DNA Tumour Viruses

Virology, Pathogenesis and Vaccines

Edited by:

Sally Roberts

Institute of Cancer and Genomic Sciences, College of Medical and Dental Sciences,
University of Birmingham, Vincent Drive, Edgbaston, Birmingham B15 2TT, United Kingdom

Caister Academic Press

www.caister.com

ISBN: 978-1-910190-79-1 (paperback)
ISBN: 978-1-910190-80-7 (ebook)

Ebooks

Cover design adapted from Figure 7.1

Contents

Current books of interest

- Cyanobacteria: Signaling and Regulation Systems 2018
- Viruses of Microorganisms 2018
- Protozoan Parasitism: From Omics to Prevention and Control 2018
- Genes, Genetics and Transgenics for Virus Resistance in Plants 2018
- Pathogenic *Escherichia coli*: Evolution, Omics, Detection and Control 2018
- Postgraduate Handbook: A Comprehensive Guide for PhD and Master's Students and their Supervisors 2018
- Enteroviruses: Omics, Molecular Biology, and Control 2018
- Molecular Biology of Kinetoplastid Parasites 2018

- Bacterial Evasion of the Host Immune System 2017
- Illustrated Dictionary of Parasitology in the Post-Genomic Era 2017
- Next-generation Sequencing and Bioinformatics for Plant Science 2017
- The CRISPR/Cas System: Emerging Technology and Application 2017
- Brewing Microbiology: Current Research, Omics and Microbial Ecology 2017
- Metagenomics: Current Advances and Emerging Concepts 2017
- *Bacillus*: Cellular and Molecular Biology (Third edition) 2017
- Cyanobacteria: Omics and Manipulation 2017

- Brain-eating Amoebae: Biology and Pathogenesis of *Naegleria fowleri* 2016
- Foot-and-Mouth Disease Virus: Current Research and Emerging Trends 2017
- *Staphylococcus*: Genetics and Physiology 2016
- Chloroplasts: Current Research and Future Trends 2016
- Microbial Biodegradation: From Omics to Function and Application 2016
- Influenza: Current Research 2016
- MALDI-TOF Mass Spectrometry in Microbiology 2016
- *Aspergillus* and *Penicillium* in the Post-genomic Era 2016
- The Bacteriocins: Current Knowledge and Future Prospects 2016
- Omics in Plant Disease Resistance 2016
- Acidophiles: Life in Extremely Acidic Environments 2016
- Climate Change and Microbial Ecology: Current Research and Future Trends 2016
- Biofilms in Bioremediation: Current Research and Emerging Technologies 2016
- Microalgae: Current Research and Applications 2016
- Gas Plasma Sterilization in Microbiology: Theory, Applications, Pitfalls and New Perspectives 2016
- Virus Evolution: Current Research and Future Directions 2016
- Arboviruses: Molecular Biology, Evolution and Control 2016
- *Shigella*: Molecular and Cellular Biology 2016

www.caister.com

Foreword

Cancers caused by human DNA tumour viruses are a considerable problem worldwide. By understanding the molecular biology of these viruses and the virus-host interactions that drive carcinogenesis, both anti-viral interventions and therapeutics have been developed for some of these viruses to reduce the burden of disease. The study of these viruses has also informed greatly on mechanisms of human tumourigenesis. The overall aim of this volume is to review the most important current research of the burden of cancers attributable to the DNA tumour viruses, their molecular virology, and the interactions between these viruses and their hosts that contribute to carcinogenesis. It covers all of the known human DNA tumour viruses and, as we have learnt much about tumour virology from the study of human adenoviruses that are not associated with any human cancer but are tumourigenic in model systems, it is very fitting to include this virus here. Epstein-Barr virus – the first human tumour virus to be discovered - and its role in epithelial carcinogenesis and lymphoma development, is covered in two separate chapters. These are complemented with an update of the development of prophylactic and therapeutic vaccines against this virus. Pathogenesis and associated therapeutics are reviewed for another herpesvirus, Kaposi sarcoma-associated herpesvirus. Hepatitis B virus, which has a partially double-stranded DNA genome, and its role in liver cancer development, is also covered in this volume. The molecular biology of small DNA tumour viruses; human papillomaviruses, and the more recently discovered Merkel cell polyomavirus – the first human polyomavirus to be associated with a human cancer, is discussed. A separate chapter addresses the development of vaccines to control infections with hepatitis B virus and human papillomavirus. Many of the DNA viruses engage in some way with the host DNA damage response pathways during their life cycle and therefore this is dealt with in a separate chapter.

I would anticipate that this volume about DNA tumour viruses would be a valuable source for all those working in the area and to those new to this field. Lastly, I am very grateful to the all the authors for their willingness to contribute to this volume.

Sally Roberts

1

Contribution of DNA Tumour Viruses to the Cancer Burden: A Global Perspective

Catherine de Martel*, Silvia Franceschi and Martyn Plummer

Infections and Cancer Epidemiology Group, International Agency for Research on Cancer, Lyon, France

**Corresponding author: demartelc@iarc.fr*

Abstract

Infections with certain infectious agents (viruses, bacteria, and parasites) have been identified as strong risk factors for specific cancers. Among those, four DNA viruses have been classified as carcinogenic to humans by IARC over the years: Hepatitis B virus (HBV) in 1994, 13 types of human papillomaviruses (HPV) in 1995, Epstein Barr virus (EBV) and Kaposi sarcoma herpes virus (KSHV) in 1997. Numerous studies in humans, animals, and mechanistic evidences support a strong causal relationship between exposure to these four viruses and human cancer in at least one site, sometimes in several sites. Together, DNA viruses have been estimated to be responsible for 1.2 million new cases of cancer worldwide in 2012, with two viruses bearing 85% of this burden: human papillomavirus (640,000 cases worldwide), and hepatitis B virus (420,000 cases). Most of the cancers due to DNA viruses occurred in Eastern Asia, Southern America, and sub-Saharan Africa, and 60% occurred in women. Merkel cell virus, another DNA virus, was discovered more recently in a rare skin cancer in humans, and was classified as probably -but not definitively- carcinogenic to humans by IARC in 2014 despite strong mechanistic evidence, due to the lack of prospective studies in humans yet available.

In this chapter we give an overview of the burden of cancers attributable to DNA viruses in the world and in eight geographical regions. Then we summarize for each infectious agent and cancer site, the body of epidemiological evidence supporting our estimates of the attributable fractions.

Introduction

Infections with certain biological agents (viruses, bacteria, and parasites) have long been identified as strong risk factors for specific cancers. In 1910 Peyton Rous, an American research worker at the Rockefeller Institute, showed that filtered blood from a hen sarcoma could propagate the tumour to healthy chickens. He concluded that a disease-causing infectious agent – known today as the retrovirus Rous sarcoma virus-1 – was present in the filtrate. For this discovery, Rous was later awarded the Nobel Prize. It took however fifty more years to discover the first human DNA oncovirus. In 1958, Burkitt lymphoma was first described by Denis Burkitt as a malignancy of the head or neck of African children. With a doubling time of 24-48 hours, the tumour is one of the fastest growing malignancies. Interestingly, the observed type of Burkitt lymphoma was confined to the malaria endemic belt, a geographical area characterized by high rainfall and a high year-round temperature. The special geographical distribution of the disease prompted the search for an infectious cause, and eventually led in 1964 to the discovery of Epstein-Barr

Table 1.1. Infectious agents that have been evaluated by the IARC Monographs and their classification according to the level of evidence of carcinogenicity.

Infectious agent	IARC Group	Year first classified
DNA viruses		
HPV, alpha types 16, 18, 31, 33, 35, 39, 45, 51, 52, 56, 58, 59	1	1995
HPV, alpha type 68	2A	1995
HPV, alpha types 26, 30, 34, 53, 66, 67, 69, 70, 73, 82, 85, 97,and beta types 5, 8	2B	1995
HPV, alpha types 6 and 11, other beta and gamma types	3	1995
EBV (HHV-4)	1	1997
KSHV (HHV-8)	1	1997
SV40	3	2012
MCV	2A	2012
BKV	2B	2012
JCV	2B	2012
HBV	1	1994
RNA viruses and retroviruses		
HCV	1	1994
HDV	3	1994
HTLV-I	1	1996
HTLV-II	3	1996
HIV-1	1	1996
HIV-2	2B	1996
Bacteria		
Helicobacter pylori	1	1994
Protozoa		
Plasmodium falciparum	2A	2012
Trematodes		
Schistosoma haematobium	1	1994
Schistosoma japonicum	2B	1994
Schistosoma mansoni	3	1994
Opistorchis viverrini	1	1994
Opistorchis felineus	3	1994
Chlonorchis sinensis	1	1994

virus (EBV) by UK scientists Anthony Epstein and Yvonne Barr (Epstein *et al.*, 1964; Young and Rickinson, 2004).

Since then, numerous studies have supported a causal relationship between specific infectious agents and human cancer. Since 1970, the International Agency for Research on Cancer (IARC) has been systematically reviewing the causes of cancer and publishing the results in the IARC Monographs on the Evaluation of Carcinogenic Risks to Humans. Each monograph is dedicated to a single agent, or a group of related agents, and contains the deliberations of an international expert working group convened by IARC.

Epidemiologists use a hierarchy of evidence to evaluate the association between a given risk factor and a disease. International variations in cancer incidence rates can give important clues to the causes of cancer, and so can time changes in incidence rates within a population, as well as cancer incidence in migrants who move from a high-risk to a low-risk country, or *vice versa*. None of these population-level approaches is conclusive, however, because there are typically multiple explanations for the same variations in risk. Case control studies allow a more detailed examination of risk factors of interest, including control for potential confounding factors. However, since risk factor exposure is assessed after disease diagnosis among cases, a case control study cannot establish a temporal relationship between exposure and disease. Prospective studies, in which exposure is assessed by measurements taken at study entry (e.g. blood samples) and participants are then followed for many years, are considered the most convincing observational data. Prospective studies are seldom available if a cancer is rare, or if it is preventable (e.g. screening for cervical cancer) as it would then be unethical to allow participants in a prospective study to develop the disease. Finally, population based intervention studies can verify that preventing exposure leads to a reduction of the disease burden in the population.

After thoroughly reviewing epidemiological studies conducted in humans, as well as studies conducted in animals and mechanistic studies, the IARC working group classifies agents according to a five-point evidence scale as follows:

- Group 1: Carcinogenic to humans
- Group 2A: Probably carcinogenic to humans
- Group 2B: Possibly carcinogenic to humans
- Group 3: Not classifiable with respect to its carcinogenicity to humans
- Group 4: Probably not carcinogenic to humans

Eleven infectious agents have so far been classified by IARC as carcinogenic to humans (Group 1) (Table 1.1). Of these, four are DNA viruses: Hepatitis B virus (HBV), 12 types of Human papillomaviruses (HPV), Epstein Barr virus (EBV) and Kaposi sarcoma herpes virus (KSHV) (IARC, 1994, 1995, 1997). Other DNA viruses are currently classified as probably carcinogenic (Group 2A), i.e. Merkel cell virus (MCV) and HPV type 68; or possibly carcinogenic (group 2B), i.e. BK polyomavirus (BKV), JC polyomavirus (JCV), and 14 other HPV types (IARC, 2011, 2012). Of note, malaria caused by an infection with the parasite *Plasmodium falciparum* in holoendemic areas has also been classified as probably carcinogenic to humans (group 2A), through reactivation of EBV, which is the known cause of endemic Burkitt lymphoma (IARC, 2012).

In this chapter, we first summarize the contribution of group 1 DNA viruses to the global burden of cancer worldwide. We then summarize for each virus the rationale that allows attributing a fraction of new cancer cases to specific DNA viruses. We review only cancer sites or types for which there is "sufficient evidence" of carcinogenicity in humans according to the IARC criteria, with the exception of a few cancer sites for which the burden of evidence has improved since they were last reviewed by the IARC monographs programme. We then briefly summarize the state of the evidence for polyomaviruses classified in groups 2A and 2B.

Global burden of cancer attributable to infectious agents

Attributable fraction: definition and calculation

The public health importance of infectious agents in causing cancer is often quantified using the attributable fraction (AF). Informally, the AF is the proportion of cancers that are caused by a particular infection in a given population. Since multiple risk factors may contribute to the development of an individual cancer case, the more formal epidemiological definition of the AF is the proportion of cancer cases that would not have occurred if the infection had been totally absent from the population.

Many infectious agents leave presence markers and mechanistic clues that allow individual cases to be attributed to the agent. Examples include the presence of EBV genome or gene products (i.e. small nuclear EBV-encoded RNAs) in malignant cells from lymphomas and nasopharyngeal carcinoma; and active replication of HPV (i.e. detection of oncoproteins E6/E7 mRNA) in anogenital cancers and head and neck cancers. In such cases, the AF is simply the prevalence of infection in cases.

In general, however, a one-to-one correspondence between the presence of the infection and the aetiology of an individual cancer cases is not possible. Prevalence must then be adjusted downwards to obtain the AF. The formula for adjusting the prevalence (Bruzzi *et al.*, 1985) is:

$$AF = p\,(R-1)/R$$

Where p is the prevalence in cases and R is the relative risk, i.e. the ratio of incidence rates in the infected and uninfected.

Two extreme cases illustrate how Bruzzi's formula works. If the relative risk is equal or close to 1 then the disease is equally common in infected and uninfected, so the presence of an infection in a cancer case is not an indication that the infection caused the disease. In this case Bruzzi's formula gives AF=0 regardless of the prevalence in cases. Conversely, if the relative risk is greater than 20, which is often the case with carcinogenic infections, then over 95% of the infected cases are attributable to the infection and the AF is only slightly lower than the prevalence. An example of the latter is liver cancer due to HBV, where the relative

Table 1.2. Number of new cancer cases occurring in 2012 attributable to DNA tumour viruses by sex and cancer sites.

INFECTIOUS AGENT AND RELATED CANCER SITE	Number of new cancer cases in 2012 attributable to infection[1]	
	Male	Female
Human papillomavirus		
Cervix uteri carcinoma	0	530 000
Vulva carcinoma	0	8 500
Anus carcinoma	17 000	18 000
Penis carcinoma	13 000	0
Vagina carcinoma	0	12 000
Oropharynx carcinoma	24 000	5 500
Oral cavity carcinoma	5 600	3 000
Larynx carcinoma	6 400	860
Hepatitis B virus		
Hepatocellular carcinoma	300 000	120 000
Epstein-Barr virus		
Hodgkin lymphoma	19 000	13 000
Burkitt lymphoma	2 900	1 900
Nasopharynx carcinoma	58 000	25 000
Kaposi sarcoma herpesvirus		
Kaposi sarcoma	29 000	15 000

risk between marker of chronic infection (i.e. HBsAg) and cancer is greater than 20 in both cohort and case control studies (Donato *et al.*, 1998; Cho *et al.*, 2011).

An alternative formula, known as Levin's formula, is often used for AF calculations for possible exposures, including lifestyle risk factors. Levin's formula is based on the prevalence in the general population, rather than in cases. We consider this to be less useful for evaluating infections since representative samples of the general population that provide accurate prevalence by age group and world regions are not available. This is particularly important when there are large changes in the infection pattern over time and serial infection prevalence surveys would be necessary.

Global burden

The global AF for all infectious agents (including viruses, bacteria, and parasites) has been estimated in a series of publication for the years 1990, 2002, 2008

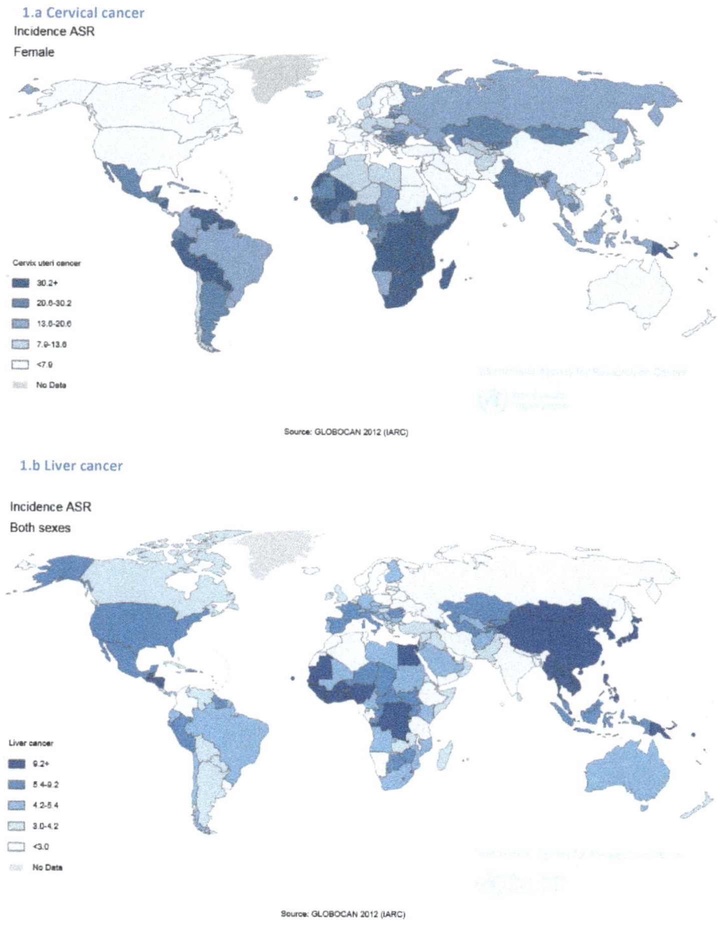

Figure 1.1. Worldwide maps of age standardized incidence rates of the two most important cancers attributable to DNA viruses: cervical cancer and liver cancer (Source: GLOBOCAN 2012, IARC).

(Pisani *et al.*, 1997; Parkin, 2006; de Martel *et al.*, 2012;) and most recently for the year 2012 (Plummer *et al.*, 2016). These estimates were based on the number of new cancer cases worldwide estimated by the GLOBOCAN project (Ferlay *et al.*, 2013). The aim of GLOBOCAN is to provide contemporary estimates of the incidence of, mortality and prevalence from 27 major types of cancer, at national level, for 184 countries of the world. The current GLOBOCAN incidence estimates are presented for the year 2012 and were used for the most recent AF calculations. Of the 14 million new cancer cases that occurred in the world in 2012, Plummer *et al.* calculated that 2.2 million were attributable to infectious agents, corresponding to a global AF of 15.4% (Plummer *et al.*, 2016). The most important infectious agents were *Helicobacter pylori* (770,000 cases worldwide, AF=5.5%, human papillomavirus (640,000 cases, AF= 4.6%), hepatitis B virus (420,000 cases, AF=3.0%), hepatitis C virus (170,000 cases, AF=1.2%) and Epstein-Barr virus (120,000 cases, AF=0.9%). Kaposi sarcoma (44,000 cases worldwide) was the second most important infectious agent in sub-Saharan Africa. Table 1.2 shows that altogether, the four DNA viruses classified as carcinogenic to humans (i.e. HPV, HBV, EBV, and KSHV, in decreasing order of global burden) were responsible for 1 200,000 new cases of cancers worldwide in 2012 (8.7% of the total cancer burden); 61% of the cases attributable to DNA viruses occurred in women (Plummer *et al.*, 2016).

Figure 1.1a, and 1.1b shows the worldwide maps of incidence rates estimated by GLOBOCAN in 2012 (Ferlay *et al.*, 2013) for the two most important cancers attributable to DNA viruses: cervical cancer, of which 100% is attributable to HPV, and liver cancer, largely dominated by hepatocellular carcinoma, of which 73% is attributable to HBV and/or HCV. For cervical cancer, 85% of the global burden occurs in less developed regions, where it accounts for almost 12% of all female cancers (Figure 1.1a). With incidence rates of 43 and 31 per 100,000, cervical cancer remains the most common cancer in women in Eastern and Middle Africa, respectively. For liver cancers, the regions of highest incidence are Eastern/South-Eastern Asia and Western Africa (Figure 1.1b).

Overall, 77% of the world cancer burden attributable to the four main oncogenic DNA viruses (0,9 million of the 1.2 million new cases occurring in 2012) falls into three geographical regions: Eastern Asia, Central Asia, and sub-Saharan Africa, in decreasing order of importance (Figure 1.2). However, the spectrum of DNA viruses responsible for cancer varies by region. For example, in Eastern Asia, and especially in China, HBV is largely preponderant over the other three viruses. In contrast, HPV largely dominates in Central Asia (especially in India). In sub-Saharan Africa, HPV is also largely dominant, but KSHV, HBV and EBV also play a non-negligible role (Plummer *et al.*, 2016).

Human papillomavirus (HPV)

Human papillomaviruses (HPV) are non-enveloped double-stranded DNA viruses that infect basal epithelial cells, causing benign and malignant lesions of the skin and the mucosa. More than 200 different genotypes (referred to as "types") of HPV have been identified to date (International Human Papillomavirus Reference Center, 2015), of which over 40 infect the genital tract. Among them, 12 types are firmly established as carcinogenic to humans by IARC (i.e. currently classified in group 1) due to their causal association with cervical cancer, other mucosal anogenital cancers, or head and neck cancers (IARC, 2011). These HPV types are typically referred to as "high-risk" types.

Cervical cancer

HPV infection is the most common sexually transmitted viral infection. Worldwide, an average of approximately 12% of women with normal cytological findings have a detectable cervical HPV infection (Bruni *et al.*, 2010). International comparisons can be affected by substantial variations in the sensitivity of the HPV assay. In IARC International HPV surveys,

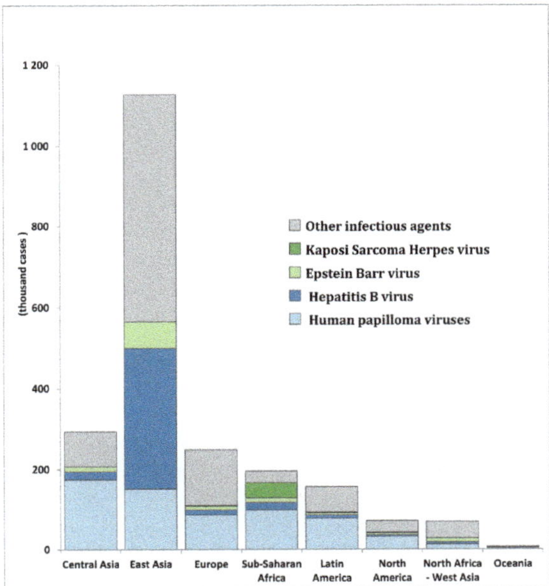

Figure 1.2. Incident cancer cases attributable to infection and to DNA viruses in eight geographical regions in 2012 (modified from Plummer *et al.*, 2016).

HPV prevalence varies greatly by region, being highest in sub-Saharan Africa (up to 50% or more), Eastern Europe and Latin America (Crosbie *et al.*, 2013). The distribution of HPV types slightly differs by region (Guan *et al.*, 2012), but overall, the high-risk types 16, 18, 31, 33, 45, 52 and 58 are the most prevalent, and HPV 16 alone is estimated to contribute around one fifth of the global prevalence in cancer free women (Bruni *et al.*, 2010). Despite its high prevalence, HPV infection is usually asymptomatic and transient, and due to a strong local cell-mediated immunity, the virus is cleared within two years in nearly all women, with an average clearance time of ten months (Plummer *et al.*, 2007; Rositch *et al.*, 2013). Only 10% of infected women will develop a persistent infection, which is a prerequisite for development of cervical cancer. HPV persistence and progression to cervical precancerous and cancerous lesions is greatly increased in immunosuppressed women, notably HIV-positive women (de Martel *et al.*, 2015b).

The absolute risk of developing cervical cancer from a persistent HPV infection is unknown, since the progression of the disease can and should be interrupted by screening. The highest grade of cervical intraepithelial neoplasia, CIN-3, represents the best surrogate endpoint to assess the risk of developing invasive cervical cancer (ICC). Using this endpoint, it has been demonstrated that cytologically normal women with persistent HPV 16 (the most aggressive of all oncogenic types) have a 47% risk of developing CIN3+ within 12 years (95% confidence interval, CI: 35%-58%) (Kjaer *et al.*, 2010). An abundance of mechanistic and retrospective epidemiological data has firmly established the chain of causality between persistent infection and cancer. One such retrospective cohort study was determinant, although it was only made possible by the existence of an unethical clinical study recruiting women between 1965 and 1974, in which treatment of CIN-3 was often withheld or delayed. The study was halted, but subsequent analyses could demonstrate that untreated CIN-3 has a considerable propensity to progress to invasive cancer, whereas the risk was very low in treated women (McCredie *et al.*, 2008). Persistent cervical infection with a carcinogenic type in the squamous columnar junction of the cervix is accompanied by expression of early viral genes E6 and E7, leading to a cascade of carcinogenic events (Quint *et al.*, 2012). Since high-risk HPV are detected by PCR in virtually all cases of cervical cancer, it is generally acknowledged that 100% cervical cancers are attributable to HPV.

The distribution of HPV types is similar in women with cervical cancer in different parts of the world, including in HIV-positive women (Clifford *et al.*, 2016). HPV 16 and 18 are the most virulent types and together are responsible for more than 70% of ICC in all regions. A meta-analysis of type-specific HPV prevalence in ICC has shown that HPV 16 prevalence ranges from 52% in Africa to 67% in Western/Central Asia, and HPV18 ranges from 14% in Eastern Asia to 20% in Oceania (Li *et al.*, 2011). The distribution of HPV types differs however by histology: HPV 16 is predominant in squamous-cell carcinoma, the most common histological type, but the contribution of HPV 16 and HPV18 is similar in adenocarcinoma.

Other anogenital cancers: anus, vulva, vagina, and penis

Anal cancer is a relatively rare malignancy with an age standardized incidence rate between 1 and 2 per 100,000 worldwide. In recent years, a few countries (i.e. USA, Scotland, Denmark, France or Australia) have reported an increase in incidence rates (Nielsen *et al.*, 2012; Shiels *et al.*, 2015). No strong risk factors are known to exist for anal cancers other than those associated with HPV transmission (i.e. number of sexual partners and anal intercourse), or immune deficiency (i.e. in HIV-positive men who have sex with men (MSM), anal cancer is one of the most commonly occurring cancers). Anal cancer increases have therefore been hypothesized to be the consequence of changing sexual practices, leading to increased prevalence of anal infection with HPV (Grulich *et al.*, 2012). The vast majority of anal cancers are SCC, but adenocarcinomas may comprise 10% to 30% of cases. For SCC, the malignancy arises from the junction between the gastrointestinal glandular mucosa and the squamous mucosa of the anus. Similar to cervical cancers, persistent infection with oncogenic types of HPV is found in more than 90% SCC cases. In the absence of any other risk factors, it is hypothesized that HPV may be a necessary cause of anal cancer as it is for cervical cancer (Alemany *et al.*, 2015). However, compared to cervical carcinoma, a stronger predominance of HPV type 16 (≥80%) is constantly reported.

HPV 16 is also relatively more predominant in vulvar, vaginal and penile cancers associated with HPV than in cervical cancer. However, in contrast to the ubiquitous presence of HPV in cervical and anal cancers, other anogenital cancer sites show a varying prevalence of HPV infection, depending upon the histological sub-type, the age distribution and the geographical area of the population studied (De Vuyst *et al.*, 2009). For vulvar cancers and penile cancers it has been shown that there are different oncogenic pathways, with or without active involvement of HPV infection (Kurman *et al.*, 1993; Chaux *et al.*, 2010; Stankiewicz *et al.*, 2011).

Vulvar cancer is a rare entity; it was estimated that for the year 2012 about 34,000 new cases were diagnosed, giving a worldwide incidence rates between 0.5 and 1.5 per 100,000 women, representing about 4% of all gynecological malignancies. As for anal cancer, incidence rates have been increasing in recent years, especially in younger women. The most current data on the prevalence of HPV in vulvar cancer cases comes from a large multinational case series of 1709 cancer cases in 39 countries (de Sanjose *et al.*, 2013). In this study, HPV diagnosis was centralized using PCR SPF-10 broad–spectrum primers and genotyping by reverse hybridization line probe essay ($LiPA_{25}$). A cancer case was attributable to HPV if both HPV-DNA and $p16^{INK4a}$ overexpression were observed simultaneously in cancer tissue. High risk HPV was more frequently associated with the warty-basaloid histological type and high risk HPV prevalence showed a strong inverse association with age. High risk HPV were present in approximately 48% (42%-54%) of vulvar cancers occurring in women aged 15 to 54, compared to 28% (23%-33%) in women aged 55 to 64, and 15% (11%-18%) in women aged 65 and older. Overall 8,500 cases worldwide (25% of all vulvar carcinoma) are attributable to HPV each year.

With about 15,000 new cases estimated worldwide in 2012, vaginal cancer is even more rare than vulvar cancer. However, in cancers of the vagina, HPV DNA is detected in the majority of cases, as well as in precursor lesions. Data from a meta-analysis of case series using PCR in tumour tissue with MY09/11 or GP5+/6+ primers (De Vuyst *et al.*, 2009) gives an average worldwide prevalence of 78% (68%-86%). This translates into about 12,000 cases attributable to HPV worldwide per year.

Invasive penile cancer is also quite rare and affects men aged 50-70 years. Globally, the annual burden for penile cancer has been estimated to be 26,000 cases in 2012. Incidence of penile cancer is higher in less developed countries, and the malignancy accounts for up to 10% of male cancers in some parts of Africa, South America and Asia. Cancers of the penis are primarily SCC (95%) and the most common penile SCC histologic sub-types are in decreasing order, keratinizing, mixed warty-basaloid, verrucous, warty, and basaloid. HPV DNA is most commonly detected in basaloid and warty tumours. Worldwide, the prevalence of HR HPV in penile cancers can be estimated from the combination of data from two reviews (Backes *et al.*, 2009; Miralles-Guri *et al.*, 2009), giving a global prevalence of 51% (47%-55%). Among HPV-related penile tumours, HPV 16 is the most common type detected, followed by HPV 18 and HPV types 6/11.

Head and neck cancer

Head and neck cancers represent a large and heterogeneous group of malignancies, for which cigarette smoking and alcohol consumption are strong risk factors. Over the last few years, worldwide cancer registry data have shown an increase of oropharyngeal cancers (OPC) - including tonsil and base of tongue- in several more developed countries, with higher and sharper increases registered among men under 60 years (Chaturvedi *et al.*, 2013). In the United States, this increase in men was accompanied by a decrease of tobacco-associated cancers at other sites, thus eliminating tobacco smoking as the main culprit (Chaturvedi *et al.*, 2011). Several molecular studies have hypothesized that HPV might be at least partially responsible for this rise, which seemed to concern a number of non-smoker and non-drinking patients. In 2007, the IARC Monograph programme re-evaluated the role of high-risk HPV types in human cancer, and the working group concluded that there was sufficient evidence for a carcinogenic role of HPV 16 infection in oropharyngeal cancers (OPC) and oral cavity cancers (IARC, 2007). They also concluded that there was limited evidence in humans for the carcinogenicity of HPV 16 in the larynx.

It is now admitted that two main distinct oncogenic pathways exist that lead to OPC, one driven by HPV, and the other by tobacco and/or alcohol. The detection of HPV in head and neck cancer by PCR-based techniques, therefore, is not enough for establishing causation, because the test cannot discriminate between transient colonisation versus persistent infection leading to transformation. Thus, relying on PCR may lead to an overestimation of the proportion of cancers caused by HPV, especially in head and neck cancer sites strongly associated with tobacco or alcohol. To overcome this issue, the detection of actively transcribing viral oncoproteins E6 and E7 (E6/E7 mRNA) in cancer tissue is now considered the gold standard to attribute a cancer case to HPV (Bishop *et al.*, 2012; Rietbergen *et al.*, 2013).

For OPC, a few case series have compared the prevalence of HPV by PCR and by E6/E7 mRNA, showing that 86.9% (95%CI 73.2-96.7) of OPC positive by PCR were also positive by E6/E7 mRNA (Ndiaye *et al.*, 2014). The fraction of OPC attributable to HPV varies greatly by world regions, being highest (30 to 50%) in more developed countries (up to 70% in the most recent studies in the USA and some North

European countries), but probably much lower and still very uncertain in most other regions of the world (Plummer *et al.*, 2016).

For cancers of the oral cavity and larynx, the prevalence of high-risk HPV was evaluated in a few small case series using E6/E7 mRNA detection methods. Most of the studies were conducted in Europe and North America, and yielded an average prevalence of 4% at both sites, i.e. a prevalence much lower than that found in oropharyngeal cancers (Plummer *et al.*, 2016). This proportion is likely to be even lower in other parts of the world. An international cross sectional study reporting on 69 cases (both sites) suggest that it is close to zero in India (Castellsague *et al.*, 2016), where the incidence of oral cavity cancer is high, mostly due to betel chewing.

Hepatitis B (HBV)

Hepatocellular carcinoma

HBV is a small DNA virus from the hepadnavirus family. It infects only humans, and can be transmitted vertically, through close contact between children and within the household, through contaminated blood and blood products, or sexually. In Asia and Africa, the virus is most often transmitted from the mother to the child at birth or between siblings at a very young age, whereas in areas of low-endemicity for HBV, the dominant routes of transmissions are contaminated blood or body fluids, or sexual intercourse. Infection with HBV is still a major global public health threat, due to its frequent severe complications including acute liver failure, cirrhosis and hepatocellular carcinoma, the main type of liver cancer. Despite the wide availability of an efficient vaccine for the last two decades, the percentage of the general adult population positive for hepatitis B virus antigen (HBAg) is estimated to vary from less than 2% in regions such as Central Latin America, North America and Western Europe to 5 to 10% in sub-Saharan Africa and Eastern Asia, leading to a prevalence of 3.7% globally, i.e. approximately 240 million chronic carriers worldwide (Ott *et al.*, 2012). Since 1990, strong declines are seen in South-Eastern Asian children, due to successful vaccination campaigns (Chang *et al.*, 2009; Hennessey *et al.*, 2013).

With 782,000 new cases estimated in 2012, liver cancer is the 7th most common cancer in the world (Ferlay *et al.*, 2013); however, it ranks third in terms of mortality figures, due to its very poor prognosis. Based on conservative assumptions, it is estimated that 420,000 cases, i.e. 54% of the world total liver cancer cases are attributable to HBV infection (Plummer *et al.*, 2016). Other main risk factors for HCC worldwide are Hepatitis C virus, male gender, diabetes, obesity, alcohol consumption, cigarette smoking, and aflatoxin-B1 exposure.

A causal association between HBV and HCC has been well established and is supported by animal studies, ecological surveys, and many sero-epidemiological studies (IARC, 1994). A meta-analysis of 20 case-control studies and two cohort studies testing all HCC patients for both HBV and HCV yielded a summary relative risk (RR) of 23.4 (95% CI: 17.2-31.7) for HBV alone (Cho *et al.*, 2011). Coinfection with HCV had an additive effect and yielded a summary RR of 51.1 (95%CI: 33.7-77.6) (Cho *et al.*, 2011). The risk of developing HCC also appears to be intimately linked to the duration of the inflammatory process, through repeated cycles of cellular necrosis and regeneration, and is increased in patients with high levels of HBV replication (Yang *et al.*, 2002). In a large Taiwanese cohort, a correlation between increased incidence of HCC and increasing levels of HBV DNA in serum was demonstrated, further supporting the role of HBV in the carcinogenesis process (Chen *et al.*, 2006).

The presence of a chronic infection with HBV in liver cancer patients is usually part of the diagnosis process and can be accurately determined using serological methods. Hence, the fraction of HBsAg seropositivity among HCC cases can be considered a proxy of HBV attributable fraction. A systematic review of case series of hepatocellular carcinoma reporting the prevalence of HBsAg and HCV antibodies reported on 119,000 HCC cases in 260 separate studies from 50 countries (de Martel *et al.*, 2015a). Overall, HBV was shown to be the dominant cause of HCC in Western/Central Asia (AF=47%), sub-Saharan Africa (AF=50%), and Eastern Asia (AF=63%), with the exception of Japan. It was much less prevalent among cases in Northern Europe (AF=8%) and North America (AF=12%).

Non-Hodgkin lymphoma (NHL)

An association between HBV and NHL has been reported in cohort studies from Europe and Korea. However, HBV has not been judged to be a well-established cause of NHL by the IARC Monograph programme (IARC, 2011).

Epstein-Barr virus (EBV)

Epstein Barr virus (EBV), or human herpes virus type 4 (HHV-4), is one of eight herpes virus known to be pathogenic to humans, the others being human herpes virus 1 and 2 (HHV-1 and -2), varicella zoster virus (HHV-3), cytomegalovirus (HHV-5), human herpes virus 6 and 7 (HHV-6 and -7), and Kaposi sarcoma herpes virus (HHV-8). Humans are the only natural hosts for EBV; after primary infection, which is usually asymptomatic and occurs early in childhood, most individuals will harbour the virus for the remainder of their life (Young and Rickinson, 2004). By adulthood, the majority of the human population (up to 90%) is infected with EBV. In less developed countries 80% of children are infected by the age of 3. In more developed countries, primary infection is increasingly delayed until adolescence or young adulthood when the more severe immune response results in a syndrome known as infectious mononucleosis.

The virus is transmitted through saliva. Once it enters the epithelium of the Waldeyer ring in the oropharynx it may infect naïve B cells in lymphoid tissue, and from there it establishes latent infection within B and T lymphocytes, and less frequently in epithelial cells. EBV has been associated with several malignancies, the strongest causal associations being with Burkitt lymphoma (BL), Hodgkin lymphoma (HL), and non-Hodgkin lymphoma (NHL) associated with post-transplant or HIV immuno-suppression (IARC, 1997). Nasopharyngeal carcinoma (NPC) is the epithelial cancer known to be induced by EBV.

Epstein–Barr encoding RNA (EBER-1 and 2) are highly expressed in all forms of EBV infection, including the three types of latent gene expression associated with BL (Latency I), NPC and HL (Latency II), and AIDS related NHL (latency III). EBER *in situ* hybridization is the gold standard method for the detection of EBV in tumour tissue, and therefore it should be used for the attribution of the malignancy to EBV.

Hodgkin lymphoma (HL)

There are over 65,000 new cases of Hodgkin lymphoma (HL) diagnosed each year worldwide. The proportion of EBV-positive HL varies by histological subtype, immunocompetence status, age, sex, and development level of the country. The classical HL type, which accounts for more than 95% of all HL cases, has been classified into four different histological subtypes; EBV is more commonly associated with the mixed cellular subtype (Stein *et al.*, 2008). In the HIV positive population, the vast majority of HL is EBV-related and mainly consists of mixed cellular subtypes (Stein *et al.*, 2008). The age distribution of the disease varies by region. In less developed areas, such as sub-Saharan Africa or central/South America where EBV infection occurs at a very young age, a first peak of incidence is often seen before the age of 15 and all pediatric HL are EBV-related. In more developed countries, the first peak of HL is seen between the ages of 15 and 35. A second peak is often seen in patients 55 years and older. The proportion of EBV-related HL in various geographic areas varies greatly from about 30% in Australia and in North America, 35% in Europe, 55% in Asia, 60% in Latin America to 75% in Africa (Lee *et al.*, 2014).

Burkitt lymphoma (BL)

As noted in the introduction, the association between the endemic form of Burkitt lymphoma (BL) and EBV has long been established (IARC, 1997). In the malaria belt of sub-Saharan Africa, the presence of clonal EBV episomes is detected in 100% of the so called "endemic" BL tumours, strongly suggesting that the tumour develops from a single infected cell, and that EBV has a role in the initiation of the neoplastic transformation. By reactivating EBV and by increasing the pool of potentially transformable B cells, the malaria parasite *Plasmodium falciparum* acts synergistically in the development of BL (IARC, 2012). The detection of EBERs or other specific EBV transcriptional gene products in tumour cells allows the lymphoma to be attributed to EBV, even in areas of lower BL incidence (Young and Rickinson, 2004).

Outside the endemic areas, EBV transcriptional gene products are detected much less frequently in BL. In Europe and North America, BL is sporadic, and is associated with EBV in only 20% of cases (IARC, 2011). Elsewhere, data are very sparse, but a few studies suggest that the rate of detection varies greatly (e.g. from 29% to 76% in Brazil alone)(Thompson and Kurzrock, 2004; Queiroga *et al.*, 2008). Worldwide, 50% of the 9,100 annual incident cases of BL estimated in 2012 can be attributed to EBV. BL is strongly associated with HIV infection (see section on lymphoma in immunosuppressed people).

Nasopharyngeal carcinoma (NPC)

With 86,500 new cases reported in 2012, NPC is also a relatively rare neoplasm worldwide, but shows a unique geographical distribution with high incidence rates (10 to 30 per 100,000 years in men) in Southern China, Singapore and Malaysia, and intermediate rates in the Arctic, Middle East, South-Eastern Asia and North Africa (Chang and Adami, 2006; Wee *et al.*, 2010). Within these regions, some ethnic groups are especially concerned, for instance the Inuits in the Artic, the Nagas tribes in the Northeast regions of India, or the descendants of the Chinese Bai Hues people in Southern China, suggesting a genetic susceptibility that could be associated with some human leukocyte antigen (HLA) class I phenotypes, or the DNA double-strand break repair pathway (Li *et al.*, 2009; Yee Ko *et al.*, 2014).

The WHO classifies NPC into three histologic types: keratinising squamous cell carcinoma (SCC) (type I); and two types of non-keratinising carcinoma characterized as differentiated (type II) or undifferentiated (type III) (Barnes *et al.*, 2005). Type III comprises over 95% of NPC in high and intermediate incidence areas, and most of the remaining 5% is type II. For the non-keratinising subtypes, EBV genome or gene products have been detected in virtually all cases, irrespective of the geographic origin, suggesting that EBV activation is a necessary step in the malignant process (Barnes *et al.*, 2005; IARC, 2011). Incidence data from GLOBOCAN 2012 show that high and intermediate incidence areas comprise about half the cases of NPC worldwide. The remaining half of NPC cases occur in low incidence areas, where type I seems to be more frequent, and may be associated with tobacco smoking, or more recently, with HPV 16 (Polesel *et al.*, 2011; Dogan *et al.*, 2014; Lin *et al.*, 2014;). In these areas, it appears that only one-third of all SCC is associated with EBV (Nicholls *et al.*, 1997; Ou *et al.*, 2007).

Hodgkin and Non-Hodgkin lymphomas in immunosuppressed people

Immunosuppressed people are at greater risk of developing EBV-related lymphoma. An increased incidence rate of NHL has long been reported in post-transplant recipients, and was noted to be particularly high among EBV-negative children undergoing EBV seroconversion, possibly via the graft (IARC, 1997). In addition to BL, diffuse large B-cell lymphomas (DLBCL), are among the earliest identified AIDS-defining illnesses (Carbone *et al.*, 2014).

DLBCL is especially frequent in the central nervous system (CNS) in severely immunosuppressed people, and includes three main morphologic variants: centroblastic, immunoblastic and anaplastic (2008). Centroblastic DLBCL predominates in the general population whereas the immunoblastic variant is the most frequent variant in severely immunosuppressed HIV-infected patients. Other very rare DLBCL types are associated with HIV/AIDS such as primary effusion lymphomas (caused by KSHV) and plasmablastic lymphomas. The vast majority of immunoblastic DLBCL in HIV-infected people with CD4 <100 are EBER-positive and so are attributable to EBV. Similarly, nearly all lymphomas of the CNS in HIV-infected people are immunoblastic DLBCL and EBER-positive. For non-CNS DLBCL, the introduction of combined antiretroviral treatment (cART) has been associated with a decrease in the proportion of EBV-associated tumours, which has been estimated around 30% (de Martel *et al.*, 2015b). For comparison, approximately 80% of all DLBCL were associated with EBV infection in the pre cART era in the HIV population (IARC, 2008).

HIV-infected people also show an excess risk of Burkitt and Hodgkin lymphoma. In 12 studies of Burkitt and Burkitt-like lymphomas occurring in the HIV population in the United States and Europe, 99 of 219 cases (45%) were EBER-positive (de Martel *et al.*, 2015b). There has been no significant change in prevalence after the introduction of cART (prevalence of 42% before 1996 versus 48% after). For comparison, the percentage of EBER-positivity in sporadic Burkitt lymphoma occurring in the general population in the United States and Europe is approximately 20%. Similarly, in 6 studies of Hodgkin lymphomas, occurring in the HIV population in the United States and Europe, 209 of 229 cases (91%) were EBER-positive with no reduction after the introduction of cART.

Other non-Hodgkin lymphoma (NHL)

In immunocompetent people, there is sufficient evidence in humans for the carcinogenicity of EBV only in some rare subtypes of non-Hodgkin lymphoma, notably in extra nodal NK/T cell lymphoma (nasal type), and in the more recently described EBV-positive DLBC lymphoma of the elderly (de Martel *et al.*, 2015b; IARC, 2011). The detection of EBV-encoded RNA by *in situ* hybridisation has also been reported in a small proportion of B-cell or T-cell lymphomas in immunocompetent adults, for instance in peripheral T cell lymphomas. Unfortunately, in spite of a relatively large number of descriptive clinico-pathological case-series, very few systematic analyses of the prevalence of EBV in different subtypes of NHL in the general population are available and the topic is still subject to large uncertainty (d'Amore *et al.*, 1996; Wang *et al.*, 2005). The estimated overall prevalence of EBV in non-Burkitt NHL varies slightly by region (range 5 to 15%), depending on the distribution of the different NHL subtypes (Teras *et al.*, 2015).

Gastric cancer

In 2014, The Cancer Genome Atlas Research Network recognized EBV-positive gastric cancer as one of the four subtypes of a new molecular classification of gastric adenocarcinoma (The Cancer Genome Atlas Research Network, 2014). A few years earlier, a meta-analysis of the international literature had shown that around 9% of all gastric cancers (5% in women and

11% in men) contain EBV DNA products in tumours cells as measured by EBER *in situ* hybridization. This fraction was higher in younger subjects and for non-antral subsites, but the analysis also showed substantial heterogeneity across studies that could not be explained by measured cofactors such as geographical area or year of publication (Murphy *et al.*, 2009).

The relationship between EBV and *H. pylori*, the strongest risk factor for gastric cancer is still unclear. In vitro studies have shown that the *H. pylori*-associated oxidant NH_2Cl may be able to convert the latent epithelial EBV infection into lytic infection with activation of the early gene (Minoura-Etoh *et al.*, 2006). A seroprevalence study of antibodies to 15 specific *H. pylori* proteins using a multiplex essay showed that the serological profile of *H. pylori* is equivalent in EBV-positive and in EBV-negative gastric biopsies obtained from 169 non-cardia gastric cancer patients in 5 countries (Camargo *et al.*, 2016). These results suggest that *H. pylori* is the main etiological factor in EBV-positive gastric cancers, and that EBV might act as a cofactor by contributing to the likelihood of malignant transformation. In 2012, EBV was re-evaluated by the IARC Monograph programme, and it was concluded that there is currently insufficient epidemiological evidence for the involvement of EBV in gastric cancer. However, the fact that the EBV genome is present in the tumour cell in a monoclonal form, and that transforming EBV proteins are expressed in the tumour cell provides a mechanistic explanation of how EBV might cause a proportion of gastric cancer (IARC, 2011).

Kaposi sarcoma herpes virus (KSHV)

Kaposi sarcoma (KS)

KSHV, also known as human herpes virus 8, is a large double stranded DNA virus from the Herpesviridae gamma subfamily. Its discovery followed years of intensive search for an infectious agent responsible for the Kaposi sarcoma (KS) epidemic occurring in AIDS patients, especially men who have sex with men (MSM) in Western countries and heterosexuals in sub-Saharan Africa. In 1993, using an elegant multi step molecular technique known as representational difference analysis, Chang *et al.* amplified unique DNA sequences found in malignant cells of KS patients but not in controls, suggestive of the presence of a new virus (Chang *et al.*, 1994). Many studies have since confirmed the near universal presence of KSHV in the characteristic spindle cells of KS lesions, both in HIV-positive and HIV-negative individuals.

Similar to EBV, humans are the natural host for KSHV, in whom it establishes a latent infection within B cells, as well as macrophages, dendritic, endothelial and epithelial cells. The routes of transmission of KSHV are not fully understood, and seem to vary by geographic area. In areas of low prevalence (<10%), such as Europe and North America, the correlation between antibody prevalence and the number of sexual partners – notably among MSM – suggests that transmission occurs mainly through oral or anal intercourse. In areas of intermediate and high prevalence (i.e. Mediterranean countries and sub-Saharan Africa, respectively), sexual transmission does not seem to be the main route (Malope *et al.*, 2008); seropositivity for KSHV is seen at a very young age and increases through childhood, suggesting mother-to-child and inter-sibling transmission, possibly via saliva (Malope *et al.*, 2007).

KSHV is a causal factor for the development of KS, and some rare lymphoproliferative disorders, including primary effusion lymphoma and multicentric Castleman disease (Du *et al.*, 2007). Although nearly all cases of KS are attributable to KSHV, four different forms of the disease have been described suggesting that other genetic or environmental factors have an important contributory role in cancer development. The classic but rare form of KS is seen in elderly Mediterranean and European men. The disease also has been often described in immuno-suppressed transplanted patients. In some regions of sub-Saharan Africa, KS has been endemic since before the HIV epidemic (Malope *et al.*, 2008). In Europe and North America, after 10 years of dramatically increasing incidence, a rapid decline of KS has followed the introduction of cART, and the incidence rate in men between 20 and 70 years old is now generally below 2 per 100,000. A study from South Africa reported a drastic reduction in the risk of developing KS of 70–80% after introduction of cART (Bohlius *et al.*, 2014). In 2012, KSHV accounts for only 2% of infection-attributable cancers worldwide, but in the less developed countries (mainly in sub-Saharan Africa), this proportion rises to 14% (Plummer *et al.*, 2016). Kaposi sarcoma mainly involves individuals below the age of 50 years and continues to be of major public health concern in African settings where KSHV is endemic and large numbers of HIV-infected individuals have late or no access to cART (Robey and Bower, 2015).

Other DNA viruses

Merkel cell virus (MCV) and Merkel cell carcinoma (MCC)

Merkel cell carcinoma (MCC) is a rare but aggressive malignancy, and accounts for less than 1% of non-melanoma skin cancers, with incidence rates ranging from 0.2 to 0.4 per 100,000 in most European and US registries; only Australia reports higher incidence rates (up to 1 per 100,000 in men), possibly due to the role played by higher UV exposure on a fair-skinned population (Agelli *et al.*, 2010; Girschik *et al.*, 2011; Kaae *et al.*, 2010). MCC generally occurs after 60 years on the sun–exposed areas of the head and the neck and is more common in men. Like Kaposi sarcoma, MCC occurs more frequently among immunosuppressed transplant and AIDS patients, a feature suggestive of an infectious origin for the tumour. In 2008, using digital transcriptome subtraction, Feng *et al.* identified a new virus integrated into cancer cell DNA in MCC tumour tissue (Feng *et al.*, 2008). Since then, Merkel cell polyomavirus has been consistently detected in MCC cells from numerous case series with prevalence in cases ranging from 59% to 100% (IARC, 2012). Human fibroblast located in the dermis –and not Merkell cells of the epidermal lineage, as originally believed- have been identified as the primary target of MCV infection and productive replication (Liu *et al.*, 2016). However, oncogenic mechanisms leading to a tumour expressing both neuroendocrine and epidermoid markers remain poorly understood. The monoclonal pattern of MCV integration into the MCC cellular genome as detected by Southern blot hybridization and PCR based analyses of MCC tissue suggests that viral infection precedes tumourigenesis and strongly argues for a causative role of the virus in carcinogenesis, yet human epidemiological studies supporting causation are sparse. Three case control studies measuring serological markers of MCV showed odds ratios around 5; the largest of these studies looked more specifically at viral early gene expression of small and large T-antigens and found odds ratios of 63.2 (95%CI 24-164) and 16.9 (95%CI 3-11), respectively (Paulson *et al.*, 2010). No prospective studies are available. Overall, the burden of evidence remains limited in humans, prompting the IARC expert working group to classify MCV as a probable (group 2A) but not definite cause of cancer in its 2012 evaluation (IARC, 2012).

Other human polyomaviruses

Merkel cell virus was the fourth polyomavirus identified in humans, after simian virus 40 (SV40), JC polyomavirus (JCV) and BK polyomavirus (BKV) the two latest viruses named after the initials of the patients from whom they were isolated. Since the discovery of JCV and BKV in 1971, a total 14 human polyomavirus species have been identified so far. Much remains unknown about the newly discovered viruses. However, SV40, JCV, and BKV have been reviewed and evaluated in 2012 by an IARC expert working group (IARC, 2012).

There is strong and consistent evidence from animal and cell-culture studies that SV40, a polyomavirus whose natural hosts are macaques, can be directly oncogenic. Because of the contamination of some polio vaccines by SV40 in the 1950s and early 1960s, millions of people have been exposed to the virus, prompting scientists to spend enormous resources and efforts in searching for SV40-induced human tumours and diseases. They have been numerous cohort studies, serological case control studies, and PCR-based case series examining potential associations between SV40 and human cancers (IARC, 2012). However, cohort studies have been largely negative, and serological cross-reactivity between SV40 and the highly prevalent BKV and JCV are likely to explain the discrepancies observed between serological studies. As for case series, results have been controversial and inconclusive due to the limitations of the PCR technique: although PCR is sensitive, it is also prone to contamination, and does not distinguish between coincidental infections in the cancer tissue and a true causal association. After a thorough review, SV40 was considered "Not classifiable with respect to its carcinogenicity to humans" (group 3) by IARC in 2012 (IARC, 2012).

BKV and JCV are very common in the human population (seroprevalence in adults of 90% and 70%, respectively) and have been associated with a few human diseases. However, carcinogenicity data in humans are sparse. For BKV, cohort studies were all negative, and a small number of case-control and case series have shown inconsistent results. For JCV, a few formal prospective and case–control studies have found no association with human cancer. Although BKV and JCV can transform cells in culture and produce tumours in inoculated animals or transgenic mice, the lack of data in human studies led the IARC working group to classify these two viruses as possibly carcinogenic to humans (group 2B) (IARC, 2012).

Cutaneous HPVs

The cutaneous HPV types that belong to the genus beta of the HPV phylogenetic tree –and among them, especially HPV-5 and HPV-8 - were first associated with the development of cutaneous squamous cell carcinoma in patients affected by a rare autosomal recessive genetic disorder called epidermodysplasia verruciformis (EV) (Tommasino, 2014). In EV patients, mutations in the transmembrane channel gene TMC6 or TMC8 create a susceptibility to infections by HPV and the development of EV-typical plane warts. In 30% to 60% of EV patients, cutaneous SCC develops at age 20 to 50 and is usually localized in sun-exposed areas of the skin.

Based on the EV carcinogenic model, several studies have implicated beta HPVs as a risk factor for cutaneous SCC in the general population. Although positive associations have been observed between cutaneous SCC and HPV of the genus beta (measured by DNA PCR in skin lesions and/or multiplex serology in plasma samples) (Bouwes Bavinck *et al.*, 2010; Iannacone *et al.*, 2014), a causal relationship has not been demonstrated yet (IARC, 2011). There is no predominant HPV type in SCC, and association with multiples types has been reported. Furthermore, HPV do not integrate into human DNA, are present in non-monoclonal form and in low copy number, and are not transcriptionally active in tumour cells (Howley and Pfister, 2015). Experimental and epidemiological evidence suggest that beta HPV may act as a co-factor in addition to UV radiation exposure in early stages of carcinogenesis, possibly through expression of oncoproteins E6 and E7 and subsequent inhibition of DNA repair mechanisms and apoptosis (Shterzer *et al.*, 2014).

The IARC expert working group stated in their 2011 evaluation that, while HPV 5 and HPV 8 are possibly carcinogenic to patients with EV (group 2B), in the general population these types and other beta types are not classifiable as to their carcinogenicity to humans (group 3) (IARC, 2011).

Conclusion and Future Trends

Approximately 1.2 million new cases of cancer occurring worldwide each year are attributable to the four DNA viruses classified as carcinogenic to humans (group 1) by IARC. Given the ubiquity of carcinogenic DNA viruses, and the fact that transmission tends to occur at young ages, primary prevention through vaccination is the best approach to prevent future cancer cases. The effectiveness of prophylactic vaccination against HBV has been clearly demonstrated and the vaccine has become relatively inexpensive. Similarly, there are three prophylactic vaccines against HPV16/18, HPV6/11/16/18 and HPV6/11/16/18/31/33/45/52/58 that are now widely considered safe and effective. For HPV vaccination, higher coverage can be achieved through the reduction of the number of doses, and development of more affordable second generation vaccines, in particular from manufacturers in less developed countries (Schiller and Muller, 2015).

Secondary prevention of DNA virus-associated cancers through antiviral treatments has improved greatly in the last decades (e.g., against HBV) but treatments are expensive and side effects sometimes severe (World Health Organization, 2015). Universal early detection of HIV and uninterrupted cART remain the most powerful tools to prevent Kaposi sarcoma and EBV-associated lymphomas in HIV-infected people, and may also substantially reduce the excess of cancer of the anus and cervix due to HPV in this population, provided that even moderate levels of immunodeficiency are avoided over the long latent phase of HPV infection. As for EBV, two independent vaccine approaches – prophylactic and therapeutic- are under development to prevent or control EBV-associated diseases. Phase 1 or phase 2 clinical studies have shown some interesting perspectives for the efficacy of both of these approaches (Balfour, 2014).

In respect to future research, lymphomas and non-melanoma skin cancer represent two priorities. The respective contribution of EBV and cutaneous HPV types to these two malignancies is ill-defined and other infectious agents, possibly including DNA viruses, may be involved.

References

Agelli, M., Clegg, L.X., Becker, J.C., and Rollison, D.E. (2010). The etiology and epidemiology of merkel cell carcinoma. Curr. Probl. Cancer 34, 14-37. doi:10.1016/j.currproblcancer.2010.01.001

Alemany, L., Saunier, M., Alvarado-Cabrero, I., Quiros, B., Salmeron, J., Shin, H.R., Pirog, E.C., Guimera, N., Hernandez-Suarez, G., Felix, A., *et al.* (2015). Human papillomavirus DNA prevalence and type distribution in anal carcinomas worldwide. Int. J. Cancer 136, 98-107. doi:10.1002/ijc.28963

Backes, D.M., Kurman, R.J., Pimenta, J.M., and Smith, J.S. (2009). Systematic review of human papillomavirus prevalence in invasive penile cancer. Cancer Causes Control 20, 449-457

Balfour, H.H., Jr. (2014). Progress, prospects, and problems in Epstein-Barr virus vaccine development. Curr. Opin. Virol. 6c, 1-5. doi:10.1016/j.coviro.2014.02.005

Barnes, L., Eveson, J.W., Reichart, P., and Sidransky, D. (2005). Pathology and Genetics of Head and Neck Tumours. In WHO Classification of Tumours (Lyon, France: IARC Press).

Bishop, J.A., Ma, X.J., Wang, H., Luo, Y., Illei, P.B., Begum, S., Taube, J.M., Koch, W.M., and Westra, W.H. (2012). Detection of transcriptionally active high-risk HPV in patients with head and neck squamous cell carcinoma as visualized by a novel E6/E7 mRNA *in situ* hybridization method. The American journal of surgical pathology 36, 1874-1882. doi:10.1097/PAS.0b013e318265fb2b

Bohlius, J., Valeri, F., Maskew, M., Prozesky, H., Garone, D., Sengayi, M., Fox, M.P., Davies, M.A., and Egger, M. (2014). Kaposi's Sarcoma in HIV-infected patients in South Africa: Multicohort study in the antiretroviral therapy era. Int. J. Cancer 135, 2644-2652. doi:10.1002/ijc.28894

Bouwes Bavinck, J.N., Neale, R.E., Abeni, D., Euvrard, S., Green, A.C., Harwood, C.A., de Koning, M.N., Naldi, L., Nindl, I., Pawlita, M., *et al.* (2010). Multicenter study of the association between betapapillomavirus infection and cutaneous squamous cell carcinoma. Cancer research 70, 9777-9786. doi:10.1158/0008-5472.can-10-0352

Bruni, L., Diaz, M., Castellsague, X., Ferrer, E., Bosch, F.X., and de Sanjose, S. (2010). Cervical human papillomavirus prevalence in 5 continents: meta-analysis of 1 million women with normal cytological findings. J. Infect. Dis. 202, 1789-1799. doi:10.1086/657321

Bruzzi, P., Green, S.B., Byar, D.P., Brinton, L.A., and Schairer, C. (1985). Estimating the population attributable risk for multiple risk factors using case-control data. Am. J. Epidem. 122, 904-914

Camargo, M.C., Kim, K.M., Matsuo, K., Torres, J., Liao, L.M., Morgan, D.R., Michel, A., Waterboer, T., Zabaleta, J., Dominguez, R.L., *et al.* (2016). Anti-*Helicobacter pylori* Antibody Profiles in Epstein-Barr virus (EBV)-Positive and EBV-Negative Gastric Cancer. Helicobacter 21, 153-157. doi:10.1111/hel.12249

Carbone, A., Vaccher, E., Gloghini, A., Pantanowitz, L., Abayomi, A., de Paoli, P., and Franceschi, S. (2014). Diagnosis and management of lymphomas and other cancers in HIV-infected patients. Nature Rev. Clin. Oncol. 11, 223-238. doi:10.1038/nrclinonc.2014.31

Castellsague, X., Alemany, L., Quer, M., Halec, G., Quiros, B., Tous, S., Clavero, O., Alos, L., Biegner, T., Szafarowski, T., *et al.* (2016). HPV Involvement in Head and Neck Cancers: Comprehensive Assessment of Biomarkers in 3680 Patients. J. Natl. Cancer Inst. 108. doi:10.1093/jnci/djv403

Chang, E.T., and Adami, H.O. (2006). The enigmatic epidemiology of nasopharyngeal carcinoma. Cancer epidemiology, biomarkers and prevention: a publication of the American Association for Cancer Research, cosponsored by the American Society of Preventive Oncology 15, 1765-1777

Chang, M.H., You, S.L., Chen, C.J., Liu, C.J., Lee, C.M., Lin, S.M., Chu, H.C., Wu, T.C., Yang, S.S., Kuo, H.S., *et al.* (2009). Decreased Incidence of Hepatocellular Carcinoma in Hepatitis B Vaccinees: A 20-Year Follow-up Study. J. Natl. Cancer Inst.

Chang, Y., Cesarman, E., Pessin, M.S., Lee, F., Culpepper, J., Knowles, D.M., and Moore, P.S. (1994). Identification of herpesvirus-like DNA sequences in AIDS-associated Kaposi's sarcoma. Science (New York, NY) 266, 1865-1869

Chaturvedi, A.K., Anderson, W.F., Lortet-Tieulent, J., Curado, M.P., Ferlay, J., Franceschi, S., Rosenberg, P.S., Bray, F., and Gillison, M.L. (2013). Worldwide trends in incidence rates for oral cavity and oropharyngeal cancers. J. Clin. Oncol.: Offical J. Am Soc. Clin. Oncol. 31, 4550-4559. doi:10.1200/jco.2013.50.3870

Chaturvedi, A.K., Engels, E.A., Pfeiffer, R.M., Hernandez, B.Y., Xiao, W., Kim, E., Jiang, B., Goodman, M.T., Sibug-Saber, M., Cozen, W., *et al.* (2011). Human papillomavirus and rising oropharyngeal cancer incidence in the United States. J. Clin. Oncol.: Offical J. Am Soc. Clin. Oncol. 29, 4294-4301. doi:10.1200/jco.2011.36.4596

Chaux, A., Lezcano, C., Cubilla, A.L., Tamboli, P., Ro, J., and Ayala, A. (2010). Comparison of subtypes of penile squamous cell carcinoma from high and low incidence geographical regions. Int J Surg Pathol 18, 268-277

Chen, C.J., Yang, H.I., Su, J., Jen, C.L., You, S.L., Lu, S.N., Huang, G.T., and Iloeje, U.H. (2006). Risk of hepatocellular carcinoma across a biological gradient of serum hepatitis B virus DNA level. Jama 295, 65-73

Cho, L.Y., Yang, J.J., Ko, K.P., Park, B., Shin, A., Lim, M.K., Oh, J.K., Park, S., Kim, Y.J., Shin, H.R., *et al.* (2011). Coinfection of hepatitis B and C viruses and risk of hepatocellular carcinoma: systematic review and meta-analysis. Int. J. Cancer 128, 176-184. doi:10.1002/ijc.25321

Clifford, G., de Vuyst, H., Tenet, V., Plummer, M., Tully, S., and Franceschi, S. (2016). Effect of HIV infection on human papillomavirus types causing invasive cervical cancer in Africa. J Acquir Immune Defic Syndr 73, 332-339. doi:10.1097/qai.0000000000001113.

Crosbie, E.J., Einstein, M.H., Franceschi, S., and Kitchener, H.C. (2013). Human papillomavirus and cervical cancer. Lancet (London, England) 382, 889-899. doi:10.1016/s0140-6736(13)60022-7

d'Amore, F., Johansen, P., Houmand, A., Weisenburger, D.D., and Mortensen, L.S. (1996). Epstein-Barr virus genome in non-Hodgkin's lymphomas occurring in immunocompetent patients: highest prevalence in nonlymphoblastic T-cell lymphoma and correlation with a poor prognosis. Danish Lymphoma Study Group, LYFO. Blood 87, 1045-1055

de Martel, C., Ferlay, J., Franceschi, S., Vignat, J., Bray, F., Forman, D., and Plummer, M. (2012). Global burden of cancers attributable to infections in 2008: a review and synthetic analysis. Lancet Oncol. 13, 607-615. doi:10.1016/s1470-2045(12)70137-7

de Martel, C., Maucort-Boulch, D., Plummer, M., and Franceschi, S. (2015a). Worldwide relative contribution of hepatitis B and C viruses in hepatocellular carcinoma. Hepatology. doi:10.1002/hep.27969

de Martel, C., Shiels, M.S., Franceschi, S., Simard, E.P., Vignat, J., Hall, H.I., Engels, E.A., and Plummer, M. (2015b). Cancers attributable to infections among adults with HIV in the United States. AIDS (London, England). doi:10.1097/qad.0000000000000808

de Sanjose, S., Alemany, L., Ordi, J., Tous, S., Alejo, M., Bigby, S.M., Joura, E.A., Maldonado, P., Laco, J., Bravo, I.G., *et al.* (2013). Worldwide human papillomavirus genotype attribution in over 2000 cases of intraepithelial and invasive lesions of the vulva. Eur. J. Cancer (Oxford, England: 1990) 49, 3450-3461. doi:10.1016/j.ejca.2013.06.033

De Vuyst, H., Clifford, G.M., Nascimento, M.C., Madeleine, M.M., and Franceschi, S. (2009). Prevalence and type distribution of human papillomavirus in carcinoma and intraepithelial neoplasia of the vulva, vagina and anus: a meta-analysis. Int. J. Cancer 124, 1626-1636

Dogan, S., Hedberg, M.L., Ferris, R.L., Rath, T.J., Assaad, A.M., and Chiosea, S.I. (2014). Human papillomavirus and Epstein-Barr virus in nasopharyngeal carcinoma in a low-incidence population. Head and neck 36, 511-516. doi:10.1002/hed.23318

Donato, F., Boffetta, P., and Puoti, M. (1998). A meta-analysis of epidemiological studies on the combined effect of hepatitis B and C virus infections in causing hepatocellular carcinoma. Int. J. Cancer 75, 347-354

Du, M.Q., Bacon, C.M., and Isaacson, P.G. (2007). Kaposi sarcoma-associated herpesvirus/human herpesvirus 8 and lymphoproliferative disorders. J. Clin. Pathol. 60, 1350-1357

Epstein, M.A., Achong, B.G., and Barr, Y.M. (1964). Virus particles in cultured lymphoblasts from Burkitt's lymphoma. Lancet (London, England) 1, 702-703

Feng, H., Shuda, M., Chang, Y., and Moore, P.S. (2008). Clonal integration of a polyomavirus in human Merkel cell carcinoma. Science (New York, NY) 319, 1096-1100

Ferlay, J., Soerjomataram, I., Ervik, M., Eser, S., Mathers, C., Rebelo, M., Parkin, D.M., and Forman, D. (2013). GLOBOCAN 2012 v1.0, Cancer Incidence and Mortality Worldwide: IARC CancerBase No. 11 [Internet].:, ed. (Lyon, France: International Agency for Research on Cancer)

Girschik, J., Thorn, K., Beer, T.W., Heenan, P.J., and Fritschi, L. (2011). Merkel cell carcinoma in Western Australia: a population-based study of incidence and survival. Brit. J. Dermatol. 165, 1051-1057. doi:10.1111/j.1365-2133.2011.10493.x

Grulich, A.E., Poynten, I.M., Machalek, D.A., Jin, F., Templeton, D.J., and Hillman, R.J. (2012). The epidemiology of anal cancer. Sexual health 9, 504-508. doi:10.1071/sh12070

Guan, P., Howell-Jones, R., Li, N., Bruni, L., de Sanjose, S., Franceschi, S., and Clifford, G.M. (2012). Human papillomavirus types in 115,789 HPV-positive women: a meta-analysis from cervical infection to cancer. Int. J. Cancer 131, 2349-2359. doi:10.1002/ijc.27485

Hennessey, K., Mendoza-Aldana, J., Bayutas, B., Lorenzo-Mariano, K.M., and Diorditsa, S. (2013). Hepatitis B control in the World Health Organization's Western Pacific Region: targets, strategies, status. Vaccine 31 Suppl 9, J85-92. doi:10.1016/j.vaccine.2012.10.082

Howley, P.M., and Pfister, H.J. (2015). Beta genus papillomaviruses and skin cancer. Virology 479-480, 290-296. doi:10.1016/j.virol.2015.02.004

Iannacone, M.R., Gheit, T., Pfister, H., Giuliano, A.R., Messina, J.L., Fenske, N.A., Cherpelis, B.S., Sondak, V.K., Roetzheim, R.G., Silling, S., *et al.* (2014). Case-control study of genus-beta human papillomaviruses in plucked eyebrow hairs and cutaneous squamous cell carcinoma. Int. J. Cancer 134, 2231-2244. doi:10.1002/ijc.28552

IARC (1994). Hepatitis viruses. Lyon, France, 8-15 June 1993. IARC Monogr Eval Carcinog Risks Hum 59, 1-255

IARC (1995). Human papillomaviruses. Lyon, France, 6-13 June 1995. IARC Monogr Eval Carcinog Risks Hum 64, 1-378

IARC (1997). Epstein-Barr Virus and Kaposi's Sarcoma Herpesvirus/Human Herpesvirus 8. Lyon, France, 17-24 June 1997. IARC Monogr Eval Carcinog Risks Hum 70, 1-524

IARC (2007). Human Papillomaviruses. Lyon, France, 15–22 February 2005. IARC Monogr Eval Carcinog Risks Hum 90, 1-670

IARC (2008). WHO Classification of Tumours of Haematopoietic and Lymphoid Tissues. (Lyon, France: IARC Press)

IARC (2011). Biological Agents. Volume 100B. A review of human carcinogens. Lyon, France, 24 February-3 March 2009. IARC Monogr Eval Carcinog Risks Hum 100B

IARC (2012). Malaria and some Polyomaviruses (SV40, BK, JC, and Merkel cell viruses)Lyon, France, 7-14 February 2012. IARC Monogr Eval Carcinog Risks Hum 104

International Human Papillomavirus Reference Center (2015). Human papillomavirus reference clones (Stockholm, Sweden: Department of Laboratory Medicine, Karolinska Institutet)

Kaae, J., Hansen, A.V., Biggar, R.J., Boyd, H.A., Moore, P.S., Wohlfahrt, J., and Melbye, M. (2010). Merkel cell carcinoma: incidence, mortality, and risk of other cancers. J. Natl. Cancer Inst. 102, 793-801. doi:10.1093/jnci/djq120

Kjaer, S.K., Frederiksen, K., Munk, C., and Iftner, T. (2010). Long-term absolute risk of cervical intraepithelial neoplasia grade 3 or worse following human papillomavirus infection: role of persistence. J. Natl. Cancer Inst. 102, 1478-1488. doi:10.1093/jnci/djq356

Kurman, R.J., Toki, T., and Schiffman, M.H. (1993). Basaloid and warty carcinomas of the vulva. Distinctive types of squamous cell carcinoma frequently associated with human papillomaviruses. Am J Surg Pathol 17, 133-145

Lee, J.H., Kim, Y., Choi, J.W., and Kim, Y.S. (2014). Prevalence and prognostic significance of Epstein-Barr virus infection in classical Hodgkin's lymphoma: a meta-analysis. Arch. Med. Res. 45, 417-431. doi:10.1016/j.arcmed.2014.06.001

Li, N., Franceschi, S., Howell-Jones, R., Snijders, P.J., and Clifford, G.M. (2011). Human papillomavirus type distribution in 30,848 invasive cervical cancers worldwide: Variation by geographical region, histological type and year of publication. Int. J. Cancer 128, 927-935

Li, X., Fasano, R., Wang, E., Yao, K.T., and Marincola, F.M. (2009). HLA associations with nasopharyngeal carcinoma. Current molecular medicine 9, 751-765

Lin, Z., Khong, B., Kwok, S., Cao, H., West, R.B., Le, Q.T., and Kong, C.S. (2014). Human papillomavirus 16 detected in nasopharyngeal carcinomas in white Americans but not in endemic Southern Chinese patients. Head and neck 36, 709-714. doi:10.1002/hed.23362

Liu, W., Yang, R., Payne, A.S., Schowalter, R.M., Spurgeon, M.E., Lambert, P.F., Xu, X., Buck, C.B., and You, J. (2016). Identifying the Target Cells and Mechanisms of Merkel Cell Polyomavirus Infection. Cell host and microbe 19, 775-787. doi:10.1016/j.chom.2016.04.024

Malope, B.I., MacPhail, P., Mbisa, G., MacPhail, C., Stein, L., Ratshikhopha, E.M., Ndhlovu, L., Sitas, F., and Whitby, D. (2008). No evidence of sexual transmission of Kaposi's sarcoma herpes virus in a heterosexual South African population. AIDS (London, England) 22, 519-526

Malope, B.I., Pfeiffer, R.M., Mbisa, G., Stein, L., Ratshikhopha, E.M., O'Connell, D.L., Sitas, F., MacPhail, P., and Whitby, D. (2007). Transmission of Kaposi sarcoma-associated herpesvirus between mothers and children in a South African population. J Acquir Immune Defic Syndr 44, 351-355

McCredie, M.R., Sharples, K.J., Paul, C., Baranyai, J., Medley, G., Jones, R.W., and Skegg, D.C. (2008). Natural history of cervical neoplasia and risk of invasive cancer in women with cervical intraepithelial neoplasia 3: a retrospective cohort study. Lancet Oncol. 9, 425-434. doi:10.1016/s1470-2045(08)70103-7

Minoura-Etoh, J., Gotoh, K., Sato, R., Ogata, M., Kaku, N., Fujioka, T., and Nishizono, A. (2006). *Helicobacter pylori*-associated oxidant monochloramine induces reactivation of Epstein-Barr virus (EBV) in gastric epithelial cells latently infected with EBV. J. Med. Microbiol. 55, 905-911. doi:10.1099/jmm.0.46580-0

Miralles-Guri, C., Bruni, L., Cubilla, A.L., Castellsague, X., Bosch, F.X., and de Sanjose, S. (2009). Human papillomavirus prevalence and type distribution in penile carcinoma. J. Clin. Pathol. 62, 870-878. doi:10.1136/jcp.2008.063149

Murphy, G., Pfeiffer, R., Camargo, M.C., and Rabkin, C.S. (2009). Meta-analysis shows that prevalence of Epstein-Barr virus-positive gastric cancer differs based on sex and anatomic location. Gastroenterology 137, 824-833. doi:10.1053/j.gastro.2009.05.001

Ndiaye, C., Mena, M., Alemany, L., Arbyn, M., Castellsague, X., Laporte, L., Bosch, F.X., de Sanjose, S., and Trottier, H. (2014). HPV DNA, E6/E7 mRNA, and $p16^{INK4a}$ detection in head and neck cancers: a systematic review and meta-analysis. Lancet Oncol. 15, 1319-1331. doi:10.1016/s1470-2045(14)70471-1

Nicholls, J.M., Agathanggelou, A., Fung, K., Zeng, X., and Niedobitek, G. (1997). The association of squamous cell carcinomas of the nasopharynx with Epstein-Barr virus shows geographical variation reminiscent of Burkitt's lymphoma. J Pathol 183, 164-168

Nielsen, A., Munk, C., and Kjaer, S.K. (2012). Trends in incidence of anal cancer and high-grade anal intraepithelial neoplasia in Denmark, 1978-2008. Int. J. Cancer 130, 1168-1173. doi:10.1002/ijc.26115

Ott, J.J., Stevens, G.A., Groeger, J., and Wiersma, S.T. (2012). Global epidemiology of hepatitis B virus infection: New estimates of age-specific HBsAg seroprevalence and endemicity. Vaccine 30, 2212-2219. doi:http://dx.doi.org/10.1016/j.vaccine.2011.12.116

Ou, S.H., Zell, J.A., Ziogas, A., and nton-Culver, H. (2007). Epidemiology of nasopharyngeal carcinoma in the United States: improved survival of Chinese patients within the keratinizing squamous cell carcinoma histology. Ann Oncol 18, 29-35

Parkin, D.M. (2006). The global health burden of infection-associated cancers in the year 2002. Int. J. Cancer 118, 3030-3044

Paulson, K.G., Carter, J.J., Johnson, L.G., Cahill, K.W., Iyer, J.G., Schrama, D., Becker, J.C., Madeleine, M.M., Nghiem, P., and Galloway, D.A. (2010). Antibodies to merkel cell polyomavirus T antigen oncoproteins reflect tumor burden in merkel cell carcinoma patients. Cancer research 70, 8388-8397. doi:10.1158/0008-5472.can-10-2128

Pisani, P., Parkin, D.M., Munoz, N., and Ferlay, J. (1997). Cancer and infection: estimates of the attributable fraction in 1990. Cancer epidemiology, biomarkers and prevention: a publication of the American Association for Cancer Research, cosponsored by the American Society of Preventive Oncology 6, 387-400

Plummer, M., de Martel, C., Vignat, J., Ferlay, J., Bray, F., and Franceschi, S. (2016). Global burden of cancer attributable to infections in 2012. Lancet Glob Health 4, e609-e616. doi:10.1016/S2214-109X(16)30143-7.

Plummer, M., Schiffman, M., Castle, P.E., Maucort-Boulch, D., and Wheeler, C.M. (2007). A 2-year prospective study of human papillomavirus persistence among women with a cytological diagnosis of atypical squamous cells of undetermined significance or low-grade squamous intraepithelial lesion. J. Infect. Dis. 195, 1582-1589. doi:10.1086/516784

Polesel, J., Franceschi, S., Talamini, R., Negri, E., Barzan, L., Montella, M., Libra, M., Vaccher, E., Franchin, G., La Vecchia, C., *et al.* (2011). Tobacco smoking, alcohol drinking, and the risk of different histological types of nasopharyngeal cancer in a low-risk population. Oral oncology 47, 541-545. doi:10.1016/j.oraloncology.2011.03.017

Queiroga, E.M., Gualco, G., Weiss, L.M., Dittmer, D.P., Araujo, I., Klumb, C.E., Harrington, W.J., Jr., and Bacchi, C.E. (2008). Burkitt lymphoma in Brazil is characterized by geographically distinct clinicopathologic features. Am J Clin Pathol 130, 946-956

Quint, W., Jenkins, D., Molijn, A., Struijk, L., van de Sandt, M., Doorbar, J., Mols, J., Van Hoof, C., Hardt, K., Struyf, F., *et al.* (2012). One virus, one lesion--individual components of CIN lesions contain a specific HPV type. J. Pathol. 227, 62-71. doi:10.1002/path.3970

Rietbergen, M.M., Leemans, C.R., Bloemena, E., Heideman, D.A., Braakhuis, B.J., Hesselink, A.T., Witte, B.I., Baatenburg de Jong, R.J., Meijer, C.J., Snijders, P.J., *et al.* (2013). Increasing prevalence rates of HPV attributable oropharyngeal squamous cell carcinomas in the Netherlands as assessed by a validated test algorithm. Int. J. Cancer 132, 1565-1571. doi:10.1002/ijc.27821

Robey, R.C., and Bower, M. (2015). Facing up to the ongoing challenge of Kaposi's sarcoma. Current opinion in infectious diseases 28, 31-40. doi:10.1097/qco.0000000000000122

Rositch, A.F., Koshiol, J., Hudgens, M.G., Razzaghi, H., Backes, D.M., Pimenta, J.M., Franco, E.L., Poole, C., and Smith, J.S. (2013). Patterns of persistent genital human papillomavirus infection among women worldwide: a literature review and meta-analysis. Int. J. Cancer 133, 1271-1285. doi:10.1002/ijc.27828

Schiller, J.T., and Muller, M. (2015). Next generation prophylactic human papillomavirus vaccines. Lancet Oncol. 16, e217-225. doi:10.1016/s1470-2045(14)71179-9

Shiels, M.S., Kreimer, A.R., Coghill, A.E., Darragh, T.M., and Devesa, S.S. (2015). Anal Cancer Incidence in the United States, 1977-2011: Distinct Patterns by Histology and Behavior. Cancer epidemiology, biomarkers and prevention: a publication of the American Association for Cancer Research, cosponsored by the American Society of Preventive Oncology 24, 1548-1556. doi: 10.1158/1055-9965.epi-15-0044

Shterzer, N., Heyman, D., Shapiro, B., Yaniv, A., Jackman, A., Serour, F., Chaouat, M., Gonen, P., Tommasino, M., and Sherman, L. (2014). Human papillomavirus types detected in skin warts and cancer differ in their transforming properties but commonly counteract UVB induced protective responses in human keratinocytes. Virology 468-470, 647-659. doi:10.1016/j.virol.2014.09.015

Stankiewicz, E., Prowse, D.M., Ng, M., Cuzick, J., Mesher, D., Hiscock, F., Lu, Y.J., Watkin, N., Corbishley, C., Lam, W., *et al.* (2011). Alternative HER/PTEN/Akt pathway activation in HPV positive and negative penile carcinomas. PLoS One 6, e17517

Stein, H., Delsol, G., Pileri, S.A., Weiss, L.M., Popperna, S., and Jaffe, E.S. (2008). Hodgkin lymphoma. In WHO classification of tumours of haematopoietic and lymphoid tissues, S.H. Swerdlow, E. Campo, N. Lee Harris, E.S. Jaffe, S.A. Pileri, H. Stein, J. Thiele, and J.W. Vardiman, eds. (Lyon, France: International Agency for Research on Cancer).

Teras, L.R., Rollison, D.E., Pawlita, M., Michel, A., Brozy, J., de Sanjose, S., Blase, J.L., and Gapstur, S.M. (2015). Epstein-Barr virus and risk of non-Hodgkin lymphoma in the cancer prevention study-II and a meta-analysis of serologic studies. Int. J. Cancer 136, 108-116. doi:10.1002/ijc.28971

The Cancer Genome Atlas Research Network (2014). The Cancer Genome Atlas Research Network: Comprehensive molecular characterization of gastric adenocarcinoma. Nature 513, 202-209. doi:10.1038/nature13480

Thompson, M.P., and Kurzrock, R. (2004). Epstein-Barr virus and cancer. Clin Cancer Res 10, 803-821

Tommasino, M. (2014). The human papillomavirus family and its role in carcinogenesis. Semin Cancer Biol 26, 13-21. doi:10.1016/j.semcancer.2013.11.002

Wang, J., Young, L., Win, W., and Taylor, C.R. (2005). Distribution and ZAP-70 expression of WHO lymphoma categories in Shanxi, China: a review of 447 cases using a tissue microarray technique. Applied immunohistochemistry and molecular morphology: AIMM / official publication of the Society for Applied Immunohistochemistry 13, 323-332

Wee, J.T., Ha, T.C., Loong, S.L., and Qian, C.N. (2010). Is nasopharyngeal cancer really a "Cantonese cancer"? Chinese J. Cancer 29, 517-526

World Health Organization (2015). Guidelines for the Prevention, Care and Treatment of Persons with Chronic Hepatitis B Infection. In (Geneva: World Health Organization).

Yang, H.I., Lu, S.N., Liaw, Y.F., You, S.L., Sun, C.A., Wang, L.Y., Hsiao, C.K., Chen, P.J., Chen, D.S., and Chen, C.J. (2002). Hepatitis B e antigen and the risk of hepatocellular carcinoma. N Engl J Med 347, 168-174. doi:10.1056/NEJMoa013215

Yee Ko, J.M., Dai, W., Wun Wong, E.H., Kwong, D., Tong Ng, W., Lee, A., Kai Cheong Ngan, R., Chung Yau, C., Tung, S., and Li Lung, M. (2014). Multigene pathway-based analyses identify nasopharyngeal carcinoma risk associations for cumulative adverse effects of TERT-CLPTM1L and DNA double-strand breaks repair. Int. J. Cancer 135, 1634-1645. doi: 10.1002/ijc.28802

Young, L.S., and Rickinson, A.B. (2004). Epstein-Barr virus: 40 years on. Nat Rev Cancer 4, 757-768.

2

Biology of the Human Papillomavirus Life Cycle

Sally Roberts*, Ieisha Pentland, Paul McCormack and Joanna L. Parish*

Institute of Cancer and Genomic Sciences, College of Medical and Dental Sciences, University of Birmingham, Birmingham, United Kingdom.

**Corresponding authors: s.roberts@bham.ac.uk, j.l.parish@bham.ac.uk*

Abstract

Human papillomaviruses (HPV) are small, double-stranded DNA viruses that replicate in squamous epithelium and cause hyperproliferative lesions, some of which are at risk of malignant transformation. The infectious HPV replication cycle is intimately linked to the differentiation program of this tissue. The virus infects undifferentiated basal keratinocytes to establish and maintain replication of the viral genome, whereas productive viral DNA amplification and formation of new progeny is restricted to differentiating keratinocytes. Successful viral replication in differentiated cells requires the virus to reprogram the post-mitotic keratinocytes in order to support viral genome replication. The use of animal PV models and cell-based tissue culture models of the virus life cycle, combined with generation of recombinant virions for infection studies, has revealed much about the complexities of the HPV life cycle. Defining the nature of the keratinocyte that HPV targets, infection strategies, and virus-host interactions necessary for virus DNA replication and virion production has contributed to the design of anti-HPV vaccines and the understanding of HPV-driven cancer development. This chapter aims to provide an overview of key HPV life cycle events and the models used to interrogate virus-host interplay.

Introduction

Human papillomaviruses (HPV) are non-enveloped, small, double-stranded circular DNA viruses that belong to the *Papillomaviridae* family. They infect squamous epithelia of the cutaneous and internal mucosal regions, namely the skin, anogenital and oropharyngeal-laryngeal tracts. Most HPV infections are asymptomatic, or form hyperproliferative but benign warty lesions and papillomas that are self-limiting. However, some infections are persistent and are at risk of progressing to cancer. HPV tropism is strictly human and it is thought that these viruses have evolved together with their human host over the past 200,000 years (Bravo and Felez-Sanchez, 2015).

The viral aetiology of warts was shown in the early part of the 20th century. It was the Italian physician Giuseppe Ciuffo who in 1907 showed that cell-free filtrates of human warts could transmit the disease suggesting a viral aetiology (Javier and Butel, 2008). This was followed in 1933 by work from Richard Shope and E. Weston Hurst who filtered soluble extracts of warts from wild cottontail rabbits and showed that this filtrate was able to transmit the disease to other wild cottontail rabbits (Shope and Hurst, 1933; Javier and Butel, 2008). The link between PV and cancer emerged when it was discovered that infection of domestic rabbits and not the natural host, led to subsequent papilloma formation that could either regress, persist without progression, or be capable of forming squamous cell carcinomas of the skin (Rous and Beard, 1935). Several other pieces of strong evidence were uncovered in the ensuing decades for the tumourigenicity of PVs culminating in the discovery in the early 1980s of a novel PV in cervical cancers by Harald zur Hausen and colleagues (Durst *et al.*, 1983; Boshart *et al.*, 1984).

Classification

To date, 209 HPV types have been identified (http://www.hpvcenter.se/html/refclones.html) and the complete genomic DNA sequences have been determined for just over 180 of these (http://pave.niad.gov). It is very likely that the number of HPV types will have expanded further by the time this chapter is published because of next generation-sequencing technologies. The major capsid *L1* gene sequence is used to classify the viruses into genera, species and types. A novel type is classified by an *L1* sequence variation of at least 10% to any other type and types can be subdivided further into variant lineages and sub-lineages, based on *L1* gene difference of 1.0-10% and 0.5-1.0% respectively (Burk *et al.*, 2013). Studies of variant genotype-phenotype relationship of carcinogenic HPV types have shown that small numbers of nucleotide changes that alter regulatory sequences or cause non-synonymous changes in protein coding sequence, can have implications for virus biology and pathogenesis (Zehbe *et al.*, 1998; Berumen *et al.*, 2001; Hildesheim *et al.*, 2001; von Knebel Doeberitz, 2016).

There are five genera (alpha-, beta-, gamma-, mu- and nu-PV) into which HPV types are classified. The mu-PV and nu-PV genera contain the smallest number of HPV types; three and one respectively, and they infect cutaneous surfaces causing benign warts - for example, HPV1 (mu-PV) induces warts and verrucae of the hands and feet. HPV204 is the most recent addition to the mu genus (Sterbenc *et al.*, 2017). The beta and gamma genera contain many more types and they infect cutaneous skin.

Beta-PVs are present on the skin and in plucked hairs of healthy immunocompetent individuals but infections are generally asymptomatic. However, patients with the genetic disorder epidermodysplasia verruciformis (EV) are highly susceptible to infections with beta-PVs (e.g. HPV5 and 8), and develop flat warts that are at high risk of progressing to cutaneous squamous cell carcinomas on sun-exposed sites. EV is classified as a primary immunodeficiency and these patients exhibit defects in both innate and cell-mediated immunity. Although the exact cause is uncertain, these individuals carry mutations in one of two genes (*EVER1/TCM6* and *EVER2/TCM8*), whose functions are linked to zinc homeostasis that is important in T cell activation (Ramoz *et al.*, 2002; Horton and Stokes, 2014). Organ transplant recipients are also susceptible to infection with this group of beta-PV and are at increased risk of non-melanoma skin cancers. The role of these viruses in the natural history of skin carcinogenesis is not resolved fully, but evidence suggests a role synergistic with UV-irradiation in the early stages of carcinogenesis, but not necessarily required for maintenance of the neoplastic phenotype (Viarisio et al., 2018). The details of the molecular mechanisms leading to beta-HPV cellular transformation are reviewed elsewhere in this volume (*See* Chapter 3: Molecular mechanisms of E6 and E7 oncoproteins from human papillomaviruses in cellular transformat-ion).

HPV types classified in the alpha genus are subdivided into those with a cutaneous tropism causing benign lesions (e.g. HPV2, a cause of common hand warts) and those that infect wet epithelial surfaces - the mucosa, and include the high-risk types identified as aetiological agents of cancer of the uterine cervix and a proportion of cancers in the vulva, vagina, anus, penis and oropharynx (tonsil and base of tongue). Other HPV types classified in this genus are HPV6 and HPV11 and, although rarely associated with malignancy, these low-risk HPV types are the cause of considerably hard to treat diseases, such as genital warts and respiratory papillomatosis.

HPV genome organization

Organization of the viral genome is relatively conserved amongst HPV types; the viral DNA consists of between 7.3 and 8.0 kilobase pairs (kbp) and contains 7-9 open reading frames (ORF) divided into early (E1, E2, E4, E5, E6, E7 and E8 [not all HPV types encode the E5 and/or E8 ORF]) and late (L1 and L2) genes (Figure 2.1A). HPV early gene proteins function primarily in viral DNA replication and gene expression, deregulation of keratinocyte proliferation and differentiation, and immune evasion. The late gene proteins constitute the viral coat, composed of both the major (L1) and minor (L2) proteins. The HPV prophylactic vaccine contains virus-like particles (VLP) assembled from the L1 protein (discussed in detail in Chapter 5 of this volume). The long control region (LCR; sometimes referred to as the upstream regulatory region) lies upstream of the early genes and contains the origin of replication (*Ori*) to which the E1 protein binds, recognition sites for a plethora of host transcription factors (Thierry, 2009) as well as binding sites for the viral regulator of replication and transcription, the E2 protein (Androphy *et al.*, 1987), and the early promoter (P^E) (Figure 2.1A and B). The virus has a complex transcription pattern, producing multiple spliced, polycistronic mRNAs synthesized from one DNA strand utilizing an early promoter (P^E), or a late promoter (P^L) that is activated as the infected

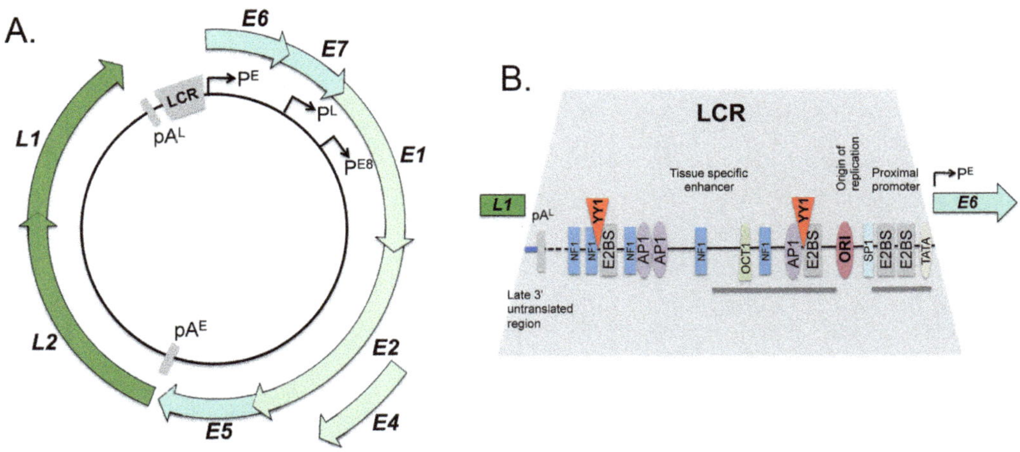

Figure 2.1. Organization of the viral genome of alpha human papillomaviruses. **(A)** Within the circular, double stranded HPV genome (~7.9 kbp), the positions of the long control region (LCR), early (E) and late (L) open reading frames, viral promoters (P^E, P^L and P^{E8}), early and late polyadenylation signals (pA^E, pA^L), are as shown. Early viral proteins E1 and E2 function in viral DNA replication and regulation of viral gene transcription. E4 is the most abundant protein expressed in the productive phase of the HPV life cycle and has been shown to disturb keratinocyte structures, including the keratin cytoskeleton. Functions of early proteins E5, E6 and E7 deregulate host pathways that control cell proliferation, cell survival and keratinocyte differentiation. For the cancer-causing HPV types, deregulation of the specific cellular pathways, including the p53 and RB tumour suppressor pathways, increase the risk of host genome instability and the acquisition of oncogenic mutations. P^{E8} drives expression of an E8^E2 protein. The late viral protein L1 and L2, expressed after initiation of HPV genome replication, make-up the viral coat, and also function in infection, virus particle internalization, trafficking of the viral DNA to the nucleus, and genome encapsidation. **(B)** The LCR lies between the early and late regions and contains the origin of replication (*ORI*), the late pA^L, E2 binding sites (E2BS) and binding sites for a large number of transcription factors (only transcription factors discussed in this chapter are shown). The solid grey bars show nucleosome positioning over the proximal promoter and tissue specific enhancer. The late 3' untranslated region (solid blue bar) contains regulatory elements important in the control of late gene expression. See main text for detail of LCR functions.

cell undergoes differentiation. A more recently discovered promoter within the E1 ORF sequence (P^{E8}) expresses the E8^E2 protein. Signals for polyadenylation of the viral early and late transcripts are present between the early and late region (pA^E), and between the late region and LCR (pA^L), respectively (Figure 2.1A).

Completion of the infectious cycle of HPV is highly dependent on the terminal differentiation programme of the keratinocyte – a feature that has limited investigations since the virus is not easily propagated in the tissue culture environment unless this differentiation process can be stimulated. This chapter examines the current understanding of the HPV life cycle, from infection through to virion progeny formation, and focuses on the different stages of the infectious cycle rather than a description of individual viral protein function. The focus is towards the cancer-causing alpha-HPV types that infect the anogenital and oropharyngeal tracts. This chapter does not delve into how these viruses have evolved to evade host immunity, and there is an excellent review that covers this area (Westrich *et al.*, 2017). Models, either animal-specific or cell-based that have supported PV propagation and have been of pivotal importance in understanding the PV life cycle and pathogenesis, are also discussed.

The HPV target cell

The HPV life cycle initiates in the basal cells of stratified epithelia that are in contact with the basal lamina that overlies the dermis (Figure 2.2). Access to the basal cells is thought to be through wounding of the multi-layered epithelia allowing the virus to gain entry into the basal cells. The target cells within the basal layer are not fully identified, but there is accumulating evidence to indicate that stem or stem cell-like cells are infected by the virus. There is also an idea that the virus may instruct reprogramming of cells to acquire stem cell-like features. A better understanding of the type of cell targeted by HPV at the different anatomical sites is beginning to shed light on why some infections with oncogenic HPV types progress to cancers and others not, as has been illustrated in HPV infections of the uterine cervix. In the cervix, the ectocervix, formed by non-keratinizing stratified epithelium, and the columnar epithelium of the

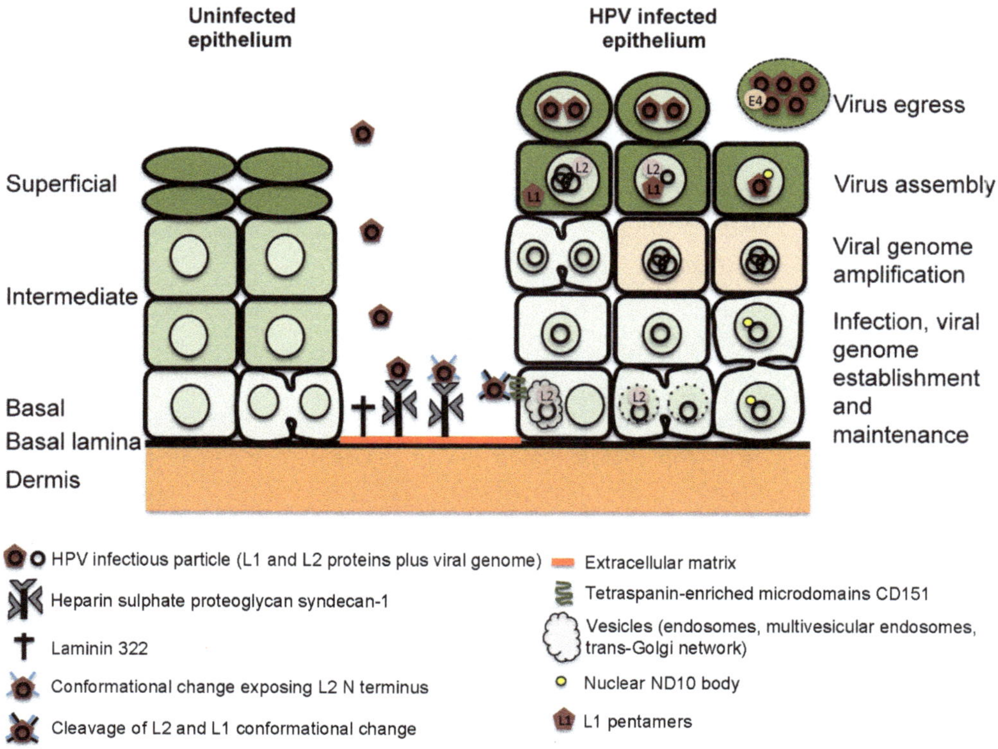

Figure 2.2. The HPV life cycle. The HPV life cycle in tightly linked to epithelial differentiation. In uninfected epithelium, mitotic activity is restricted to the basal cell layers. As cells leave this layer they exit the cell cycle and complete differentiation. HPV infectious particles gain entry to the basal cells *via* wounding, bind to receptors on the basal lamina and following a number of capsid conformational changes binds to the basal keratinocyte. Following internalization, the L2-viral DNA complex is trafficked through the cytoplasm and gains access to the nucleoplasm in order to initiate establishment of the viral DNA once the infected cells undergoes mitosis. Differentiation is delayed as the infected cell migrates upwards and host cell cycle is activated in these cells (shaded light green) to maintain the viral genomes. Viral genome amplification occurs as the cells move into the G2 phase (shaded pink). Differentiation enables expression of the late proteins and L1 pentamers assemble in the cytoplasm (shaded dark green). L2 mediates viral genome encapsidation and assembly of complete particles occurs. Cells containing HPV particles are sloughed off the surface and released into the environment; remodelling of keratinocyte structural integrity by E4 probably plays a role in this final stage. Refer to the main text for full details of the different stages of the HPV life cycle.

endocervix, is separated by the squamo-columnar junction (SCJ). The SCJ contains squamous metaplastic epithelia, which differentiates from a thin layer of immature epithelium, formed by the proliferation of sub-columnar reserve cells, to mature stratified squamous metaplastic tissue, a process that remodels the cervical epithelia. The boundary between immature and mature metaplasia is known as the transformation zone (TZ). The cervical TZ is the region from which the majority of high-grade premalignant lesions and cancers arise. Elegant work from Michael Herfs, Christopher Crum and colleagues of the cellular makeup of this region showed that precursor lesions and cancers are linked to a small and discrete population of non-stratified cuboidal cells of embryonic origin within the SCJ (Herfs *et al.*, 2012). These cells express a junction cell specific expression profile (cytokeratin 7 (CK7), anterior gradient protein 2 (AGR2), cluster differentiation 63 (CD63), matrix metalloproteinase 7 (MMP7) and guanine deaminase (GDA)) that also marks embryonic cervical epithelium. The junction cells manufacture basal located reserve cells, which continue to express the junction cell specific biomarkers, and subsequently differentiate downwards into squamous metaplasia that is negative for the junction cell proteins (Herfs *et al.*, 2013). Such a 'top down' differentiation program would lead to exposure of the cuboidal junction cells to HPV infection and if these cells are unable to support a productive HPV infection, then an abortive or non-productive infection whereby viral gene expression is deregulated might be at higher risk of malignant conversion (Egawa *et al.*, 2015). The idea that these TZ junction cells are progenitors of cervical cancers is supported by the phenotypic similarities between these cells and high-grade premalignant lesions and cancers (Herfs *et al.*, 2012; Herfs *et al.*, 2013; Mirkovic *et al.*, 2015; Herfs *et al.*, 2017).

Since many HPV positive anal cancers arise at the TZ of the anal canal it was hypothesized that a similar population of SCJ cells at this site could be vulnerable to infection by high-risk HPV types (Herfs *et al.*, 2017). While the two sites exhibit anatomical similarities, namely squamous metaplasia that originates from CK7 positive cells, only the cervix has a single layer of CK7-positive junction cells (Yang *et al.*, 2015). Also, in contrast to the cervical disease, far fewer anal cancers and precursors were CK7–positive. It may be that the anal CK7 cells are less susceptible to HPV infection or that these cells have a reduced risk of neoplastic conversion (Yang *et al.*, 2015).

But what of the tonsil - another site of HPV driven carcinogenesis and primarily HPV16 positive? The outer surface of the palatine tonsil is covered by a continuous sheet of stratified squamous epithelium and is separated from the underlying stroma by the basement membrane. The epithelium that extends down into the tonsil crypts is discontinuous, broken up by the dense lymphoid infiltrate and, unlike the surface epithelium shows a loss of normal maturation. HPV tonsil cancers consistently arise from the tonsil crypt suggesting that oncogenic infections within the crypt are at higher risk of malignant progression than those of surface epithelium (Westra, 2012). The target cells within the crypts may be more accessible to HPV infection given the discontinuous nature of the epithelium and being an immune-privileged tissue, infections may escape immune detection and elimination. However, the target cell(s) within the tonsil have not been identified; in fact so far there is no evidence of a productive HPV16 infection in tonsils; in the crypt or at the surface, although isolated tonsil keratinocytes transfected with complete HPV16 genomes support virus replication and the production of infectious particles (Palmer *et al.*, 2014; Israr *et al.*, 2016). Although little is understood of HPV biology at this anatomical site, progenitor epithelial cells (expressing high levels of CD44 and nerve growth factor receptor) are present in the surface and crypt epithelia, but those in the crypt have a more primitive phenotype (Kang *et al.*, 2015). Expression of HPV16 E6 and E7 oncogenes in this population of cells isolated either from the crypt or the surface epithelium robustly depressed stratification and differentiation in the crypt progenitor cells, but less so in the surface epithelium progenitor cells, following growth in three dimensional (3-D) organotypic culture, suggesting that virus biology is not equivocal in the two progenitor populations.

Identification of the target cells in the cutaneous skin has been difficult and it is only in the past few years that stem cells can be marked and their progeny tracked (Blanpain and Fuchs, 2014). A single master stem cell population presiding over all skin lineages (hair follicles, sebaceous and sweat glands, interfollicular epidermis (IFE)) is unlikely and the epidermis actually harbours multiple stem cell niches (Blanpain and Fuchs, 2014).

Upon wounding of epithelial tissue, normal tissue homeostasis is disturbed and stem cells are mobilized to sites of repair, where they may be more accessible to infection by HPV. The notion that stem cells in the skin are permissive for PV infection is strongly supported by analysis of CRPV. In a non-wounding CRPV infection of domestic rabbits, early viral transcripts were localized to epidermal cells that possessed stem cell-like characteristics in the bulb and outer epithelial root sheet of the hair follicle (Schmitt *et al.*, 1996). Upon infection, the virus might reprogram stem cells by altering proliferation and migration capacity. For example, expression of HPV16 E6 and E7 in murine stem cells, isolated from the hair follicle bulge, leads to an increase in proliferation and mobilization of the cells (Michael *et al.*, 2013). In a transgenic mouse model, HPV16 E6 and E7 expressing stem cells accumulated in dysplastic regions suggesting that this might also have a role in early stages of neoplastic development (da Silva-Diz *et al.*, 2013). The basal layer in the IFE contains stem cells and transient progenitors and these cells undergo symmetrical cell division to produce two long-lived daughter cells or asymmetrically to produce two daughter cells with dissimilar fates of proliferation and differentiation. Upon infection of basal cells the virus has a balancing act to perform, on one hand to enable persistence of viral episomes in the basal compartment through symmetrical division, but also to disturb the fate of cells undergoing asymmetrical division. The position of the mitotic spindle has an important role to play in the symmetry of division and the E6 and E7 proteins might cooperate to alter mitotic cell polarity by targeting the host factors involved in spindle positioning, for viral persistence and vegetative replication (Nguyen and Munger, 2009; Marsh *et al.*, 2017).

The site of HPV infection and the target cell infected at these sites clearly are going to be determining factors in the outcome of infection – whether a productive, non-productive/abortive or latent infection. Consequently, if infection is by a high-risk HPV type, then there is a risk of malignant progression (Egawa *et al.*, 2015).

Latent HPV infections

The idea of latent PV infections has been around for decades - defined by the presence of viral DNA in the absence of visible sign of disease, although the molecular makeup of a latent infection is not clear (reviewed in Maglennon and Doorbar, 2012). Animal PV have proved to be informative models of latency. In an animal model of rabbit oral papillomavirus (ROPV), a naturally occurring infection of domestic rabbits that leads to the formation of benign oral papillomas, Maglennon and colleagues showed that following regression of productive infections in the tongue, low copy ROPV DNA was detectable in a proportion of basal cells long after visible disease had gone, suggesting a latent infection had been established (Maglennon *et al.*, 2011). Latent infection was associated with low-level expression of early viral transcripts encoding E1 and E2, and E6 and E7, suggesting that for the viral episomes to persist there is a requirement for the function of these proteins. Asymptomatic infection occurred following infection of rabbits with low doses of CRPV and was associated with low level expression of E1 encoding RNA transcripts (Amella *et al.*, 1994). Clinical outcomes of HPV-associated disease have also supported a latent or quiescent state for HPV replication. For example, anogenital warts and recurrent respiratory papillomatosis, both caused by the low-risk HPV types 6 and 11, have high recurrence rates which makes effective clinical management very difficult. These patients cannot mount effective immune responses to clear the infections and they are likely to become dormant or latent, and thus sensitive to reactivation (Bonagura *et al.*, 2010).

In cattle infected with bovine PV (BPV) the virus could persist in a latent state and the infection reactivated by exposure to immune-suppressants, other carcinogens and/or physical trauma (Campo *et al.*, 1994a; Campo *et al.*, 1994b). Similarly immunosuppression in the ROPV model caused reactivation of ROPV DNA replication at the previously infected sites (Maglennon *et al.*, 2014). Coming back to clinical situations in humans, reactivation of latent cervical HPV infections in organ transplant patients using immunosuppressive drugs may account for the increase in HPV cervical infections in these patients. Thus, the model of PV latency is one of low-level viral genomes in basal stem or stem-like cells requiring a low level of early gene activity (the specific early viral function requirements may vary depending on the cell the latent infection has established in) and under immunological control, that could be induced to reactivate.

Whilst reactivation of a latent infection to a productive infection would require persistence of viral episomes, another scenario would be reactivation of quiescent integrated viral DNA potentially leading to neoplastic disease. A recent analysis identified a reservoir of apparently transcriptionally silent, E2 disrupted, HPV16 DNA in morphologically normal cervices of older women, removed during hysterectomy operation for reasons unrelated to cervical dysplasia (Leonard *et al.*, 2016). Reactivation of E6 and E7 oncoprotein expression from these 'latent' HPV16 integrants could explain why an increase in the risk of cervical cancer continues to increase throughout a woman's life-time. But what would cause reactivation of the silent integrants? Treatment of the cervical cancer cell line CaSki with the demethylating agent 5-azacytidine activates previously transcriptionally silent HPV16 integrants (Van Tine *et al.*, 2004). The host genome-wide demethylation changes, which occur naturally as part of the ageing process, could be a mechanism of reactivation of such silent integrated HPV DNA in older women (Leonard *et al.*, 2016).

Mechanisms of HPV infection

HPV virion structure

The virus particle has not been defined at an atomic scale but native bovine papillomavirus type 1, and HPV16 pseudovirions (PsV) or quasivirus have been solved using high-resolution cryo-EM, the highest resolution of these being 4Å (Cardone *et al.*, 2014; Guan *et al.*, 2017). The T-7d icosahedral capsid is composed of 360 copies of the major capsid protein L1 arranged into 72 capsomers, and each capsomer is composed of 5 copies of L1. The amount of the minor capsid protein L2 in the capsid is unclear and studies have indicated 72 L2 molecules per capsid, but L2 amounts may be variable and could range from 12 to 72 molecules. Twelve of the capsomers lie on the icosahedral 5 fold vertices - pentavalent coordination; the remaining 60 capsomers are each surrounded by six other capsomers - hexavalent coordination. The C-terminus of L1 links capsomers together and enables disulphide bond formation between neighbouring L1 proteins (HPV16 Cys175 and Cys428). The precise localization of L2 in the capsid is missing, but most likely one L2 molecule occludes the centre of each capsomer with a small amount of N-terminal region extending outwards to the capsid surface (Cerqueira and Schiller, 2017; Guan *et al.*, 2017). The capsid forms around one chromatinized viral episome (Zhao *et al.*, 1998).

Infection

Deciphering the mechanisms of HPV infection was hampered by a lack of cell culture systems in which to propagate infectious virus, although native virus can be purified from some HPV infections (Doorbar and Gallimore, 1987) and from tissue culture models of propagation (McLaughlin-Drubin *et al.*, 2003; Cruz *et al.*, 2015). The development of cell-based systems to generate PsV; VLP composed of the L1 protein alone or together with L2, that do not contain the viral DNA, but instead can be packaged with a reporter plasmid has been a significant step forward in understanding the infection mechanisms. The activity of the reporter acts as a surrogate readout for HPV entry. In addition, the development of an *in vivo* murine cervicovaginal challenge model combined with the widespread use of PsV has provided insight into the early events of HPV infection (Roberts *et al.*, 2007). The current understanding is that virus particles gain access to extracellular matrix (ECM) and basal keratinocytes following wounding of the epithelium. Infection, which is a remarkably slow and asynchronous process (Giroglou *et al.*, 2001; Day *et al.*, 2004), involves a cascade of conformational changes in both capsid proteins and multiple receptor interactions that enable final binding to the basal keratinocytes and internalization (Figure 2.2).

Virus attachment (primary receptor)

It is well recognised that initial HPV binding occurs via heparin sulphate proteoglycans (HSPGs), a family of glycoproteins containing one or more covalently bonded heparin sulphate chains (glycosaminoglycans) which are modified by acetylation and sulphonation, extending heterogeneity (Sarrazin *et al.*, 2011). HSPGs are located on the cell surfaces of basal keratinocytes and bound to the ECM, but also secreted into the extracellular milieu upon epithelial wounding. Early studies showed that treatment of keratinocytes with heparinase caused a reduction in HPV11 VLP binding (Joyce *et al.*, 1999), and pre-incubation with heparin inhibited HPV16 and HPV33 PsV infection (Giroglou *et al.*, 2001). Using HPV16 PsV it was shown that the virus particle binds to the highly sulphonated glycan (Cerqueira *et al.*, 2013). Primary keratinocytes are difficult to infect with different HPV types (Ozbun, 2002; Day *et al.*, 2008) and this is thought to be due to the pattern of post-translational modification of HSPGs, such as N- and O-linked sulphonation (Schowalter *et al.*, 2011). Differences in sulphonation patterns occur during would healing so this may explain why, as shown in the murine cervicovaginal tract model, microwounding was required for PV infection (Kines *et al.*, 2009). HPV types from different genera seem to have different HSPG affinities dictated by post-translational modification and this could perhaps explain the differences in anatomical tropism of these viruses (Johnson *et al.*, 2009). The HSPG syndecan-1 is recognised as the ECM attachment receptor to which virus particles bind (Shafti-Keramat *et al.*, 2003) (Figure 2.2).

The HPV-HSPG interaction depends on L1 and involves at least two HSPG binding sites on the protein via electrostatic and polar interactions. This interaction leads to conformational changes in the capsid exposing L1 residues for cleavage by kallikrein-8 (a secreted serine protease) (Cerqueira *et al.*, 2015b) and in the L2 protein resulting in exposure of the amino terminus (Dasgupta *et al.*, 2011; Richards *et al.*, 2013). Following this, the host chaperone protein cyclophilin B binds to the exposed portion of L2 causing a further conformational change at the amino terminus to reveal a furin [a proprotein convertase (PC)] cleavage site (Richards *et al.*, 2006; Bienkowska-Haba *et al.*, 2009). This cascade of conformational changes reduces affinity of the virus particle for the HSPG receptor and exposes a binding site on L1 that facilitates the transfer of the virus particle to a secondary receptor complex responsible for virus entry on basal keratinocytes (Figure 2.2). Pharmacological inhibition of the convertase prevents efficient infection of HPV16 PsV (Richards *et al.*, 2006). PsV can infect cells lacking HSPGs but only if they had been pre-cleaved by furin, suggesting that the initial binding event to HSPGs is important for furin-mediated L2 cleavage (Day *et al.*, 2008). However, Craig Meyers showed that native HPV16 particles could infect cells independently of cellular furin (either in the presence of a furin inhibitor or infection of furin deficient cells) (Cruz *et al.*, 2015). In fact, the L2 protein was cleaved by furin during virion morphogenesis in organotypic raft cultures of foreskin or cervical cells harbouring the HPV16 genome. Furin independent infection was noted also for HPV45 particles, but not HPV18 or 31, suggesting that some alpha types could bypass the initial HSPG binding, permitting attachment directly to the functional entry receptor (Cruz *et al.*, 2015). Also, HPV16 PsV infection of human keratinocytes (HaCaT) did not require extracellular furin and was shown not to regulate the release of virus particles from ECM, whereas cleavage of HSPG chains using heparinase releases infectious particles from the ECM (Surviladze *et al.*, 2015). Moreover, cells infected with virus and treated with the furin inhibitor (decanoyl-RVKR-CMK), show retention of L2 and the viral genome in late endosomes (see later) and escape from the endosomes is dependent on cleavage at the furin site in L2 (Richards *et al.*, 2006). Whether furin cleavage occurs during virion morphogenesis in infected tissue or during the infection process, inhibitors of furin and related PCs could be an effective method of controlling genital HPV infections. Indeed, human α-defensin 5, an innate effector peptide secreted by genital tract epithelial cells, has been shown to bind to HPV16 PsV preventing furin cleavage of L2 and infection (Wiens and Smith, 2015). Similarly, the seaweed derived sulphated polysaccharide carrageenan blocks HPV16 PsV infection *in vitro* and cervical infection in mice by binding to the capsid and blocking the attachment to HPSG receptor as well as occluding viral surfaces involved in other non-HPSG interactions required for infection (Buck *et al.*, 2006; Roberts *et al.*, 2007).

The different tropisms of the viruses could indicate the ability of the virus to infect using non-HSPG receptors. ECM-bound laminin 332 (LN332, formerly LN5) has been shown to act as a non-HSPG receptor for HPV16 and closely related types (Culp *et al.*, 2006; Richards *et al.*, 2014). It has been suggested that LN332 acts as a transient receptor for the virus particle (Figure 2.2). However, HPV18 PsV was unable to utilize ECM-bound LN332, suggesting that it might use a different non-HSPG receptor and such specificity may contribute to the anatomical tropisms of the different alpha species (Richards *et al.*, 2014).

Virus attachment (secondary receptor)

A plethora of candidates have been suggested as receptors for attachment and entry into keratinocytes, including integrins, tetraspanins, annexin A2 and growth factor receptors, all highly expressed in the target cell - basal keratinocytes. Early studies using monoclonal antibodies directed against integrins indicated that the α6 integrin was involved in binding of HPV6 VLPs to cells (Evander *et al.*, 1997) and several other studies have also implicated this integrin in HPV infection (Abban and Meneses, 2010; Scheffer *et al.*, 2013). There is evidence that the viral particles are recruited to discrete sites on the cell membrane known as tetraspanin-enriched domains (TERM) and specifically those containing the tetraspanin CD151 in association with integrins, including the α6 integrin (Scheffer *et al.*, 2013). Moreover, CD151 motifs involved in integrin activity but not those associated with sorting into clathrin coated pits were important for HPV16 internalization (Scheffer *et al.*, 2013). There is also evidence that following attachment of the virus particle (HPV16 PsV) to the primary HSPG receptor, the virus is released to the surface of the keratinocyte in complex with the HSPG receptor through normal HSPG shedding process from the ECM by the action of sheddases, including matrix metalloproteinases (Surviladze *et al.*, 2012). The shed high molecular weight complexes contain virus-HSPG-HS and other bioactive molecules, including growth factors that enable binding to growth factor receptors on the surface of the keratinocyte. Indeed, both the epidermal growth factor (EGF) and keratinocyte GF (KGF) pathways (and possibly other GF pathways) are activated upon HPV16 PsV binding to human keratinocytes and GF receptor inhibitors block infection (Surviladze *et al.*, 2015).

Supporting these models is the finding that signal transduction pathways associated with these different receptors are activated upon HPV infection of human keratinocytes. Activation of the focal adhesion kinase (FAK), a pathway downstream of α6 integrin, and the phosphoinositide 3-kinase (PI3K), downstream of EGFR, are important for HPV16 PsV infection of human keratinocytes (Abban and Meneses, 2010; Surviladze *et al.*, 2015).

Internalization

The next stage in the HPV infection cycle is virus uptake and virus disassembly (Figure 2.2), and both events are poorly understood. The virus is endocytosed and this process occurs over many hours in an asynchronous fashion (Selinka *et al.*, 2002; Day *et al.*, 2003; Culp and Christensen, 2004). However, the speed of internalization following infection may depend on the cell cycle phase of the target cell. Broniarczyk and colleagues showed that if cells are infected in late S to early G2/M phase, HPV16 PsV reporter DNA gene expression occurs within 8 hours of infection (Broniarczyk *et al.*, 2015), indicating very rapid infection. Non-enveloped viruses undergo endocytosis generally by one of two pathways, either mediated by clathrin or caveolae (Pelkmans and Helenius, 2003; Smith and Helenius, 2004). However, HPV16 entry is clathrin-, caveolin-, cholesterol- and flotillin- independent, but dependent on actin polymerisation (Yang *et al.*, 2003; Spoden *et al.*, 2008; Schelhaas *et al.*, 2012) and attachment to the TERM (Spoden *et al.*, 2008; Scheffer *et al.*, 2013), and this is a common pathways amongst several high risk HPV types (Spoden *et al.*, 2013). Following internalization, the virus particles are trafficked to multivesicular endosomes (MVE), a process dependent on another tetraspanin CD63 and its cytoplasmic partner syntenin-1 (Grassel *et al.*, 2016). This study showed that the syntenin-1 partner ALIX (ESCRT-associated adaptor protein) was critical for this process and identified CD63-syntenin-1-ALIX as a novel pathway for regulating post-endocytic delivery of HPV particles to MVE (Grassel *et al.*, 2016).

Disassembly and separation of the virus particle into L1 and an L2-viral DNA complex initiates during endosomal trafficking and requires endosomal acidification, and this uncoating step involves cyclophilins (Bienkowska-Haba *et al.*, 2012). While the majority of L1 is routed to the lysosomes, L2 remains tethered to the viral genome and is delivered along the microtubule filaments to the trans-Golgi network (TGN) (Florin *et al.*, 2006; Day *et al.*, 2013). However, it is possible that not all L1 molecules are routed to the lysosomes since some L1 protein has been shown to also traffic to the TGN (Lipovsky *et al.*, 2013). Interferon-γ has been shown to inhibit HPV16 L1 and L2 separation during internalization and prevent trafficking of the L2-genome complex from the late endosomes (Day *et al.*, 2017). Interferon inhibition is mediated through the L2 protein but this action did not extend to all alpha HPV types tested suggesting type-specific differences in this stage of infection. A genome-wide siRNA screen revealed that the escape of L2-genome complexes from vesicles to the TGN was dependent on the retromer, a major component of the retrograde transport pathway interacting with viral particles at endosomal membranes (Lipovsky *et al.*, 2013). Inhibition of this pathway blocked HPV16 PsV infection of cervical cancer cells by about 80% indicating that this is a major (but perhaps not exclusive) intracellular trafficking pathway utilised by HPV. Silencing expression of the cargo recognition core of the retromer (VPS26, VPS29, VPS35) also inhibited PsV infection, and colocalization studies indicated a physical association between L2 and the retromer (Lipovsky *et al.*, 2013). Sorting nexins determine fate of cargoes and L2 has been shown to bind SNX17 and SNX27; however the nature of the association is dissimilar between the two molecules suggesting different roles in HPV infection (Bergant and Banks, 2013; Pim *et al.*, 2015).

The final step in the infection pathway is nuclear entry and this stage is dependent on the infected cell entering mitosis (Figure 2.2). Leaving the TGN, the L2-DNA complex within a membrane bound vesicle locates to the vicinity of the centrioles and when the nuclear envelope breaks down during mitosis, the complex gains access to the condensed host chromatin, alleviating the need for nuclear import factors (Aydin *et al.*, 2014; DiGiuseppe *et al.*, 2016; Calton *et al.*, 2017). A chromatin-binding region in L2 (residues 188-334) is necessary to promote the L2 association with mitotic chromatin (Aydin *et al.*, 2014). The L2-viral genome complex is finally delivered to the sub-nuclear domain 10 (ND10) bodies whereupon viral transcription is initiated (Day *et al.*, 2004).

Our understanding of the infectious process is being updated constantly and outcomes appear often to depend on the nature of the reagents used (PsV *vs* virons derived from infected tissues) and infection systems (tissue culture grown cells *vs* animal models).

HPV gene transcription

Correct timing of expression of the different HPV proteins is crucial for the successful replication of the virus in differentiating epithelia. Regulation of activity of the viral promoters is one level of control (discussed below), but polyadenylation signalling and the regulation of alternative RNA splicing also play significant roles in controlling HPV gene expression (these areas are covered by two excellent reviews (Johansson and Schwartz, 2013; Graham and Faizo, 2017)).

Initiation of papillomavirus gene transcription is mediated by the co-ordinated action of the viral transcriptional regulator E2 and a plethora of cellular transcriptional regulators. These viral and cellular factors generally bind to specific sites within the viral LCR, which contains the early promoter, origin of replication, tissue specific transcription enhancer and the late 3′ untranslated region, which contains the late polyadenylation signal (Figure 2.1B). The early viral promoter is a canonical RNA polymerase II-dependent promoter with a consensus TATA box, which recruits the general transcription factor TFIID, immediately upstream of the transcriptional start site of the E6 ORF. Although the HPV E2 protein does not harbour intrinsic enzymatic activity, it is essential for many important processes in the virus life cycle. E2 binds as a homodimer to specific sequences within the viral LCR with the general sequence ACCG(N)4CGGT (Androphy *et al.*, 1987), and in doing so is able to recruit factors required for replication and transcription control. Multiple E2 binding sites exist in the LCR of all HPV types, with a bias for four consensus binding sites in the alpha-HPV genus (McBride, 2013). E2 can either activate or repress transcription depending on the context of the occupied binding sites and the factors that are recruited to the LCR by direct interaction with E2 (Bernard *et al.*, 1989; Romanczuk *et al.*, 1990; Bouvard *et al.*, 1994). The occupancy of each E2 binding site within the LCR is in part regulated by differential affinity of E2 for the exact sequence of each binding site (Thain *et al.*, 1997; Dell *et al.*, 2003). Two tandem E2 binding sites are situated proximal to the early promoter region and overlap the TATA box and an SP1 transcription factor binding site. E2 occupancy of these sites represses transcription of E6/E7 encoding transcripts due to exclusion of TFIID and SP1 by steric hindrance (Dong *et al.*, 1994; Tan *et al.*, 1994; Hou *et al.*, 2000). However, these promoter proximal binding sites are bound by E2 with relative low affinity and are therefore occupied at high E2 concentrations (Demeret *et al.*, 1997; Steger and Corbach, 1997; Thain *et al.*, 1997), creating a rheostatic control mechanism for the regulation of early transcript production coupled to keratinocyte differentiation. As infected cells differentiate, E2 protein levels increase (Xue *et al.*, 2010), resulting in repression of the early promoter and a reduction in E6 and E7 expression, which further promotes terminal differentiation. A truncated E2 protein is also expressed from a separate promoter within the E1 ORF upstream of the E8 ORF (P^{E8}) (Straub *et al.*, 2015) (Figure 2.1A). Splicing of this transcript into the E2 ORF results in a message that encodes 12-16 amino acids from E8 fused to the hinge and C-terminal DNA binding domain and dimerization domain of E2, termed E8^E2 (reviewed by Dreer *et al.*, 2017). E8^E2 is able to bind to the E2 binding sites within the viral LCR and in doing so strongly represses transcription (Stubenrauch *et al.*, 2000; Lace *et al.*, 2008; Dreer *et al.*, 2016).

E2 is also able to control virus transcription by the recruitment of key cellular transcriptional regulators through direct interaction. Bromodomain containing protein 4 (BRD4) has been shown to play an important role in transcriptional activation and repression functions of E2. E2 binds directly to the C-terminal domain of BRD4 (Abbate *et al.*, 2006; Wu *et al.*, 2016) and competes with cellular transcriptional regulators including the positive transcription elongation factor (P-TEFb) and the general co-factor MED1, which is part of the Mediator complex, resulting in repression of transcription (Smith *et al.*, 2010; Yan *et al.*, 2010; Wu *et al.*, 2016). Conversely, BRD4 interaction has been shown to be essential for E2-dependent transcription activation (McPhillips *et al.*, 2006; Schweiger *et al.*, 2006; Lee and Chiang, 2009) and presumably stimulates E2-dependent transcription activation via its intrinsic histone acetyltransferase (HAT) activity, which induces nucleosome remodelling and chromatin decompaction and results in transcriptional activation (Devaiah *et al.*, 2016).

However, an E2-independent role for the activation of early transcription by BRD4 immediately after infection has also been demonstrated. Using HPV18 quasivirus, McKinney and colleagues elegantly demonstrated that BRD4 activates both transcription and replication independent of E2 expression following virus entry (McKinney *et al.*, 2016). BRD4 also plays a role in late gene expression and recruitment to the late promoter is induced by cellular differentiation. In this context, BRD4 recruits cyclin dependent kinase 8 (CDK8), a subunit of the Mediator complex, which in turn recruits CDK9 to stimulate late gene transcription elongation (Songock *et al.*, 2017).

E2 has also been shown to bind and recruit the aceyltransferase cAMP response element binding protein (CBP) and its homologue p300 to activate transcription (Lee *et al.*, 2000; Peng *et al.*, 2000; Muller *et al.*, 2002; Kruppel *et al.*, 2008). It is also interesting to note that p300 can acetylate bovine PV E2 protein to stimulate transcription activity (Quinlan *et al.*, 2013).

Viral LCR activity is further regulated by several cellular transcription factors independent of E2 binding (Figure 2.1B). The viral enhancer activity is restricted to keratinocytes, which is thought to confer cell specificity to HPV infections. This is in part due to the central role of the AP1 transcriptional activator in activating viral transcription. The AP1 family of transcription factors are heterodimers of the proto-oncogenes JUN and FOS. The AP1 heterodimers responsible for the stimulation of HPV enhancer activity are JUNB and FRA2 heterodimers (Bouallaga *et al.*, 2000), which are specifically upregulated in differentiating keratinocytes (Kyo *et al.*, 1997; Mehic *et al.*, 2005). AP1 dependent stimulation of viral transcription is attenuated by the Ying-yang 1 (YY1) transcriptional regulator, which strongly represses early transcription production when recruited to the tissue specific enhancer (O'Connor *et al.*, 1996). However, YY1 can also stimulate HPV transcription when bound to the extreme 5′ region of the viral LCR with the transcriptional co-activator TEF-1 (Kanaya *et al.*, 1997). Host factors have also been shown to restrict differentiation-dependent HPV transcription. The ND10 associated protein Sp100 binds to the HPV31 genome and restricts both the initial activation of virus transcription and late virus transcription in differentiated cells (Stepp *et al.*, 2013; Stepp *et al.*, 2017).

Epigenetic control of HPV transcription

HPV gene transcription is further regulated by epigenetic modifications. The majority of E2 binding sites contain CpG dinucleotides, which can become methylated to prevent E2 binding (Thain *et al.*, 1997; Kim *et al.*, 2003). This is particularly significant in HPV-driven cancers that contain episomal forms of the viral genomes; CpG methylation of episomes is increased during carcinogenesis, which prevents E2-mediated repression of E6 and E7 (Kalantari *et al.*, 2004; Kalantari *et al.*, 2008; Chaiwongkot *et al.*, 2013; Simanaviciene *et al.*, 2015). CpG methylation within the viral enhancer provides a mechanism for the attenuation of E6/E7 expression by keeping the chromatin in a closed state and unresponsive to transcription activation (De-Castro Arce *et al.*, 2012). HPV transcription is also regulated by epigenetic modification of histones associated with viral genomes in persistent viral infections. Using the cervical carcinoma cell line CaSki, which contains over 500 copies of integrated head-to-tail concatemers of HPV16, it was demonstrated that nucleosomes are specifically positioned over the viral enhancer and early promoter (Stunkel and Bernard, 1999) (Figure 2.1B). These nucleosomes contribute to the repression of early transcription; treatment of W12 cells (which contain HPV16 episomes) with the histone deacetylase trichostatin A renders the chromatin more accessible to transcriptional regulators and increases E2-dependent E6/E7 repression (Bechtold *et al.*, 2003). More recently, it was shown that acetylation of intragenic histones within the early gene region of HPV16 correlates with enhanced E6 and E7 expression (Johansson *et al.*, 2015). The mechanisms underpinning nucleosome phasing and epigenetic regulation of HPV episomes is not understood, but it has been demonstrated that the cellular chromatin organisation protein CCCTC binding factor (CTCF) associates with HPV episomes (Mehta *et al.*, 2015; Paris *et al.*, 2015). The binding of CTCF within the E2 ORF of HPV18 regulates the activity of the early promoter in viral episome containing cells (Paris *et al.*, 2015). Given the central role of CTCF in insulation of epigenetic boundaries, nucleosome phasing and chromatin looping, it is likely that CTCF binding to the HPV genome controls nucleosome positioning and epigenetic transcription control.

Epigenetic modifications also regulate HPV gene expression control during cellular differentiation. Evidence suggests that a switch in epigenetic modification of the viral LCR upon cellular differentiation is in part responsible for activation of the viral late promoter. A loss of CpG methylation within the LCR, which was most dramatic within the 5′ enhancer region, was observed upon differentiation of cells containing HPV16 episomes (Kalantari *et al.*, 2008). This loss of CpG methylation coupled to a dramatic increase in dimethylated H3 lysine 4 and acetylated H3 shown in HPV31 genomes (Wooldridge and Laimins, 2008) indicates a differentiation-dependent increase in transcription activity through epigenetic modulation. Moreover, these changes in histone modifications coincide with changes in transcription factor binding to the HPV31 promoters, including increased YY1 recruitment to the tissue specific enhancer, and C/EBP-alpha and SP1 recruitment to the late promoter within the E7 ORF (Wooldridge and Laimins, 2008).

HPV genome replication

Following entry into the nucleus, HPV DNA is maintained as a chromatinized, supercoiled circular DNA molecule, or episome. Replication of HPV episomes has three distinct phases in the virus life cycle - establishment, maintenance and amplification (Figure 2.2). Although the precise mechanisms responsible for the transition between these phases are not understood, regulation is in part governed by the differentiation status of the infected cell. As such, the replication cycle of papillomaviruses takes around three weeks to complete, as this is the time it takes for keratinocytes to fully differentiate. All phases of HPV replication rely on host cell polymerases and regulatory factors since the virus does not encode proteins that can fully support replication.

Establishment replication

The first phase of replication follows virus entry in the basal cells, which is immediately preceded by transcription of early genes including those which encode the E1 and E2 viral proteins (Ozbun, 2002). Both E1 and E2 proteins are required for the initial burst of viral replication which serves to amplify viral genomes to about 50 to 100 copies per cell, based on the episome copy number in cell lines isolated from clinical lesions (Stanley *et al.*, 1989; Bedell *et al.*, 1991) and in animal models of papillomavirus infection (Maglennon *et al.*, 2011). Once E2 binds to the binding site found adjacent to an A-T rich patch known to be the viral origin of replication (*Ori*), it is able to recruit of E1 protein molecules (Ustav *et al.*, 1991; Ustav *et al.*, 1993; Stenlund, 2003). E1 dimers then assemble in a co-operative fashion into a double hexameric helicase (Mohr *et al.*, 1990; Sanders and Stenlund, 2000) which catalyses the melting of the DNA at the replication origin (Sedman and Stenlund, 1998). The assembly of the E1 helicase on *Ori* by E2 is strongly inhibited by the E8^E2 repressor protein, which binds to the E2 binding sites, but is unable to recruit E1 (Lace *et al.*, 2008; Straub *et al.*, 2014). E1 binds to and recruits cellular replication factors including DNA polymerase α/primase, replication protein A (RPA) and topoisomerase I (Han *et al.*, 1999b; Clower *et al.*, 2006; Hu *et al.*, 2006), which are considered important in the establishment of an E1-dependent bidirectional replication fork (Orav *et al.*, 2015). Other host cell factors that are important in the initiation of early replication events include BRD4 and topoisomerase binding protein 1 (TOPBP1), which are both recruited to the origin of replication by the E1-E2 viral replication proteins (Gauson *et al.*, 2015). It has been elegantly demonstrated that BRD4 is colocalised with E1 and E2 proteins in discrete nuclear foci in the absence of viral DNA and this complex is required for efficient initiation of replication (Sakakibara *et al.*, 2013a). However, amplification of viral DNA results in a displacement of BRD4 from replication factories. It is therefore thought that BRD4 plays an important role in the initiation of viral DNA replication by confining E1 and E2 replication centres to specific regions of the host nucleus through interaction with host chromatin (Sakakibara *et al.*, 2013a). Interestingly, BRD4 and E2 have been shown to co-localise in defined regions of the host chromatin that are in close proximity to common fragile sites (Jang *et al.*, 2014), which are frequently found to contain integrated HPV DNA in HPV-driven cancers (Thorland *et al.*, 2003).

Use of the viral helicase rather than the cellular replication helicases, such as the mini-chromosome maintenance (MCM) proteins allows the initial replication phase to be uncoupled from highly restricted cellular replication to facilitate amplification of viral episomes in the infected cell. However, it has also been shown that this initial burst of replication activates the cellular DNA damage response (DDR) (Reinson *et al.*, 2013), which may restrict viral amplification in favour of persistent infection (reviewed by Porter *et al.*, 2017). Indeed, strict regulation of the E1 helicase appears to be essential for efficient viral genome amplification since nuclear E1 accumulation can block S-phase entry and activate the ataxia-telangiectasia (ATM) - dependent DDR pathway by inducing double strand breaks in the host genome (Fradet-Turcotte *et al.*, 2011; Reinson *et al.*, 2013). Nuclear-cytoplasmic shuttling of E1 is controlled by cyclin E/A-CDK2-mediated phosphorylation of E1 (Fradet-Turcotte *et al.*, 2010) and E1 activity is in part regulated by interaction with the cellular WD repeat-containing factor UAF1 and associated deubiquitinating enzymes such as USP1, 12 and 46 (Cote-Martin *et al.*, 2008; Lehoux *et al.*, 2012; Lehoux *et al.*, 2014). Thus, it appears that tight regulation of the activity and localisation of the viral helicase is essential for establishment of infection.

Maintenance replication

The second phase of PV replication is characterised by the stable maintenance of episome copy number, and is thought to proceed *via* unidirectional replication fork movement that does not require the viral origin of replication and is therefore presumed to be independent of E1 helicase (Orav *et al.*, 2015). In support of this, HPV genomes are able to replicate independently of E1 protein in human keratinocytes (Angeletti *et al.*, 2002; Kim and Lambert, 2002; Egawa *et al.*, 2012). How the switch from early amplification to maintenance replication is controlled is not fully understood. It is known that the stable maintenance of virus copy number is controlled by utilisation of cellular replication factors rather than the viral E1 helicase. Also, replication in this phase of the virus life cycle has been shown to occur either by the replication of each viral episome once per cell cycle alongside the cellular DNA, or by the over replication of a few episomes to maintain episomal load (Hoffmann *et al.*, 2006). Maintenance replication also requires the cellular DNA repair pathways and it has been shown that inhibition of the ATM and ATM-related (ATR) pathways results in viral DNA degradation and rapid elimination of episomes from cells (Edwards *et al.*, 2013a; Edwards *et al.*, 2013b).

Although the E1 and E2 proteins are not required for episomal replication during the stable maintenance phase, E2 is vital for the persistence of viral genomes in the proliferating infected cells by physically tethering viral episomes to host cell chromatin during mitosis (Feeney and Parish, 2009). By E2 binding to consensus binding sites within the viral LCR and simultaneously making high affinity contacts with host chromatin, episomes are tethered to host cell DNA during mitotic chromosomal segregation (Lehman and Botchan, 1998; Skiadopoulos and McBride, 1998; Ilves *et al.*, 1999; Bastien and McBride, 2000). The E2 proteins from different PV types associate with chromatin through different mechanisms (Oliveira *et al.*, 2006) and several cellular proteins have been reported to be essential for tethering of viral genomes through interaction with E2, including BRD4, TOPBP1 and the DNA helicase CHLR1 (You *et al.*, 2004; Baxter *et al.*, 2005; Abbate *et al.*, 2006; Parish *et al.*, 2006; Donaldson *et al.*, 2007). From these extensive studies, it seems likely that BRD4 is the *bona fide* chromatin receptor for E2, which physically attaches E2-bound episomes to chromosomes during mitosis, while evidence suggests that CHLR1 plays a functional role in loading E2-genome complexes onto BRD4 bound regions during DNA replication (Feeney *et al.*, 2011; Harris *et al.*, 2017). While different PV types appear to have evolved slightly different mechanisms for episome tethering to support persistence, it seems that this is a fundamental requirement for the persistence of all PV infections.

Amplification replication

The third and final phase of replication is characterised by a dramatic amplification of viral episomes in the upper differentiated layers of the epithelium. This productive replication phase coincides with a significant rise in E1 and E2 expression and is therefore thought to be E1 and E2 dependent (Burnett *et al.*, 1990; Penrose and McBride, 2000; Xue *et al.*,

2010; Paris *et al.*, 2015). Both the E1 and E2 proteins independently activate the cellular DDR (Kadaja *et al.*, 2007; Moody and Laimins, 2009; Fradet-Turcotte *et al.*, 2011; Reinson *et al.*, 2013; Johnson *et al.*, 2017) and it has been demonstrated that several DDR proteins are recruited to E1-E2 dependent viral replication *foci* including NBS1, MRE11, BRCA1, RAD51 and the RAD50 interacting protein RINT1 (Gillespie *et al.*, 2012; Anacker *et al.*, 2014; Chappell *et al.*, 2015; Campos-Leon *et al.*, 2017). Evidence suggests a switch from bidirectional replication in the maintenance phase to recombination-mediated replication in the amplification phase (Flores and Lambert, 1997; Sakakibara *et al.*, 2013b). The productive phase of HPV replication which occurs in cells that have progressed to a more G2–like state, requires activation of the ATM/ATR-dependent homologous recombination DNA repair pathway to support viral genome amplification (Moody and Laimins, 2009; Chappell *et al.*, 2015; Hong *et al.*, 2015; Anacker *et al.*, 2016). The interaction of HPV with host DNA repair pathways is also discussed in Chapter 11 of this volume.

Induction of cellular milieu supportive of viral DNA amplification

Studies using keratinocyte based HPV life cycle models have shown that E6 and E7 functions are essential for the maintenance replication and amplification of the cancer causing alpha-HPV genomes (Thomas *et al.*, 1999; Flores *et al.*, 2000; Park and Androphy, 2002; Lee and Laimins, 2004; Wang *et al.*, 2009; Nicolaides *et al.*, 2011; Delury *et al.*, 2013). The proteins are expressed together in the virus life cycle, and are transcribed as a bicistronic transcript initiated from the early promoter. The overall picture of their role is one whereby they work in a synergistic fashion to generate a host cell environment that supports viral DNA synthesis in the differentiating epithelium – promoting cell cycle activity, cell survival and delaying differentiation. Of course, the expression of E6 and E7 proteins of the cancer causing alpha-HPV types is retained in cancers and many of the host interactions made during the virus life cycle contribute to the oncogenicity of these proteins. The nature of these interactions, and their function is discussed in more depth in Chapter 3 of this book.

A major function of E7 is to induce host cell DNA replication machinery. Indeed, expression of E7 alone simulates S phase activity in post-mitotic differentiating keratinocytes through the actions of E7 upon RB family members p105, p107 and p130 - p130 functioning in differentiating cells (Dyson *et al.*, 1989; Munger *et al.*, 1989; Banerjee *et al.*, 2006; Genovese *et al.*, 2008). The interaction between E7 and RB proteins disrupts the association with the E2F family of transcription factors whose primary function is to activate transcription of genes necessary for S phase entry (Cheng *et al.*, 1995). E7 has also been shown to degrade RB by binding the cellular ubiquitin ligase CUL2 and this might be particularly relevant for destabilizing p130 in the infected differentiating keratinocyte (Huh *et al.*, 2007). Interaction with the RB pocket proteins is not the only interaction E7 makes – it deregulates many other transcription factors to promote a proliferative phenotype (Moody and Laimins, 2010). It has been shown that E7 regulates the expression of various microRNAs (miR) and the regulation of some of these has been shown to be important for viral genome replication during the virus life cycle (Gunasekharan and Laimins, 2013; Gunasekharan *et al.*, 2016). One example is miR-203, which is down regulated by HPV31 E7 *via* interference with MAPK activation of the microRNA (Melar-New and Laimins, 2010). MiR-203 is a negative regulator of proliferation upon keratinocyte differentiation and its expression has been shown to be inhibitory towards HPV DNA amplification. Downstream targets of miR-203 include p63, a family of transcription factors involved in controlling the balance between proliferation and differentiation. In normal differentiation, p63 levels decrease, partly due to increased levels of miR-203. However, in differentiating HPV31 genome containing CIN612 cells, high levels of p63 are maintained thus helping to keep these cells in cycle during the viral DNA amplification phase (Melar-New and Laimins, 2010). Also, E5 functions likely promote host cell cycle progression of the HPV infected cells. E5 is a membrane-integrated protein that leads to an enhancement of EGFR signalling in cells that have moved up from the basal layer to the suprabasal compartment (Straight *et al.*, 1995). Interestingly, it has recently been shown in an HPV18 life cycle model that E5 potentiation of EGFR signalling suppresses the activity of the keratinocyte growth factor receptor pathway and this contributes to the delay in differentiation of cells undergoing viral genome amplification (Wasson *et al.*, 2017).

The E7-induced premature cell cycle progression leads to activation of host defence pathways including the upregulation of the p53 protein, which is degraded by E6 in complex with the ubiquitin ligase E6AP, abrogating p53 mediated control over cell cycle arrest, apoptosis, senescence and DNA repair (Scheffner *et al.*, 1990). Indeed, in organotypic raft cultures of primary keratinocytes transfected with HPV18 E6 null genomes, very few suprabasal cells contained amplified viral DNA compared to structures established from cells containing wild type viral genomes (Wang *et al.*, 2009). Moreover, p53 accumulated to high levels in the E6 null mutant cells and the reintroduction of full-length E6 or knockdown of p53 expression in these cells partially restored genome amplification (Wang *et al.*, 2009; Kho *et al.*, 2013). However, these studies also implicated E6 functions in addition to p53 inactivation in aiding viral DNA amplification. One such function could be the ability of E6 proteins to degrade a subset of proteins containing PDZ protein interaction domains; many of which are involved in establishing and maintaining cell polarity pathways (e.g. epithelial cell polarity, mitotic spindle polarity) (James and Roberts, 2016). Abrogation of this E6 activity in HPV life cycle model systems reduces viral DNA replication (Lee and Laimins, 2004; Delury *et al.*, 2013). Precisely how this E6 function supports viral DNA amplification is not fully understood. However, expansion of S-phase-like competent cells was suppressed in the upper layers of rafts formed from cells harbouring HPV18 genomes defective in PDZ protein binding, thus reducing the number of cells able to move into a G2-like phase in order for viral DNA amplification to occur to high levels (Delury *et al.*, 2013). Indeed, this phenotype correlates with a reduction in mitotic cells in the suprabasal layers compared to the epithelia formed from cells harbouring wild type HPV18 genomes (Marsh *et al.*, 2017).

Several studies have reported that E4, which is abundantly expressed in cells supporting viral DNA amplification, may contribute to the G2-like phenotype of suprabasal cells. The E4 proteins of both benign and high-risk HPV types are strong inducers of G2 arrest of keratinocytes, but mechanisms vary between the virus types (Davy *et al.*, 2002; Knight *et al.*, 2006; Davy and Doorbar, 2007). However, in an HPV18 life cycle model, abrogation of the G2 arrest function of E4 did not adversely affect viral DNA amplification, whereas a similar E4 mutation in an HPV16 system reduced viral DNA amplification, suggesting that there are type specific differences in the role of E4 in the virus life cycle (Knight *et al.*, 2011; Egawa *et al.*, 2017).

Virion assembly and transmission

Assembly of viral progeny

Assembly of new infectious virions takes place in the late phase of the HPV life cycle that is restricted to the most differentiated region of the infected epithelia (Figure 2.2). Keratinocyte based models of the virus life cycle have facilitated the unravelling of this late phase, and virion formation probably undergoes multiple phases from initial assembly, superassembly, encapsidation and final maturation. Expression of the capsid proteins is restricted to the differentiating keratinocytes and occurs after E4 expression, avoiding immune recognition of the highly immunogenic particles by the host. A network of inter-pentameric disulphide linkages is critical for the maturation of highly stable infectious particles (Cardone *et al.*, 2014). Even though virion assembly takes place largely in the nucleus, most likely initiating at ND10 bodies (Day *et al.*, 1998), virion assembly begins in the cytoplasm where L1 assembles into pentamers prior to nuclear import by karyopherin *via* binding the nuclear localization signal at the C-terminus of the L1 protein (Merle *et al.*, 1999; Nelson *et al.*, 2002). Karyopherin $\alpha2\beta1$ likely also serves as a chaperone to prevent premature assembly of the cytoplasmic pentamers into capsids, but once displaced from the nuclear located pentamer full assembly can take place (Bird *et al.*, 2008). Other cellular chaperones such as heat shock protein 70 family (HSP70) might be also involved in the regulation of this process - from subassembly status to full assembly (Chromy *et al.*, 2006; Song *et al.*, 2010). These molecular chaperones have been implicated to facilitate L2 incorporation into the capsid. HSP70 is required for transport of L2 to the nucleus, and along with the cochaperone HSP40, associates with assembled L1 and L2 containing VLP (Florin *et al.*, 2004). HSP70 binds to the exposed C-terminus of the L2 protein. HSP70 can also bind to sites in the N-terminus of L2 and this association probably assists in L2 protein folding, most likely to protect the C-terminus from protein interactions that would hinder virion morphogenesis. However, the chaperone is displaced from the L1-L2 VLP by encapsidation of reporter DNA suggesting that the chaperone is dislodged from the particle upon viral genome encapsidation (Florin *et al.*, 2004). Similarly nucleophosmin interacts with L2 and this interaction may have a critical role in establishing the correct interactions between L2 and L1 during capsid assembly (Day *et al.*, 2015). Detail of the mechanism of HPV genome encapsidation is not understood, but is likely to involve factors that aid stabilization of the virion. There is no evidence for distinct packaging signals within the viral genome, but perhaps size matters here. The Schiller group recently reported the use of a cell-free assembly system to study PV assembly; consisting of a nuclear extract, HPV16 capsomers (disassembled capsids) and a GFP reporter plasmid to produce infectious HPV16 particles (Cerqueira *et al.*, 2015a). They found evidence that in the nucleus the viral particle undergoes assembly and disassembly during which it captures DNA of different sizes, a process that is energy (ATP) dependent. Only when it packages an episome of the correct size (8 kbp) is the particle stabilized to endow resistance to disassembly. It is hypothesized that a 'size discrimination mechanism' of packaging would be able to enable specificity in viral DNA packaging, without the need for a sequence dependent mechanism. Particles that have not undergone the final stage of stabilization could disassemble and reassemble until packaging episomal DNA of the correct size is achieved (Cerqueira *et al.*, 2015a). It will be important to establish if this is a common mechanism of viral genome packaging amongst different HPV types and whether this process occurs in the presence of a chromatinized viral genome.

Once assembled and packaged the naturally occurring redox gradient of the most superficial terminally differentiated cell layers allows maturation of the virus particles as the cells move up to a more oxidized environment in superficial epithelia (Conway *et al.*, 2009). This slow process enables the production of more ordered and stable infectious virus particles stabilized by L1 disulphide covalent interactions formed between pentamers (Ishii *et al.*, 2003; Conway *et al.*, 2009; Ryndock *et al.*, 2014).

Egress from infected tissues

The manner of release of the infectious particles from the superficial parakeratotic squames is thought to be a passive process whereby the virus packed squames are sloughed off the epithelia surface and the cells break open to release the virus particles (Figure 2.2). These cells may be made more fragile through the actions of the E4 protein on destabilization of the structural integrity of the cytokeratin network and a weakening of the cornified cell envelopes of differentiated squames (Doorbar *et al.*, 1991; Roberts *et al.*, 1993; Brown *et al.*, 2006). The action of E4 involves E4 assembly into amyloid fibres that disrupt the cytokeratin and deplete keratins through increased phosphorylation and ubiquitinylation of the keratin proteins (McIntosh *et al.*, 2008; McIntosh *et al.*, 2010).

Anogenital transmission of HPV is believed to occur by the shed virus gaining access to the basal epithelium *via* intercourse-induced microabrasions in the uninfected epithelium. In a mouse cervicovaginal infection model, mechanical abrasion of the genital epithelium permitted HPV16 PsV infection whereas untreated epithelium was resistant to infection (Roberts *et al.*, 2007). Chemical disruption of the epithelium using nonoxynol-9 (N-9), a non-ionic membrane active surfactant promoted PsV infection (Roberts *et al.*, 2007). Visualization of the capsids showed binding to the basement membrane in areas of the mucosa epithelium disrupted and wounded by physical and chemical induced abrasion. Worryingly, N-9 is widely used in over-the-counter spermicide and use of such products may be a risk factor for genital HPV infection. Moreover, the columnar epithelium next to the transformation zone, the site where cervical cancers arise, was particularly sensitive to N-9 induced wounding. On the other hand, the vaginal lubricant carrageenan protected against PsV infection following mechanical and N-9 wounding. Therefore, carrageenan could be used as a safe carrier for N-9 since the spermicide activity of N-9 is preserved (Roberts *et al.*, 2007).

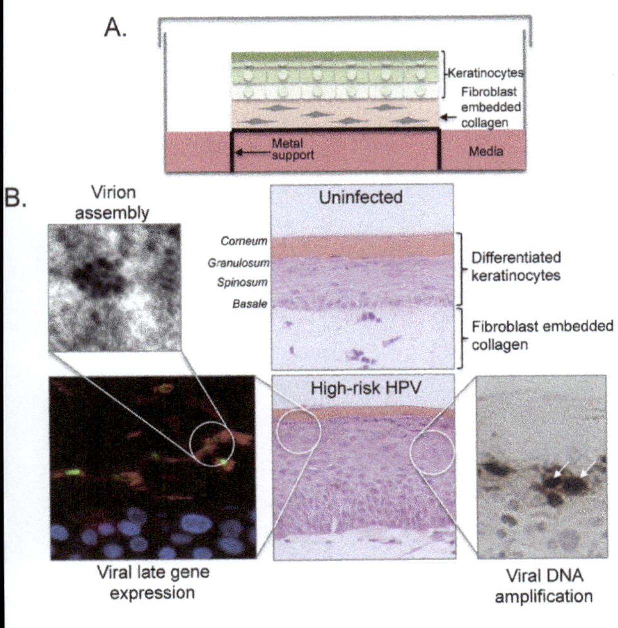

Figure 2.3. Modelling the HPV life cycle in 3-D. (A) Keratinocytes that maintain the HPV genome as episomes are grown in organotypic raft culture. The keratinocytes are seeded onto a fibroblast-embedded collagen matrix and grown at the air-liquid (media) interface by mounting on a solid support. Keratinocytes complete differentiation in between two to three weeks. (B) Haematoxylin and eosin stained raft sections of uninfected and cells harbouring high-risk HPV episomes show morphological differentiation of the keratinocytes; the different epithelial layers are indicated. Viral genome amplification (brown stain) is detected in the nuclei of cells (white arrows) by *in situ* hybridization using a high-risk HPV DNA probe. Viral late protein expression (L1, green) is detected in nuclei of cells in the upper suprabasal layer. These cells also express the HPV E4 protein (red). Electron microscopy detects arrays of HPV particles (~ 55 nm) in L1 positive cells.

Table 2. 1. Animal PV model systems.

Animal PV (host animal used)	Type of lesion	Main role in PV research
Bovine PV types 1, 2 and 4 (cattle, horses)	Cutaneous fibropapillomas (BPV1, 2), bladder cancers (BPV1, 2,), upper gastrointestinal tract cancers (BPV 4),	Latent infection, carcinogenesis, cofactor involvement, immune evasion, prophylactic and therapeutic vaccination
Canine oral PV type 1 (beagles)	Oral cavity papillomas	Host immune responses to PV, prophylactic and therapeutic immunization
Rhesus PV type 1 (macaques)	Squamous cell carcinoma of the cervix	Vaccine delivery, virus transmission, virus protein functions, host immune response to PV and vaccines
Cottontail PV type 1- Shope (domestic rabbits)	Cutaneous carcinomas	Carcinogenesis, latent infection, host immunity, virus protein function and virus life cycle, therapeutic vaccination, antiviral therapies
Rabbit oral PV (domestic rabbits)	Benign lesions on tongue and oral cavity, experimental infection of genital tissues	Latent infection, virus protein function, host immune response, prophylactic vaccination
Mouse (*Mus musculus*) PV type 1 (laboratory bred mice)	Cutaneous and mucosal lesions (anogenital and base of tongue), squamous cell carcinomas	Host immunity to virus infection, virus tissue tropism, virus transmission

Modelling the virus life cycle

The difficulty of propagation of PV in tissue culture has been a serious problem to researchers over the decades. There have been several different approaches used to overcome this limitation, ranging from animal models of PV replication and transformation, xenograft models using human tissues to keratinocyte cell lines that maintain the viral DNA as an un-integrated episome - either isolated from natural infections, or established by transfection of keratinocytes with cloned HPV genomes. The latter, when combined with the organotypic raft culture system (Asselineau and Prunieras, 1984), where upon the keratinocytes are grown on a fibroblast-embedded collagen matrix that is supported at the air-liquid interface, permits the formation of a fully differentiated stratified 3-D structure that supports the viral replication cycle (Figure 2.3).

Animal Models

PV are ubiquitous in the animal kingdom and infect birds, reptiles and mammals causing cutaneous papillomas, and oral and genital warts. In some cases, the animal PV induces malignant transformation: for example, BPV4 and squamous cell carcinomas of the upper gastrointestinal (GI) tract of bracken-fed cattle (Jarrett *et al.*, 1978; Campo *et al.*, 1994b), rhesus PV type 1 (RHPV1) and cervical neoplasia in the macaque (Ostrow *et al.*, 1990) and the induction of squamous cell carcinomas of the skin following experimental infection of domestic rabbits with the Shope CRPV1 (Rous and Beard, 1935). As discussed earlier in this chapter, several animal PV have been used to model virus latency. A summary of the different models and their use in PV research is given in Table 2.1.

Studies on the animal PV laid the foundations for the future development of vaccination against HPV. It was demonstrated that CRPV driven papilloma development in the skin coincided with the production of serum neutralizing antibodies (Shope and Hurst, 1933; Kidd *et al.*, 1936), which were capable of protecting against high dose viral challenge (Shope, 1937). The specificity of the neutralizing effect to PVs was shown when infection of domestic rabbits with ROPV could induce specific neutralizing antibodies, but the host was still susceptible to CRPV infection and *vice versa*. Long-lasting protective immunity against CRPV infection was established using VLP containing CRPV L1 (Breitburd *et al.*, 1995; Christensen *et al.*, 1996). Campo showed in the late 1990s that BPV1 neutralizing antibodies were elicited in response to prophylactic vaccination with BPV1 virions and BPV4 VLPs (including those formed by L1 alone), and provided protective immunity against BPV4 infection (Kirnbauer *et al.*, 1996). The N-terminus of the BPV4 L2 protein was also an effective prophylactic vaccine in cattle (Campo *et al.*, 1997). Beagles infected with canine oral papillomavirus type 1 have also proved a useful model to study prophylactic and therapeutic vaccination for papillomas at a mucosal epithelial site (Stanley *et al.*, 2001; Moore *et al.*, 2003). Further technical developments to induce viral DNA mediated papilloma formation and improved infection protocols in rabbits (Ito and Evans, 1961) permitted a better understanding of viral genetics and the potential of pre-clinical anti viral therapies. Christensen *et al.* demonstrated that the rabbit model could be an effective system to test nucleoside analogues for their anti-viral potential, either through topical application or *via* intralesional injections (Christensen *et al.*, 2000; Christensen *et al.*, 2014). The model has also been used to study the potential of viral gene based vaccines. The Christensen laboratory showed that combinatorial vaccination with E1, E2, E6, and E7 CRPV genes displayed protection against viral challenge through a T cell mediated immune response (Han *et al.*, 1999a) and these DNA based vaccinations slowed carcinoma development (Han *et al.*, 2000).

Whilst the investigations of the different animal models have proven very informative in the different areas of PV biology, pathogenesis and immunity, the pattern of life cycle organization is heterogeneous; amongst themselves and when compared to the HPV types that cause malignancies at the mucosa (Peh *et al.*, 2002). In fact, the animal PV with an epithelial tropism and life cycle organization that resembles most closely that of HPV16 is the ROPV (Peh *et al.*, 2002).

The discovery of a mouse papillomavirus (MusPV1 – relabelled Mus musculus (Mmu) PV1) that could infect laboratory-bred mice was an important milestone in the development of small animal models of PV infection (Ingle *et al.*, 2011; Joh *et al.*, 2011; Hu *et al.*, 2017; Uberoi and Lambert, 2017). Importantly, MmuPV1 infections form virus-producing productive lesions at cutaneous and mucosal epithelial sites, with the vagina, cervical, anal and base of tongue (mice do not have tonsils) epithelia being susceptible to infection (Sundberg *et al.*, 2014; Cladel *et al.*, 2015; Hu *et al.*, 2015; Cladel *et al.*, 2016). There is also evidence that the virus may have oncogenic potential; dysplasia can be detected at some of the infected sites and skin tumour formation has also been reported (Cladel *et al.*, 2013; Handisurya *et al.*, 2013). The future potential of this animal model is far-reaching in areas of host immunity and vaccine development, virus transmission, and cancer development at the skin and mucosa.

Transgenic mouse models have been of particular strength in the characterization of the properties of the viral oncogenes of the alpha-HPV (E6, E7 and E5) *in vivo*. The models utilize the basal specific promoter cassette of keratin 14 to drive expression in the basal cells of stratified squamous epithelium and enabling analysis of the contribution of the oncogenes to skin and cervical carcinogenesis (Herber *et al.*, 1996; Song *et al.*, 1999). The role of the viral oncogenes in carcinogenesis was found to be site-specific - in the skin, E5 and E6 are the most potent oncogenes, whereas in the cervix E7 is more potent (Song *et al.*, 2000; Maufort *et al.*, 2007) in driving carcinogenesis. However, the potency of E5 and E6 in the cervix was modified upon lengthy treatment of the mice with oestrogen leading to more severe neoplastic disease (E5) or cancers (E6) (Shai *et al.*, 2007; Maufort *et al.*, 2010). Similar analyses have been extended to other sites of HPV-induced carcinogenesis, including the anal canal and head and neck epithelia (Strati *et al.*, 2006; Strati and Lambert, 2007; Jabbar *et al.*, 2010; Thomas *et al.*, 2011). Dissection of the role of the individual functions of the oncoproteins to carcinogenesis showed that whilst inactivation of the p53 and RB tumour suppressor pathways were involved, so too were other E6 and E7 functions, including the ability of E6 to bind to PDZ domain-containing proteins (Simonson *et al.*, 2005; Balsitis *et al.*, 2006).

Human tissue xenografts

Xenograft models were initially developed to circumvent the issue of PV propagation outside of the natural host. One of the first models for PV research utilized athymic (nude) mice grafted with human tissue skin (foreskin and cervix), which was infected *in vitro* with a cell-free extract derived from HPV11 positive vulvar condylomata (known as the Hershey isolate) (Kreider *et al.*, 1985; Kreider *et al.*, 1987a). These xenografts supported the production of large amounts of the HPV11 virions and were morphologically and histologically indistinguishable from the naturally occurring condylomas. As well as a source of infectious virus, the model allowed for the first time analysis of the time course of the virus life cycle using HPV11 DNA and RNA probes (Stoler *et al.*, 1990). Increased cellular proliferation and other morphologic features of the condylomata could be attributed directly to the marked upregulation of early viral gene activity and viral DNA (Stoler *et al.*, 1990). This technique not only permitted the growth of HPV infected human tissue outside of the human host, it was the first experiment to demonstrate HPV driven morphological transformation of human tissues (Kreider *et al.*, 1985; Kreider *et al.*, 1986; Kreider *et al.*, 1987b). Neil Christensen and John Kreider used the athymic mouse model to demonstrate antibody-mediated neutralization of an HPV (HPV11) infection (Christensen and Kreider, 1990; Christensen *et al.*, 1990). Bonnez *et al.* further adapted this xenograft by using severe combined immunodeficiency mice (SCID), and considered this model to be more effective than the nude mouse model for overall growth of HPV infected tissue (Bonnez *et al.*, 1993). This model was also deemed advantageous over the nude model as human immune cells could be administered to the mouse, paving the way for future studies to dissect the host immunological response to HPV infection.

Since HPV16 is the most prevalent type in cervical cancers, the ability to model this high-risk PV life cycle became a prime target. The athymic model was used for grafting HPV16 containing W12 cells (derived from a low-grade cervical intraepithelial lesion and maintains HPV16 genomes as unintegrated episomes; see below for more detail about this cell line) onto the flank of mice, allowing epithelial differentiation and viral particle formation in the terminally differentiated cells (Sterling *et al.*, 1990). A gene gun approach to deliver HPV16 DNA (derived from W12 cell line) into foreskin grafted on to SCID mice also successfully yielded papilloma formation (Brandsma *et al.*, 1995). However propagation and passaging of this virus was not demonstrated in uninfected cells or tissues. Nevertheless, Bonnez *et al.* successfully demonstrated the isolation and passaging of HPV16 virus in the SCID xenograft model using human foreskin tissue infected with clinically isolated virus preparations (Bonnez *et al.*, 1998). This model has been used to demonstrate that antisera raised against HPV16 VLPs block HPV16 infection (White *et al.*, 1998).

Both the animal PV models and the HPV infected xenografts have provided a strong rationale for VLP vaccinations to control PV infection, and have provided key evidence for the direct contributiion of HPV in causing neoplasia, as well as serving as a platform for the search for antiviral therapies.

HPV genome containing cells that support virus production

The establishment of cell lines from patients' low-grade neoplastic anogenital tissues that are immortal but not tumourigenic and, when grown in organotypic raft culture, formed a reconstituted epithelium resembling the low-grade cervical lesion from which they were derived, opened up new avenues of investigation into the virus life cycle. The two most well-researched cell lines are W12, harbouring HPV16 (Stanley *et al.*, 1989) and CIN612 carrying the HPV31 subtype 31b (Rader *et al.*, 1990), and as in the natural lesion, the cells retain the viral genomes as unintegrated episomes at similar copy numbers (both approximately 50 copies per cell). In CIN612 rafts, the complete synthesis of HPV31b virions was achieved only upon addition of the phorbol ester 12-O-tetradecanoylphorbol-13-acetate (TPA) to the media (Meyers *et al.*, 1992). TPA activates protein kinase C signalling that stimulates the full epidermal differentiation programme required for expression of late viral transcripts (Hummel *et al.*, 1995). These cell lines have proven powerful in gaining insight into many aspects of the virus life cycle, including the identification of the differentiation-dependent late promoter in the *E7* gene (Hummel *et al.*, 1992) and the complexity of viral transcriptome (Doorbar *et al.*, 1990; Milligan *et al.*, 2007), as just a few examples, and they are still valued models today for the study of life cycle events (Anacker *et al.*, 2016; Spriggs and Laimins, 2017).

Notably, upon extended cell culture the viral genomes in the W12 and CIN612 cells become integrated into the cellular DNA. In 3-D growth, these cells formed structures with a more severe dysplastic higher-grade phenotype, showing atypical basal-like cells throughout the epithelium, high number of mitotic (normal and abnormal) cells and evidence of invasion into the underlying collagen (Bedell *et al.*, 1991; De Geest *et al.*, 1993). These cell lines have therefore proved a good model in which to study HPV driven cervical disease progression. High-risk HPV DNA integration was demonstrated to be a key event in progression. In W12 cells, not only did integration of the HPV16 genome give rise to cells that had a selective growth advantage over cells harbouring episomes, correlating with increased stability of the host-viral fusion transcripts encoding the E6 and E7 oncogenes, but also the W12 clones with integrated DNA eventually acquired numerical and structural chromosomal abnormalities, linking HPV integration to host genomic instability (Jeon *et al.*, 1995; Jeon and Lambert, 1995). But why do HPV DNA integrant containing cells emerge in an episome-containing population? Using the W12 model, Pett and Coleman showed that latent (transcriptionally silent) integrants become active once episomes have been cleared by an interferon driven antiviral response. They proposed that the cells in response to the host cell activity rapidly lose the remaining episomes, relieving episome-mediated inhibition (by E2) of the integrant; deregulated E6 and E7 expression from the integrant then confers a selective growth advantage (Pett *et al.*, 2006). The W12 cell line continues to be a key model for studying the early events in cervical neoplastic progression (Scarpini *et al.*, 2014).

As yet, no similar precancerous cell lines have been established from HPV infected tissues in the head and neck – primarily base of tongue and the tonsil. In fact, HPV precancerous lesions at this body site have so far not been found and this might be because the virus driven carcinogenic pathway at this site is different to the cervix (Brakenhoff *et al.*, 2017).

Clearly, whilst the biopsy-derived cell lines are invaluable models of cervical disease, they are limited in their use to investigate individual HPV protein function. A breakthrough came when the Laimins laboratory established stable cell lines from normal human foreskin keratinocytes transfected with cloned HPV31 genomes (Frattini and Laimins, 1994). In these cells, the HPV31 genome was maintained as a stable extrachromosomal episome at a copy number similar to the biopsy derived cell line CIN612. In organotypic raft culture, the cells supported the complete HPV31 life cycle, including the formation of progeny virus without the need for treatment with protein kinase C activators. This model enabled the genetic analysis of papillomavirus biology by introducing mutations into the viral genome (Frattini and Laimins, 1994). Primary keratinocyte based models were subsequently established for other high-risk HPV types, including the most prevalent types in cervical cancer HPV16 and 18 (Frattini *et al.*, 1997; Meyers *et al.*, 1997; Wilson and Laimins, 2005) (Figure 2.3). Through the use of these models, the field has gained molecular insight into viral gene function, the role of regulatory sequences and detailed viral transcription profiles in the complete differentiation-dependent virus life cycle. Our own laboratories has used this physiological model widely to investigate host-virus events in HPV driven pathogenesis and to probe E2, E4, E5 and E6 functions in the high-risk HPV life cycle (Wilson *et al.*, 2007; Knight *et al.*, 2011; Leonard *et al.*, 2012; Delury *et al.*, 2013; Paris *et al.*, 2015; Harris *et al.*, 2017; Marsh *et al.*, 2017; Wasson *et al.*, 2017). In relation to HPV biology in the tonsil, the use of primary keratinocytes from the tonsil has shown that while these cells support the HPV16 life cycle (the most common type (> 90%) found in tonsil cancers) the final conformation of the virions produced in the organotypic rafts was different to those produced in anogenital derived keratinocytes suggesting assembly and maturation events are different between the two tissues (Israr *et al.*, 2016). Therefore, the tonsil-HPV models could prove to be very useful to understanding HPV biology and pathogenesis at this anatomical site (Lace *et al.*, 2009; Langsfeld and Laimins, 2016; Evans *et al.*, 2017).

There is however a number of limitations to the primary keratinocyte based models. The success rate of colony formation following HPV genome transfection is often very low, and given that extension of the life span of the cells and eventual immortalization is required for long-term viral genome maintenance it is not possible to generate lines using non-oncogenic HPV genomes for long-term culture, although short-term maintenance of low-risk HPV11 genomes is possible in primary cells (Thomas *et al.*, 2001). Furthermore, the study of early life cycle events such as viral genome establishment and critical virus-host interactions that occur prior to immortalization is also precluded in the primary cell model. To overcome some of these limitations, immortalized keratinocytes have been used. The two most commonly used cell lines are NIKS (normal immortalised keratinocytes) that arose spontaneously from normal human neonate foreskin keratinocytes (Allen-Hoffmann *et al.*, 2000), these are near diploid (contain an extra isochromosome of the long arm of chromosome 8), and primary human foreskin keratinocytes immortalized using the catalytic subunit of human telomerase (N-TERT) (Dickson *et al.*, 2000). Both cell lines support episome maintenance of cloned HPV genomes, including low-risk (e.g. HPV11) and high-risk (e.g. HPV16, HPV18) (Flores *et al.*, 1999; Fang *et al.*, 2006; Egawa *et al.*, 2017), and upon rafting the stratified cells produce infectious virus particles. Recombinant HPV genomes have also been used to establish experimental systems (Lee *et al.*, 2004; Wang *et al.*, 2009; Van Doorslaer *et al.*, 2016). Michelle Ozbun's group (Lee *et al.*, 2004) generated a recombinant replication deficient adenovirus type 5 containing the complete HPV16 genome flanked by LoxP sites. Infection of primary cervical keratinocytes with the recombinant Ad/HPV16 virus alongside an adenovirus containing the Cre recombinase enabled recombination and the HPV16 episomes were maintained at between 50-80 copies per cell. Because the recombinant Ad/HPV16 virus encodes eGFP that will be expressed only upon recombination, infected cells can be easily separated from the non-infected population thereby enabling very early life cycle events to be studied. In another study, transfection of primary keratinocytes with a plasmid containing a recombinant HPV18 genome with LoxP sites in the URR, together with a Cre recombinase expressing plasmid yielded cells harbouring 2000-5000 copies (much higher levels than in natural HPV infections) of the viral episomes and the complete virus life cycle recapitulated by growth in 3-D (Wang *et al.*, 2009). This system is not dependent on an immortalization function and has been used to define a role for E6 in viral DNA amplification using an immortalization defective mutant HPV18 genome (Wang *et al.*, 2009).

Future perspectives

The intimate link between PV propagation and epithelial differentiation is an aspect of PV biology that has made it difficult to replicate the PV life cycle in tissue culture. However, the use of animal PV models, cells derived from natural PV infections, and cells transfected with cloned HPV genomes have facilitated the study of the PV life cycle on a molecular level. These physiological models, combined with the production of VLP, PsV and isolation of virus particles assembled in keratinocytes grown in 3-D cultures have unquestionably pushed this field far. There are still many unanswered questions around how these viruses complete their life cycle in our squamous epithelia, including whether regulation of the life cycle is different between anatomical sites. Also, proteomic investigations of the HPV E6 and E7 ‘interactome’ show that they bind to a very large number of host proteins and as yet the function of many of these remains to be explained. A far greater understanding of the epigenetic control of the virus life cycle is needed, including the epigenetic regulation of the early to late gene expression switch. HPV infection leads to the production of many alternatively spliced transcripts and some have the potential to encode novel HPV proteins, which currently have no function assigned to them. Many events that occur in the late stages of the HPV life cycle remain speculative, including the role of E4 in virus release and transmission.

Several animal PV-based models have been exploited, from the large (BPV; cattle) to the small (CRPV1, ROPV1; rabbits), to yield a greater understanding of virus-host interactions, including host immune response to infection, latent PV infection, pathways to cancer, and many different aspects of PV vaccination. The discovery of MmuPV1 that infects laboratory bred mice and induces virus-producing lesions and dysplasia at the same anatomical epithelial sites as the human virus, including the base of tongue is an important advance in animal PV model development. Now, with the increasing incidence of HPV positive cancers in the oropharynx (tonsil, base of tongue) and the emerging evidence that the mechanisms of HPV driven carcinogenesis in the oropharynx and the cervix are different, indicates that there is a need for a better understanding of the regulation of the HPV life cycle in the oropharynx. Naturally derived HPV-positive “premalignant” oropharyngeal cells have as yet not been isolated, but primary tonsil keratinocytes transfected with HPV genomes support the HPV life cycle and they will aid the identification of HPV-tonsil interactions critical for HPV replication at this body site.

Acknowledgements

We thank all present and past members of the Roberts and Parish groups for their support. We apologize most sincerely to those authors whose PV research could not be cited due to space limitations.

References

Abban, C.Y., and Meneses, P.I. (2010). Usage of heparan sulfate, integrins, and FAK in HPV16 infection. Virology 403, 1-16. http://dx.doi.org/10.1016/j.virol.2010.04.007

Abbate, E.A., Voitenleitner, C., and Botchan, M.R. (2006). Structure of the papillomavirus DNA-tethering complex E2:Brd4 and a peptide that ablates HPV chromosomal association. Mol Cell 24, 877-889. http://dx.doi.org/10.1016/j.molcel.2006.11.002

Allen-Hoffmann, B.L., Schlosser, S.J., Ivarie, C.A., Sattler, C.A., Meisner, L.F., and O'Connor, S.L. (2000). Normal growth and differentiation in a spontaneously immortalized near-diploid human keratinocyte cell line, NIKS. J Invest Dermatol 114, 444-455. http://dx.doi.org/10.1046/j.1523-1747.2000.00869.x

Amella, C.A., Lofgren, L.A., Ronn, A.M., Nouri, M., Shikowitz, M.J., and Steinberg, B.M. (1994). Latent infection induced with cottontail rabbit papillomavirus. A model for human papillomavirus latency. Am J Pathol 144, 1167-1171.

Anacker, D.C., Aloor, H.L., Shepard, C.N., Lenzi, G.M., Johnson, B.A., Kim, B., and Moody, C.A. (2016). HPV31 utilizes the ATR-Chk1 pathway to maintain elevated RRM2 levels and a replication-competent environment in differentiating Keratinocytes. Virology 499, 383-396. http://dx.doi.org/10.1016/j.virol.2016.09.028

Anacker, D.C., Gautam, D., Gillespie, K.A., Chappell, W.H., and Moody, C.A. (2014). Productive replication of human papillomavirus 31 requires DNA repair factor Nbs1. J Virol 88, 8528-8544. http://dx.doi.org/10.1128/JVI.00517-14

Androphy, E.J., Lowy, D.R., and Schiller, J.T. (1987). Bovine papillomavirus E2 trans-activating gene product binds to specific sites in papillomavirus DNA. Nature 325, 70-73. http://dx.doi.org/10.1038/325070a0

Angeletti, P.C., Kim, K., Fernandes, F.J., and Lambert, P.F. (2002). Stable replication of papillomavirus genomes in Saccharomyces cerevisiae. J Virol 76, 3350-3358.

Asselineau, D., and Prunieras, M. (1984). Reconstruction of 'simplified' skin: control of fabrication. Br J Dermatol 111 Suppl 27, 219-222.

Aydin, I., Weber, S., Snijder, B., Samperio Ventayol, P., Kuhbacher, A., Becker, M., Day, P.M., Schiller, J.T., Kann, M., Pelkmans, L., *et al.* (2014). Large scale RNAi reveals the requirement of nuclear envelope breakdown for nuclear import of human papillomaviruses. PLoS Pathog 10, e1004162. http://dx.doi.org/10.1371/journal.ppat.1004162

Balsitis, S., Dick, F., Dyson, N., and Lambert, P.F. (2006). Critical roles for non-pRb targets of human papillomavirus type 16 E7 in cervical carcinogenesis. Cancer Res 66, 9393-9400. http://dx.doi.org/10.1158/0008-5472.CAN-06-0984

Banerjee, N.S., Genovese, N.J., Noya, F., Chien, W.M., Broker, T.R., and Chow, L.T. (2006). Conditionally activated E7 proteins of high-risk and low-risk human papillomaviruses induce S phase in postmitotic, differentiated human keratinocytes. J Virol 80, 6517-6524. http://dx.doi.org/10.1128/JVI.02499-05

Bastien, N., and McBride, A.A. (2000). Interaction of the papillomavirus E2 protein with mitotic chromosomes. Virology 270, 124-134. http://dx.doi.org/10.1006/viro.2000.0265

Baxter, M.K., McPhillips, M.G., Ozato, K., and McBride, A.A. (2005). The mitotic chromosome binding activity of the papillomavirus E2 protein correlates with interaction with the cellular chromosomal protein, Brd4. J Virol 79, 4806-4818. http://dx.doi.org/10.1128/JVI.79.8.4806-4818.2005

Bechtold, V., Beard, P., and Raj, K. (2003). Human papillomavirus type 16 E2 protein has no effect on transcription from episomal viral DNA. J Virol 77, 2021-2028.

Bedell, M.A., Hudson, J.B., Golub, T.R., Turyk, M.E., Hosken, M., Wilbanks, G.D., and Laimins, L.A. (1991). Amplification of human papillomavirus genomes *in vitro* is dependent on epithelial differentiation. J Virol 65, 2254-2260.

Bergant, M., and Banks, L. (2013). SNX17 facilitates infection with diverse papillomavirus types. J Virol 87, 1270-1273. http://dx.doi.org/10.1128/JVI.01991-12

Bernard, B.A., Bailly, C., Lenoir, M.C., Darmon, M., Thierry, F., and Yaniv, M. (1989). The human papillomavirus type 18 (HPV18) E2 gene product is a repressor of the HPV18 regulatory region in human keratinocytes. J Virol 63, 4317-4324.

Berumen, J., Ordonez, R.M., Lazcano, E., Salmeron, J., Galvan, S.C., Estrada, R.A., Yunes, E., Garcia-Carranca, A., Gonzalez-Lira, G., and Madrigal-de la Campa, A. (2001). Asian-American variants of human papillomavirus 16 and risk for cervical cancer: a case-control study. J Natl Cancer Inst 93, 1325-1330.

Bienkowska-Haba, M., Patel, H.D., and Sapp, M. (2009). Target cell cyclophilins facilitate human papillomavirus type 16 infection. PLoS pathogens 5, e1000524. http://dx.doi.org/10.1371/journal.ppat.1000524

Bienkowska-Haba, M., Williams, C., Kim, S.M., Garcea, R.L., and Sapp, M. (2012). Cyclophilins facilitate dissociation of the human papillomavirus type 16 capsid protein L1 from the L2/DNA complex following virus entry. J Virol 86, 9875-9887. http://dx.doi.org/10.1128/JVI.00980-12

Bird, G., O'Donnell, M., Moroianu, J., and Garcea, R.L. (2008). Possible role for cellular karyopherins in regulating polyomavirus and papillomavirus capsid assembly. J Virol 82, 9848-9857. http://dx.doi.org/10.1128/JVI.01221-08

Blanpain, C., and Fuchs, E. (2014). Stem cell plasticity. Plasticity of epithelial stem cells in tissue regeneration. Science 344, 1242281. http://dx.doi.org/10.1126/science.1242281

Bonagura, V.R., Hatam, L.J., Rosenthal, D.W., de Voti, J.A., Lam, F., Steinberg, B.M., and Abramson, A.L. (2010). Recurrent respiratory papillomatosis: a complex defect in immune responsiveness to human papillomavirus-6 and -11. APMIS 118, 455-470. http://dx.doi.org/10.1111/j.1600-0463.2010.02617.x

Bonnez, W., DaRin, C., Borkhuis, C., de Mesy Jensen, K., Reichman, R.C., and Rose, R.C. (1998). Isolation and propagation of human papillomavirus type 16 in human xenografts implanted in the severe combined immunodeficiency mouse. J Virol 72, 5256-5261.

Bonnez, W., Rose, R.C., Da Rin, C., Borkhuis, C., de Mesy Jensen, K.L., and Reichman, R.C. (1993). Propagation of human papillomavirus type 11 in human xenografts using the severe combined immunodeficiency (SCID) mouse and comparison to the nude mouse model. Virology 197, 455-458. http://dx.doi.org/10.1006/viro.1993.1611

Boshart, M., Gissmann, L., Ikenberg, H., Kleinheinz, A., Scheurlen, W., and zur Hausen, H. (1984). A new type of papillomavirus DNA, its presence in genital cancer biopsies and in cell lines derived from cervical cancer. EMBO J 3, 1151-1157.

Bouallaga, I., Massicard, S., Yaniv, M., and Thierry, F. (2000). An enhanceosome containing the Jun B/Fra-2 heterodimer and the HMG-I(Y) architectural protein controls HPV 18 transcription. EMBO Rep 1, 422-427. http://dx.doi.org/10.1093/embo-reports/kvd091

Bouvard, V., Storey, A., Pim, D., and Banks, L. (1994). Characterization of the human papillomavirus E2 protein: evidence of trans-activation and trans-repression in cervical keratinocytes. EMBO J 13, 5451-5459.

Brakenhoff, R.H., Wagner, S., and Klussmann, J.P. (2017). Molecular Patterns and Biology of HPV-Associated HNSCC. Recent Results Cancer Res 206, 37-56. http://dx.doi.org/10.1007/978-3-319-43580-0_3

Brandsma, J.L., Brownstein, D.G., Xiao, W., and Longley, B.J. (1995). Papilloma formation in human foreskin xenografts after inoculation of human papillomavirus type 16 DNA. J Virol 69, 2716-2721.

Bravo, I.G., and Felez-Sanchez, M. (2015). Papillomaviruses: Viral evolution, cancer and evolutionary medicine. Evol Med Public Health 2015, 32-51. http://dx.doi.org/10.1093/emph/eov003

Breitburd, F., Kirnbauer, R., Hubbert, N.L., Nonnenmacher, B., Trin-Dinh-Desmarquet, C., Orth, G., Schiller, J.T., and Lowy, D.R. (1995). Immunization with viruslike particles from cottontail rabbit papillomavirus (CRPV) can protect against experimental CRPV infection. J Virol 69, 3959-3963.

Broniarczyk, J., Massimi, P., Bergant, M., and Banks, L. (2015). Human Papillomavirus Infectious Entry and Trafficking Is a Rapid Process. J Virol 89, 8727-8732. http://dx.doi.org/10.1128/JVI.00722-15

Brown, D.R., Kitchin, D., Qadadri, B., Neptune, N., Batteiger, T., and Ermel, A. (2006). The human papillomavirus type 11 E1--E4 protein is a transglutaminase 3 substrate and induces abnormalities of the cornified cell envelope. Virology 345, 290-298. http://dx.doi.org/10.1016/j.virol.2005.09.048

Buck, C.B., Thompson, C.D., Roberts, J.N., Muller, M., Lowy, D.R., and Schiller, J.T. (2006). Carrageenan is a potent inhibitor of papillomavirus infection. PLoS Pathog 2, e69. http://dx.doi.org/10.1371/journal.ppat.0020069

Burk, R.D., Harari, A., and Chen, Z. (2013). Human papillomavirus genome variants. Virology 445, 232-243. http://dx.doi.org/10.1016/j.virol.2013.07.018

Burnett, S., Strom, A.C., Jareborg, N., Alderborn, A., Dillner, J., Moreno-Lopez, J., Pettersson, U., and Kiessling, U. (1990). Induction of bovine papillomavirus E2 gene expression and early region transcription by cell growth arrest: correlation with viral DNA amplification and evidence for differential promoter induction. J Virol 64, 5529-5541.

Calton, C.M., Bronnimann, M.P., Manson, A.R., Li, S., Chapman, J.A., Suarez-Berumen, M., Williamson, T.R., Molugu, S.K., Bernal, R.A., and Campos, S.K. (2017). Translocation of the papillomavirus L2/vDNA complex across the limiting membrane requires the onset of mitosis. PLoS Pathog 13, e1006200. http://dx.doi.org/10.1371/journal.ppat.1006200

Campo, M.S., Jarrett, W.F., O'Neil, W., and Barron, R.J. (1994a). Latent papillomavirus infection in cattle. Res Vet Sci 56, 151-157.

Campo, M.S., O'Neil, B.W., Barron, R.J., and Jarrett, W.F. (1994b). Experimental reproduction of the papilloma-carcinoma complex of the alimentary canal in cattle. Carcinogenesis 15, 1597-1601.

Campo, M.S., O'Neil, B.W., Grindlay, G.J., Curtis, F., Knowles, G., and Chandrachud, L. (1997). A peptide encoding a B-cell epitope from the N-terminus of the capsid protein L2 of bovine papillomavirus-4 prevents disease. Virology 234, 261-266. http://dx.doi.org/10.1006/viro.1997.8649

Campos-Leon, K., Wijendra, K., Siddiqa, A., Pentland, I., Feeney, K.M., Knapman, A., Davies, R., Androphy, E.J., and Parish, J.L. (2017). Association of Human Papillomavirus 16 E2 with Rad50-Interacting Protein 1 Enhances Viral DNA Replication. J Virol 91. http://dx.doi.org/10.1128/JVI.02305-16

Cardone, G., Moyer, A.L., Cheng, N., Thompson, C.D., Dvoretzky, I., Lowy, D.R., Schiller, J.T., Steven, A.C., Buck, C.B., and Trus, B.L. (2014). Maturation of the human papillomavirus 16 capsid. MBio 5, e01104-01114. http://dx.doi.org/10.1128/mBio.01104-14

Cerqueira, C., Liu, Y., Kuhling, L., Chai, W., Hafezi, W., van Kuppevelt, T.H., Kuhn, J.E., Feizi, T., and Schelhaas, M. (2013). Heparin increases the infectivity of Human Papillomavirus type 16 independent of cell surface proteoglycans and induces L1 epitope exposure. Cell Microbiol 15, 1818-1836. http://dx.doi.org/10.1111/cmi.12150

Cerqueira, C., Pang, Y.Y., Day, P.M., Thompson, C.D., Buck, C.B., Lowy, D.R., and Schiller, J.T. (2015a). A Cell-Free Assembly System for Generating Infectious Human Papillomavirus 16 Capsids Implicates a Size Discrimination Mechanism for Preferential Viral Genome Packaging. J Virol 90, 1096-1107. http://dx.doi.org/10.1128/JVI.02497-15

Cerqueira, C., Samperio Ventayol, P., Vogeley, C., and Schelhaas, M. (2015b). Kallikrein-8 Proteolytically Processes Human Papillomaviruses in the Extracellular Space To Facilitate Entry into Host Cells. J Virol 89, 7038-7052. http://dx.doi.org/10.1128/JVI.00234-15

Cerqueira, C., and Schiller, J.T. (2017). Papillomavirus assembly: An overview and perspectives. Virus Res 231, 103-107. http://dx.doi.org/10.1016/j.virusres.2016.11.010

Chaiwongkot, A., Pientong, C., Ekalaksananan, T., Vinokurova, S., Kongyingyoes, B., Chumworathayi, B., Patarapadungkit, N., Siriaunkgul, S., and von Knebel Doeberitz, M. (2013). Detection of the human papillomavirus 58 physical state using the amplification of papillomavirus oncogene transcripts assay. J Virol Methods 189, 290-298. http://dx.doi.org/10.1016/j.jviromet.2013.02.011

Chappell, W.H., Gautam, D., Ok, S.T., Johnson, B.A., Anacker, D.C., and Moody, C.A. (2015). Homologous Recombination Repair Factors Rad51 and BRCA1 Are Necessary for Productive Replication of Human Papillomavirus 31. J Virol 90, 2639-2652. http://dx.doi.org/10.1128/JVI.02495-15

Cheng, S., Schmidt-Grimminger, D.C., Murant, T., Broker, T.R., and Chow, L.T. (1995). Differentiation-dependent up-regulation of the human papillomavirus E7 gene reactivates cellular DNA replication in suprabasal differentiated keratinocytes. Genes Dev 9, 2335-2349.

Christensen, N.D., Cladel, N.M., Hu, J., and Balogh, K.K. (2014). Formulation of cidofovir improves the anti-papillomaviral activity of topical treatments in the CRPV/rabbit model. Antiviral Res 108, 148-155. http://dx.doi.org/10.1016/j.antiviral.2014.06.001

Christensen, N.D., and Kreider, J.W. (1990). Antibody-mediated neutralization *in vivo* of infectious papillomaviruses. J Virol 64, 3151-3156.

Christensen, N.D., Kreider, J.W., Cladel, N.M., Patrick, S.D., and Welsh, P.A. (1990). Monoclonal antibody-mediated neutralization of infectious human papillomavirus type 11. J Virol 64, 5678-5681.

Christensen, N.D., Pickel, M.D., Budgeon, L.R., and Kreider, J.W. (2000). *in vivo* anti-papillomavirus activity of nucleoside analogues including cidofovir on CRPV-induced rabbit papillomas. Antiviral Res 48, 131-142.

Christensen, N.D., Reed, C.A., Cladel, N.M., Han, R., and Kreider, J.W. (1996). Immunization with viruslike particles induces long-term protection of rabbits against challenge with cottontail rabbit papillomavirus. J Virol 70, 960-965.

Chromy, L.R., Oltman, A., Estes, P.A., and Garcea, R.L. (2006). Chaperone-mediated *in vitro* disassembly of polyoma- and papillomaviruses. J Virol 80, 5086-5091. http://dx.doi.org/10.1128/JVI.80.10.5086-5091.2006

Cladel, N.M., Budgeon, L.R., Balogh, K.K., Cooper, T.K., Hu, J., and Christensen, N.D. (2015). A novel pre-clinical murine model to study the life cycle and progression of cervical and anal papillomavirus infections. PLoS One 10, e0120128. http://dx.doi.org/10.1371/journal.pone.0120128

Cladel, N.M., Budgeon, L.R., Balogh, K.K., Cooper, T.K., Hu, J., and Christensen, N.D. (2016). Mouse papillomavirus MmuPV1 infects oral mucosa and preferentially targets the base of the tongue. Virology 488, 73-80. http://dx.doi.org/10.1016/j.virol.2015.10.030

Cladel, N.M., Budgeon, L.R., Cooper, T.K., Balogh, K.K., Hu, J., and Christensen, N.D. (2013). Secondary infections, expanded tissue tropism, and evidence for malignant potential in immunocompromised mice infected with Mus musculus papillomavirus 1 DNA and virus. J Virol 87, 9391-9395. http://dx.doi.org/10.1128/JVI.00777-13

Clower, R.V., Fisk, J.C., and Melendy, T. (2006). Papillomavirus E1 protein binds to and stimulates human topoisomerase I. J Virol 80, 1584-1587. http://dx.doi.org/10.1128/JVI.80.3.1584-1587.2006

Conway, M.J., Alam, S., Ryndock, E.J., Cruz, L., Christensen, N.D., Roden, R.B., and Meyers, C. (2009). Tissue-spanning redox gradient-dependent assembly of native human papillomavirus type 16 virions. J Virol 83, 10515-10526. http://dx.doi.org/10.1128/JVI.00731-09

Cote-Martin, A., Moody, C., Fradet-Turcotte, A., D'Abramo, C.M., Lehoux, M., Joubert, S., Poirier, G.G., Coulombe, B., Laimins, L.A., and Archambault, J. (2008). Human papillomavirus E1 helicase interacts with the WD repeat protein p80 to promote maintenance of the viral genome in keratinocytes. J Virol 82, 1271-1283. http://dx.doi.org/10.1128/JVI.01405-07

Cruz, L., Biryukov, J., Conway, M.J., and Meyers, C. (2015). Cleavage of the HPV16 Minor Capsid Protein L2 during Virion Morphogenesis Ablates the Requirement for Cellular Furin during De Novo Infection. Viruses 7, 5813-5830. http://dx.doi.org/10.3390/v7112910

Culp, T.D., Budgeon, L.R., and Christensen, N.D. (2006). Human papillomaviruses bind a basal extracellular matrix component secreted by keratinocytes which is distinct from a membrane-associated receptor. Virology 347, 147-159. http://dx.doi.org/10.1016/j.virol.2005.11.025

Culp, T.D., and Christensen, N.D. (2004). Kinetics of *in vitro* adsorption and entry of papillomavirus virions. Virology 319, 152-161. http://dx.doi.org/10.1016/j.virol.2003.11.004

da Silva-Diz, V., Sole-Sanchez, S., Valdes-Gutierrez, A., Urpi, M., Riba-Artes, D., Penin, R.M., Pascual, G., Gonzalez-Suarez, E., Casanovas, O., Vinals, F., *et al.* (2013). Progeny of Lgr5-expressing hair follicle stem cell contributes to papillomavirus-induced tumor development in epidermis. Oncogene 32, 3732-3743. http://dx.doi.org/10.1038/onc.2012.375

Dasgupta, J., Bienkowska-Haba, M., Ortega, M.E., Patel, H.D., Bodevin, S., Spillmann, D., Bishop, B., Sapp, M., and Chen, X.S. (2011). Structural basis of oligosaccharide receptor recognition by human papillomavirus. J. Biol. Chem. 286, 2617-2624. http://dx.doi.org/10.1074/jbc.M110.160184

Davy, C., and Doorbar, J. (2007). G2/M cell cycle arrest in the life cycle of viruses. Virology 368, 219-226. http://dx.doi.org/10.1016/j.virol.2007.05.043

Davy, C.E., Jackson, D.J., Wang, Q., Raj, K., Masterson, P.J., Fenner, N.F., Southern, S., Cuthill, S., Millar, J.B., and Doorbar, J. (2002). Identification of a G(2) arrest domain in the E1 wedge E4 protein of human papillomavirus type 16. J Virol 76, 9806-9818.

Day, P.M., Baker, C.C., Lowy, D.R., and Schiller, J.T. (2004). Establishment of papillomavirus infection is enhanced by promyelocytic leukemia protein (PML) expression. Proc Natl Acad Sci USA 101, 14252-14257. http://dx.doi.org/10.1073/pnas.0404229101

Day, P.M., Lowy, D.R., and Schiller, J.T. (2003). Papillomaviruses infect cells via a clathrin-dependent pathway. Virology 307, 1-11. http://dx.doi.org/10.1016/S0042-6822(02)00143-5

Day, P.M., Lowy, D.R., and Schiller, J.T. (2008). Heparan sulfate-independent cell binding and infection with furin-precleaved papillomavirus capsids. J. Virol. 82, 12565-12568. http://dx.doi.org/10.1128/JVI.01631-08

Day, P.M., Roden, R.B., Lowy, D.R., and Schiller, J.T. (1998). The papillomavirus minor capsid protein, L2, induces localization of the major capsid protein, L1, and the viral transcription/replication protein, E2, to PML oncogenic domains. J Virol 72, 142-150.

Day, P.M., Thompson, C.D., Lowy, D.R., and Schiller, J.T. (2017). Interferon Gamma Prevents Infectious Entry of Human Papillomavirus 16 via an L2-Dependent Mechanism. J Virol 91. http://dx.doi.org/10.1128/JVI.00168-17

Day, P.M., Thompson, C.D., Pang, Y.Y., Lowy, D.R., and Schiller, J.T. (2015). Involvement of Nucleophosmin (NPM1/B23) in Assembly of Infectious HPV16 Capsids. Papillomavirus Res 1, 74-89. http://dx.doi.org/10.1016/j.pvr.2015.06.005

Day, P.M., Thompson, C.D., Schowalter, R.M., Lowy, D.R., and Schiller, J.T. (2013). Identification of a role for the trans-Golgi network in human papillomavirus 16 pseudovirus infection. J Virol 87, 3862-3870. http://dx.doi.org/10.1128/JVI.03222-12

De Geest, K., Turyk, M.E., Hosken, M.I., Hudson, J.B., Laimins, L.A., and Wilbanks, G.D. (1993). Growth and differentiation of human papillomavirus type 31b positive human cervical cell lines. Gynecol Oncol 49, 303-310. http://dx.doi.org/10.1006/gyno.1993.1131

De-Castro Arce, J., Gockel-Krzikalla, E., and Rosl, F. (2012). Silencing of multi-copy HPV16 by viral self-methylation and chromatin occlusion: a model for epigenetic virus-host interaction. Hum Mol Genet 21, 1693-1705. http://dx.doi.org/10.1093/hmg/ddr604

Dell, G., Wilkinson, K.W., Tranter, R., Parish, J., Leo Brady, R., and Gaston, K. (2003). Comparison of the structure and DNA-binding properties of the E2 proteins from an oncogenic and a non-oncogenic human papillomavirus. J Mol Biol 334, 979-991.

Delury, C.P., Marsh, E.K., James, C.D., Boon, S.S., Banks, L., Knight, G.L., and Roberts, S. (2013). The role of protein kinase A regulation of the E6 PDZ-binding domain during the differentiation-dependent life cycle of human papillomavirus type 18. J Virol 87, 9463-9472. http://dx.doi.org/10.1128/JVI.01234-13

Demeret, C., Desaintes, C., Yaniv, M., and Thierry, F. (1997). Different mechanisms contribute to the E2-mediated transcriptional repression of human papillomavirus type 18 viral oncogenes. J Virol 71, 9343-9349.

Devaiah, B.N., Case-Borden, C., Gegonne, A., Hsu, C.H., Chen, Q., Meerzaman, D., Dey, A., Ozato, K., and Singer, D.S. (2016). BRD4 is a histone acetyltransferase that evicts nucleosomes from chromatin. Nat Struct Mol Biol 23, 540-548. http://dx.doi.org/10.1038/nsmb.3228

Dickson, M.A., Hahn, W.C., Ino, Y., Ronfard, V., Wu, J.Y., Weinberg, R.A., Louis, D.N., Li, F.P., and Rheinwald, J.G. (2000). Human keratinocytes that express *hTERT* and also bypass a p16(INK4a)-enforced mechanism that limits life span become immortal yet retain normal growth and differentiation characteristics. Mol Cell Biol 20, 1436-1447.

DiGiuseppe, S., Luszczek, W., Keiffer, T.R., Bienkowska-Haba, M., Guion, L.G., and Sapp, M.J. (2016). Incoming human papillomavirus type 16 genome resides in a vesicular compartment throughout mitosis. Proc Natl Acad Sci USA 113, 6289-6294. http://dx.doi.org/10.1073/pnas.1600638113

Donaldson, M.M., Boner, W., and Morgan, I.M. (2007). TopBP1 regulates human papillomavirus type 16 E2 interaction with chromatin. J Virol 81, 4338-4342. http://dx.doi.org/10.1128/JVI.02353-06

Dong, G., Broker, T.R., and Chow, L.T. (1994). Human papillomavirus type 11 E2 proteins repress the homologous E6 promoter by interfering with the binding of host transcription factors to adjacent elements. J Virol 68, 1115-1127.

Doorbar, J., Ely, S., Sterling, J., McLean, C., and Crawford, L. (1991). Specific interaction between HPV-16 E1-E4 and cytokeratins results in collapse of the epithelial cell intermediate filament network. Nature 352, 824-827. http://dx.doi.org/10.1038/352824a0

Doorbar, J., and Gallimore, P.H. (1987). Identification of proteins encoded by the L1 and L2 open reading frames of human papillomavirus 1a. J Virol 61, 2793-2799.

Doorbar, J., Parton, A., Hartley, K., Banks, L., Crook, T., Stanley, M., and Crawford, L. (1990). Detection of novel splicing patterns in a HPV16-containing keratinocyte cell line. Virology 178, 254-262.

Dreer, M., Fertey, J., van de Poel, S., Straub, E., Madlung, J., Macek, B., Iftner, T., and Stubenrauch, F. (2016). Interaction of NCOR/SMRT Repressor Complexes with Papillomavirus E8^E2C Proteins Inhibits Viral Replication. PLoS Pathog 12, e1005556. http://dx.doi.org/10.1371/journal.ppat.1005556

Dreer, M., van de Poel, S., and Stubenrauch, F. (2017). Control of viral replication and transcription by the papillomavirus E8^E2 protein. Virus Res 231, 96-102. http://dx.doi.org/10.1016/j.virusres.2016.11.005

Durst, M., Gissmann, L., Ikenberg, H., and zur Hausen, H. (1983). A papillomavirus DNA from a cervical carcinoma and its prevalence in cancer biopsy samples from different geographic regions. Proc Natl Acad Sci USA 80, 3812-3815.

Dyson, N., Howley, P.M., Munger, K., and Harlow, E. (1989). The human papilloma virus-16 E7 oncoprotein is able to bind to the retinoblastoma gene product. Science 243, 934-937.

Edwards, T.G., Helmus, M.J., Koeller, K., Bashkin, J.K., and Fisher, C. (2013a). Human papillomavirus episome stability is reduced by aphidicolin and controlled by DNA damage response pathways. J Virol 87, 3979-3989. http://dx.doi.org/10.1128/JVI.03473-12

Edwards, T.G., Vidmar, T.J., Koeller, K., Bashkin, J.K., and Fisher, C. (2013b). DNA damage repair genes controlling human papillomavirus (HPV) episome levels under conditions of stability and extreme instability. PLoS One 8, e75406. http://dx.doi.org/10.1371/journal.pone.0075406

Egawa, N., Egawa, K., Griffin, H., and Doorbar, J. (2015). Human Papillomaviruses; Epithelial Tropisms, and the Development of Neoplasia. Viruses 7, 3863-3890. http://dx.doi.org/10.3390/v7072802

Egawa, N., Nakahara, T., Ohno, S., Narisawa-Saito, M., Yugawa, T., Fujita, M., Yamato, K., Natori, Y., and Kiyono, T. (2012). The E1 protein of human papillomavirus type 16 is dispensable for maintenance replication of the viral genome. J Virol 86, 3276-3283. http://dx.doi.org/10.1128/JVI.06450-11

Egawa, N., Wang, Q., Griffin, H.M., Murakami, I., Jackson, D., Mahmood, R., and Doorbar, J. (2017). HPV16 and 18 genome amplification show different E4-dependence, with 16E4 enhancing E1 nuclear accumulation and replicative efficiency via its cell cycle arrest and kinase activation functions. PLoS Pathog 13, e1006282. http://dx.doi.org/10.1371/journal.ppat.1006282

Evander, M., Frazer, I.H., Payne, E., Qi, Y.M., Hengst, K., and McMillan, N.A. (1997). Identification of the alpha6 integrin as a candidate receptor for papillomaviruses. J. Virol. 71, 2449-2456.

Evans, M.R., James, C.D., Loughran, O., Nulton, T.J., Wang, X., Bristol, M.L., Windle, B., and Morgan, I.M. (2017) An oral keratinocyte life cycle model identifies novel host genome regulation by human papillomavirus 16 relevant to HPV positive head and neck cancer. Oncotarget 8, 81892-81909. http://dx.doi.org/10.18632/oncotarget.18328

Fang, L., Meyers, C., Budgeon, L.R., and Howett, M.K. (2006). Induction of productive human papillomavirus type 11 life cycle in epithelial cells grown in organotypic raft cultures. Virology 347, 28-35. http://dx.doi.org/10.1016/j.virol.2005.10.043

Feeney, K.M., and Parish, J.L. (2009). Targeting mitotic chromosomes: a conserved mechanism to ensure viral genome persistence. Proc Biol Sci 276, 1535-1544. http://dx.doi.org/10.1098/rspb.2008.1642

Feeney, K.M., Saade, A., Okrasa, K., and Parish, J.L. (2011). *in vivo* analysis of the cell cycle dependent association of the bovine papillomavirus E2 protein and ChlR1. Virology 414, 1-9. http://dx.doi.org/10.1016/j.virol.2011.03.015

Flores, E.R., Allen-Hoffmann, B.L., Lee, D., and Lambert, P.F. (2000). The human papillomavirus type 16 E7 oncogene is required for the productive stage of the viral life cycle. J Virol 74, 6622-6631.

Flores, E.R., Allen-Hoffmann, B.L., Lee, D., Sattler, C.A., and Lambert, P.F. (1999). Establishment of the human papillomavirus type 16 (HPV-16) life cycle in an immortalized human foreskin keratinocyte cell line. Virology 262, 344-354. http://dx.doi.org/10.1006/viro.1999.9868

Flores, E.R., and Lambert, P.F. (1997). Evidence for a switch in the mode of human papillomavirus type 16

DNA replication during the viral life cycle. J Virol 71, 7167-7179.

Florin, L., Becker, K.A., Lambert, C., Nowak, T., Sapp, C., Strand, D., Streeck, R.E., and Sapp, M. (2006). Identification of a dynein interacting domain in the papillomavirus minor capsid protein l2. J Virol 80, 6691-6696. http://dx.doi.org/10.1128/JVI.00057-06

Florin, L., Becker, K.A., Sapp, C., Lambert, C., Sirma, H., Muller, M., Streeck, R.E., and Sapp, M. (2004). Nuclear translocation of papillomavirus minor capsid protein L2 requires Hsc70. J Virol 78, 5546-5553. http://dx.doi.org/10.1128/JVI.78.11.5546-5553.2004

Fradet-Turcotte, A., Bergeron-Labrecque, F., Moody, C.A., Lehoux, M., Laimins, L.A., and Archambault, J. (2011). Nuclear accumulation of the papillomavirus E1 helicase blocks S-phase progression and triggers an ATM-dependent DNA damage response. J Virol 85, 8996-9012. http://dx.doi.org/10.1128/JVI.00542-11

Fradet-Turcotte, A., Moody, C., Laimins, L.A., and Archambault, J. (2010). Nuclear export of human papillomavirus type 31 E1 is regulated by Cdk2 phosphorylation and required for viral genome maintenance. J Virol 84, 11747-11760. http://dx.doi.org/10.1128/JVI.01445-10

Frattini, M.G., and Laimins, L.A. (1994). Binding of the human papillomavirus E1 origin-recognition protein is regulated through complex formation with the E2 enhancer-binding protein. Proc Natl Acad Sci USA 91, 12398-12402.

Frattini, M.G., Lim, H.B., Doorbar, J., and Laimins, L.A. (1997). Induction of human papillomavirus type 18 late gene expression and genomic amplification in organotypic cultures from transfected DNA templates. J Virol 71, 7068-7072.

Gauson, E.J., Donaldson, M.M., Dornan, E.S., Wang, X., Bristol, M., Bodily, J.M., and Morgan, I.M. (2015). Evidence supporting a role for TopBP1 and Brd4 in the initiation but not continuation of human papillomavirus 16 E1/E2-mediated DNA replication. J Virol 89, 4980-4991. http://dx.doi.org/10.1128/JVI.00335-15

Genovese, N.J., Banerjee, N.S., Broker, T.R., and Chow, L.T. (2008). Casein kinase II motif-dependent phosphorylation of human papillomavirus E7 protein promotes p130 degradation and S-phase induction in differentiated human keratinocytes. J Virol 82, 4862-4873. http://dx.doi.org/10.1128/JVI.01202-07

Gillespie, K.A., Mehta, K.P., Laimins, L.A., and Moody, C.A. (2012). Human papillomaviruses recruit cellular DNA repair and homologous recombination factors to viral replication centers. J Virol 86, 9520-9526. http://dx.doi.org/10.1128/JVI.00247-12

Giroglou, T., Florin, L., Schäfer, F., Streeck, R.E., and Sapp, M. (2001). Human papillomavirus infection requires cell surface heparan sulfate. J. Virol. 75, 1565-1570. http://dx.doi.org/10.1128/JVI.75.3.1565-1570.2001

Graham, S.V., and Faizo, A.A. (2017). Control of human papillomavirus gene expression by alternative splicing. Virus Res 231, 83-95. http://dx.doi.org/10.1016/j.virusres.2016.11.016

Grassel, L., Fast, L.A., Scheffer, K.D., Boukhallouk, F., Spoden, G.A., Tenzer, S., Boller, K., Bago, R., Rajesh, S., Overduin, M., *et al.* (2016). The CD63-Syntenin-1 Complex Controls Post-Endocytic Trafficking of Oncogenic Human Papillomaviruses. Sci Rep 6, 32337. http://dx.doi.org/10.1038/srep32337

Guan, J., Bywaters, S.M., Brendle, S.A., Ashley, R.E., Makhov, A.M., Conway, J.F., Christensen, N.D., and Hafenstein, S. (2017). Cryoelectron Microscopy Maps of Human Papillomavirus 16 Reveal L2 Densities and Heparin Binding Site. Structure 25, 253-263. http://dx.doi.org/10.1016/j.str.2016.12.001

Gunasekharan, V., and Laimins, L.A. (2013). Human papillomaviruses modulate microRNA 145 expression to directly control genome amplification. J Virol 87, 6037-6043. http://dx.doi.org/10.1128/JVI.00153-13

Gunasekharan, V.K., Li, Y., Andrade, J., and Laimins, L.A. (2016). Post-Transcriptional Regulation of KLF4 by High-Risk Human Papillomaviruses Is Necessary for the Differentiation-Dependent Viral Life Cycle. PLoS Pathog 12, e1005747. http://dx.doi.org/10.1371/journal.ppat.1005747

Han, R., Cladel, N.M., Reed, C.A., Peng, X., Budgeon, L.R., Pickel, M., and Christensen, N.D. (2000). DNA vaccination prevents and/or delays carcinoma development of papillomavirus-induced skin papillomas on rabbits. J Virol 74, 9712-9716.

Han, R., Cladel, N.M., Reed, C.A., Peng, X., and Christensen, N.D. (1999a). Protection of rabbits from viral challenge by gene gun-based intracutaneous vaccination with a combination of cottontail rabbit papillomavirus E1, E2, E6, and E7 genes. J Virol 73, 7039-7043.

Han, Y., Loo, Y.M., Militello, K.T., and Melendy, T. (1999b). Interactions of the papovavirus DNA replication initiator proteins, bovine papillomavirus type 1 E1 and simian virus 40 large T antigen, with human replication protein A. J Virol 73, 4899-4907.

Handisurya, A., Day, P.M., Thompson, C.D., Buck, C.B., Pang, Y.Y., Lowy, D.R., and Schiller, J.T. (2013). Characterization of Mus musculus papillomavirus 1 infection *in situ* reveals an unusual pattern of late gene expression and capsid protein localization. J Virol 87, 13214-13225. http://dx.doi.org/10.1128/JVI.02162-13

Harris, L., McFarlane-Majeed, L., Campos-Leon, K., Roberts, S., and Parish, J.L. (2017). The Cellular DNA Helicase ChlR1 Regulates Chromatin and Nuclear Matrix Attachment of the Human Papillomavirus 16 E2 Protein and High-Copy-Number Viral Genome Establishment. J Virol 91. http://dx.doi.org/10.1128/JVI.01853-16

Herber, R., Liem, A., Pitot, H., and Lambert, P.F. (1996). Squamous epithelial hyperplasia and carcinoma in mice transgenic for the human papillomavirus type 16 E7 oncogene. J Virol 70, 1873-1881.

Herfs, M., Soong, T.R., Delvenne, P., and Crum, C.P. (2017). Deciphering the Multifactorial Susceptibility of Mucosal Junction Cells to HPV Infection and Related Carcinogenesis. Viruses 9. http://dx.doi.org/10.3390/v9040085

Herfs, M., Vargas, S.O., Yamamoto, Y., Howitt, B.E., Nucci, M.R., Hornick, J.L., McKeon, F.D., Xian, W., and Crum, C.P. (2013). A novel blueprint for 'top down' differentiation defines the cervical squamocolumnar junction during development, reproductive life, and neoplasia. J Pathol 229, 460-468. http://dx.doi.org/10.1002/path.4110

Herfs, M., Yamamoto, Y., Laury, A., Wang, X., Nucci, M.R., McLaughlin-Drubin, M.E., Munger, K., Feldman, S., McKeon, F.D., Xian, W., *et al.* (2012). A discrete population of squamocolumnar junction cells implicated in the pathogenesis of cervical cancer. Proc Natl Acad Sci USA 109, 10516-10521. http://dx.doi.org/10.1073/pnas.1202684109

Hildesheim, A., Schiffman, M., Bromley, C., Wacholder, S., Herrero, R., Rodriguez, A., Bratti, M.C., Sherman, M.E., Scarpidis, U., Lin, Q.Q., *et al.* (2001). Human papillomavirus type 16 variants and risk of cervical cancer. J Natl Cancer Inst 93, 315-318.

Hoffmann, R., Hirt, B., Bechtold, V., Beard, P., and Raj, K. (2006). Different modes of human papillomavirus DNA replication during maintenance. J Virol 80, 4431-4439. http://dx.doi.org/10.1128/JVI.80.9.4431-4439.2006

Hong, S., Cheng, S., Iovane, A., and Laimins, L.A. (2015). STAT-5 Regulates Transcription of the Topoisomerase IIbeta-Binding Protein 1 (TopBP1) Gene To Activate the ATR Pathway and Promote Human Papillomavirus Replication. MBio 6, e02006-02015. http://dx.doi.org/10.1128/mBio.02006-15

Horton, J.S., and Stokes, A.J. (2014). The transmembrane channel-like protein family and human papillomaviruses: Insights into epidermodysplasia verruciformis and progression to squamous cell carcinoma. Oncoimmunology 3, e28288. http://dx.doi.org/10.4161/onci.28288

Hou, S.Y., Wu, S.Y., Zhou, T., Thomas, M.C., and Chiang, C.M. (2000). Alleviation of human papillomavirus E2-mediated transcriptional repression via formation of a TATA binding protein (or TFIID)-TFIIB-RNA polymerase II-TFIIF preinitiation complex. Mol Cell Biol 20, 113-125.

Hu, J., Budgeon, L.R., Cladel, N.M., Balogh, K., Myers, R., Cooper, T.K., and Christensen, N.D. (2015). Tracking vaginal, anal and oral infection in a mouse papillomavirus infection model. J Gen Virol 96, 3554-3565. http://dx.doi.org/10.1099/jgv.0.000295

Hu, Y., Clower, R.V., and Melendy, T. (2006). Cellular topoisomerase I modulates origin binding by bovine papillomavirus type 1 E1. J Virol 80, 4363-4371. http://dx.doi.org/10.1128/JVI.80.9.4363-4371.2006

Hu, J., Cladel., N.M., Budgeon, L.R., Balogh, K.K., and Christensen, N.D. (2017) The mouse papillomavirus infection model. Viruses 9, E246. http://dx.doi.org/10.3390/v9090246.

Huh, K., Zhou, X., Hayakawa, H., Cho, J.Y., Libermann, T.A., Jin, J., Harper, J.W., and Munger, K. (2007). Human papillomavirus type 16 E7 oncoprotein associates with the cullin 2 ubiquitin ligase complex, which contributes to degradation of the retinoblastoma tumor suppressor. J Virol 81, 9737-9747. http://dx.doi.org/10.1128/JVI.00881-07

Hummel, M., Hudson, J.B., and Laimins, L.A. (1992). Differentiation-induced and constitutive transcription of human papillomavirus type 31b in cell lines containing viral episomes. J Virol 66, 6070-6080.

Hummel, M., Lim, H.B., and Laimins, L.A. (1995). Human papillomavirus type 31b late gene expression is regulated through protein kinase C-mediated changes in RNA processing. J Virol 69, 3381-3388.

Ilves, I., Kivi, S., and Ustav, M. (1999). Long-term episomal maintenance of bovine papillomavirus type 1 plasmids is determined by attachment to host chromosomes, which Is mediated by the viral E2 protein and its binding sites. J Virol 73, 4404-4412.

Ingle, A., Ghim, S., Joh, J., Chepkoech, I., Bennett Jenson, A., and Sundberg, J.P. (2011). Novel laboratory mouse papillomavirus (MusPV) infection. Vet Pathol 48, 500-505. http://dx.doi.org/10.1177 /0300985810377186

Ishii, Y., Tanaka, K., and Kanda, T. (2003). Mutational analysis of human papillomavirus type 16 major capsid protein L1: the cysteines affecting the intermolecular bonding and structure of L1-capsids. Virology 308, 128-136.

Israr, M., Biryukov, J., Ryndock, E.J., Alam, S., and Meyers, C. (2016). Comparison of human papillomavirus type 16 replication in tonsil and foreskin epithelia. Virology 499, 82-90. http://dx.doi.org/10.1016/j.virol.2016.09.004

Ito, Y., and Evans, C.A. (1961). Induction of Tumors in Domestic Rabbits with Nucleic Acid Preparations from Partially Purified Shope Papilloma Virus and from Extracts of the Papillomas of Domestic and Cottontail Rabbits. J Exp Med 114, 485-500.

Jabbar, S., Strati, K., Shin, M.K., Pitot, H.C., and Lambert, P.F. (2010). Human papillomavirus type 16 E6 and E7 oncoproteins act synergistically to cause head and neck cancer in mice. Virology 407, 60-67. http://dx.doi.org/10.1016/j.virol.2010.08.003

James, C.D., and Roberts, S. (2016). Viral Interactions with PDZ Domain-Containing Proteins-An Oncogenic Trait? Pathogens 5. http://dx.doi.org/10.3390/pathogens5010008

Jang, M.K., Shen, K., and McBride, A.A. (2014). Papillomavirus genomes associate with BRD4 to replicate at fragile sites in the host genome. PLoS Pathog 10, e1004117. http://dx.doi.org/10.1371/journal.ppat.1004117

Jarrett, W.F., McNeil, P.E., Grimshaw, W.T., Selman, I.E., and McIntyre, W.I. (1978). High incidence area of cattle cancer with a possible interaction between an environmental carcinogen and a papilloma virus. Nature 274, 215-217.

Javier, R.T., and Butel, J.S. (2008). The history of tumor virology. Cancer Res 68, 7693-7706. http://dx.doi.org/10.1158/0008-5472.CAN-08-3301

Jeon, S., Allen-Hoffmann, B.L., and Lambert, P.F. (1995). Integration of human papillomavirus type 16 into the human genome correlates with a selective growth advantage of cells. J Virol 69, 2989-2997.

Jeon, S., and Lambert, P.F. (1995). Integration of human papillomavirus type 16 DNA into the human genome leads to increased stability of E6 and E7 mRNAs: implications for cervical carcinogenesis. Proc Natl Acad Sci USA 92, 1654-1658.

Joh, J., Jenson, A.B., King, W., Proctor, M., Ingle, A., Sundberg, J.P., and Ghim, S.J. (2011). Genomic analysis of the first laboratory-mouse papillomavirus. J Gen Virol 92, 692-698. http://dx.doi.org/10.1099/vir.0.026138-0

Johansson, C., Jamal Fattah, T., Yu, H., Nygren, J., Mossberg, A.K., and Schwartz, S. (2015). Acetylation of intragenic histones on HPV16 correlates with enhanced HPV16 gene expression. Virology 482, 244-259. http://dx.doi.org/10.1016/j.virol.2015.02.053

Johansson, C., and Schwartz, S. (2013). Regulation of human papillomavirus gene expression by splicing and polyadenylation. Nat Rev Microbiol 11, 239-251. http://dx.doi.org/10.1038/nrmicro2984

Johnson, B.A., Aloor, H.L., and Moody, C.A. (2017). The Rb binding domain of HPV31 E7 is required to maintain high levels of DNA repair factors in infected cells. Virology 500, 22-34. http://dx.doi.org/10.1016/j.virol.2016.09.029

Johnson, K.M., Kines, R.C., Roberts, J.N., Lowy, D.R., Schiller, J.T., and Day, P.M. (2009). Role of heparan sulfate in attachment to and infection of the murine female genital tract by human papillomavirus. J. Virol. 83, 2067-2074. http://dx.doi.org/10.1128/JVI.02190-08

Joyce, J.G., Tung, J.S., Przysiecki, C.T., Cook, J.C., Lehman, E.D., Sands, J.A., Jansen, K.U., and Keller, P.M. (1999). The L1 major capsid protein of human papillomavirus type 11 recombinant virus-like particles interacts with heparin and cell-surface glycosaminoglycans on human keratinocytes. J. Biol. Chem. 274, 5810-5822.

Kadaja, M., Sumerina, A., Verst, T., Ojarand, M., Ustav, E., and Ustav, M. (2007). Genomic instability of the host cell induced by the human papillomavirus replication machinery. EMBO J 26, 2180-2191. http://dx.doi.org/10.1038/sj.emboj.7601665

Kalantari, M., Calleja-Macias, I.E., Tewari, D., Hagmar, B., Lie, K., Barrera-Saldana, H.A., Wiley, D.J., and Bernard, H.U. (2004). Conserved methylation patterns of human papillomavirus type 16 DNA in asymptomatic infection and cervical neoplasia. J Virol 78, 12762-12772. http://dx.doi.org/10.1128/JVI.78.23.12762-12772.2004

Kalantari, M., Lee, D., Calleja-Macias, I.E., Lambert, P.F., and Bernard, H.U. (2008). Effects of cellular differentiation, chromosomal integration and 5-aza-2'-deoxycytidine treatment on human papillomavirus-16 DNA methylation in cultured cell lines. Virology 374, 292-303. http://dx.doi.org/10.1016/j.virol.2007.12.016

Kanaya, T., Kyo, S., and Laimins, L.A. (1997). The 5' region of the human papillomavirus type 31 upstream regulatory region acts as an enhancer which augments viral early expression through the action of YY1. Virology 237, 159-169. http://dx.doi.org/10.1006/viro.1997.8771

Kang, S.Y., Kannan, N., Zhang, L., Martinez, V., Rosin, M.P., and Eaves, C.J. (2015). Characterization of Epithelial Progenitors in Normal Human Palatine Tonsils and Their HPV16 E6/E7-Induced Perturbation. Stem Cell Reports 5, 1210-1225. http://dx.doi.org/10.1016/j.stemcr.2015.09.020

Kho, E.Y., Wang, H.K., Banerjee, N.S., Broker, T.R., and Chow, L.T. (2013). HPV-18 E6 mutants reveal p53 modulation of viral DNA amplification in organotypic cultures. Proc Natl Acad Sci USA 110, 7542-7549. http://dx.doi.org/10.1073/pnas.1304855110

Kidd, J.G., Beard, J.W., and Rous, P. (1936). Serological Reactions with a Virus Causing Rabbit Papillomas Which Become Cancerous: I. Tests of the Blood of Animals Carrying the Papilloma. J Exp Med 64, 63-77.

Kim, K., Garner-Hamrick, P.A., Fisher, C., Lee, D., and Lambert, P.F. (2003). Methylation patterns of papillomavirus DNA, its influence on E2 function, and implications in viral infection. J Virol 77, 12450-12459.

Kim, K., and Lambert, P.F. (2002). E1 protein of bovine papillomavirus 1 is not required for the maintenance of viral plasmid DNA replication. Virology 293, 10-14. http://dx.doi.org/10.1006/viro.2001.1305

Kines, R.C., Thompson, C.D., Lowy, D.R., Schiller, J.T., and Day, P.M. (2009). The initial steps leading to papillomavirus infection occur on the basement membrane prior to cell surface binding. Proc Natl Acad Sci USA 106, 20458-20463. http://dx.doi.org/10.1073/pnas.0908502106

Kirnbauer, R., Chandrachud, L.M., O'Neil, B.W., Wagner, E.R., Grindlay, G.J., Armstrong, A., McGarvie, G.M., Schiller, J.T., Lowy, D.R., and Campo, M.S. (1996). Virus-like particles of bovine papillomavirus type 4 in prophylactic and therapeutic immunization. Virology 219, 37-44. http://dx.doi.org/10.1006/viro.1996.0220

Knight, G.L., Pugh, A.G., Yates, E., Bell, I., Wilson, R., Moody, C.A., Laimins, L.A., and Roberts, S. (2011). A cyclin-binding motif in human papillomavirus type 18 (HPV18) E1^E4 is necessary for association with CDK-cyclin complexes and G2/M cell cycle arrest of keratinocytes, but is not required for differentiation-dependent viral genome amplification or L1 capsid protein expression. Virology 412, 196-210. http://dx.doi.org/10.1016/j.virol.2011.01.007

Knight, G.L., Turnell, A.S., and Roberts, S. (2006). Role for Wee1 in inhibition of G2-to-M transition through the cooperation of distinct human papillomavirus type 1 E4 proteins. J Virol 80, 7416-7426. http://dx.doi.org/10.1128/JVI.00196-06

Kreider, J.W., Howett, M.K., Leure-Dupree, A.E., Zaino, R.J., and Weber, J.A. (1987a). Laboratory production *in vivo* of infectious human papillomavirus type 11. J Virol 61, 590-593.

Kreider, J.W., Howett, M.K., Lill, N.L., Bartlett, G.L., Zaino, R.J., Sedlacek, T.V., and Mortel, R. (1986). *In vivo* transformation of human skin with human papillomavirus type 11 from condylomata acuminata. J Virol 59, 369-376.

Kreider, J.W., Howett, M.K., Stoler, M.H., Zaino, R.J., and Welsh, P. (1987b). Susceptibility of various human tissues to transformation *in vivo* with human papillomavirus type 11. Int J Cancer 39, 459-465.

Kreider, J.W., Howett, M.K., Wolfe, S.A., Bartlett, G.L., Zaino, R.J., Sedlacek, T., and Mortel, R. (1985). Morphological transformation *in vivo* of human uterine cervix with papillomavirus from condylomata acuminata. Nature 317, 639-641.

Kruppel, U., Muller-Schiffmann, A., Baldus, S.E., Smola-Hess, S., and Steger, G. (2008). E2 and the co-activator p300 can cooperate in activation of the human papillomavirus type 16 early promoter. Virology 377, 151-159. http://dx.doi.org/10.1016/j.virol.2008.04.006

Kyo, S., Klumpp, D.J., Inoue, M., Kanaya, T., and Laimins, L.A. (1997). Expression of AP1 during cellular differentiation determines human papillomavirus E6/E7 expression in stratified epithelial cells. J Gen Virol 78 (Pt 2), 401-411. http://dx.doi.org/10.1099/0022-1317-78-2-401

Lace, M.J., Anson, J.R., Klingelhutz, A.J., Lee, J.H., Bossler, A.D., Haugen, T.H., and Turek, L.P. (2009). Human papillomavirus (HPV) type 18 induces extended growth in primary human cervical, tonsillar, or foreskin keratinocytes more effectively than other high-risk mucosal HPVs. J Virol 83, 11784-11794. http://dx.doi.org/10.1128/JVI.01370-09

Lace, M.J., Anson, J.R., Thomas, G.S., Turek, L.P., and Haugen, T.H. (2008). The E8--E2 gene product of human papillomavirus type 16 represses early transcription and replication but is dispensable for viral plasmid persistence in keratinocytes. J Virol 82, 10841-10853. http://dx.doi.org/10.1128/JVI.01481-08

Langsfeld, E., and Laimins, L.A. (2016). Human papillomaviruses: research priorities for the next decade. Trends Cancer 2, 234-240. http://dx.doi.org/10.1016/j.trecan.2016.04.001

Lee, A.Y., and Chiang, C.M. (2009). Chromatin adaptor Brd4 modulates E2 transcription activity and protein stability. J Biol Chem 284, 2778-2786. http://dx.doi.org/10.1074/jbc.M805835200

Lee, C., and Laimins, L.A. (2004). Role of the PDZ domain-binding motif of the oncoprotein E6 in the pathogenesis of human papillomavirus type 31. J Virol 78, 12366-12377. http://dx.doi.org/10.1128/JVI.78.22.12366-12377.2004

Lee, D., Lee, B., Kim, J., Kim, D.W., and Choe, J. (2000). cAMP response element-binding protein-binding protein binds to human papillomavirus E2 protein and activates E2-dependent transcription. J Biol Chem 275, 7045-7051.

Lee, J.H., Yi, S.M., Anderson, M.E., Berger, K.L., Welsh, M.J., Klingelhutz, A.J., and Ozbun, M.A. (2004). Propagation of infectious human papillomavirus type 16 by using an adenovirus and Cre/LoxP mechanism. Proc Natl Acad Sci USA 101, 2094-2099. http://dx.doi.org/10.1073/pnas.0308615100

Lehman, C.W., and Botchan, M.R. (1998). Segregation of viral plasmids depends on tethering to chromosomes and is regulated by phosphorylation. Proc Natl Acad Sci USA 95, 4338-4343.

Lehoux, M., Fradet-Turcotte, A., Lussier-Price, M., Omichinski, J.G., and Archambault, J. (2012). Inhibition of human papillomavirus DNA replication by an E1-derived p80/UAF1-binding peptide. J Virol 86, 3486-3500. http://dx.doi.org/10.1128/JVI.07003-11

Lehoux, M., Gagnon, D., and Archambault, J. (2014). E1-mediated recruitment of a UAF1-USP deubiquitinase complex facilitates human papillomavirus DNA replication. J Virol 88, 8545-8555. http://dx.doi.org/10.1128/JVI.00379-14

Leonard, S.M., Pereira, M., Roberts, S., Cuschieri, K., Nuovo, G., Athavale, R., Young, L., Ganesan, R., and Woodman, C.B. (2016). Evidence of disrupted high-risk human papillomavirus DNA in morphologically normal cervices of older women. Sci Rep 6, 20847. http://dx.doi.org/10.1038/srep20847

Leonard, S.M., Wei, W., Collins, S.I., Pereira, M., Diyaf, A., Constandinou-Williams, C., Young, L.S., Roberts, S., and Woodman, C.B. (2012). Oncogenic human papillomavirus imposes an instructive pattern of DNA methylation changes which parallel the natural history of cervical HPV infection in young women. Carcinogenesis 33, 1286-1293. http://dx.doi.org/10.1093/carcin/bgs157

Lipovsky, A., Popa, A., Pimienta, G., Wyler, M., Bhan, A., Kuruvilla, L., Guie, M.A., Poffenberger, A.C., Nelson, C.D., Atwood, W.J., *et al*. (2013). Genome-wide siRNA screen identifies the retromer as a cellular entry factor for human papillomavirus. Proc Natl Acad Sci USA 110, 7452-7457. http://dx.doi.org/10.1073/pnas.1302164110

Maglennon, G.A., and Doorbar, J. (2012). The biology of papillomavirus latency. Open Virol J 6, 190-197. http://dx.doi.org/10.2174/1874357901206010190

Maglennon, G.A., McIntosh, P., and Doorbar, J. (2011). Persistence of viral DNA in the epithelial basal layer suggests a model for papillomavirus latency following immune regression. Virology 414, 153-163. http://dx.doi.org/10.1016/j.virol.2011.03.019

Maglennon, G.A., McIntosh, P.B., and Doorbar, J. (2014). Immunosuppression facilitates the reactivation of latent papillomavirus infections. J Virol 88, 710-716. http://dx.doi.org/10.1128/JVI.02589-13

Marsh, E.K., Delury, C.P., Davies, N.J., Weston, C.J., Miah, M.A.L., Banks, L., Parish, J.L., Higgs, M.R., and Roberts, S. (2017). Mitotic control of human papillomavirus genome-containing cells is regulated by the function of the PDZ-binding motif of the E6 oncoprotein. Oncotarget 8, 19491-19506. http://dx.doi.org/10.18632/oncotarget.14469

Maufort, J.P., Shai, A., Pitot, H.C., and Lambert, P.F. (2010). A role for HPV16 E5 in cervical carcinogenesis. Cancer Res 70, 2924-2931. http://dx.doi.org/10.1158/0008-5472.CAN-09-3436

Maufort, J.P., Williams, S.M., Pitot, H.C., and Lambert, P.F. (2007). Human papillomavirus 16 E5 oncogene contributes to two stages of skin carcinogenesis. Cancer Res 67, 6106-6112. http://dx.doi.org/10.1158/0008-5472.CAN-07-0921

McBride, A.A. (2013). The papillomavirus E2 proteins. Virology 445, 57-79. http://dx.doi.org/10.1016/j.virol.2013.06.006

McIntosh, P.B., Laskey, P., Sullivan, K., Davy, C., Wang, Q., Jackson, D.J., Griffin, H.M., and Doorbar, J. (2010). E1--E4-mediated keratin phosphorylation and ubiquitylation: a mechanism for keratin depletion in HPV16-infected epithelium. J Cell Sci 123, 2810-2822. http://dx.doi.org/10.1242/jcs.061978

McIntosh, P.B., Martin, S.R., Jackson, D.J., Khan, J., Isaacson, E.R., Calder, L., Raj, K., Griffin, H.M., Wang, Q., Laskey, P., *et al*. (2008). Structural analysis reveals an amyloid form of the human papillomavirus type 16

E1--E4 protein and provides a molecular basis for its accumulation. J Virol 82, 8196-8203. http://dx.doi.org/10.1128/JVI.00509-08

McKinney, C.C., Kim, M.J., Chen, D., and McBride, A.A. (2016). Brd4 Activates Early Viral Transcription upon Human Papillomavirus 18 Infection of Primary Keratinocytes. MBio 7. http://dx.doi.org/10.1128/mBio.01644-16

McLaughlin-Drubin, M.E., Wilson, S., Mullikin, B., Suzich, J., and Meyers, C. (2003). Human papillomavirus type 45 propagation, infection, and neutralization. Virology 312, 1-7.

McPhillips, M.G., Oliveira, J.G., Spindler, J.E., Mitra, R., and McBride, A.A. (2006). Brd4 is required for e2-mediated transcriptional activation but not genome partitioning of all papillomaviruses. J Virol 80, 9530-9543. http://dx.doi.org/10.1128/JVI.01105-06

Mehic, D., Bakiri, L., Ghannadan, M., Wagner, E.F., and Tschachler, E. (2005). Fos and jun proteins are specifically expressed during differentiation of human keratinocytes. J Invest Dermatol 124, 212-220. http://dx.doi.org/10.1111/j.0022-202X.2004.23558.x

Mehta, K., Gunasekharan, V., Satsuka, A., and Laimins, L.A. (2015). Human papillomaviruses activate and recruit SMC1 cohesin proteins for the differentiation-dependent life cycle through association with CTCF insulators. PLoS Pathog 11, e1004763. http://dx.doi.org/10.1371/journal.ppat.1004763

Melar-New, M., and Laimins, L.A. (2010). Human papillomaviruses modulate expression of microRNA 203 upon epithelial differentiation to control levels of p63 proteins. J Virol 84, 5212-5221. http://dx.doi.org/10.1128/JVI.00078-10

Merle, E., Rose, R.C., LeRoux, L., and Moroianu, J. (1999). Nuclear import of HPV11 L1 capsid protein is mediated by karyopherin alpha2beta1 heterodimers. J Cell Biochem 74, 628-637.

Meyers, C., Frattini, M.G., Hudson, J.B., and Laimins, L.A. (1992). Biosynthesis of human papillomavirus from a continuous cell line upon epithelial differentiation. Science 257, 971-973.

Meyers, C., Mayer, T.J., and Ozbun, M.A. (1997). Synthesis of infectious human papillomavirus type 18 in differentiating epithelium transfected with viral DNA. J Virol 71, 7381-7386.

Michael, S., Lambert, P.F., and Strati, K. (2013). The HPV16 oncogenes cause aberrant stem cell mobilization. Virology 443, 218-225. http://dx.doi.org/10.1016/j.virol.2013.04.008

Milligan, S.G., Veerapraditsin, T., Ahamet, B., Mole, S., and Graham, S.V. (2007). Analysis of novel human papillomavirus type 16 late mRNAs in differentiated W12 cervical epithelial cells. Virology 360, 172-181. http://dx.doi.org/10.1016/j.virol.2006.10.012

Mirkovic, J., Howitt, B.E., Roncarati, P., Demoulin, S., Suarez-Carmona, M., Hubert, P., McKeon, F.D., Xian, W., Li, A., Delvenne, P., *et al.* (2015). Carcinogenic HPV infection in the cervical squamo-columnar junction. J Pathol 236, 265-271. http://dx.doi.org/10.1002/path.4533

Mohr, I.J., Clark, R., Sun, S., Androphy, E.J., MacPherson, P., and Botchan, M.R. (1990). Targeting the E1 replication protein to the papillomavirus origin of replication by complex formation with the E2 transactivator. Science 250, 1694-1699.

Moody, C.A., and Laimins, L.A. (2009). Human papillomaviruses activate the ATM DNA damage pathway for viral genome amplification upon differentiation. PLoS Pathog 5, e1000605. http://dx.doi.org/10.1371/journal.ppat.1000605

Moody, C.A., and Laimins, L.A. (2010). Human papillomavirus oncoproteins: pathways to transformation. Nat Rev Cancer 10, 550-560. http://dx.doi.org/10.1038/nrc2886

Moore, R.A., Walcott, S., White, K.L., Anderson, D.M., Jain, S., Lloyd, A., Topley, P., Thomsen, L., Gough, G.W., and Stanley, M.A. (2003). Therapeutic immunisation with COPV early genes by epithelial DNA delivery. Virology 314, 630-635.

Muller, A., Ritzkowsky, A., and Steger, G. (2002). Cooperative activation of human papillomavirus type 8 gene expression by the E2 protein and the cellular coactivator p300. J Virol 76, 11042-11053.

Munger, K., Werness, B.A., Dyson, N., Phelps, W.C., Harlow, E., and Howley, P.M. (1989). Complex formation of human papillomavirus E7 proteins with the retinoblastoma tumor suppressor gene product. EMBO J 8, 4099-4105.

Nelson, L.M., Rose, R.C., and Moroianu, J. (2002). Nuclear import strategies of high risk HPV16 L1 major capsid protein. J Biol Chem 277, 23958-23964. http://dx.doi.org/10.1074/jbc.M200724200

Nguyen, C.L., and Munger, K. (2009). Human papillomavirus E7 protein deregulates mitosis via an association with nuclear mitotic apparatus protein 1. J Virol 83, 1700-1707. http://dx.doi.org/10.1128/JVI.01971-08

Nicolaides, L., Davy, C., Raj, K., Kranjec, C., Banks, L., and Doorbar, J. (2011). Stabilization of HPV16 E6 protein by PDZ proteins, and potential implications for genome maintenance. Virology 414, 137-145. http://dx.doi.org/10.1016/j.virol.2011.03.017

O'Connor, M.J., Tan, S.H., Tan, C.H., and Bernard, H.U. (1996). YY1 represses human papillomavirus type 16 transcription by quenching AP-1 activity. J Virol 70, 6529-6539.

Oliveira, J.G., Colf, L.A., and McBride, A.A. (2006). Variations in the association of papillomavirus E2 proteins with mitotic chromosomes. Proc Natl Acad Sci USA 103, 1047-1052. http://dx.doi.org/10.1073/pnas.0507624103

Orav, M., Geimanen, J., Sepp, E.M., Henno, L., Ustav, E., and Ustav, M. (2015). Initial amplification of the HPV18 genome proceeds via two distinct replication mechanisms. Sci Rep 5, 15952. http://dx.doi.org/10.1038/srep15952

Ostrow, R.S., McGlennen, R.C., Shaver, M.K., Kloster, B.E., Houser, D., and Faras, A.J. (1990). A rhesus monkey model for sexual transmission of a papillomavirus isolated from a squamous cell carcinoma. Proc Natl Acad Sci USA 87, 8170-8174.

Ozbun, M.A. (2002). Human papillomavirus type 31b infection of human keratinocytes and the onset of early transcription. J. Virol. 76, 11291-11300. http://dx.doi.org/10.1128/JVI.76.22.11291-11300.2002

Palmer, E., Newcombe, R.G., Green, A.C., Kelly, C., Noel Gill, O., Hall, G., Fiander, A.N., Pirotte, E., Hibbitts, S.J., Homer, J., *et al.* (2014). Human papillomavirus infection is rare in nonmalignant tonsil tissue in the UK: implications for tonsil cancer precursor lesions. Int J Cancer 135, 2437-2443. http://dx.doi.org/10.1002/ijc.28886

Paris, C., Pentland, I., Groves, I., Roberts, D.C., Powis, S.J., Coleman, N., Roberts, S., and Parish, J.L. (2015). CCCTC-binding factor recruitment to the early region of the human papillomavirus 18 genome regulates viral oncogene expression. J Virol 89, 4770-4785. http://dx.doi.org/10.1128/JVI.00097-15

Parish, J.L., Bean, A.M., Park, R.B., and Androphy, E.J. (2006). ChlR1 is required for loading papillomavirus E2 onto mitotic chromosomes and viral genome maintenance. Mol Cell 24, 867-876. http://dx.doi.org/10.1016/j.molcel.2006.11.005

Park, R.B., and Androphy, E.J. (2002). Genetic analysis of high-risk e6 in episomal maintenance of human papillomavirus genomes in primary human keratinocytes. J Virol 76, 11359-11364.

Peh, W.L., Middleton, K., Christensen, N., Nicholls, P., Egawa, K., Sotlar, K., Brandsma, J., Percival, A., Lewis, J., Liu, W.J., *et al.* (2002). Life cycle heterogeneity in animal models of human papillomavirus-associated disease. J Virol 76, 10401-10416.

Pelkmans, L., and Helenius, A. (2003). Insider information: what viruses tell us about endocytosis. Curr. Opin. Cell Biol.15, 414-422.

Peng, Y.C., Breiding, D.E., Sverdrup, F., Richard, J., and Androphy, E.J. (2000). AMF-1/Gps2 binds p300 and enhances its interaction with papillomavirus E2 proteins. J Virol 74, 5872-5879.

Penrose, K.J., and McBride, A.A. (2000). Proteasome-mediated degradation of the papillomavirus E2-TA protein is regulated by phosphorylation and can modulate viral genome copy number. J Virol 74, 6031-6038.

Pett, M.R., Herdman, M.T., Palmer, R.D., Yeo, G.S., Shivji, M.K., Stanley, M.A., and Coleman, N. (2006). Selection of cervical keratinocytes containing integrated HPV16 associates with episome loss and an endogenous antiviral response. Proc Natl Acad Sci USA 103, 3822-3827. http://dx.doi.org/10.1073/pnas.0600078103

Pim, D., Broniarczyk, J., Bergant, M., Playford, M.P., and Banks, L. (2015). A Novel PDZ Domain Interaction Mediates the Binding between Human Papillomavirus 16 L2 and Sorting Nexin 27 and Modulates Virion Trafficking. J Virol 89, 10145-10155. http://dx.doi.org/10.1128/JVI.01499-15

Porter, S.S., Stepp, W.H., Stamos, J.D., and McBride, A.A. (2017). Host cell restriction factors that limit transcription and replication of human papillomavirus. Virus Res 231, 10-20. http://dx.doi.org/10.1016/j.virusres.2016.11.014

Quinlan, E.J., Culleton, S.P., Wu, S.Y., Chiang, C.M., and Androphy, E.J. (2013). Acetylation of conserved lysines in bovine papillomavirus E2 by p300. J Virol 87, 1497-1507. http://dx.doi.org/10.1128/JVI.02771-12

Rader, J.S., Golub, T.R., Hudson, J.B., Patel, D., Bedell, M.A., and Laimins, L.A. (1990). In vitro differentiation of epithelial cells from cervical neoplasias resembles *in vivo* lesions. Oncogene 5, 571-576.

Ramoz, N., Rueda, L.A., Bouadjar, B., Montoya, L.S., Orth, G., and Favre, M. (2002). Mutations in two adjacent novel genes are associated with epidermodysplasia verruciformis. Nat Genet 32, 579-581. http://dx.doi.org/10.1038/ng1044

Reinson, T., Toots, M., Kadaja, M., Pipitch, R., Allik, M., Ustav, E., and Ustav, M. (2013). Engagement of the ATR-dependent DNA damage response at the human papillomavirus 18 replication centers during the initial amplification. J Virol 87, 951-964. http://dx.doi.org/10.1128/JVI.01943-12

Richards, K.F., Bienkowska-Haba, M., Dasgupta, J., Chen, X.S., and Sapp, M. (2013). Multiple heparan sulfate binding site engagements are required for the infectious entry of human papillomavirus type 16. J Virol 87, 11426-11437. http://dx.doi.org/10.1128/jvi.01721-13

Richards, K.F., Mukherjee, S., Bienkowska-Haba, M., Pang, J., and Sapp, M. (2014). Human papillomavirus species-specific interaction with the basement membrane-resident non-heparan sulfate receptor. Viruses 6, 4856-4879. http://dx.doi.org/10.3390/v6124856

Richards, R.M., Lowy, D.R., Schiller, J.T., and Day, P.M. (2006). Cleavage of the papillomavirus minor capsid protein, L2, at a furin consensus site is necessary for infection. Proc Natl Acad Sci USA 103, 1522-1527. http://dx.doi.org/10.1073/pnas.0508815103

Roberts, J.N., Buck, C.B., Thompson, C.D., Kines, R., Bernardo, M., Choyke, P.L., Lowy, D.R., and Schiller, J.T. (2007). Genital transmission of HPV in a mouse model is potentiated by nonoxynol-9 and inhibited by carrageenan. Nat Med 13, 857-861. http://dx.doi.org/10.1038/nm1598

Roberts, S., Ashmole, I., Johnson, G.D., Kreider, J.W., and Gallimore, P.H. (1993). Cutaneous and mucosal human papillomavirus E4 proteins form intermediate filament-like structures in epithelial cells. Virology 197, 176-187. http://dx.doi.org/10.1006/viro.1993.1578

Romanczuk, H., Thierry, F., and Howley, P.M. (1990). Mutational analysis of *cis* elements involved in E2 modulation of human papillomavirus type 16 P97 and type 18 P105 promoters. J Virol 64, 2849-2859.

Rous, P., and Beard, J.W. (1935). The Progression to Carcinoma of Virus-Induced Rabbit Papillomas (Shope). J Exp Med 62, 523-548.

Ryndock, E.J., Conway, M.J., Alam, S., Gul, S., Murad, S., Christensen, N.D., and Meyers, C. (2014). Roles for human papillomavirus type 16 l1 cysteine residues 161, 229, and 379 in genome encapsidation and capsid stability. PLoS One 9, e99488. http://dx.doi.org/10.1371/journal.pone.0099488

Sakakibara, N., Chen, D., Jang, M.K., Kang, D.W., Luecke, H.F., Wu, S.Y., Chiang, C.M., and McBride, A.A. (2013a). Brd4 is displaced from HPV replication factories as they expand and amplify viral DNA. PLoS

Pathog 9, e1003777. http://dx.doi.org/10.1371/journal.ppat.1003777

Sakakibara, N., Chen, D., and McBride, A.A. (2013b). Papillomaviruses use recombination-dependent replication to vegetatively amplify their genomes in differentiated cells. PLoS Pathog 9, e1003321. http://dx.doi.org/10.1371/journal.ppat.1003321

Sanders, C.M., and Stenlund, A. (2000). Transcription factor-dependent loading of the E1 initiator reveals modular assembly of the papillomavirus origin melting complex. J Biol Chem 275, 3522-3534.

Sarrazin, S., Lamanna, W.C., and Esko, J.D. (2011). Heparan sulfate proteoglycans. Cold Spring Harb Perspect Biol 3. http://dx.doi.org/10.1101/cshperspect.a004952

Scarpini, C.G., Groves, I.J., Pett, M.R., Ward, D., and Coleman, N. (2014). Virus transcript levels and cell growth rates after naturally occurring HPV16 integration events in basal cervical keratinocytes. J Pathol 233, 281-293. http://dx.doi.org/10.1002/path.4358

Scheffer, K.D., Gawlitza, A., Spoden, G.a., Zhang, X.a., Lambert, C., Berditchevski, F., and Florin, L. (2013). Tetraspanin CD151 mediates papillomavirus type 16 endocytosis. J. Virol. 87, 3435-3446. http://dx.doi.org/10.1128/JVI.02906-12

Scheffner, M., Werness, B.A., Huibregtse, J.M., Levine, A.J., and Howley, P.M. (1990). The E6 oncoprotein encoded by human papillomavirus types 16 and 18 promotes the degradation of p53. Cell 63, 1129-1136.

Schelhaas, M., Shah, B., Holzer, M., Blattmann, P., Kühling, L., Day, P.M., Schiller, J.T., and Helenius, A. (2012). Entry of human papillomavirus type 16 by actin-dependent, clathrin- and lipid raft-independent endocytosis. PLoS pathogens 8, e1002657. http://dx.doi.org/10.1371/journal.ppat.1002657

Schmitt, A., Rochat, A., Zeltner, R., Borenstein, L., Barrandon, Y., Wettstein, F.O., and Iftner, T. (1996). The primary target cells of the high-risk cottontail rabbit papillomavirus colocalize with hair follicle stem cells. J Virol 70, 1912-1922.

Schowalter, R.M., Pastrana, D.V., and Buck, C.B. (2011). Glycosaminoglycans and sialylated glycans sequentially facilitate Merkel cell polyomavirus infectious entry. PLoS pathogens 7, e1002161. http://dx.doi.org/10.1371/journal.ppat.1002161

Schweiger, M.R., You, J., and Howley, P.M. (2006). Bromodomain protein 4 mediates the papillomavirus E2 transcriptional activation function. J Virol 80, 4276-4285. http://dx.doi.org/10.1128/JVI.80.9.4276-4285.2006

Sedman, J., and Stenlund, A. (1998). The papillomavirus E1 protein forms a DNA-dependent hexameric complex with ATPase and DNA helicase activities. J Virol 72, 6893-6897.

Selinka, H.-C., Giroglou, T., and Sapp, M. (2002). Analysis of the infectious entry pathway of human papillomavirus type 33 pseudovirions. Virology 299, 279-287.

Shafti-Keramat, S., Handisurya, A., Kriehuber, E., Meneguzzi, G., Slupetzky, K., and Kirnbauer, R. (2003). Different heparan sulfate proteoglycans serve as cellular receptors for human papillomaviruses. J Virol 77, 13125-13135.

Shai, A., Brake, T., Somoza, C., and Lambert, P.F. (2007). The human papillomavirus E6 oncogene dysregulates the cell cycle and contributes to cervical carcinogenesis through two independent activities. Cancer Res 67, 1626-1635. http://dx.doi.org/10.1158/0008-5472.CAN-06-3344

Shope, R.E. (1937). Immunization of Rabbits to Infectious Papillomatosis. J Exp Med 65, 219-231.

Shope, R.E., and Hurst, E.W. (1933). Infectious Papillomatosis of Rabbits: With a Note on the Histopathology. J Exp Med 58, 607-624.

Simanaviciene, V., Popendikyte, V., Gudleviciene, Z., and Zvirbliene, A. (2015). Different DNA methylation pattern of HPV16, HPV18 and HPV51 genomes in asymptomatic HPV infection as compared to cervical neoplasia. Virology 484, 227-233. http://dx.doi.org/10.1016/j.virol.2015.06.008

Simonson, S.J., Difilippantonio, M.J., and Lambert, P.F. (2005). Two distinct activities contribute to human papillomavirus 16 E6's oncogenic potential. Cancer Res 65, 8266-8273. http://dx.doi.org/10.1158/0008-5472.CAN-05-1651

Skiadopoulos, M.H., and McBride, A.A. (1998). Bovine papillomavirus type 1 genomes and the E2 transactivator protein are closely associated with mitotic chromatin. J Virol 72, 2079-2088.

Smith, A.E., and Helenius, A. (2004). How viruses enter animal cells. Science (New York, NY) 304, 237-242. http://dx.doi.org/10.1126/science.1094823

Smith, J.A., White, E.A., Sowa, M.E., Powell, M.L., Ottinger, M., Harper, J.W., and Howley, P.M. (2010). Genome-wide siRNA screen identifies SMCX, EP400, and Brd4 as E2-dependent regulators of human papillomavirus oncogene expression. Proc Natl Acad Sci USA 107, 3752-3757. http://dx.doi.org/10.1073/pnas.0914818107

Song, H., Moseley, P.L., Lowe, S.L., and Ozbun, M.A. (2010). Inducible heat shock protein 70 enhances HPV31 viral genome replication and virion production during the differentiation-dependent life cycle in human keratinocytes. Virus Res 147, 113-122. http://dx.doi.org/10.1016/j.virusres.2009.10.019

Song, S., Liem, A., Miller, J.A., and Lambert, P.F. (2000). Human papillomavirus types 16 E6 and E7 contribute differently to carcinogenesis. Virology 267, 141-150. http://dx.doi.org/10.1006/viro.1999.0106

Song, S., Pitot, H.C., and Lambert, P.F. (1999). The human papillomavirus type 16 E6 gene alone is sufficient to induce carcinomas in transgenic animals. J Virol 73, 5887-5893.

Songock, W.K., Scott, M.L., and Bodily, J.M. (2017). Regulation of the human papillomavirus type 16 late promoter by transcriptional elongation. Virology 507, 179-191. http://dx.doi.org/10.1016/j.virol.2017.04.021

Spoden, G., Freitag, K., Husmann, M., Boller, K., Sapp, M., Lambert, C., and Florin, L. (2008). Clathrin- and caveolin-independent entry of human papillomavirus type 16--involvement of tetraspanin-enriched microdomains (TEMs). PloS one 3, e3313. http://dx.doi.org/10.1371/journal.pone.0003313

Spoden, G., Kühling, L., Cordes, N., Frenzel, B., Sapp, M., Boller, K., Florin, L., and Schelhaas, M. (2013). Human papillomavirus types 16, 18, and 31 share similar endocytic requirements for entry. J. Virol. 87, 7765-7773. http://dx.doi.org/10.1128/JVI.00370-13

Spriggs, C.C., and Laimins, L.A. (2017). FANCD2 Binds Human Papillomavirus Genomes and Associates with a Distinct Set of DNA Repair Proteins to Regulate Viral Replication. MBio 8. http://dx.doi.org/10.1128/mBio.02340-16

Stanley, M.A., Browne, H.M., Appleby, M., and Minson, A.C. (1989). Properties of a non-tumorigenic human cervical keratinocyte cell line. Int J Cancer 43, 672-676.

Stanley, M.A., Moore, R.A., Nicholls, P.K., Santos, E.B., Thomsen, L., Parry, N., Walcott, S., and Gough, G. (2001). Intra-epithelial vaccination with COPV L1 DNA by particle-mediated DNA delivery protects against mucosal challenge with infectious COPV in beagle dogs. Vaccine 19, 2783-2792.

Steger, G., and Corbach, S. (1997). Dose-dependent regulation of the early promoter of human papillomavirus type 18 by the viral E2 protein. J Virol 71, 50-58.

Stenlund, A. (2003). E1 initiator DNA binding specificity is unmasked by selective inhibition of non-specific DNA binding. EMBO J 22, 954-963. http://dx.doi.org/10.1093/emboj/cdg091

Stepp, W.H., Meyers, J.M., and McBride, A.A. (2013). Sp100 provides intrinsic immunity against human papillomavirus infection. MBio 4, e00845-00813. http://dx.doi.org/10.1128/mBio.00845-13

Stepp, W.H., Stamos, J.D., Khurana, S., Warburton, A., and McBride, A.A. (2017). Sp100 colocalizes with HPV replication foci and restricts the productive stage of the infectious cycle. PLoS Pathog 13, e1006660. http://dx.doi.org/10.1371/journal.ppat.1006660

Sterbenc, A., Hosnjak, L., Chouhy, D., Bolatti, E.M., Ostrbenk, A., Seme, K., Kocjan, B.J., Luzar, B., Giri, A.A., and Poljak, M. (2017). Molecular characterization, tissue tropism, and genetic variability of the novel Mupapillomavirus type HPV204 and phylogenetically related types HPV1 and HPV63. PLoS One 12, e0175892. http://dx.doi.org/10.1371/journal.pone.0175892

Sterling, J., Stanley, M., Gatward, G., and Minson, T. (1990). Production of human papillomavirus type 16 virions in a keratinocyte cell line. J Virol 64, 6305-6307.

Stoler, M.H., Whitbeck, A., Wolinsky, S.M., Broker, T.R., Chow, L.T., Howett, M.K., and Kreider, J.W. (1990). Infectious cycle of human papillomavirus type 11 in human foreskin xenografts in nude mice. J Virol 64, 3310-3318.

Straight, S.W., Herman, B., and McCance, D.J. (1995). The E5 oncoprotein of human papillomavirus type 16 inhibits the acidification of endosomes in human keratinocytes. J Virol 69, 3185-3192.

Strati, K., and Lambert, P.F. (2007). Role of Rb-dependent and Rb-independent functions of papillomavirus E7 oncogene in head and neck cancer. Cancer Res 67, 11585-11593. http://dx.doi.org/10.1158/0008-5472.CAN-07-3007

Strati, K., Pitot, H.C., and Lambert, P.F. (2006). Identification of biomarkers that distinguish human papillomavirus (HPV)-positive versus HPV-negative head and neck cancers in a mouse model. Proc Natl Acad Sci USA 103, 14152-14157. http://dx.doi.org/10.1073/pnas.0606698103

Straub, E., Dreer, M., Fertey, J., Iftner, T., and Stubenrauch, F. (2014). The viral E8^E2C repressor limits productive replication of human papillomavirus 16. J Virol 88, 937-947. http://dx.doi.org/10.1128/JVI.02296-13

Straub, E., Fertey, J., Dreer, M., Iftner, T., and Stubenrauch, F. (2015). Characterization of the Human Papillomavirus 16 E8 Promoter. J Virol 89, 7304-7313. http://dx.doi.org/10.1128/JVI.00616-15

Stubenrauch, F., Hummel, M., Iftner, T., and Laimins, L.A. (2000). The E8E2C protein, a negative regulator of viral transcription and replication, is required for extrachromosomal maintenance of human papillomavirus type 31 in keratinocytes. J Virol 74, 1178-1186.

Stunkel, W., and Bernard, H.U. (1999). The chromatin structure of the long control region of human papillomavirus type 16 represses viral oncoprotein expression. J Virol 73, 1918-1930.

Sundberg, J.P., Stearns, T.M., Joh, J., Proctor, M., Ingle, A., Silva, K.A., Dadras, S.S., Jenson, A.B., and Ghim, S.J. (2014). Immune status, strain background, and anatomic site of inoculation affect mouse papillomavirus (MmuPV1) induction of exophytic papillomas or endophytic trichoblastomas. PLoS One 9, e113582. http://dx.doi.org/10.1371/journal.pone.0113582

Surviladze, Z., Dziduszko, A., and Ozbun, M.A. (2012). Essential roles for soluble virion-associated heparan sulfonated proteoglycans and growth factors in human papillomavirus infections. PLoS Pathog 8, e1002519. http://dx.doi.org/10.1371/journal.ppat.1002519

Surviladze, Z., Sterkand, R.T., and Ozbun, M.A. (2015). Interaction of human papillomavirus type 16 particles with heparan sulfate and syndecan-1 molecules in the keratinocyte extracellular matrix plays an active role in infection. J. Gen. Virol. 96, 2232-2241. http://dx.doi.org/10.1099/vir.0.000147

Tan, S.H., Leong, L.E., Walker, P.A., and Bernard, H.U. (1994). The human papillomavirus type 16 E2 transcription factor binds with low cooperativity to two flanking sites and represses the E6 promoter through displacement of Sp1 and TFIID. J Virol 68, 6411-6420.

Thain, A., Webster, K., Emery, D., Clarke, A.R., and Gaston, K. (1997). DNA binding and bending by the human papillomavirus type 16 E2 protein. Recognition of an extended binding site. J Biol Chem 272, 8236-8242.

Thierry, F. (2009). Transcriptional regulation of the papillomavirus oncogenes by cellular and viral transcription factors in cervical carcinoma. Virology 384, 375-379. http://dx.doi.org/10.1016/j.virol.2008.11.014

Thomas, J.T., Hubert, W.G., Ruesch, M.N., and Laimins, L.A. (1999). Human papillomavirus type 31 oncoproteins E6 and E7 are required for the maintenance of episomes during the viral life cycle in normal human keratinocytes. Proc Natl Acad Sci USA 96, 8449-8454.

Thomas, J.T., Oh, S.T., Terhune, S.S., and Laimins, L.A. (2001). Cellular changes induced by low-risk human papillomavirus type 11 in keratinocytes that stably maintain viral episomes. J Virol 75, 7564-7571. http://dx.doi.org/10.1128/JVI.75.16.7564-7571.2001

Thomas, M.K., Pitot, H.C., Liem, A., and Lambert, P.F. (2011). Dominant role of HPV16 E7 in anal carcinogenesis. Virology 421, 114-118. http://dx.doi.org/10.1016/j.virol.2011.09.018

Thorland, E.C., Myers, S.L., Gostout, B.S., and Smith, D.I. (2003). Common fragile sites are preferential targets for HPV16 integrations in cervical tumors. Oncogene 22, 1225-1237. http://dx.doi.org/10.1038/sj.onc.1206170

Uberoi, A., and Lambert, P.F. (2017). Rodent papillomaviruses. Viruses 9, E362. http://dx.doi.org/10.3390/v9120362.

Ustav, E., Ustav, M., Szymanski, P., and Stenlund, A. (1993). The bovine papillomavirus origin of replication requires a binding site for the E2 transcriptional activator. Proc Natl Acad Sci USA 90, 898-902.

Ustav, M., Ustav, E., Szymanski, P., and Stenlund, A. (1991). Identification of the origin of replication of bovine papillomavirus and characterization of the viral origin recognition factor E1. EMBO J 10, 4321-4329.

Van Doorslaer, K., Porter, S., McKinney, C., Stepp, W.H., and McBride, A.A. (2016). Novel recombinant papillomavirus genomes expressing selectable genes. Sci Rep 6, 37782. http://dx.doi.org/10.1038/srep37782

Van Tine, B.A., Kappes, J.C., Banerjee, N.S., Knops, J., Lai, L., Steenbergen, R.D., Meijer, C.L., Snijders, P.J., Chatis, P., Broker, T.R., *et al.* (2004). Clonal selection for transcriptionally active viral oncogenes during progression to cancer. J Virol 78, 11172-11186. http://dx.doi.org/10.1128/JVI.78.20.11172-11186.2004

Viarisio, D., Muller-Decker, K., Accardi, R., Robitaille, A., Durst, M., Beer, K., Jansen, L., Flechtenmacher, C., Bozza, M., Harbottle, R., Voegele, C., Ardin, M., Zavadil, J., Caldeira, S., Gissmann, L., and Tommasino, M. (2018). Beta HPV38 oncoproteins act with a hit-and-run mechanism in ultraviolet radiation-induced skin carcinogenesis in mice. PLoS Pathog 14, e1006783. http://dx.doi.org/10.1371/journal.ppat.1006783.

von Knebel Doeberitz, M. (2016). HPV 16 and its variants: minor genome variations make a big difference. J Natl Cancer Inst 108. http://dx.doi.org/10.1093/jnci/djw123

Wang, H.K., Duffy, A.A., Broker, T.R., and Chow, L.T. (2009). Robust production and passaging of infectious HPV in squamous epithelium of primary human keratinocytes. Genes Dev 23, 181-194. http://dx.doi.org/10.1101/gad.1735109

Wasson, C.W., Morgan, E.L., Müller, M., Ross, R.L., Hartley, M., Roberts, S., and Macdonald, A. (2017). Human papillomavirus type 18 E5 oncogene supports cell cycle progression and impairs epithelial differentiation by modulating growth factor receptor signalling during the virus life cycle. Oncotarget 8:103581-103600. http://dx.doi.org/10.18632/oncotarget.21658

Westra, W.H. (2012). The morphologic profile of HPV-related head and neck squamous carcinoma: implications for diagnosis, prognosis, and clinical management. Head Neck Pathol 6 Suppl 1, S48-54. http://dx.doi.org/10.1007/s12105-012-0371-6

Westrich, J.A., Warren, C.J., and Pyeon, D. (2017). Evasion of host immune defenses by human papillomavirus. Virus Res 231, 21-33. http://dx.doi.org/10.1016/j.virusres.2016.11.023

White, W.I., Wilson, S.D., Bonnez, W., Rose, R.C., Koenig, S., and Suzich, J.A. (1998). In vitro infection and type-restricted antibody-mediated neutralization of authentic human papillomavirus type 16. J Virol 72, 959-964.

Wiens, M.E., and Smith, J.G. (2015). Alpha-defensin HD5 inhibits furin cleavage of human papillomavirus 16 L2 to block infection. J Virol 89, 2866-2874. http://dx.doi.org/10.1128/JVI.02901-14

Wilson, R., and Laimins, L.A. (2005). Differentiation of HPV-containing cells using organotypic "raft" culture or methylcellulose. Methods Mol Med 119, 157-169. http://dx.doi.org/10.1385/1-59259-982-6:157

Wilson, R., Ryan, G.B., Knight, G.L., Laimins, L.A., and Roberts, S. (2007). The full-length E1E4 protein of human papillomavirus type 18 modulates differentiation-dependent viral DNA amplification and late gene expression. Virology 362, 453-460. http://dx.doi.org/10.1016/j.virol.2007.01.005

Wooldridge, T.R., and Laimins, L.A. (2008). Regulation of human papillomavirus type 31 gene expression during the differentiation-dependent life cycle through histone modifications and transcription factor binding. Virology 374, 371-380. http://dx.doi.org/10.1016/j.virol.2007.12.011

Wu, S.Y., Nin, D.S., Lee, A.Y., Simanski, S., Kodadek, T., and Chiang, C.M. (2016). BRD4 Phosphorylation Regulates HPV E2-Mediated Viral Transcription, Origin Replication, and Cellular MMP-9 Expression. Cell Rep 16, 1733-1748. http://dx.doi.org/10.1016/j.celrep.2016.07.001

Xue, Y., Bellanger, S., Zhang, W., Lim, D., Low, J., Lunny, D., and Thierry, F. (2010). HPV16 E2 is an immediate early marker of viral infection, preceding E7 expression in precursor structures of cervical carcinoma. Cancer Res 70, 5316-5325. http://dx.doi.org/10.1158/0008-5472.CAN-09-3789

Yan, J., Li, Q., Lievens, S., Tavernier, J., and You, J. (2010). Abrogation of the Brd4-positive transcription elongation factor B complex by papillomavirus E2 protein contributes to viral oncogene repression. J Virol 84, 76-87. http://dx.doi.org/10.1128/JVI.01647-09

Yang, E.J., Quick, M.C., Hanamornroongruang, S., Lai, K., Doyle, L.A., McKeon, F.D., Xian, W., Crum, C.P., and Herfs, M. (2015). Microanatomy of the cervical and anorectal squamocolumnar junctions: a proposed model for anatomical differences in HPV-related cancer risk. Mod Pathol 28, 994-1000. http://dx.doi.org/10.1038/modpathol.2015.54

Yang, R., Yutzy, W.H., Viscidi, R.P., and Roden, R.B.S. (2003). Interaction of L2 with -Actin Directs Intracellular Transport of Papillomavirus and Infection. J. Biol. Chem. 278, 12546-12553. http://dx.doi.org/10.1074/jbc.M208691200

You, J., Croyle, J.L., Nishimura, A., Ozato, K., and Howley, P.M. (2004). Interaction of the bovine papillomavirus E2 protein with Brd4 tethers the viral DNA to host mitotic chromosomes. Cell 117, 349-360.

Zehbe, I., Voglino, G., Delius, H., Wilander, E., and Tommasino, M. (1998). Risk of cervical cancer and geographical variations of human papillomavirus 16 E6 polymorphisms. Lancet 352, 1441-1442. http://dx.doi.org/10.1016/S0140-6736(05)61263-9

Zhao, K.N., Sun, X.Y., Frazer, I.H., and Zhou, J. (1998). DNA packaging by L1 and L2 capsid proteins of bovine papillomavirus type 1. Virology 243, 482-491. http://dx.doi.org/10.1006/viro.1998.9091

3

Molecular Mechanisms of E6 and E7 Oncoproteins From Human Papillomaviruses in Cellular Transformation

Massimo Tommasino*

Infections and Cancer Biology Group, International Agency for Research on Cancer, World Health Organization, 69372 Lyon Cedex 08, France

**Corresponding author: tommasinom@iarc.fr*

Abstract

The small double-stranded DNA human papillomaviruses (HPVs) form a family of approximately 200 viruses. They can infect the epithelia of the genital and upper respiratory tracts and the skin and are classified into several genera based on their DNA sequence. A small number belonging to genus alpha, referred to as mucosal high-risk (HR) HPV types, are clearly associated with human carcinogenesis. In particular, HR HPV16 and HPV18 are the most carcinogenic types, responsible for a large proportion of HPV-positive anogenital cancers. A vast number of studies have demonstrated that the products of two early genes, E6 and E7, play a key role in cellular transformation and cancer development. Both oncoproteins are very small molecules and do not have any enzymatic activities, but exert their biological properties by interacting with a large number of cellular proteins. These interactions result in modification of the activities of the cellular proteins, which, in some cases, are targeted to degradation via the proteasome pathway. The most studied examples are the ability of HR HPV E6 and E7 to induce degradation of p53 and pRb, respectively. Inactivation of these key tumour suppressors strongly alters cellular gene expression, leading to chromosomal instability and cellular transformation.

Despite the great knowledge of the biology of mucosal HR HPV types, additional research is needed to characterize the biology of the majority of HPV types that have been poorly investigated so far, with a final aim of evaluating their potential roles in other human diseases.

The human papillomavirus family and its association with human diseases

Approximately 200 human papillomavirus (HPV) types have been identified so far. They are double-stranded circular DNA viruses with a highly conserved organization of the genome (Van Doorslaer *et al.*, 2013). The HPV phylogenetic tree has been generated considering the nucleotide sequence of the L1 major capsid protein (Bernard *et al.*, 2010; de Villiers, 2013). It is subdivided into genera, which in turn include different HPV species. The genera alpha, beta, and gamma comprise the majority of the HPV types. The known associations between specific subgroups of HPV types and human diseases are shown in Table 3.1.

Alpha HPV types

The most studied HPV types so far belong to genus alpha, since they were among the first ones to be associated with human diseases. Specifically, the majority of the alpha HPV types infect the epithelia of

Table 3.1. HPV types from different genera and their association with human diseases.

Genus	HPV type[a]	Disease (% of attributed cases to infection)[b]
Alpha	16	Cervical squamous cell carcinoma (~50) Cervical adenocarcinoma (~35) Oropharyngeal cancer (~25)
	18	Cervical squamous cell carcinoma (~20) Cervical adenocarcinoma (~35)
	31, 33, 35, 39, 45, 51, 52, 56, 58, 59	Cervical squamous cell carcinoma (~30)
	6, 11	Benign genital lesions Respiratory papillomatosis
	13, 32	Oral focal epithelial hyperplasia
	2,3, 27, 57	Skin warts
Mu	1	Skin warts
Beta	5, 8	Non melanoma skin cancer (NMSC) in EV patients
	HPV9, 12, 14, 15, 17, 19-25, 36-38, 47, 49, 75, 76, 80, 92, 93, 96, 98-100, 104, 105, 107, 110, 111, 113, 115, 118, 120, 122, 124, 143, 145, 150- 152, 159	Suspected to be associated with NMSC in EV patients as well as immuno-compromised and immuno-competent individuals
Gamma	Approximately 80 different HPV types	Unknown

[a] Many of the identified HPV types that belong to different genera (i.e. alpha, beta, gamma, and mu) of the HPV phylogenetic tree are shown. [b] In addition, the main diseases that have been associated with different HPV types are indicated.

the mucosa of the genital and upper digestive tracts; they are classified as low-risk (LR) and high-risk (HR) HPV types, according to their ability to induce the development of benign or malignant lesions, respectively.

These viruses are sexually transmitted, and approximately 70–80% of sexually active women develop an HPV infection during their lifetime. The HR HPV types have been clearly associated with cervical cancers and a subset of other genital tract cancers, i.e. anal, vulvar, vaginal, and penile cancers (Plummer *et al.*, 2016). Twelve different HR HPV types, namely 16, 18, 31, 33, 35, 39, 45, 51, 52, 56, 58, and 59, have been classified as carcinogenic to humans (Bouvard *et al.*, 2009). Worldwide epidemiological studies have clearly demonstrated that HPV16 and HPV18 are the most oncogenic viruses in the alpha mucosal HR HPV group; together, they are responsible for approximately 70% of cervical squamous cell carcinoma (SCC) and adenocarcinoma (ADC) (Munoz *et al.*, 2003; Li *et al.*, 2011). However, the ratio between these two HPV types in SCC and ADC differs substantially. In ADC, HPV16 and HPV18 show similar oncogenicity; they are equally represented in approximately 35% of cases. In contrast, in SCC, HPV16 and HPV18 are responsible for about 50% and 20% of the invasive cancers, respectively (Munoz *et al.*, 2003; Li *et al.*, 2011). For the other cancers of the genital tract, approximately 90%, 75%, 25%, and 35% of invasive anal, vaginal, vulvar, and penile carcinomas, respectively, are attributable to HR HPV infection (de Sanjose *et al.*, 2013; Alemany *et al.*, 2014; Alemany *et al.*, 2015; Alemany *et al.*, 2016;). Most HR HPV infections are cleared by the immune system in 12–14 months without leading to any clinical manifestations. In a small percentage of cases, the infection can persist and, after a period of latency, evolve into premalignant lesions, which may still regress, or progress to an invasive cervical carcinoma.

LR HPV types (e.g. types 6 and 11) are responsible for genital warts, or condylomata acuminata. The majority of genital lesions induced by HPV6 or HPV11 are cleared by the immune system within 2 years of infection. However, recurrence of the disease can be observed in a minority of cases. LR HPV6 and HPV11 are also associated with a rare disease in children and adults, laryngeal papillomatosis, also referred to as recurrent respiratory papillomatosis (Carifi *et al.*, 2015). In children, LR HPV infection is acquired at birth during passage through an HPV-infected birth canal. Although these tumours can recur frequently, they do not usually progress to malignancy. However, recurrent respiratory papillomatosis has been rarely associated with pulmonary malignancies (Can *et al.*, 2015; Karatayli-Ozgursoy *et al.*, 2016). Interestingly, an HPV11 genome with increased size (10.4 kb) has been isolated from a malignant lung lesion of a patient with recurrent respiratory papillomatosis. Sequence analysis revealed that the HPV11 genome isolated from the pulmonary cancer cells contained a duplication of approximately 2,500 bp that comprises the entire long control region and the E6 and E7 oncogenes (Yuan *et al.*, 2013). Thus, the greater oncogenic potential of this divergent HPV11 genome could be explained by elevated expression levels of E6 and E7 genes. These findings highlight the potential oncogenicity of the LR HPV types. However, whether they are involved in other human malignancies remains to be confirmed. Interestingly, it has recently been reported that HPV6 was detected in 3.7% of penile carcinomas worldwide (Alemany *et al.*, 2016).

During the past two decades, it has become clear that the HR HPV types are also involved in head and neck cancers (HNC). The major HNC site associated with these viruses is the oropharynx. Approximately 30% of oropharyngeal SCCs worldwide are linked to HR HPV infections (Ndiaye *et al.*, 2014; Castellsague *et al.*, 2016). The other HNC sites are marginally affected by HPV infection, although the causal role of the HR HPV types in a small percentage of oral cavity and laryngeal cancers has been consistently shown in independent studies (Marur *et al.*, 2010; Halec *et al.*, 2013; Castellsague *et al.*, 2016). In contrast to cervical cancer, the majority of oropharyngeal carcinomas (85–95%) are associated with HPV16, whereas the other HR HPV types appear to have only a marginal role (Castellsague *et al.*, 2016). This different percentage of HPV16-positive cancers in the oropharynx and cervix may be explained by the greater efficiency of HPV16 compared with the other HR HPV types in evading the immune system (Schiffman *et al.*, 2005; Hasan *et al.*, 2007). Since the immune system is more active in the oropharynx than in the genital tract, it is likely that infections of other HR HPV types are better cleared compared with HPV16 infections.

Additional alpha HPV types have a cutaneous or mucosal tropism and have been associated with the development of benign lesions, such as skin warts (e.g. HPV2, 27, and 57) and oral mucosal lesions called focal epithelial hyperplasia (HPV13 and 32).

Beta HPV types

Approximately 50 different beta HPV types have been characterized so far, but it is likely that many more exist. The first beta HPV types, HPV5 and 8, were isolated in the skin of individuals suffering from a rare autosomal recessive genetic disorder called epidermodysplasia verruciformis (EV) (reviewed in Bouvard *et al.*, 2006). These patients have an elevated susceptibility to skin beta HPV infection compared with the general population. As a consequence, they develop disseminated pityriasis versicolor-like lesions and flat warts, which in approximately 30–60% of cases progress to multifocal SCC on sun-exposed areas of the body. The majority of EV patients harbour mutations in *EVER1* (*TMC6*) or *EVER2* (*TMC8*), two adjacent genes that synthesize proteins involved in the control of intracellular levels of zinc in keratinocytes and immune cells (Lazarczyk *et al.*, 2009; Lazarczyk *et al.*, 2012). It has also been shown that EV patients can harbour mutations in genes other than *EVER*s, leading to the same high susceptibility to beta HPV infection (Crequer *et al.*, 2012; S.L. Li *et al.*, 2016). A recent study provided evidence that other viruses interfere with the *EVER1*/*EVER2* network. The human gamma herpesvirus Epstein–Barr virus (EBV), via the oncoprotein LMP1, strongly downregulates *EVER1* and *EVER2* expression. Ectopic expression of *EVER1* impairs the ability of EBV to infect B cells (Frecha *et al.*, 2015).

Since the discovery of HPV5 and 8, many more beta HPV types have been isolated, as a result of the development of highly sensitive polymerase chain reaction (PCR)-based methods (Van Doorslaer *et al.*, 2013). They are subdivided into five different species: beta-1, -2, -3, -4, and -5, and are abundantly present in the skin of normal individuals without inducing clear clinical manifestations (Astori *et al.*, 1998; de Koning *et al.*, 2009). However, biological and epidemiological studies support their role in the development of non-melanoma skin cancer (NMSC), together with ultraviolet (UV) light (reviewed in Tommasino, 2014). Individuals with a history of skin SCC show a higher positivity for viral DNA in the skin and/or antibodies against the L1 major capsid protein (Forslund *et al.*, 2007; Waterboer *et al.*, 2008; Bouwes Bavinck *et al.*,

2010; Karagas *et al.*, 2010; Iannacone *et al.*, 2012; Farzan *et al.*, 2013; Iannacone *et al.*, 2014). A recent meta-analysis showed that beta-1 HPV5, 8, and 20 and beta-2 HPV17 and 38 are weakly but significantly associated with the risk of NMSC in the general population, with an odds ratio never higher than 2.0 (Chahoud *et al.*, 2015). The role of some of these beta HPV types in skin SCC is also supported by studies in patients with invasive melanoma under treatment with a BRAF inhibitor (BRAFi) (Cohen *et al.*, 2015; Falchook *et al.*, 2013). These patients develop, as side effects, skin SCC and keratoacanthomas (Flaherty *et al.*, 2010; Kefford *et al.*, 2010; Chapman *et al.*, 2011). Analysis of the skin SCC of melanoma patients under BRAFi treatment revealed the presence of the beta-2 HPV types 17, 38, and 111.

Although together these data support a link between beta HPV infection and NMSC, the scenario appears to differ substantially from that of the well-established oncogenic viruses, such the alpha HR HPV types. For instance, cervical cancer cells contain at least one copy of the viral genome, which is often integrated into the host DNA. Beta HPV DNA does not usually integrate into the host genome and appears to be lost in cancer cells. Quantitative PCR experiments revealed that in skin SCC, the ratio between viral and cellular DNA is much less than 1 (Weissenborn *et al.*, 2005), indicating that most of the cancer cells do not contain the viral DNA. In agreement with this possibility, a study showed no correlation between beta HPV DNA-positivity in primary cutaneous SCC and the corresponding lymph node metastasis (Toll *et al.*, 2014). In addition, the beta HPV DNA load is higher in actinic keratosis, which is considered the premalignant lesion of SCC, than in SCC (Vasiljevic *et al.*, 2008; Weissenborn *et al.*, 2005). Low gene expression for E6 and E7 from E6/E7 transcripts of HPV8, 9, and 15 was detected in actinic keratosis and SCC, but not in healthy tissue, suggesting an active role of the viruses in these pathological conditions (Dang *et al.*, 2006). Thus, all these findings support the concept that beta HPV types play a role at an early stage of carcinogenesis, facilitating the DNA damage induced by UV irradiation, which is one of the key risk factors for NMSC. At a late stage, after the accumulation of many cancer-driving DNA mutations, the expression of the viral genes may no longer be required. This model is consistent with the epidemiological and biological findings. The biological findings demonstrate, as described below, the ability of the viral proteins to deregulate machineries involved in cellular proliferation, DNA repair, and apoptosis in several experimental models.

In addition, beta HPV types have been detected in the oral mucosal epithelium, eyebrow hairs, and penile and external genital samples. However there is no clear evidence for their involvement in any lesions at these anatomical sites (Bottalico *et al.*, 2011; Pierce Campbell *et al.*, 2013; Torres *et al.*, 2015; Dona *et al.*, 2016). A recent prospective study indicated that detection of some beta HPV types in the oral cavity is associated with the risk of HNC (Agalliu *et al.*, 2016). Another study showed that the beta-3 species, which includes HPV types 49, 75, 76, and 115, is more prevalent in the nasal cavity than in the skin (Forslund *et al.*, 2013). In addition, biological studies showed that beta-3 HPV49 E6 and E7 have a functional similarity to the oncoproteins of mucosal HR HPV types (Cornet *et al.*, 2012; Viarisio *et al.*, 2016). Thus,

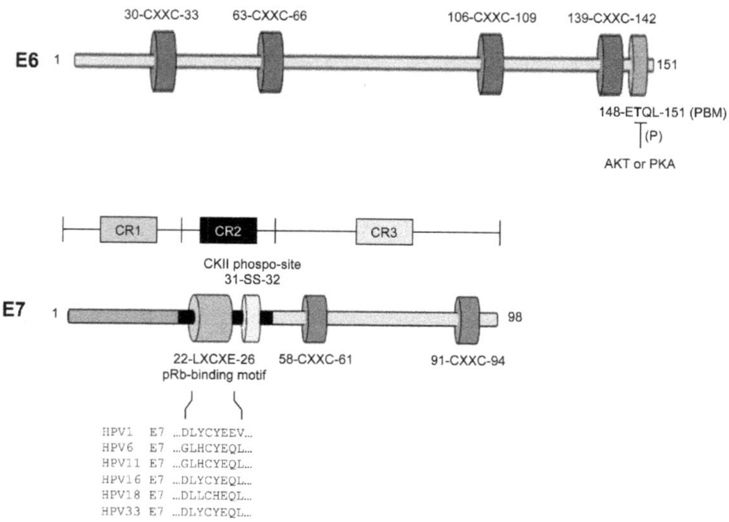

Figure 3.1. The structures of HPV16 E6 and E7. The schematic structures of HPV16 E6 and E7 are shown, with the position of several amino acid motifs that are important for their structure and functions. Both oncoproteins contain CXXC motifs (four in E6 and two in E7) that are able to form zinc complexes. In addition, E6 has a consensus PDZ-binding motif (PBM) (ETQL) at the C-terminus, that mediates binding to PDZ proteins. The T within the PBM can be phosphorylated (P) by the kinases AKT or PKA, negatively regulating the interactions with the PDZ proteins. E7 includes three regions that are homologous to adenovirus E1A (conserved regions 1–3, CR1–3). CR2 comprises the pRb-binding motif (LXCXE) and two serines (31 and 32) that are phosphorylated by casein kinase II (CKII). A few sequences of the pRb-binding motif of different E7 with different pRB binding affinity are also shown.

it is likely that beta-3 HPV types may also have a mucosal tropism.

Gamma HPV types

Genus gamma includes more HPV types than the other genera (Van Doorslaer *et al.*, 2013). Approximately 80 gamma HPV types have been isolated from a variety of human specimens, including (i) healthy tissue from the oral cavity, skin, and female genital tract; (ii) benign skin lesions from normal and immune-suppressed individuals; (iii) SCCs from immune-suppressed individuals; and (iv) benign or premalignant lesions from the female genital tract. Interestingly, three gamma HPV types, 101, 103, and 108, isolated from cervicovaginal cells of healthy women or women with cervical lesions revealed an unusual genome organization (Chen *et al.*, 2007; Nobre *et al.*, 2009). All three genomes lack the E6 open reading frame (ORF).

Currently, very little is known about the epidemiology and biology of the gamma HPV types. The different HPV types appear to have a cutaneous and/or a mucosal tropism. Importantly, their potential role in human diseases is largely unexplored so far.

The biological properties of HPV E6 and E7 oncoproteins

A vast number of studies have demonstrated that the products of two early genes, E6 and E7 from the mucosal HR HPV types, exert key functions in the viral life cycle, hijacking many cellular machineries and guaranteeing progeny production. An important and well-characterized property of the viral proteins is the ability to drive quiescent cells into a proliferative state, enabling efficient replication of the HPV genome. In addition, these viral proteins have developed several mechanisms to circumvent immune surveillance in order to persist as long as possible in the host and produce more progeny (Einstein *et al.*, 2009).

E6 and E7 are very small proteins, with approximately 150 and 100 amino acids, respectively (Figure 3.1). They have in common the presence in the primary structure of CXXC motifs that bind zinc, specifically four in E6 and two in E7. This feature is conserved in E6 and E7 from all HPV types isolated so far. Based on this similarity, it has been proposed that the proteins may have the same ancestral origin (Cole and Danos, 1987). The fact that the two CXXC motifs of E6 and at the C-terminus of E7 from most, if not all, HPV types are separated by the same number of amino acids (n=29) further supports this hypothesis. Findings on E7 protein from the rhesus monkey papillomavirus 1 (MmPV1) are in line with the evolution of E6 and E7 from a common ancestor. MmPV1 E7 has a PDZ binding motif at the C-terminus, which is a feature of E6 from the mucosal HR HPV types (Tomaic *et al.*, 2009).

E6 and E7 proteins do not have any enzymatic activities, and exert their functions by interacting with a large number of cellular proteins, with consequent alterations or loss of the biological properties of the cellular proteins (reviewed in Roman and Munger, 2013; Vande Pol and Klingelhutz, 2013; Tommasino, 2014). The high adaptability of E6 and E7 in forming complexes with different cellular proteins can be explained by their flexible structure, which can be modified according to the interacting partners (Calcada *et al.*, 2013; Trave and Zanier, 2016).

In addition to their ability to sustain the viral life cycle, some E6 and E7 proteins act as viral oncoproteins and are able to promote malignant transformation of infected cells. Since HR HPV16 and HPV18 were the first to be associated with human carcinogenesis, their E6 and E7 proteins have been extensively studied. Early studies on cervical cancer-derived cell lines, e.g. SiHa and CaSki, show that the viral DNA was found to be randomly integrated into the host genome, with only preservation of E6 and E7 expression (Schwarz *et al.*, 1985). It has subsequently been shown that E6 and E7 expression is necessary for the maintenance of the malignant phenotype. Silencing of both viral genes in cervical cancer-derived cell lines results in rapid loss of the proliferative state (Chen *et al.*, 1995; Hamada *et al.*, 1996; Butz *et al.*, 2003). Similar findings were obtained *in vivo*, using transgenic animal models (Jabbar *et al.*, 2009; Jabbar *et al.*, 2012). The oncogenic functions of HR HPV E6 and E7 oncoproteins have been demonstrated in a vast number of *in vitro* and *in vivo* experimental models (Lambert *et al.*, 2006; Tommasino, 2014). The classical *in vitro* transformation assay for E6 and E7 that has been widely used is the immortalization of the primary human keratinocytes (HPKs), which represent the natural host of the virus. HPKs from different tissues are immortalized by E6 and E7 from HPV16 as well as from other HR HPV types (Durst *et al.*, 1987; Pirisi *et al.*, 1987; Bedell *et al.*, 1989; Sedman *et al.*, 1991). In contrast, LR HPV E6 and E7 did not show any immortalization activity in this assay (Barbosa *et al.*, 1991; Halbert *et al.*, 1992). A few studies showed that also some beta HPV E6 and E7 can immortalize HPKs, supporting their possible role in carcinogenesis. For instance, E6 and E7 from the cutaneous beta-2 HPV38 are able to efficiently immortalize HPKs, although in a different manner than E6 and E7 from the HR HPV types (Caldeira *et al.*, 2003; Gabet *et al.*,

2008; Muench *et al*., 2010). HPK immortalization by HPV38 E6/E7 is always preceded by an apparent 3–6-week lag phase, in which cells enlarge and remain growth-arrested. Thereafter, islands of small proliferating cells appear and continue to grow indefinitely (Gabet *et al*., 2008). In contrast to HPV38, E6 and E7 from beta-3 HPV49 immortalize HPKs in a similar manner to mucosal HR HPV oncoproteins (Cornet *et al*., 2012). Indeed, expression of HPV49 E6 and E7 in HPKs resulted in constant growth, and immortalization is not preceeded by a crisis as observed in HPV38 E6 and E7 HPKs. Very little is known about the ability of E6 and E7 from other beta HPV types to immortalize HPKs. E6 and E7 from beta-1 HPV14, 24, and 36 and beta-2 HPV22 and 23 can only, with different efficiencies, extend the life span of HPKs, without leading to immortalization (Cornet *et al*., 2012), whereas E6 and E7 from beta-1 HPV8 have low immortalizing activity in the same cells (Schmitt *et al*., 1994).

Experiments in three-dimensional (3D) organotypic raft culture, which closely mimic the epithelium stratification (Shamir and Ewald, 2014), showed that E6 and E7 from beta HPV types 5, 12, 15, 17, 20, and 38 alter keratinocyte differentiation (Boxman *et al*., 2001; Kazem *et al*., 2012).

Very limited information is available about the transforming properties of E6 and E7 from gamma HPV types. A study on gamma HPV108 that lacks the E6 gene showed that E7 protein stimulates proliferation in 3D organotypic cultured HPKs (Nobre *et al*., 2009).

Immortalization of primary cells is normally associated with crucial pathways controlled by tumour suppressors, such as p53 and the product of the retinoblastoma gene (pRb). Different HPV types have developed distinct mechanisms to alter the regulation of these pathways, of which some will be described in the following sections.

Features and transforming properties of E6 protein

E6 is a small and basic protein of approximately 150 amino acids. As already mentioned, a conserved structural feature among all E6 proteins is the presence of two pairs of zinc-binding motifs, present at the N- and C-terminal regions, referred to as E6N and E6C, respectively (Figure 3.1). The structures of isolated E6N and E6C, as well as of the entire E6 protein, have been determined (Nomine *et al*., 2006; Zanier *et al*., 2012; Zanier *et al*., 2013; Martinez-Zapien *et al*., 2016). These structural studies identified the regions

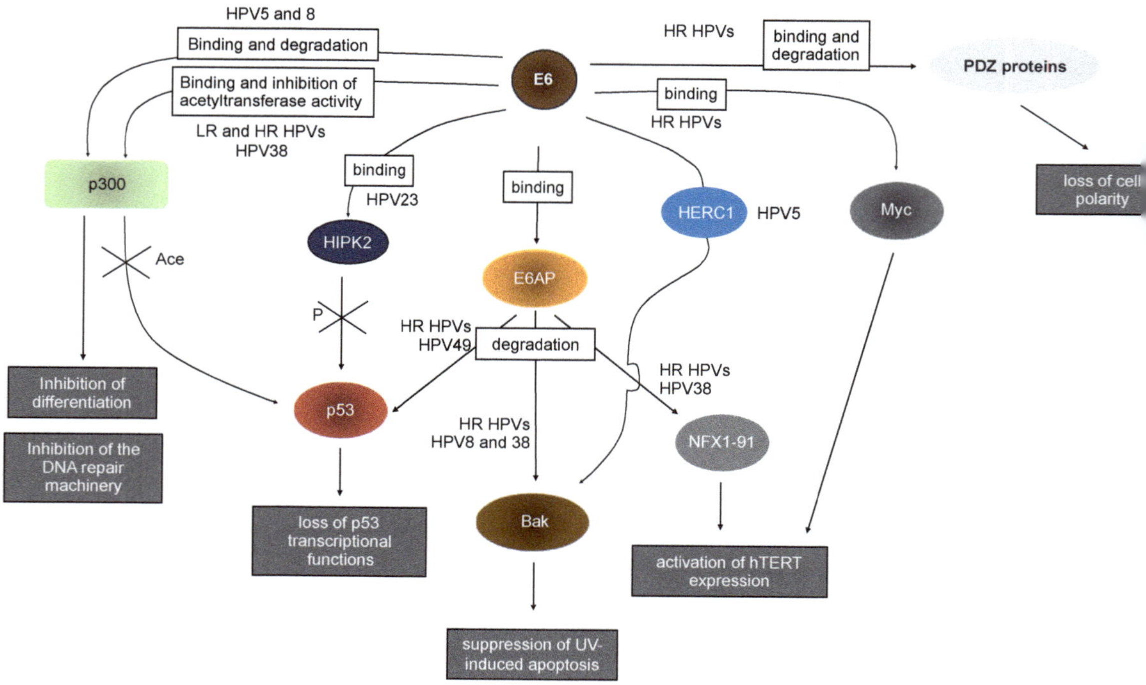

Figure 3.2. Interactions between E6 and cellular targets. Many of the targets of E6 from different HPV types and the consequence of the E6/cellular protein interactions are shown, as inhibition of p53 acetylation (Ace) or phosphorylation (P). The dark grey boxes describe the events induced by the different E6 oncoproteins.

involved in the binding of a motif, LXXLL, that is present in many E6-interacting cellular proteins, such as the ubiquitin ligase E3 enzyme E6AP, the notch co-activator MAML1, the interferon regulatory factor-3, and the focal adhesion protein paxillin.

E6 oncoproteins, p53, and deregulation of apoptosis

The best-characterized HPV16 E6 interaction is one with E6AP that mediates the formation of a trimeric complex with the tumour suppressor p53 (Martinez-Zapien *et al.*, 2016) (Figure 3.2). The formation of this complex requires, first, the interaction of E6 with the LXXLL motif of E6AP, which induces conformational changes, generating a cleft for the binding of p53. The consequence of E6/E6AP/p53 interaction is the rapid degradation of p53 via the proteasome pathway (Scheffner *et al.*, 1990; Huibregtse *et al.*, 1991; Scheffner *et al.*, 1993). A key function of p53 is to orchestrate the response to cellular stresses by regulating the expression of a vast number of cellular genes, leading to different outcomes, e.g. temporary cell-cycle arrest and DNA repair or apoptosis. When cells are exposed to different types of stress, p53 is rapidly stabilized by post-translational modifications, e.g. phosphorylation, methylation, and acetylation, which in turn confer on p53 the specificity to activate different sets of cellular genes (MacLaine and Hupp, 2011; Marouco *et al.*, 2013). The final aim of p53 is to preserve genome integrity and to prevent the production of potentially transformed progeny. In the context of HPV infection, p53 is activated by the stress induced by expression of the viral oncogenes. Thus, E6 needs to inactivate p53 in order to maintain the proliferative state of the infected cells. As a side effect, HPV-infected cells become highly susceptible to the accumulation of DNA damage and to cellular transformation.

Because the viral life cycle is highly dependent on p53 inactivation, HR HPV types have developed redundant mechanisms for targeting p53 functions. HR HPV E6 proteins and, to a lesser extent, LR HPV E6 proteins are able to inhibit the expression of p53-regulated genes by directly interacting with the histone acetyltransferases, CBP and p300, that act as transcriptional co-activators for p53 and other transcription factors (Figure 3.2) (Patel *et al.*, 1999; Zimmermann *et al.*, 1999). E6 proteins from some beta HPV types also have the property of inactivating p300 (Muench *et al.*, 2010; Howie *et al.*, 2011; Wallace *et al.*, 2012; Wallace *et al.*, 2014; Wallace *et al.*, 2015). Beta-1 HPV5 and HPV8 E6 bind and promote p300 degradation via the proteasome pathway (Howie *et al.*, 2011). Although beta HPV5 and HPV8 E6 proteins weakly interact with E6AP (Bedard *et al.*, 2008), p300 degradation induced by these two viral proteins is independent of this ubiquitin ligase, but rather is linked to a decrease of p300 phosphorylation at a specific site of the C-terminus (S1834) that is important for protein stability (Howie *et al.*, 2011). Loss of p300 in E6-expressing cells results in the inhibition of calcium-mediated differentiation as well as a decrease of the efficiency in repairing UV-induced DNA damage, showing as the accumulation of thymine dimers and double-strand breaks (Wallace *et al.*, 2012, Wallace *et al.*, 2015). The accumulation of double-strand breaks is linked to inhibition of expression of key proteins involved in the DNA repair machinery, such as ATR, BRAC1, and BRAC2, as well as a decrease of p53 activation by phosphorylation and acetylation (Wallace *et al.*, 2012, Wallace *et al.*, 2015). These findings are in agreement with a previous study that showed the ability of HPV5 and HR HPV18 E6 to delay the UVB-induced thymine dimer repair (Giampieri and Storey, 2004). Beta-2 HPV38 E6 weakly interacts with p300, without promoting its degradation. As a consequence, the accumulation of UV-induced DNA damage is much less pronounced in HPV38 E6 HPKs than in HPV5 or HPV8 E6 cells (Howie *et al.*, 2011; Wallace *et al.*, 2012, Wallace *et al.*, 2015). However, it is important to underline that all these experiments have been performed with cells expressing the single HPV oncogene E6. It is likely that the presence of E7 could lead to alteration of the same pathways with different mechanisms. For instance, HPV38 E7 counteracts the transcriptional functions of p53 by inducing ΔNp73α, as described in detail below (Accardi *et al.*, 2006; Accardi *et al.*, 2011). Another study highlighted the importance of HPV38 E6/p300 interaction in cellular transformation (Muench *et al.*, 2010). Indeed, HPV38 E6, via its interaction with p300, counteracts the pro-apoptotic functions of p53. Importantly, an HPV38 E6 mutant unable to associate with p300 loses the capacity to prevent p53 acetylation, p53-dependent transcription, and apoptosis induction. In addition, this HPV38 E6 mutant does not cooperate with HPV38 E7 in the immortalization of HPKs (Muench *et al.*, 2010). Accordingly, E6 from the beta types HPV5, 8, and 38 can prevent p53 stabilization in response to aberrant mitosis and dysregulated centrosome duplication (Wallace *et al.*, 2014).

Another study, on E6 from beta-2 HPV23, has highlighted another mechanism to block the p53 activation induced by UV irradiation. HIPK2 is a cellular kinase responsible for p53 phosphorylation at serine 46 upon UV irradiation. HPV23 E6 directly interacts with HIPK2, with consequent inhibition of the kinase activity of HIPK2 (Muschik *et al.*, 2011).

Together, these findings fully support the model of skin carcinogenesis in which beta HPV types exacerbate the accumulation of UV-induced damage, affecting p53 and the DNA repair machinery. In agreement with these conclusions, transgenic animal models expressing beta HPV E6 and E7 in skin keratinocytes highlighted the cooperation of UV and viral oncoproteins in promoting skin lesions (Marcuzzi *et al.*, 2009; Viarisio *et al.*, 2011). Interestingly, transgenic mouse models expressing E6 and E7 genes from beta-3 HPV49, in agreement with *in vitro* studies (Cornet *et al.*, 2012), highlighted functional similarities of HPV16 and HPV49 (Viarisio *et al.*, 2016).

It has also been well demonstrated that alpha and beta HPV types have the ability to inhibit the apoptotic response via additional mechanisms (Figure 3.2). LR and HR mucosal HPV types, as well as E6 from beta and other cutaneous HPV types, are able to interact with the pro-apoptotic protein BAK, a member of the Bcl-2 family, promoting its degradation (Jackson *et al.*, 2000; Underbrink *et al.*, 2008). In normal cells, apoptotic stimuli induce changes in BAK post-translational modifications, leading to oligomerization at the mitochondrial membrane. This activated form of BAK promotes the formation of pores, facilitating the release of cytochrome C from the mitochondria, followed by induction of caspase apoptotic cascades (Dewson *et al.*, 2008; Dewson *et al.*, 2009). In the case of HR HPV E6 proteins, the BAK degradation appears to be mediated by E6AP and the ubiquitin pathway. Also, beta HPV8 E6 is able, at least in part, to promote BAK degradation via the interaction with E6AP (Underbrink *et al.*, 2008). A recent study on beta HPV5 E6 has further characterized the mechanism of BAK degradation. The viral protein targets for degradation the late activated form of BAK via interaction with the E3 ubiquitin ligase HERC1 and not E6AP (Holloway *et al.*, 2015). Whether this is a mechanism shared with E6 oncoproteins from other beta HPV types remains to be evaluated.

E6 oncoproteins and hTERT

The interaction of HR HPV E6 proteins with E6AP is also involved in the maintenance of telomere length, which is an essential requirement for cellular immortalization. Somatic cells have very little or no telomerase activity and telomeres shorten as a function of cellular division to finally reach a critical size, which promotes chromosome aggregation and replicative senescence (reviewed in Blackburn *et al.*, 2015). HR HPV E6 proteins are able to activate the transcription of the *hTERT* (human telomerase reverse transcriptase) gene, preventing the shortening of telomeres after each cell division, facilitating the immortalization of HPKs (Figure 3.2). This E6 activity was initially described for HPV16 E6 (Klingelhutz *et al.*, 1996). It is likely that this property is conserved in E6 proteins from many HR HPV types, which are able, together with the corresponding E7, to efficiently immortalize HPKs (Hiller *et al.*, 2006). As mentioned above, the HPV-mediated activation of *hTERT* transcription is in part due to the E6/E6AP interaction, which results in the association with and degradation of a negative transcriptional regulator, NFX1-91, which is a spliced variant of the *NFX1* gene (Gewin *et al.*, 2004). The other spliced variant of the *NFX1* gene is NFX1-123, which positively regulates *hTERT* expression (Gewin *et al.*, 2004). NFX1-123 also interacts with HPV16 E6; however, in contrast to NFX1-91, it is not degraded by the viral protein and its interaction with E6 leads to an increase in *hTERT* expression by a post-transcriptional mechanism (Katzenellenbogen *et al.*, 2007). Interestingly, NFX1-91 can also act as an activator of the transcription on other promoters. Indeed, it has been shown that NFX1-91 activates the expression of p105, the precursor of p50, which negatively regulates the transcriptional activity of NF-κB. HPV16 E6 antagonizes these functions of NFX1-91, repressing the expression of p105 and altering the regulation of the NF-κB pathway (Xu *et al.*, 2010).

In agreement with the concept that HR HPV types have multiple mechanisms to target the same cellular pathway, it has been shown that HPV16 E6 is able to directly associate with the transcription factor Myc, which in turn acquires the property of efficiently activating the *hTERT* promoter (Veldman *et al.*, 2003). Findings indicate that this additional E6 mechanism is independent of E6/E6AP interaction (Sekaric *et al.*, 2008).

Also beta HPV E6 oncoproteins can activate *hTERT* transcription (Bedard *et al.*, 2008). According to the ability to immortalize HPKs, beta-2 HPV38 and beta-3 HPV49 efficiently activate *hTERT* transcription (Bedard *et al.*, 2008; Gabet *et al.*, 2008; Cornet *et al.*, 2012). In contrast, E6 proteins from beta-1 HPV5 and 20, and beta-2 HPV22 have a much reduced activity compared with HPV38 (Bedard *et al.*, 2008; Cornet *et al.*, 2012). Findings indicate that the mechanism by which beta HPV E6 activates *hTERT* expression involves E6AP and NFX1-91. A comparative study among E6 oncoprotein from HPV types 5, 20, 22, and 38 showed a close correlation between the efficiency of the different E6 proteins in interacting with E6AP or NFX1-91 and in activating telomerase (Bedard *et al.*, 2008; Cornet *et al.*, 2012). In addition, HPV38, through the E7 oncoprotein, activates *hTERT* transcription by an alternative mechanism, which is mediated by the p53-related protein ΔNp73α (Gabet *et al.*, 2008).

E6 oncoproteins and the PDZ proteins

Another important feature of E6 is the ability to interact with a family of cellular proteins that contain a domain of approximately 80–90 amino acids, termed PDZ (Figure 3.2). The analysis of the human proteome revealed that more than 150 PDZ proteins exist, which may contain multiple motifs (Luck *et al.*, 2012). They promote the formation of multimeric protein complexes and regulate several signalling pathways, controlling different cellular events, including cell–cell contact and cell polarity (reviewed in Ganti *et al.*, 2015).

The E6 oncoproteins interact with the PDZ proteins via a binding motif (PBM) at the C-terminus, X-S/T-X-Φ, where X is any amino acid and Φ is a hydrophobic amino acid (Figure 3.1). All E6 from the mucosal HR HPV types contain a PBM motif (reviewed in Ganti *et al.*, 2015). No such sequence has been identified in most E6 proteins from LR HPV types, with the exception of LR HPV40 E6, which also appears to target some of the PDZ proteins (Van Doorslaer *et al.*, 2015; Thomas *et al.*, 2016). For beta and gamma HPV types, there are no lines of evidence that E6 from these viruses contains a PBM and is able to target any PDZ protein. However, it has been shown that beta HPV8 E6 is able to inhibit the expression of a gene encoding the PDZ protein syntenin-2 (Lazic *et al.*, 2012), supporting the hypothesis that these HPV types have developed alternative mechanisms to neutralize the functions of some PDZ proteins (Figure 3.2).

The fact that a PBM motif is present in E6 from the most oncogenic HPV types known so far supports the concept that this property is intimately linked to virus-mediated carcinogenesis. Accordingly, experiments in *in vitro* and *in vivo* experimental models demonstrate the role of E6 PBM in cellular transformation (reviewed in Ganti *et al.*, 2015). Expression of ΔPBM HPV18 E6 in human keratinocytes results in a much attenuated transformed phenotype compared with the wild-type viral oncoprotein (Watson *et al.*, 2003). Transgenic mice expressing HR HPV16 E6 in the basal layer of the epidermis under the control of the cytokeratin K14 promoter have an elevated proliferation in the skin and mucosa, which strongly decreased upon deletion of the PBM (Nguyen *et al.*, 2003). In addition, long-term exposure of these transgenic mice to oestrogen treatment leads to the development of cervical cancer that is strictly dependent on the integrity of the PBM (Shai *et al.*, 2007).

Several studies showed that the E6 oncoproteins from different mucosal HR HPV types have a different efficiency in binding specific PDZ proteins. For instance, HPV16 and HPV18 E6 differ in their binding affinity for two PDZ proteins, i.e. Disc Large (Dlg) and Scribble (Scrib) proteins. These two viral proteins have almost identical PDZ-binding motifs (148ETQL151 in HPV16 E6 and 155ETQV158 in HPV18 E6). However, the last L or V residues in the HPV16 or HPV18 E6 PBM confer preferential binding for Scrib or Dlg, respectively. Swapping the two amino acids in the two PBMs of HPV16 and HPV18 E6 inverts the Scrib or Dlg binding efficiency of HPV16 and HPV18 E6 (Thomas *et al.*, 2005). Interestingly, a comparative analysis of E6 PBM from a broad spectrum of mucosal HPV types revealed that the total number of PDZ proteins targeted by the different E6 oncoproteins increases according to their oncogenicity (Thomas *et al.*, 2016).

Although the ability to target the cellular PDZ proteins is an exclusive property of the mucosal HPV E6 oncoproteins, E7 can facilitate such interaction. HPV16 E6 associates with and promotes the degradation of the PDZ protein Na(+)/H(+) exchange regulatory factor 1 (NHERF-1). HPV16 E7, through its positive impact on the cell, promotes the accumulation of the phosphorylated form of NHERF-1, which is more efficiently targeted by E6 compared with the un-phosphorylated form (Figure 3.2) (Accardi *et al.*, 2011).

Several studies have shown that the E6/PDZ protein interaction is negatively modulated by phosphorylation of the S/T within the E6 PBM. Protein kinase A (PKA) phosphorylates the HPV18 E6 PBM, abolishing the interaction with Dlg (Kuhne *et al.*, 2000). In the case of HPV16 E6, the PBM can be phosphorylated by PKA as well as AKT (Figure 3.2) (Boon and Banks, 2013). Many findings support the concept that this phosphorylation is more likely linked to the viral life cycle rather than the HPV-driven cellular transformation.

Features and transforming properties of E7 protein

Similar to E6, E7 is a small protein, of approximately 100 amino acids, that contains zinc-binding motifs and has the ability to interact with a vast number of cellular proteins (Figure 3.1). Based on several structural and functional similarities to E1A, E7 primary structures can be separated into three conserved regions: CR1–3

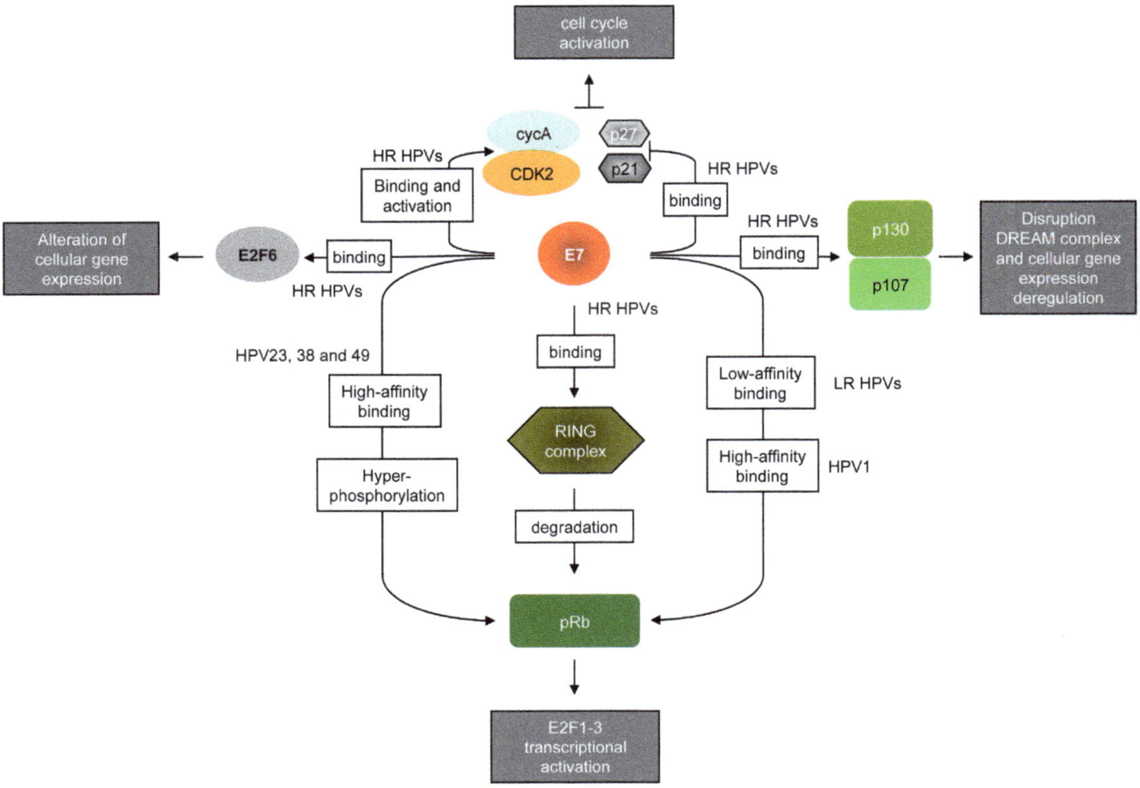

Figure 3.3. Interactions between E7 and cellular targets. Many of the properties of E7 from different HPV types and their induced events are shown. The dark grey boxes describe the events induced by the E7 oncoproteins.

(Figure 3.1) (Phelps *et al.*, 1988). The majority of the E7 proteins from cutaneous and mucosal HPV types have in CR2 a motif (LXCXE) that mediates interaction with the tumour suppressor gene product retinoblastoma (pRb1) and its related proteins, p107 and p130. In many E7 proteins, CR2 also includes a phosphorylation site of casein kinase 2 (CK2), which is located immediately after the pocket protein-binding motif (Figure 3.1) (Firzlaff *et al.*, 1989; Barbosa *et al.*, 1990). Studies of HR HPV E7 showed that deletion of this domain hampers its transforming properties in *in vitro* experimental models, without altering the interaction with pRb (Barbosa *et al.*, 1990; Firzlaff *et al.*, 1991). CR3, similarly to the N- and C-domain of E6, contains two zinc-binding motifs, CXXC (Barbosa *et al.*, 1989; McIntyre *et al.*, 1993), which are separated by 29 amino acids. Studies of E7 oncoproteins from some mucosal HR HPV types showed that CR3 mediates the formation of homodimers and/or tetramers (McIntyre *et al.*, 1993; Clemens *et al.*, 1995; Zwerschke *et al.*, 1996). Point mutations in one of the two CXXC motifs of E7 severely impair its ability to homodimerize and affect the transforming properties in rodent cells (McIntyre *et al.*, 1993). However, it appears that multimerization does not interfere with the interaction of E7 with pRb (Alonso *et al.*, 2006). Replacement of E7 CR3 with the zinc-binding motif of E6 or with CR3 of adenovirus E1A does not significantly alter the transforming properties of E7 CR1 and CR2 (Mavromatis *et al.*, 1997; Braspenning *et al.*, 1998).

E7, the pocket proteins, and deregulation of the cell cycle

One of the most studied properties of E7 oncoprotein is the ability to target pRb, p107, and p130, also termed the "pocket proteins" (Figure 3.3. These proteins play a key role in the control of the cell cycle and exert their activity mainly by regulating the expression of different sets of cellular genes. In quiescent cells, pRb is bound to members of the E2F transcription factor family, E2F1–3, repressing their activities. When cells are exposed to mitogenic stimuli, activation of cyclin-dependent kinases (CDKs) results in phosphorylation of pRb, which in turn loses the ability to bind E2F transcription factors. Free and active E2Fs induce the expression of genes encoding proteins involved in promoting the G1/S transition, such as cyclin A and cyclin E (Cobrinik, 2005; Dimova and Dyson, 2005). HPV E7 binds the hypo-phosphorylated form of pRb, disrupting the complex with E2F. Thus, E7, mimicking the inactivation of pRb by phosphorylation, constitutively activates the cell cycle in a manner independent of external stimuli. Many studies have shown that E7 oncoproteins from different HPV types retain the ability to bind pRb, although with different affinity (Munger *et al.*, 1989; Ciccolini *et al.*, 1994; Schmitt *et al.*, 1994; Caldeira *et al.*, 2000; Caldeira *et al.*, 2003; Cornet *et al.*, 2012;). E7 from the mucosal LR HPV types 6 and 11 binds pRb with a lower affinity than HR HPV16 (Munger *et al.*, 1989). This is explained by subtle difference in the region containing the pRb binding motif, LXCXE (Figure 3.1). In all HR HPV E7 proteins, the LXCXE motif is preceded by an aspartic acid residue (D21), whereas HPV6 E7 has in the same position a glycine (G22). G22D mutation in HPV6 E7 significantly enhanced its affinity for pRb and its *in vitro* transforming properties (Heck *et al.*, 1992; Sang and Barbosa, 1992). However, the D residue immediately before the LXCXE motif is not present only in E7 from HR HPV; it can also be present in E7 from HPV types that have been clearly linked to benign lesions. For instance, E7 from cutaneous HPV1 from genus mu, associated with skin warts, has a DLXCXE motif and binds pRb with a similar efficiency to HPV16 E7 (Ciccolini *et al.*, 1994; Schmitt *et al.*, 1994). The consequence of the interaction of HPV16 E7 with pRb is the rapid degradation of pRb via the proteasome pathway (Berezutskaya *et al.*, 1997; Jones and Munger, 1997). HPV1 E7 is not able to destabilize pRb and is less efficient than HPV16 E7 in deregulating the cell cycle (Giarre *et al.*, 2001; Gonzalez *et al.*, 2001). HPV16 E7 promotes pRb degradation via the proteasome pathway by associating with a cullin 2-RING ubiquitin ligase complex (Huh *et al.*, 2007; White *et al.*, 2012). In addition to the LXCXE motif in CR2, it has been shown that HPV16 E7 CR3 contributes to the interaction with and stabilization of pRb (Todorovic *et al.*, 2012).

For the beta HPV types, only a small number of studies have evaluated the ability of E7 to bind pRb. Beta-1 HPV5 or 8 E7 binds pRb with much reduced affinity compared with HPV16 E7 (Yamashita *et al.*, 1993). Beta-2 HPV38 E7 can efficiently associate with pRb and its expression in immortalized rodent fibroblasts (NIH 3T3) leads to pRb destabilization (Caldeira *et al.*, 2003). However, in HPKs, HPV38 E7 induces pRb hyper-phosphorylation and stabilization (Caldeira *et al.*, 2003). A more recent study compared the ability of E7 from beta-1 (HPV14, 24, and 36), beta-2 (HPV23 and 38), and beta-3 (HPV49) to target pRb (Cornet *et al.*, 2012). All these E7 oncoproteins bind pRb and induce its destabilization or hyper-phosphorylation/stabilization in NIH 3T3 or HPKs, respectively, as observed for HPV38 E7 (Caldeira *et al.*, 2003; Cornet *et al.*, 2012). The reasons why these beta HPV E7 oncoproteins induce opposite effects in human and rodent cells are still unknown.

Also the other two pocket proteins, p107 and p130, are bound and destabilized by the HR HPV E7 oncoproteins (Jones and Munger, 1997; Gonzalez *et al.*, 2001; Genovese *et al.*, 2008; Helt and Galloway, 2001). p107 and p130 form complexes with other members of the E2F family, E2F4 and E2F5, that are involved in controlling additional cell-cycle checkpoints. Both these pocket proteins regulate the G0/G1 transition; in addition, p107 plays a role in the G2 phase (Cobrinik, 2005; Dimova and Dyson, 2005). In contrast to E2F1–3, which positively regulate cellular gene expression, E2F4 and E2F5 are part of inhibitory complexes of transcription, together with the hypo-phosphorylated forms of p107 or p130 and histone deacetylase enzymes. In particular, recent studies led to the characterization of a multimeric complex (DREAM: DP (dimerization partner), RB-like, E2F4, and MuvB) (Muller *et al.*, 2012; Sadasivam and DeCaprio, 2013), which represses promoters containing regions that are conserved in transcriptionally regulated cell-cycle genes (Muller and Engeland, 2010). HPV16 E7, by binding the hypo-phosphorylated forms of p107 and p130, releases the transcriptional repression of these complexes (Figure 3.3; Nor Rashid *et al.*, 2011; Fischer *et al.*, 2014).

The E2F family includes three additional members, E2F6–8, which lack a transactivation domain and act as repressors of transcription in a pRb-independent manner (Dimova and Dyson, 2005; Lammens *et al.*, 2009). E2F6 is part of a multimeric complex together with several polycomb proteins, leading to epigenetic changes associated with chromatin remodelling and gene silencing, e.g. histone 3 lysine 27 trimethylation (H3K27me3) (Cartwright *et al.*, 1998; Gaubatz *et al.*, 1998; Trimarchi *et al.*, 1998; Attwooll *et al.*, 2005). It has been shown that HPV16 E7 interacts with E2F6 via CR3, interfering with the inhibitory function of E2F6 in cellular gene expression (Figure 3.3 (McLaughlin-Drubin *et al.*, 2008). Interestingly, E7 from LR HPV types as well as E1A and LT oncoproteins, from adenovirus and simian virus 40 (SV40), respectively, have the ability to associate with E2F6. The fact that LR HPV E7 retains this property indicates that the E7/E2F6 interaction is not directly linked to cellular transformation, but it may facilitate the viral life cycle, reprogramming cellular gene expression.

Additional mechanisms of E7 in deregulating the cell cycle

As observed for E6, which has redundant mechanisms in targeting p53, the E7 oncoprotein deregulates the cell cycle in multiple ways. It has been reported by several independent studies that HPV16 E7 is able to associate with CDK complexes, i.e. cyclin A/CDK2 and cyclin E/CDK2 (Figure 3.3 (Arroyo *et al.*, 1993; Davies *et al.*, 1993; Tommasino *et al.*, 1993; McIntyre *et al.*, 1996). Since p107 can form a complex with cyclin A/CDK2 via a motif that is located outside the regions involved in the interaction with the viral oncoproteins (Ewen *et al.*, 1992), the E7/cyclin association is at least in part mediated by p107 (McIntyre *et al.*, 1996). However, there is also clear evidence that HPV16 E7 can directly interact with cyclin A (He *et al.*, 2003; Nguyen and Munger, 2008; Tommasino *et al.*, 1993). This property is also conserved in E7 from the LR HPV types (Nguyen and Munger, 2008). Interestingly, it has been shown that purified recombinant E7 proteins from HR HPV16 and HPV31, and LR HPV6b activate cyclin E and cyclin A/CDK2 complexes (He *et al.*, 2003).

Additional studies have shown that HPV16 E7 can also stimulate CDK activity by directly binding the two CDK inhibitors $p21^{WAF1/CIP1}$ and $p27^{KIP1}$, neutralizing their inhibitory effects on the cell cycle (Zerfass-Thome *et al.*, 1996; Funk *et al.*, 1997; Jones *et al.*, 1997).

No data are available on the possible impact of beta HPV E7 oncoproteins on cell-cycle deregulation with mechanisms independent of the pocket proteins.

Conclusions

The biological studies have clearly demonstrated the oncogenic activities of E6 and E7 oncoproteins, which deregulate fundamental cellular events by redundant mechanisms, such as the cell cycle, apoptosis, differentiation, senescence, cell polarity, and immune-response-related pathways. Only some of these mechanisms have been reviewed in this chapter. Importantly, the studies on the biological properties of E6 and E7 oncoproteins also led to the discovery of important cellular mechanisms, increasing our understanding of cell biology.

The limited knowledge of the biology of E6 and E7 from the majority of HPV types known so far means that future studies are needed to establish new links between HPV types and human diseases, such as NMSC.

Acknowledgements

I apologize to those authors whose important contributions to HPV research could not be cited or adequately discussed due to space limitations. I thank all members of the Infections and Cancer Biology Group at IARC for their constant support, Dr Karen Müller for the editing, and Nicole Suty for the preparation of this manuscript.

References

Accardi, R., Dong, W., Smet, A., Cui, R., Hautefeuille, A., Gabet, A.S., Sylla, B.S., Gissmann, L., Hainaut, P., and Tommasino, M. (2006). Skin human papillomavirus type 38 alters p53 functions by accumulation of deltaNp73. EMBO reports 7, 334-340. http://dx.doi.org/10.1038/sj.embor.7400615

Accardi, R., Rubino, R., Scalise, M., Gheit, T., Shahzad, N., Thomas, M., Banks, L., Indiveri, C., Sylla, B.S., Cardone, R.A., *et al.* (2011). E6 and E7 from human papillomavirus type 16 cooperate to target the PDZ protein Na/H exchange regulatory factor 1. J. Virol. 85, 8208-8216. http://dx.doi.org/10.1128/jvi.00114-11

Accardi, R., Scalise, M., Gheit, T., Hussain, I., Yue, J., Carreira, C., Collino, A., Indiveri, C., Gissmann, L., Sylla, B.S., *et al.* (2011). IkappaB kinase beta promotes cell survival by antagonizing p53 functions through DeltaNp73alpha phosphorylation and stabilization. Mol. Cell. Biol. 31, 2210-2226. http://dx.doi.org/10.1128/mcb.00964-10

Agalliu, I., Gapstur, S., Chen, Z., Wang, T., Anderson, R.L., Teras, L., Kreimer, A.R., Hayes, R.B., Freedman, N.D., and Burk, R.D. (2016). Associations of Oral alpha-, beta-, and gamma-Human Papillomavirus Types With Risk of Incident Head and Neck Cancer. JAMA oncology. http://dx.doi.org/10.1001/jamaoncol.2015.5504

Alemany, L., Cubilla, A., Halec, G., Kasamatsu, E., Quiros, B., Masferrer, E., Tous, S., Lloveras, B., Hernandez-Suarez, G., Lonsdale, R., *et al.* (2016). Role of Human Papillomavirus in Penile Carcinomas Worldwide. European urology 69, 953-961. http://dx.doi.org/10.1016/j.eururo.2015.12.007

Alemany, L., Saunier, M., Alvarado-Cabrero, I., Quiros, B., Salmeron, J., Shin, H.R., Pirog, E.C., Guimera, N., Hernandez-Suarez, G., Felix, A., *et al.* (2015). Human papillomavirus DNA prevalence and type distribution in anal carcinomas worldwide. Int. J. Cancer 136, 98-107. http://dx.doi.org/10.1002/ijc.28963

Alemany, L., Saunier, M., Tinoco, L., Quiros, B., Alvarado-Cabrero, I., Alejo, M., Joura, E.A., Maldonado, P., Klaustermeier, J., Salmeron, J., *et al.* (2014). Large contribution of human papillomavirus in vaginal neoplastic lesions: a worldwide study in 597 samples. Eur. J. Cancer (Oxford, England: 1990) 50, 2846-2854. http://dx.doi.org/10.1016/j.ejca.2014.07.018

Alonso, L.G., Smal, C., Garcia-Alai, M.M., Chemes, L., Salame, M., and de Prat-Gay, G. (2006). Chaperone holdase activity of human papillomavirus E7 oncoprotein. Biochemistry 45, 657-667. http://dx.doi.org/10.1021/bi0522549

Arroyo, M., Bagchi, S., and Raychaudhuri, P. (1993). Association of the human papillomavirus type 16 E7 protein with the S-phase-specific E2F-cyclin A complex. Mol. Cell. Biol. 13, 6537-6546.

Astori, G., Lavergne, D., Benton, C., Hockmayr, B., Egawa, K., Garbe, C., and de Villiers, E.M. (1998). Human papillomaviruses are commonly found in normal skin of immunocompetent hosts. J. Invest. Dermatol. 110, 752-755. http://dx.doi.org/10.1046/j.1523-1747.1998.00191.x

Attwooll, C., Oddi, S., Cartwright, P., Prosperini, E., Agger, K., Steensgaard, P., Wagener, C., Sardet, C., Moroni, M.C., and Helin, K. (2005). A novel repressive E2F6 complex containing the polycomb group protein, EPC1, that interacts with EZH2 in a proliferation-specific manner. J. Biol. Chem. 280, 1199-1208. http://dx.doi.org/10.1074/jbc.M412509200

Barbosa, M.S., Edmonds, C., Fisher, C., Schiller, J.T., Lowy, D.R., and Vousden, K.H. (1990). The region of the HPV E7 oncoprotein homologous to adenovirus E1a and Sv40 large T antigen contains separate domains for Rb binding and casein kinase II phosphorylation. The EMBO journal 9, 153-160.

Barbosa, M.S., Lowy, D.R., and Schiller, J.T. (1989). Papillomavirus polypeptides E6 and E7 are zinc-binding proteins. J. Virol. 63, 1404-1407.

Barbosa, M.S., Vass, W.C., Lowy, D.R., and Schiller, J.T. (1991). In vitro biological activities of the E6 and E7 genes vary among human papillomaviruses of different oncogenic potential. J. Virol. 65, 292-298.

Bedard, K.M., Underbrink, M.P., Howie, H.L., and Galloway, D.A. (2008). The E6 oncoproteins from human betapapillomaviruses differentially activate telomerase through an E6AP-dependent mechanism and prolong the lifespan of primary keratinocytes. J. Virol. 82, 3894-3902. http://dx.doi.org/10.1128/jvi.01818-07

Bedell, M.A., Jones, K.H., Grossman, S.R., and Laimins, L.A. (1989). Identification of human papillomavirus type 18 transforming genes in immortalized and primary cells. J. Virol. 63, 1247-1255.

Berezutskaya, E., Yu, B., Morozov, A., Raychaudhuri, P., and Bagchi, S. (1997). Differential regulation of the pocket domains of the retinoblastoma family proteins by the HPV16 E7 oncoprotein. Cell growth and differentiation: the molecular biology journal of the American Association for Cancer Research 8, 1277-1286.

Bernard, H.U., Burk, R.D., Chen, Z., van Doorslaer, K., zur Hausen, H., and de Villiers, E.M. (2010). Classification of papillomaviruses (PVs) based on 189 PV types and proposal of taxonomic amendments. Virology 401, 70-79. http://dx.doi.org/10.1016/j.virol.2010.02.002

Blackburn, E.H., Epel, E.S., and Lin, J. (2015). Human telomere biology: A contributory and interactive factor in aging, disease risks, and protection. Science (New York, NY) 350, 1193-1198. http://dx.doi.org/10.1126/science.aab3389

Boon, S.S., and Banks, L. (2013). High-risk human papillomavirus E6 oncoproteins interact with 14-3-3zeta in a PDZ binding motif-dependent manner. J. Virol. 87, 1586-1595. http://dx.doi.org/10.1128/jvi.02074-12

Bottalico, D., Chen, Z., Dunne, A., Ostoloza, J., McKinney, S., Sun, C., Schlecht, N.F., Fatahzadeh, M., Herrero, R., Schiffman, M., *et al.* (2011). The oral cavity contains abundant known and novel human papillomaviruses from the Betapapillomavirus and Gammapapillomavirus genera. J. Infect. Dis. 204, 787-792. http://dx.doi.org/10.1093/infdis/jir383

Bouvard, V., Baan, R., Straif, K., Grosse, Y., Secretan, B., El Ghissassi, F., Benbrahim-Tallaa, L., Guha, N., Freeman, C., Galichet, L., *et al.* (2009). A review of human carcinogens--Part B: biological agents. Lancet Oncol. 10, 321-322.

Bouvard, V., Gabet, A.S., Accardi, R., Sylla, B.S., and Tommasino, M. (2006). The Cutaneous Human Papillomavirus Types and Non-melanoma Skin Cancer. In Papillomavirus Research: From Natural History To Vaccines and Beyond, M. Saveria Campo, ed. (Division of Pathological Sciences, Institute of Comparative Medicine University of Glasgow, Glasgow G61 1QH, UK: Caister Academic Press), pp. 269-277. ISBN 978-1-904455-04-2

Bouwes Bavinck, J.N., Neale, R.E., Abeni, D., Euvrard, S., Green, A.C., Harwood, C.A., de Koning, M.N., Naldi, L., Nindl, I., Pawlita, M., *et al.* (2010). Multicenter study of the association between betapapillomavirus infection and cutaneous squamous cell carcinoma. Cancer research 70, 9777-9786. http://dx.doi.org/10.1158/0008-5472.can-10-0352

Boxman, I.L., Mulder, L.H., Noya, F., de Waard, V., Gibbs, S., Broker, T.R., ten Kate, F., Chow, L.T., and ter Schegget, J. (2001). Transduction of the E6 and E7 genes of epidermodysplasia-verruciformis-associated human papillomaviruses alters human keratinocyte growth and differentiation in organotypic cultures. J. Invest. Dermatol. 117, 1397-1404. http://dx.doi.org/10.1046/j.0022-202x.2001.01602.x

Braspenning, J., Marchini, A., Albarani, V., Levy, L., Ciccolini, F., Cremonesi, C., Ralston, R., Gissmann, L., and Tommasino, M. (1998). The CXXC Zn binding motifs of the human papillomavirus type 16 E7 oncoprotein are not required for its *in vitro* transforming activity in rodent cells. Oncogene 16, 1085-1089. http://dx.doi.org/10.1038/sj.onc.1201617

Butz, K., Ristriani, T., Hengstermann, A., Denk, C., Scheffner, M., and Hoppe-Seyler, F. (2003). siRNA targeting of the viral E6 oncogene efficiently kills human papillomavirus-positive cancer cells. Oncogene 22, 5938-5945. http://dx.doi.org/10.1038/sj.onc.1206894

Calcada, E.O., Felli, I.C., Hosek, T., and Pierattelli, R. (2013). The heterogeneous structural behavior of E7 from HPV16 revealed by NMR spectroscopy. Chembiochem: a European journal of chemical biology 14, 1876-1882. http://dx.doi.org/10.1002/cbic.201300172

Caldeira, S., de Villiers, E.M., and Tommasino, M. (2000). Human papillomavirus E7 proteins stimulate proliferation independently of their ability to associate with retinoblastoma protein. Oncogene 19, 821-826. http://dx.doi.org/10.1038/sj.onc.1203375

Caldeira, S., Zehbe, I., Accardi, R., Malanchi, I., Dong, W., Giarre, M., de Villiers, E.M., Filotico, R., Boukamp, P., and Tommasino, M. (2003). The E6 and E7 proteins of the cutaneous human papillomavirus type 38 display transforming properties. J. Virol. 77, 2195-2206.

Can, N.T., Tretiakova, M.S., and Taxy, J.B. (2015). Natural History and Malignant Transformation in Recurrent Respiratory Papillomatosis: Human Papillomavirus (HPV), Dysplasia and an Autopsy Review. Fetal and pediatric pathology 34, 80-90. http://dx.doi.org/10.3109/15513815.2014.968271

Carifi, M., Napolitano, D., Morandi, M., and Dall'Olio, D. (2015). Recurrent respiratory papillomatosis: current and future perspectives. Therapeut. Clin. Risk Manag. 11, 731-738. http://dx.doi.org/10.2147/tcrm.s81825

Cartwright, P., Muller, H., Wagener, C., Holm, K., and Helin, K. (1998). E2F-6: a novel member of the E2F family is an inhibitor of E2F-dependent transcription. Oncogene 17, 611-623. http://dx.doi.org/10.1038/sj.onc.1201975

Castellsague, X., Alemany, L., Quer, M., Halec, G., Quiros, B., Tous, S., Clavero, O., Alos, L., Biegner, T., Szafarowski, T., *et al.* (2016). HPV Involvement in Head and Neck Cancers: Comprehensive Assessment of Biomarkers in 3680 Patients. J. Natl. Cancer Inst. 108, djv403. http://dx.doi.org/10.1093/jnci/djv403

Chahoud, J., Semaan, A., Chen, Y., Cao, M., Rieber, A.G., Rady, P., and Tyring, S.K. (2015). Association Between beta-Genus Human Papillomavirus and Cutaneous Squamous Cell Carcinoma in Immunocompetent Individuals-A Meta-analysis. JAMA dermatology. http://dx.doi.org/10.1001/jamadermatol.2015.4530

Chapman, P.B., Hauschild, A., Robert, C., Haanen, J.B., Ascierto, P., Larkin, J., Dummer, R., Garbe, C., Testori, A., Maio, M., *et al.* (2011). Improved survival with vemurafenib in melanoma with BRAF V600E mutation. New Engl. J. Med. 364, 2507-2516. http://dx.doi.org/10.1056/NEJMoa1103782

Chen, Z., Kamath, P., Zhang, S., Weil, M.M., and Shillitoe, E.J. (1995). Effectiveness of three ribozymes for cleavage of an RNA transcript from human papillomavirus type 18. Cancer gene therapy 2, 263-271.

Chen, Z., Schiffman, M., Herrero, R., Desalle, R., and Burk, R.D. (2007). Human papillomavirus (HPV) types 101 and 103 isolated from cervicovaginal cells lack an E6 open reading frame (ORF) and are related to gamma-papillomaviruses. Virology 360, 447-453. http://dx.doi.org/10.1016/j.virol.2006.10.022

Ciccolini, F., Di Pasquale, G., Carlotti, F., Crawford, L., and Tommasino, M. (1994). Functional studies of E7 proteins from different HPV types. Oncogene 9, 2633-2638.

Clemens, K.E., Brent, R., Gyuris, J., and Munger, K. (1995). Dimerization of the human papillomavirus E7 oncoprotein *in vivo*. Virology 214, 289-293. http://dx.doi.org/10.1006/viro.1995.9926

Cobrinik, D. (2005). Pocket proteins and cell cycle control. Oncogene 24, 2796-2809. http://dx.doi.org/10.1038/sj.onc.1208619

Cohen, D.N., Lawson, S.K., Shaver, A.C., Du, L., Nguyen, H.P., He, Q., Johnson, D.B., Lumbang, W.A., Moody, B.R., Prescott, J.L., *et al.* (2015). Contribution of Beta-HPV Infection and UV Damage to Rapid-Onset Cutaneous Squamous Cell Carcinoma during BRAF-Inhibition Therapy. Clinical cancer research: an official journal of the American Association for Cancer Research 21, 2624-2634. http://dx.doi.org/10.1158/1078-0432.ccr-14-2667

Cole, S.T., and Danos, O. (1987). Nucleotide sequence and comparative analysis of the human papillomavirus type 18 genome. Phylogeny of papillomaviruses and repeated structure of the E6 and E7 gene products. Journal of molecular biology 193, 599-608.

Cornet, I., Bouvard, V., Campo, M.S., Thomas, M., Banks, L., Gissmann, L., Lamartine, J., Sylla, B.S., Accardi, R., and Tommasino, M. (2012). Comparative analysis of transforming properties of E6 and E7 from different beta human papillomavirus types. J. Virol. 86, 2366-2370. http://dx.doi.org/10.1128/jvi.06579-11

Crequer, A., Troeger, A., Patin, E., Ma, C.S., Picard, C., Pedergnana, V., Fieschi, C., Lim, A., Abhyankar, A., Gineau, L., *et al.* (2012). Human RHOH deficiency causes T cell defects and susceptibility to EV-HPV infections. J. Clin. Invest. 122, 3239-3247. http://dx.doi.org/10.1172/jci62949

Dang, C., Koehler, A., Forschner, T., Sehr, P., Michael, K., Pawlita, M., Stockfleth, E., and Nindl, I. (2006). E6/E7 expression of human papillomavirus types in cutaneous squamous cell dysplasia and carcinoma in immunosuppressed organ transplant recipients. Brit. J. Dermatol. 155, 129-136. http://dx.doi.org/10.1111/j.1365-2133.2006.07378.x

Davies, R., Hicks, R., Crook, T., Morris, J., and Vousden, K. (1993). Human papillomavirus type 16 E7 associates with a histone H1 kinase and with p107 through sequences necessary for transformation. J. Virol. 67, 2521-2528.

de Koning, M.N., Weissenborn, S.J., Abeni, D., Bouwes Bavinck, J.N., Euvrard, S., Green, A.C., Harwood, C.A., Naldi, L., Neale, R., Nindl, I., *et al.* (2009). Prevalence and associated factors of betapapillomavirus infections in individuals without cutaneous squamous cell carcinoma. J. Gen. Virol. 90, 1611-1621. http://dx.doi.org/10.1099/vir.0.010017-0

de Sanjose, S., Alemany, L., Ordi, J., Tous, S., Alejo, M., Bigby, S.M., Joura, E.A., Maldonado, P., Laco, J., Bravo, I.G., *et al.* (2013). Worldwide human papillomavirus genotype attribution in over 2000 cases of intraepithelial and invasive lesions of the vulva. Eur. J. Cancer (Oxford, England: 1990) 49, 3450-3461. http://dx.doi.org/10.1016/j.ejca.2013.06.033

de Villiers, E.M. (2013). Cross-roads in the classification of papillomaviruses. Virology 445, 2-10. http://dx.doi.org/10.1016/j.virol.2013.04.023

Dewson, G., Kratina, T., Czabotar, P., Day, C.L., Adams, J.M., and Kluck, R.M. (2009). Bak activation for apoptosis involves oligomerization of dimers via their alpha6 helices. Molecular cell 36, 696-703. http://dx.doi.org/10.1016/j.molcel.2009.11.008

Dewson, G., Kratina, T., Sim, H.W., Puthalakath, H., Adams, J.M., Colman, P.M., and Kluck, R.M. (2008). To trigger apoptosis, Bak exposes its BH3 domain and homodimerizes via BH3:groove interactions. Molecular cell 30, 369-380. http://dx.doi.org/10.1016/j.molcel.2008.04.005

Dimova, D.K., and Dyson, N.J. (2005). The E2F transcriptional network: old acquaintances with new faces. Oncogene 24, 2810-2826. http://dx.doi.org/10.1038/sj.onc.1208612

Dona, M.G., Gheit, T., Vescio, M.F., Latini, A., Moretto, D., Benevolo, M., Cristaudo, A., Tommasino, M., and Giuliani, M. (2016). Incidence, clearance and duration of cutaneous beta and gamma human papillomavirus anal infection. The Journal of infection 73, 380-383. http://dx.doi.org/10.1016/j.jinf.2016.07.006

Durst, M., Dzarlieva-Petrusevska, R.T., Boukamp, P., Fusenig, N.E., and Gissmann, L. (1987). Molecular and cytogenetic analysis of immortalized human primary keratinocytes obtained after transfection with human papillomavirus type 16 DNA. Oncogene 1, 251-256.

Einstein, M.H., Schiller, J.T., Viscidi, R.P., Strickler, H.D., Coursaget, P., Tan, T., Halsey, N., and Jenkins, D. (2009). Clinician's guide to human papillomavirus immunology: knowns and unknowns. The Lancet Infectious diseases 9, 347-356. http://dx.doi.org/10.1016/s1473-3099(09)70108-2

Ewen, M.E., Faha, B., Harlow, E., and Livingston, D.M. (1992). Interaction of p107 with cyclin A independent of complex formation with viral oncoproteins. Science (New York, NY) 255, 85-87.

Falchook, G.S., Rady, P., Hymes, S., Nguyen, H.P., Tyring, S.K., Prieto, V.G., Hong, D.S., and Kurzrock, R. (2013). Merkel cell polyomavirus and HPV-17 associated with cutaneous squamous cell carcinoma arising in a patient with melanoma treated with the BRAF inhibitor dabrafenib. JAMA dermatology 149, 322-326. http://dx.doi.org/10.1001/jamadermatol.2013.2023

Farzan, S.F., Waterboer, T., Gui, J., Nelson, H.H., Li, Z., Michael, K.M., Perry, A.E., Spencer, S.K., Demidenko, E., Green, A.C., *et al.* (2013). Cutaneous alpha, beta and gamma human papillomaviruses in relation to squamous cell carcinoma of the skin: a population-based study. Int. J. Cancer 133, 1713-1720. http://dx.doi.org/10.1002/ijc.28176

Firzlaff, J.M., Galloway, D.A., Eisenman, R.N., and Luscher, B. (1989). The E7 protein of human papillomavirus type 16 is phosphorylated by casein kinase II. The New biologist 1, 44-53.

Firzlaff, J.M., Luscher, B., and Eisenman, R.N. (1991). Negative charge at the casein kinase II phosphorylation site is important for transformation but not for Rb protein binding by the E7 protein of human papillomavirus type 16. Proc Natl Acad Sci USA 88, 5187-5191.

Fischer, M., Quaas, M., Wintsche, A., Muller, G.A., and Engeland, K. (2014). Polo-like kinase 4 transcription is activated via CRE and NRF1 elements, repressed by DREAM through CDE/CHR sites and deregulated by HPV E7 protein. Nucleic acids research 42, 163-180. http://dx.doi.org/10.1093/nar/gkt849

Flaherty, K.T., Puzanov, I., Kim, K.B., Ribas, A., McArthur, G.A., Sosman, J.A., O'Dwyer, P.J., Lee, R.J., Grippo, J.F., Nolop, K., *et al.* (2010). Inhibition of mutated, activated BRAF in metastatic melanoma. New Engl. J. Med. 363, 809-819. http://dx.doi.org/10.1056/NEJMoa1002011

Forslund, O., Iftner, T., Andersson, K., Lindelof, B., Hradil, E., Nordin, P., Stenquist, B., Kirnbauer, R., Dillner, J., and de Villiers, E.M. (2007). Cutaneous human papillomaviruses found in sun-exposed skin: Beta-papillomavirus species 2 predominates in squamous cell carcinoma. J. Infect. Dis. 196, 876-883. http://dx.doi.org/10.1086/521031

Forslund, O., Johansson, H., Madsen, K.G., and Kofoed, K. (2013). The nasal mucosa contains a large spectrum of human papillomavirus types from the Betapapillomavirus and Gammapapillomavirus genera. J. Infect. Dis. 208, 1335-1341. http://dx.doi.org/10.1093/infdis/jit326

Frecha, C., Chevalier, S.A., van Uden, P., Rubio, I., Siouda, M., Saidj, D., Cohen, C., Lomonte, P., Accardi, R., and Tommasino, M. (2015). Expression of the epidermodysplasia verruciformis-associated genes EVER1 and EVER2 is activated by exogenous DNA and inhibited by LMP1 oncoprotein from Epstein-Barr virus. J. Virol. 89, 1461-1467. http://dx.doi.org/10.1128/jvi.02936-14

Funk, J.O., Waga, S., Harry, J.B., Espling, E., Stillman, B., and Galloway, D.A. (1997). Inhibition of CDK activity and PCNA-dependent DNA replication by p21 is blocked by interaction with the HPV-16 E7 oncoprotein. Genes and development 11, 2090-2100.

Gabet, A.S., Accardi, R., Bellopede, A., Popp, S., Boukamp, P., Sylla, B.S., Londono-Vallejo, J.A., and Tommasino, M. (2008). Impairment of the telomere/telomerase system and genomic instability are associated with keratinocyte immortalization induced by the skin human papillomavirus type 38. FASEB journal: official publication of the Federation of American Societies for Experimental Biology 22, 622-632. http://dx.doi.org/10.1096/fj.07-8389com

Ganti, K., Broniarczyk, J., Manoubi, W., Massimi, P., Mittal, S., Pim, D., Szalmas, A., Thatte, J., Thomas, M., Tomaic, V., *et al.* (2015). The Human Papillomavirus E6 PDZ Binding Motif: From Life Cycle to Malignancy. Viruses 7, 3530-3551. http://dx.doi.org/10.3390/v7072785

Gaubatz, S., Wood, J.G., and Livingston, D.M. (1998). Unusual proliferation arrest and transcriptional control properties of a newly discovered E2F family member, E2F-6. Proc Natl Acad Sci USA 95, 9190-9195.

Genovese, N.J., Banerjee, N.S., Broker, T.R., and Chow, L.T. (2008). Casein kinase II motif-dependent phosphorylation of human papillomavirus E7 protein promotes p130 degradation and S-phase induction in differentiated human keratinocytes. J. Virol. 82, 4862-4873. http://dx.doi.org/10.1128/jvi.01202-07

Gewin, L., Myers, H., Kiyono, T., and Galloway, D.A. (2004). Identification of a novel telomerase repressor that interacts with the human papillomavirus type-16

E6/E6-AP complex. Genes and development 18, 2269-2282. http://dx.doi.org/10.1101/gad.1214704

Giampieri, S., and Storey, A. (2004). Repair of UV-induced thymine dimers is compromised in cells expressing the E6 protein from human papillomaviruses types 5 and 18. Brit. J. Cancer 90, 2203-2209. http://dx.doi.org/10.1038/sj.bjc.6601829

Giarre, M., Caldeira, S., Malanchi, I., Ciccolini, F., Leao, M.J., and Tommasino, M. (2001). Induction of pRb degradation by the human papillomavirus type 16 E7 protein is essential to efficiently overcome p16^{INK4a}-imposed G1 cell cycle Arrest. J. Virol. 75, 4705-4712. http://dx.doi.org/10.1128/jvi.75.10.4705-4712.2001

Gonzalez, S.L., Stremlau, M., He, X., Basile, J.R., and Munger, K. (2001). Degradation of the retinoblastoma tumor suppressor by the human papillomavirus type 16 E7 oncoprotein is important for functional inactivation and is separable from proteasomal degradation of E7. J. Virol. 75, 7583-7591. http://dx.doi.org/10.1128/jvi.75.16.7583-7591.2001

Halbert, C.L., Demers, G.W., and Galloway, D.A. (1992). The E6 and E7 genes of human papillomavirus type 6 have weak immortalizing activity in human epithelial cells. J. Virol. 66, 2125-2134.

Halec, G., Holzinger, D., Schmitt, M., Flechtenmacher, C., Dyckhoff, G., Lloveras, B., Hofler, D., Bosch, F.X., and Pawlita, M. (2013). Biological evidence for a causal role of HPV16 in a small fraction of laryngeal squamous cell carcinoma. Brit. J. Cancer 109, 172-183. http://dx.doi.org/10.1038/bjc.2013.296

Hamada, K., Sakaue, M., Alemany, R., Zhang, W.W., Horio, Y., Roth, J.A., and Mitchell, M.F. (1996). Adenovirus-mediated transfer of HPV 16 E6/E7 antisense RNA to human cervical cancer cells. Gynecologic oncology 63, 219-227. http://dx.doi.org/10.1006/gyno.1996.0310

Hasan, U.A., Bates, E., Takeshita, F., Biliato, A., Accardi, R., Bouvard, V., Mansour, M., Vincent, I., Gissmann, L., Iftner, T., *et al.* (2007). TLR9 expression and function is abolished by the cervical cancer-associated human papillomavirus type 16. J. Immunol. (Baltimore, Md: 1950) 178, 3186-3197.

He, W., Staples, D., Smith, C., and Fisher, C. (2003). Direct activation of cyclin-dependent kinase 2 by human papillomavirus E7. J. Virol. 77, 10566-10574.

Heck, D.V., Yee, C.L., Howley, P.M., and Munger, K. (1992). Efficiency of binding the retinoblastoma protein correlates with the transforming capacity of the E7 oncoproteins of the human papillomaviruses. Proc Natl Acad Sci USA 89, 4442-4446.

Helt, A.M., and Galloway, D.A. (2001). Destabilization of the retinoblastoma tumor suppressor by human papillomavirus type 16 E7 is not sufficient to overcome cell cycle arrest in human keratinocytes. J. Virol. 75, 6737-6747. http://dx.doi.org/10.1128/jvi.75.15.6737-6747.2001

Hiller, T., Poppelreuther, S., Stubenrauch, F., and Iftner, T. (2006). Comparative analysis of 19 genital human papillomavirus types with regard to p53 degradation, immortalization, phylogeny, and epidemiologic risk classification. Cancer epidemiology, biomarkers and prevention: a publication of the American Association for Cancer Research, cosponsored by the American Society of Preventive Oncology 15, 1262-1267. http://dx.doi.org/10.1158/1055-9965.epi-05-0778

Holloway, A., Simmonds, M., Azad, A., Fox, J.L., and Storey, A. (2015). Resistance to UV-induced apoptosis by beta-HPV5 E6 involves targeting of activated BAK for proteolysis by recruitment of the HERC1 ubiquitin ligase. Int. J. Cancer 136, 2831-2843. http://dx.doi.org/10.1002/ijc.29350

Howie, H.L., Koop, J.I., Weese, J., Robinson, K., Wipf, G., Kim, L., and Galloway, D.A. (2011). Beta-HPV 5 and 8 E6 promote p300 degradation by blocking AKT/p300 association. PLoS pathogens 7, e1002211. http://dx.doi.org/10.1371/journal.ppat.1002211

Huh, K., Zhou, X., Hayakawa, H., Cho, J.Y., Libermann, T.A., Jin, J., Harper, J.W., and Munger, K. (2007). Human papillomavirus type 16 E7 oncoprotein associates with the cullin 2 ubiquitin ligase complex, which contributes to degradation of the retinoblastoma tumor suppressor. J. Virol. 81, 9737-9747. http://dx.doi.org/10.1128/jvi.00881-07

Huibregtse, J.M., Scheffner, M., and Howley, P.M. (1991). A cellular protein mediates association of p53 with the E6 oncoprotein of human papillomavirus types 16 or 18. The EMBO journal 10, 4129-4135.

Iannacone, M.R., Gheit, T., Pfister, H., Giuliano, A.R., Messina, J.L., Fenske, N.A., Cherpelis, B.S., Sondak, V.K., Roetzheim, R.G., Silling, S., *et al.* (2014). Case-control study of genus-beta human papillomaviruses in plucked eyebrow hairs and cutaneous squamous cell carcinoma. Int. J. Cancer 134, 2231-2244. http://dx.doi.org/10.1002/ijc.28552

Iannacone, M.R., Gheit, T., Waterboer, T., Giuliano, A.R., Messina, J.L., Fenske, N.A., Cherpelis, B.S., Sondak, V.K., Roetzheim, R.G., Michael, K.M., *et al.* (2012). Case-control study of cutaneous human papillomaviruses in squamous cell carcinoma of the skin. Cancer epidemiology, biomarkers and prevention: a publication of the American Association for Cancer Research, cosponsored by the American Society of Preventive Oncology 21, 1303-1313. http://dx.doi.org/10.1158/1055-9965.epi-12-0032

Jabbar, S.F., Abrams, L., Glick, A., and Lambert, P.F. (2009). Persistence of high-grade cervical dysplasia and cervical cancer requires the continuous expression of the human papillomavirus type 16 E7 oncogene. Cancer research 69, 4407-4414. http://dx.doi.org/10.1158/0008-5472.can-09-0023

Jabbar, S.F., Park, S., Schweizer, J., Berard-Bergery, M., Pitot, H.C., Lee, D., and Lambert, P.F. (2012). Cervical cancers require the continuous expression of the human papillomavirus type 16 E7 oncoprotein even in the presence of the viral E6 oncoprotein. Cancer research 72, 4008-4016. http://dx.doi.org/10.1158/0008-5472.can-11-3085

Jackson, S., Harwood, C., Thomas, M., Banks, L., and Storey, A. (2000). Role of Bak in UV-induced apoptosis in skin cancer and abrogation by HPV E6 proteins. Genes and development 14, 3065-3073.

Jones, D.L., Alani, R.M., and Munger, K. (1997). The human papillomavirus E7 oncoprotein can uncouple cellular differentiation and proliferation in human keratinocytes by abrogating p21[Cip1]-mediated inhibition of cdk2. Genes and development 11, 2101-2111.

Jones, D.L., and Munger, K. (1997). Analysis of the p53-mediated G1 growth arrest pathway in cells expressing the human papillomavirus type 16 E7 oncoprotein. J. Virol. 71, 2905-2912.

Karagas, M.R., Waterboer, T., Li, Z., Nelson, H.H., Michael, K.M., Bavinck, J.N., Perry, A.E., Spencer, S.K., Daling, J., Green, A.C., *et al.* (2010). Genus beta human papillomaviruses and incidence of basal cell and squamous cell carcinomas of skin: population based case-control study. BMJ (Clinical research ed) 341, c2986. http://dx.doi.org/10.1136/bmj.c2986

Karatayli-Ozgursoy, S., Bishop, J.A., Hillel, A., Akst, L., and Best, S.R. (2016). Risk Factors for Dysplasia in Recurrent Respiratory Papillomatosis in an Adult and Pediatric Population. The Annals of otology, rhinology, and laryngology 125, 235-241. http://dx.doi.org/10.1177/0003489415608196

Katzenellenbogen, R.A., Egelkrout, E.M., Vliet-Gregg, P., Gewin, L.C., Gafken, P.R., and Galloway, D.A. (2007). NFX1-123 and poly(A) binding proteins synergistically augment activation of telomerase in human papillomavirus type 16 E6-expressing cells. J. Virol. 81, 3786-3796. http://dx.doi.org/10.1128/jvi.02007-06

Kazem, S., van der Meijden, E., Struijk, L., de Gruijl, F.R., and Feltkamp, M.C. (2012). Human papillomavirus 8 E6 disrupts terminal skin differentiation and prevents pro-Caspase-14 cleavage. Virus research 163, 609-616. http://dx.doi.org/10.1016/j.virusres.2011.12.015

Kefford, R., Arkenau, H., Brown, M.P., Millward, M., Infante, J.R., Long, G.V., Ouellet, D., Curtis, M., Lebowitz, P.F., and Falchook, G.S. (2010). Phase I/II study of GSK2118436, a selective inhibitor of oncogenic mutant BRAF kinase, in patients with metastatic melanoma and other solid tumors. J. Clin. Oncol. 28.

Klingelhutz, A.J., Foster, S.A., and McDougall, J.K. (1996). Telomerase activation by the E6 gene product of human papillomavirus type 16. Nature 380, 79-82. http://dx.doi.org/10.1038/380079a0

Kuhne, C., Gardiol, D., Guarnaccia, C., Amenitsch, H., and Banks, L. (2000). Differential regulation of human papillomavirus E6 by protein kinase A: conditional degradation of human discs large protein by oncogenic E6. Oncogene 19, 5884-5891.

Lambert, P.F., Balsitis, S.J., Shai, A., Simonson, S.J.S., and Williams, S.M.G. (2006). Transgenic Mouse Models for the In Vivo Analysis of Papillomavirus Oncogene Function. In Papillomavirus Research: From Natural History To Vaccines and Beyond, M. Saveria Campo, ed. (Division of Pathological Sciences, Institute of Comparative Medicine University of Glasgow, Glasgow G61 1QH, UK: Caister Academic Press), pp. 213-228. ISBN 978-1-904455-04-2

Lammens, T., Li, J., Leone, G., and De Veylder, L. (2009). Atypical E2Fs: new players in the E2F transcription factor family. Trends in cell biology 19, 111-118. http://dx.doi.org/10.1016/j.tcb.2009.01.002

Lazarczyk, M., Cassonnet, P., Pons, C., Jacob, Y., and Favre, M. (2009). The EVER proteins as a natural barrier against papillomaviruses: a new insight into the pathogenesis of human papillomavirus infections. Microbiology and molecular biology reviews: MMBR 73, 348-370. http://dx.doi.org/10.1128/mmbr.00033-08

Lazarczyk, M., Dalard, C., Hayder, M., Dupre, L., Pignolet, B., Majewski, S., Vuillier, F., Favre, M., and Liblau, R.S. (2012). EVER proteins, key elements of the natural anti-human papillomavirus barrier, are regulated upon T-cell activation. PloS one 7, e39995. http://dx.doi.org/10.1371/journal.pone.0039995

Lazic, D., Hufbauer, M., Zigrino, P., Buchholz, S., Kazem, S., Feltkamp, M.C., Mauch, C., Steger, G., Pfister, H., and Akgul, B. (2012). Human papillomavirus type 8 E6 oncoprotein inhibits transcription of the PDZ protein syntenin-2. J. Virol. 86, 7943-7952. http://dx.doi.org/10.1128/jvi.00132-12

Li, N., Franceschi, S., Howell-Jones, R., Snijders, P.J., and Clifford, G.M. (2011). Human papillomavirus type distribution in 30,848 invasive cervical cancers worldwide: Variation by geographical region, histological type and year of publication. Int. J. Cancer 128, 927-935. http://dx.doi.org/10.1002/ijc.25396

Li, S.L., Duo, L.N., Wang, H.J., Dai, W., Zhou, E.H., Xu, Y.N., Zhao, T., Xiao, Y.Y., Xia, L., Yang, Z.H., *et al.* (2016). Identification of LCK mutation in a family with atypical epidermodysplasia verruciformis with T-cell defects and virus-induced squamous cell carcinoma. Brit. J. Dermatol.. http://dx.doi.org/10.1111/bjd.14679

Luck, K., Charbonnier, S., and Trave, G. (2012). The emerging contribution of sequence context to the specificity of protein interactions mediated by PDZ domains. FEBS letters 586, 2648-2661. http://dx.doi.org/10.1016/j.febslet.2012.03.056

MacLaine, N.J., and Hupp, T.R. (2011). How phosphorylation controls p53. Cell cycle (Georgetown, Tex) 10, 916-921. http://dx.doi.org/10.4161/cc.10.6.15076

Marcuzzi, G.P., Hufbauer, M., Kasper, H.U., Weissenborn, S.J., Smola, S., and Pfister, H. (2009). Spontaneous tumour development in human papillomavirus type 8 E6 transgenic mice and rapid induction by UV-light exposure and wounding. J. Gen. Virol. 90, 2855-2864. http://dx.doi.org/10.1099/vir.0.012872-0

Marouco, D., Garabadgiu, A.V., Melino, G., and Barlev, N.A. (2013). Lysine-specific modifications of p53: a matter of life and death? Oncotarget 4, 1556-1571. http://dx.doi.org/10.18632/oncotarget.1436

Martinez-Zapien, D., Ruiz, F.X., Poirson, J., Mitschler, A., Ramirez, J., Forster, A., Cousido-Siah, A., Masson, M., Vande Pol, S., Podjarny, A., *et al.* (2016). Structure of the E6/E6AP/p53 complex required for HPV-mediated degradation of p53. Nature 529, 541-545. http://dx.doi.org/10.1038/nature16481

Marur, S., D'Souza, G., Westra, W.H., and Forastiere, A.A. (2010). HPV-associated head and neck cancer: a virus-related cancer epidemic. Lancet Oncol. 11, 781-789. http://dx.doi.org/10.1016/s1470-2045(10)70017-6

Mavromatis, K.O., Jones, D.L., Mukherjee, R., Yee, C., Grace, M., and Munger, K. (1997). The carboxyl-terminal zinc-binding domain of the human papillomavirus E7 protein can be functionally replaced

by the homologous sequences of the E6 protein. Virus research 52, 109-118.
McIntyre, M.C., Frattini, M.G., Grossman, S.R., and Laimins, L.A. (1993). Human papillomavirus type 18 E7 protein requires intact Cys-X-X-Cys motifs for zinc binding, dimerization, and transformation but not for Rb binding. J. Virol. 67, 3142-3150.
McIntyre, M.C., Ruesch, M.N., and Laimins, L.A. (1996). Human papillomavirus E7 oncoproteins bind a single form of cyclin E in a complex with cdk2 and p107. Virology 215, 73-82. http://dx.doi.org/10.1006/viro.1996.0008
McLaughlin-Drubin, M.E., Huh, K.W., and Munger, K. (2008). Human papillomavirus type 16 E7 oncoprotein associates with E2F6. J. Virol. 82, 8695-8705. http://dx.doi.org/10.1128/jvi.00579-08
Muench, P., Probst, S., Schuetz, J., Leiprecht, N., Busch, M., Wesselborg, S., Stubenrauch, F., and Iftner, T. (2010). Cutaneous papillomavirus E6 proteins must interact with p300 and block p53-mediated apoptosis for cellular immortalization and tumorigenesis. Cancer research 70, 6913-6924. http://dx.doi.org/10.1158/0008-5472.can-10-1307
Muller, G.A., and Engeland, K. (2010). The central role of CDE/CHR promoter elements in the regulation of cell cycle-dependent gene transcription. The FEBS journal 277, 877-893. http://dx.doi.org/10.1111/j.1742-4658.2009.07508.x
Muller, G.A., Quaas, M., Schumann, M., Krause, E., Padi, M., Fischer, M., Litovchick, L., DeCaprio, J.A., and Engeland, K. (2012). The CHR promoter element controls cell cycle-dependent gene transcription and binds the DREAM and MMB complexes. Nucleic acids research 40, 1561-1578. http://dx.doi.org/10.1093/nar/gkr793
Munger, K., Werness, B.A., Dyson, N., Phelps, W.C., Harlow, E., and Howley, P.M. (1989). Complex formation of human papillomavirus E7 proteins with the retinoblastoma tumor suppressor gene product. The EMBO journal 8, 4099-4105.
Munoz, N., Bosch, F.X., de Sanjose, S., Herrero, R., Castellsague, X., Shah, K.V., Snijders, P.J., and Meijer, C.J. (2003). Epidemiologic classification of human papillomavirus types associated with cervical cancer. New Engl. J. Med. 348, 518-527. http://dx.doi.org/10.1056/NEJMoa021641
Muschik, D., Braspenning-Wesch, I., Stockfleth, E., Rosl, F., Hofmann, T.G., and Nindl, I. (2011). Cutaneous HPV23 E6 prevents p53 phosphorylation through interaction with HIPK2. PloS one 6, e27655. http://dx.doi.org/10.1371/journal.pone.0027655
Ndiaye, C., Mena, M., Alemany, L., Arbyn, M., Castellsague, X., Laporte, L., Bosch, F.X., de Sanjose, S., and Trottier, H. (2014). HPV DNA, E6/E7 mRNA, and $p16^{INK4a}$ detection in head and neck cancers: a systematic review and meta-analysis. Lancet Oncol. 15, 1319-1331. http://dx.doi.org/10.1016/s1470-2045(14)70471-1
Nguyen, C.L., and Munger, K. (2008). Direct association of the HPV16 E7 oncoprotein with cyclin A/CDK2 and cyclin E/CDK2 complexes. Virology 380, 21-25. http://dx.doi.org/10.1016/j.virol.2008.07.017
Nguyen, M.L., Nguyen, M.M., Lee, D., Griep, A.E., and Lambert, P.F. (2003). The PDZ ligand domain of the human papillomavirus type 16 E6 protein is required for E6's induction of epithelial hyperplasia *in vivo*. J. Virol. 77, 6957-6964.
Nobre, R.J., Herraez-Hernandez, E., Fei, J.W., Langbein, L., Kaden, S., Grone, H.J., and de Villiers, E.M. (2009). E7 oncoprotein of novel human papillomavirus type 108 lacking the E6 gene induces dysplasia in organotypic keratinocyte cultures. J. Virol. 83, 2907-2916. http://dx.doi.org/10.1128/jvi.02490-08
Nomine, Y., Masson, M., Charbonnier, S., Zanier, K., Ristriani, T., Deryckere, F., Sibler, A.P., Desplancq, D., Atkinson, R.A., Weiss, E., *et al.* (2006). Structural and functional analysis of E6 oncoprotein: insights in the molecular pathways of human papillomavirus-mediated pathogenesis. Molecular cell 21, 665-678. http://dx.doi.org/10.1016/j.molcel.2006.01.024
Nor Rashid, N., Yusof, R., and Watson, R.J. (2011). Disruption of repressive p130-DREAM complexes by human papillomavirus 16 E6/E7 oncoproteins is required for cell-cycle progression in cervical cancer cells. J. Gen. Virol. 92, 2620-2627. http://dx.doi.org/10.1099/vir.0.035352-0
Patel, D., Huang, S.M., Baglia, L.A., and McCance, D.J. (1999). The E6 protein of human papillomavirus type 16 binds to and inhibits co-activation by CBP and p300. The EMBO journal 18, 5061-5072. http://dx.doi.org/10.1093/emboj/18.18.5061
Phelps, W.C., Yee, C.L., Munger, K., and Howley, P.M. (1988). The human papillomavirus type 16 E7 gene encodes transactivation and transformation functions similar to those of adenovirus *E1A*. Cell 53, 539-547.
Pierce Campbell, C.M., Messina, J.L., Stoler, M.H., Jukic, D.M., Tommasino, M., Gheit, T., Rollison, D.E., Sichero, L., Sirak, B.A., Ingles, D.J., *et al.* (2013). Cutaneous human papillomavirus types detected on the surface of male external genital lesions: a case series within the HPV Infection in Men Study. Journal of clinical virology: the official publication of the Pan American Society for Clinical Virology 58, 652-659. http://dx.doi.org/10.1016/j.jcv.2013.10.011
Pirisi, L., Yasumoto, S., Feller, M., Doniger, J., and DiPaolo, J.A. (1987). Transformation of human fibroblasts and keratinocytes with human papillomavirus type 16 DNA. J. Virol. 61, 1061-1066.
Plummer, M., de Martel, C., Vignat, J., Ferlay, J., Bray, F., and Franceschi, S. (2016). Global burden of cancers attributable to infections in 2012: a synthetic analysis. The Lancet Global health 4, e609-616. http://dx.doi.org/10.1016/s2214-109x(16)30143-7
Roman, A., and Munger, K. (2013). The papillomavirus E7 proteins. Virology 445, 138-168. http://dx.doi.org/10.1016/j.virol.2013.04.013
Sadasivam, S., and DeCaprio, J.A. (2013). The DREAM complex: master coordinator of cell cycle-dependent gene expression. Nature reviews Cancer 13, 585-595. http://dx.doi.org/10.1038/nrc3556
Sang, B.C., and Barbosa, M.S. (1992). Single amino acid substitutions in "low-risk" human papillomavirus (HPV) type 6 E7 protein enhance features characteristic

of the "high-risk" HPV E7 oncoproteins. Proc Natl Acad Sci USA 89, 8063-8067.

Scheffner, M., Huibregtse, J.M., Vierstra, R.D., and Howley, P.M. (1993). The HPV-16 E6 and E6-AP complex functions as a ubiquitin-protein ligase in the ubiquitination of p53. Cell 75, 495-505.

Scheffner, M., Werness, B.A., Huibregtse, J.M., Levine, A.J., and Howley, P.M. (1990). The E6 oncoprotein encoded by human papillomavirus types 16 and 18 promotes the degradation of p53. Cell 63, 1129-1136.

Schiffman, M., Herrero, R., Desalle, R., Hildesheim, A., Wacholder, S., Rodriguez, A.C., Bratti, M.C., Sherman, M.E., Morales, J., Guillen, D., *et al.* (2005). The carcinogenicity of human papillomavirus types reflects viral evolution. Virology 337, 76-84. http://dx.doi.org/10.1016/j.virol.2005.04.002

Schmitt, A., Harry, J.B., Rapp, B., Wettstein, F.O., and Iftner, T. (1994). Comparison of the properties of the E6 and E7 genes of low- and high-risk cutaneous papillomaviruses reveals strongly transforming and high Rb-binding activity for the E7 protein of the low-risk human papillomavirus type 1. J. Virol. 68, 7051-7059.

Schwarz, E., Freese, U.K., Gissmann, L., Mayer, W., Roggenbuck, B., Stremlau, A., and zur Hausen, H. (1985). Structure and transcription of human papillomavirus sequences in cervical carcinoma cells. Nature 314, 111-114.

Sedman, S.A., Barbosa, M.S., Vass, W.C., Hubbert, N.L., Haas, J.A., Lowy, D.R., and Schiller, J.T. (1991). The full-length E6 protein of human papillomavirus type 16 has transforming and trans-activating activities and cooperates with E7 to immortalize keratinocytes in culture. J. Virol. 65, 4860-4866.

Sekaric, P., Cherry, J.J., and Androphy, E.J. (2008). Binding of human papillomavirus type 16 E6 to E6AP is not required for activation of *hTERT*. J. Virol. 82, 71-76. http://dx.doi.org/10.1128/jvi.01776-07

Shai, A., Brake, T., Somoza, C., and Lambert, P.F. (2007). The human papillomavirus E6 oncogene dysregulates the cell cycle and contributes to cervical carcinogenesis through two independent activities. Cancer research 67, 1626-1635. http://dx.doi.org/10.1158/0008-5472.can-06-3344

Shamir, E.R., and Ewald, A.J. (2014). Three-dimensional organotypic culture: experimental models of mammalian biology and disease. Nat. Rev. Mol. Cell Biol. 15, 647-664. http://dx.doi.org/10.1038/nrm3873

Thomas, M., Massimi, P., Navarro, C., Borg, J.P., and Banks, L. (2005). The hScrib/Dlg apico-basal control complex is differentially targeted by HPV-16 and HPV-18 E6 proteins. Oncogene 24, 6222-6230. http://dx.doi.org/10.1038/sj.onc.1208757

Thomas, M., Myers, M.P., Massimi, P., Guarnaccia, C., and Banks, L. (2016). Analysis of Multiple HPV E6 PDZ Interactions Defines Type-Specific PDZ Fingerprints That Predict Oncogenic Potential. PLoS pathogens 12, e1005766. http://dx.doi.org/10.1371/journal.ppat.1005766

Todorovic, B., Hung, K., Massimi, P., Avvakumov, N., Dick, F.A., Shaw, G.S., Banks, L., and Mymryk, J.S. (2012). Conserved region 3 of human papillomavirus 16 E7 contributes to deregulation of the retinoblastoma tumor suppressor. J. Virol. 86, 13313-13323. http://dx.doi.org/10.1128/jvi.01637-12

Toll, A., Lloveras, B., Masferrer, E., Ferrandiz-Pulido, C., Garcia-Patos, V., Gheit, T., Pujol, R.M., and Tommasino, M. (2014). Human beta papillomavirus DNA study in primary cutaneous squamous cell carcinomas and their corresponding metastases. Archives of dermatological research 306, 93-95. http://dx.doi.org/10.1007/s00403-013-1424-8

Tomaic, V., Gardiol, D., Massimi, P., Ozbun, M., Myers, M., and Banks, L. (2009). Human and primate tumour viruses use PDZ binding as an evolutionarily conserved mechanism of targeting cell polarity regulators. Oncogene 28, 1-8. http://dx.doi.org/10.1038/onc.2008.365

Tommasino, M. (2014). The human papillomavirus family and its role in carcinogenesis. Seminars in Cancer Biology 26, 13-21. http://dx.doi.org/10.1016/j.semcancer.2013.11.002

Tommasino, M., Adamczewski, J.P., Carlotti, F., Barth, C.F., Manetti, R., Contorni, M., Cavalieri, F., Hunt, T., and Crawford, L. (1993). HPV16 E7 protein associates with the protein kinase p33CDK2 and cyclin A. Oncogene 8, 195-202.

Torres, M., Gheit, T., McKay-Chopin, S., Rodriguez, C., Romero, J.D., Filotico, R., Dona, M.G., Ortiz, M., and Tommasino, M. (2015). Prevalence of beta and gamma human papillomaviruses in the anal canal of men who have sex with men is influenced by HIV status. Journal of clinical virology: the official publication of the Pan American Society for Clinical Virology 67, 47-51. http://dx.doi.org/10.1016/j.jcv.2015.04.005

Trave, G., and Zanier, K. (2016). HPV-mediated inactivation of tumor suppressor p53. Cell cycle (Georgetown, Tex) 15, 2231-2232. http://dx.doi.org/10.1080/15384101.2016.1191257

Trimarchi, J.M., Fairchild, B., Verona, R., Moberg, K., Andon, N., and Lees, J.A. (1998). E2F-6, a member of the E2F family that can behave as a transcriptional repressor. Proc Natl Acad Sci USA 95, 2850-2855.

Underbrink, M.P., Howie, H.L., Bedard, K.M., Koop, J.I., and Galloway, D.A. (2008). E6 proteins from multiple human betapapillomavirus types degrade Bak and protect keratinocytes from apoptosis after UVB irradiation. J. Virol. 82, 10408-10417. http://dx.doi.org/10.1128/jvi.00902-08

Van Doorslaer, K., DeSalle, R., Einstein, M.H., and Burk, R.D. (2015). Degradation of Human PDZ-Proteins by Human Alphapapillomaviruses Represents an Evolutionary Adaptation to a Novel Cellular Niche. PLoS pathogens 11, e1004980. http://dx.doi.org/10.1371/journal.ppat.1004980

Van Doorslaer, K., Tan, Q., Xirasagar, S., Bandaru, S., Gopalan, V., Mohamoud, Y., Huyen, Y., and McBride, A.A. (2013). The Papillomavirus Episteme: a central resource for papillomavirus sequence data and analysis. Nucleic acids research 41, D571-578. http://dx.doi.org/10.1093/nar/gks984

Vande Pol, S.B., and Klingelhutz, A.J. (2013). Papillomavirus E6 oncoproteins. Virology 445, 115-137. http://dx.doi.org/10.1016/j.virol.2013.04.026

Vasiljevic, N., Hazard, K., Dillner, J., and Forslund, O. (2008). Four novel human betapapillomaviruses of species 2 preferentially found in actinic keratosis. J. Gen. Virol. 89, 2467-2474. http://dx.doi.org/10.1099/vir.0.2008/001925-0

Veldman, T., Liu, X., Yuan, H., and Schlegel, R. (2003). Human papillomavirus E6 and Myc proteins associate *in vivo* and bind to and cooperatively activate the telomerase reverse transcriptase promoter. Proc Natl Acad Sci USA 100, 8211-8216. http://dx.doi.org/10.1073/pnas.1435900100

Viarisio, D., Mueller-Decker, K., Kloz, U., Aengeneyndt, B., Kopp-Schneider, A., Grone, H.J., Gheit, T., Flechtenmacher, C., Gissmann, L., and Tommasino, M. (2011). E6 and E7 from beta HPV38 cooperate with ultraviolet light in the development of actinic keratosis-like lesions and squamous cell carcinoma in mice. PLoS pathogens 7, e1002125. http://dx.doi.org/10.1371/journal.ppat.1002125

Viarisio, D., Muller-Decker, K., Zanna, P., Kloz, U., Aengeneyndt, B., Accardi, R., Flechtenmacher, C., Gissmann, L., and Tommasino, M. (2016). Novel ss-HPV49 Transgenic Mouse Model of Upper Digestive Tract Cancer. Cancer research. http://dx.doi.org/10.1158/0008-5472.can-16-0370

Wallace, N.A., Robinson, K., and Galloway, D.A. (2014). Beta human papillomavirus E6 expression inhibits stabilization of p53 and increases tolerance of genomic instability. J. Virol. 88, 6112-6127. http://dx.doi.org/10.1128/jvi.03808-13

Wallace, N.A., Robinson, K., Howie, H.L., and Galloway, D.A. (2012). HPV 5 and 8 E6 abrogate ATR activity resulting in increased persistence of UVB induced DNA damage. PLoS pathogens 8, e1002807. http://dx.doi.org/10.1371/journal.ppat.1002807

Wallace, N.A., Robinson, K., Howie, H.L., and Galloway, D.A. (2015). beta-HPV 5 and 8 E6 disrupt homology dependent double strand break repair by attenuating BRCA1 and BRCA2 expression and foci formation. PLoS pathogens 11, e1004687. http://dx.doi.org/10.1371/journal.ppat.1004687

Waterboer, T., Abeni, D., Sampogna, F., Rother, A., Masini, C., Sehr, P., Michael, K.M., and Pawlita, M. (2008). Serological association of beta and gamma human papillomaviruses with squamous cell carcinoma of the skin. Brit. J. Dermatol. 159, 457-459. http://dx.doi.org/10.1111/j.1365-2133.2008.08621.x

Watson, R.A., Thomas, M., Banks, L., and Roberts, S. (2003). Activity of the human papillomavirus E6 PDZ-binding motif correlates with an enhanced morphological transformation of immortalized human keratinocytes. Journal of cell science 116, 4925-4934. http://dx.doi.org/10.1242/jcs.00809

Weissenborn, S.J., Nindl, I., Purdie, K., Harwood, C., Proby, C., Breuer, J., Majewski, S., Pfister, H., and Wieland, U. (2005). Human papillomavirus-DNA loads in actinic keratoses exceed those in non-melanoma skin cancers. J. Invest. Dermatol. 125, 93-97. http://dx.doi.org/10.1111/j.0022-202X.2005.23733.x

White, E.A., Sowa, M.E., Tan, M.J., Jeudy, S., Hayes, S.D., Santha, S., Munger, K., Harper, J.W., and Howley, P.M. (2012). Systematic identification of interactions between host cell proteins and E7 oncoproteins from diverse human papillomaviruses. Proc Natl Acad Sci USA 109, E260-267. http://dx.doi.org/10.1073/pnas.1116776109

Xu, M., Katzenellenbogen, R.A., Grandori, C., and Galloway, D.A. (2010). NFX1 plays a role in human papillomavirus type 16 E6 activation of NFkappaB activity. J. Virol. 84, 11461-11469. http://dx.doi.org/10.1128/jvi.00538-10

Yamashita, T., Segawa, K., Fujinaga, Y., Nishikawa, T., and Fujinaga, K. (1993). Biological and biochemical activity of E7 genes of the cutaneous human papillomavirus type 5 and 8. Oncogene 8, 2433-2441.

Yuan, H., Zhou, D., Wang, J., and Schlegel, R. (2013). Divergent human papillomavirus associated with recurrent respiratory papillomatosis with lung involvement. Genome announcements 1. http://dx.doi.org/10.1128/genomeA.00474-13

Zanier, K., Charbonnier, S., Sidi, A.O., McEwen, A.G., Ferrario, M.G., Poussin-Courmontagne, P., Cura, V., Brimer, N., Babah, K.O., Ansari, T., *et al.* (2013). Structural basis for hijacking of cellular LxxLL motifs by papillomavirus E6 oncoproteins. Science (New York, NY) 339, 694-698. http://dx.doi.org/10.1126/science.1229934

Zanier, K., ould M'hamed ould Sidi, A., Boulade-Ladame, C., Rybin, V., Chappelle, A., Atkinson, A., Kieffer, B., and Trave, G. (2012). Solution structure analysis of the HPV16 E6 oncoprotein reveals a self-association mechanism required for E6-mediated degradation of p53. Structure (London, England: 1993) 20, 604-617. http://dx.doi.org/10.1016/j.str.2012.02.001

Zerfass-Thome, K., Zwerschke, W., Mannhardt, B., Tindle, R., Botz, J.W., and Jansen-Durr, P. (1996). Inactivation of the cdk inhibitor $p27^{KIP1}$ by the human papillomavirus type 16 E7 oncoprotein. Oncogene 13, 2323-2330.

Zimmermann, H., Degenkolbe, R., Bernard, H.U., and O'Connor, M.J. (1999). The human papillomavirus type 16 E6 oncoprotein can down-regulate p53 activity by targeting the transcriptional coactivator CBP/p300. J. Virol. 73, 6209-6219.

Zwerschke, W., Joswig, S., and Jansen-Durr, P. (1996). Identification of domains required for transcriptional activation and protein dimerization in the human papillomavirus type-16 E7 protein. Oncogene 12, 213-220.

4

Hepatitis B Virus: Molecular Biology, Carcinogenic Mechanisms, and Therapies

Nicholas Duchemin[1], Sumedha Bagga[2] and Michael J. Bouchard[2*]

[1]Graduate School of Biomedical Sciences and Professional Studies, Drexel University College of Medicine, Philadelphia, PA, USA
[2]Department of Biochemistry and Molecular Biology, Drexel University College of Medicine, Philadelphia, PA, USA

**Corresponding author: mjb93@drexel.edu*

Abstract

A chronic hepatitis B virus (HBV) infection can cause liver cancer, specifically hepatocellular carcinoma (HCC). A variety of mechanisms that can lead from HBV infection to development of HCC have been proposed and could include prolonged expression of HBV proteins that alter cellular signaling pathways, recurrent anti-HBV inflammatory responses associated with death of HBV-infected hepatocytes and concomitant liver regeneration, and integration of the HBV genome into the host cell chromosome, which could alter expression of cellular genes. Because it typically requires decades for a chronic HBV infection to cause liver cancer, it has been difficult to identify a single HBV-related effect that is carcinogenic, and it is likely that subtle HBV-associated effects over a prolonged period of time cause HCC. Anti-HBV therapies, such as interferon or nucleoside analogs can successfully control HBV replication and delay development of diseases associated with an HBV infection; however, these treatments are typically not curative. In this chapter, mechanisms that have been proposed to contribute to the progression and development of HBV-associated HCC, as well as established and emerging anti-HBV therapeutic strategies, will be discussed.

Introduction and Overview

The hepatitis B virus (HBV) is a small enveloped DNA virus that primarily infects hepatocytes of the human liver. An HBV infection can have various consequences for the liver, including the development of liver cancer following years of a chronic infection. The development of HBV-associated liver cancer, specifically hepatocellular carcinoma (HCC), is a complex molecular process that has not been narrowed to one precise causative event, and various putative mechanisms have been suggested to influence the development and progression of HBV-associated HCC. Three potential mechanisms that can lead from HBV infection to development of HBV-associated HCC have been frequently proposed: 1) prolonged expression of HBV proteins that result in alterations in cellular signaling pathways and that stimulate changes within hepatocytes to promote a more oncogenic environment, 2) chronic immune-mediated inflammation and concomitant liver regeneration that can lead to fibrosis and cirrhosis and ultimately progress to HCC, and 3) the integration of the HBV genome into the host cell chromosome, which could alter expression of endogenous cellular genes that could promote oncogenesis. Due to the typical requirement for a decades-long chronic HBV infection prior to development of HCC, it remains difficult to identify a single HBV-related effect that is carcinogenic, and it is likely that subtle HBV-associated effects over a prolonged period of time eventually influence and lead to HCC development. Moreover, it is also possible that combinations of the three potential mechanisms described above play contributing roles in promoting the development of HBV-associated HCC. Antiviral HBV therapies, such as interferon or nucleoside analogs, have shown successful control of viral replication, causing a delay in the development of diseases associated with an HBV infection; however, these treatments are typically not curative due to the persistent presence of HBV covalently closed circular DNA (cccDNA) replicative intermediates that resides within the nuclei of HBV-infected hepatocytes. Mechanisms that have been proposed to contribute to the progression and development of HBV-associated HCC, as well as established and emerging anti-HBV therapeutic strategies, will be the focus of this chapter.

Global epidemiology of HBV infection

HCC, which constitutes ~85-90% of primary liver cancers, is the fifth most common cancer worldwide and has the second highest mortality rate of any cancer (El-Serag and Rudolph, 2007; El-Serag, 2012; Jeang, 2014; Sunbul, 2014; Trepo, 2014). Globally, major risk factors for HCC vary by region. For example, China, Africa, and South East Asia are high-risk regions for HBV-associated HCC; whereas, HBV-associated HCC is less common in the United States and Canada, where chronic hepatitis C virus (HCV) infection, alcohol consumption, and obesity are more commonly linked to the development of HCC (El-Serag and Rudolph, 2007; El-Serag, 2012). Other risk factors for developing HCC include toxin exposure, diabetes, non-alcoholic fatty liver disease, and tobacco use; combinations of multiple risk factors, such as combined alcohol use in the context of a chronic HBV infection, can dramatically increase the risk for HCC development (El-Serag and Rudolph, 2007; El-Serag, 2012). Many liver carcinogenic factors can cause persistent liver damage that is associated with hepatitis, fibrosis, cirrhosis, and continuous immune-mediated destruction of hepatocytes, the major cell type in the liver and direct target of an HBV or HCV infection (El-Serag and Rudolph, 2007; El-Serag, 2012; Jeang, 2014; Sunbul, 2014).

Although many factors have been linked to the development of HCC, globally, a chronic HBV infection remains the leading cause of HCC, and despite the existence of an effective HBV vaccine (see accompanying chapter on HBV and HPV vaccines), there are an estimated 2 billion individuals worldwide who show serological evidence of HBV infection (El-Serag, 2012; Sunbul, 2014; WHO, 2016). The most recent estimate from the World Health Organization (WHO) is that there are over 240 million individuals chronically infected by HBV worldwide (WHO, 2017). Areas with high incidence of HBV infection, such as Asia and Africa, have an estimated 5% - 10% of the total adult population displaying evidence of a chronic HBV infections (WHO, 2016). Moreover, an estimated 600,000 people die every year due to complications resulting from HBV infection; these complications include hepatitis, fulminant hepatitis, liver fibrosis, liver cirrhosis, and primary liver cancer (HCC) (Sunbul, 2014; WHO, 2016). Epidemiological studies have identified a direct correlation between areas of the world with a high incidence of chronic HBV infection and areas of the world with a high incidence of HCC. For example, only about 10% of HCC in Western Europe and North America, which have a low incidence of chronic HBV infections, can be attributed to HBV infection; while roughly 70% of HCC cases in China, Africa, and South East Asia, which have high incidences of chronic HBV-infection, are attributed to HBV infection (El-Serag, 2012; Sunbul, 2014; WHO, 2016). A global study of mortality rates over the past 2 to 3 decades demonstrated that chronic HBV infection was linked to over half of liver cancer-associated mortalities in 2010 alone (Lozano, 2012). Furthermore, this study also demonstrated that over the time period analyzed, the global mortality rate for liver cancer has increased by over 50% (Lozano, 2012). Many of these cases of liver cancer were directly attributable to HBV infection, although infection with HCV and a rise in obesity-associated diseases, such as non-alcoholic fatty liver disease, were also contributors to the increased global incidence of liver cancer observed during this time period (El-Serag and Rudolph, 2007; El-Serag, 2012).

The human hepatitis B virus (HBV)

Hepadnaviruses

Inflammation of the liver, or hepatitis, is most commonly caused by a viral infection. There are five main hepatitis viruses, referred to as the hepatitis A, B, C, D, and E viruses (WHO, 2008). Each of these viruses belongs to a distinct viral family, with the only shared characteristic being that they all infect hepatocytes in the liver. HBV is a member of the *Hepadnaviridae* family; this viral family is comprised of two distinct genera: the orthohepadnaviruses, which infect mammals, and the avihepadnaviruses, which infect birds. While there is significant DNA sequence similarity within each genera, there is almost no homology between the two genera, and membership in the family is instead related to common characteristics such as virion structure, genome organization, the narrow host range of viruses within this family, and the mode of genome replication. The orthohepadnaviruses exhibit a maximum of 40% sequence divergence among members of this genera, compared to about 20% among the avihepadnaviruses (Schaefer, 2007; Seeger *et al.*, 2007; Seeger, 2013). The orthohepadnaviruses include the human hepatitis B virus (HBV), the Woodchuck Hepatitis Virus (WHV), the Ground Squirrel Hepatitis Virus (GSHV), and the Woolly Monkey Hepatitis B Virus (WMHV), while the avihepadnaviruses include the Duck Hepatitis B Virus (DHBV) and the Heron Hepatitis B Virus (HHBV), to name a few (Schaefer, 2007; Seeger *et al.*, 2007; Seeger, 2013). To date, only chronic infections with orthohepadnaviruses have been linked to the development of HCC in their respective hosts.

Four HBV serotypes, adw, ayw, ayr, and adr, and eight HBV genotypes, A to H, have been identified. Classification of HBV into the four serotypes is based on the antigenic determinants of a shared region of the HBV envelope protein (HBsAg), designated region "A"; this "A" region of HBsAg is the basis for the cross-serotype protection afforded by the current HBV vaccine (Carman, 1997; Kay and Zoulim, 2007; Mina *et al.*, 2015). HBV classification into eight distinct genotypes is based on an intergenotypic divergence of 8% or more in the complete nucleotide sequence (Bancroft *et al.*, 1972; Schaefer, 2007; Locarnini *et al.*, 2013; Seeger *et al.*, 2013; Pourkarim *et al.*, 2014). Recently, two new genotypes, I and J, have also been described. The different genotypes of HBV exhibit distinct geographical distribution patterns and diverse clinical manifestations (Kurbanov *et al.*, 2008; Tran *et al.*, 2008; Tatematsu *et al.*, 2009; Seeger *et al.*, 2013).

The HBV life cycle

The blood of an HBV-infected individual includes both infectious HBV virions that contain the viral genome and subviral HBV particles that are composed only of viral envelope proteins; subviral particles are present at significantly higher levels than infectious virions (Seeger, 2013). The infectious HBV virion is also referred to as the Dane particle in recognition of D. S. Dane who first visualized and described the infectious virion (Dane, 1970). The infectious virion contains a partially double-stranded DNA genome that is encapsidated by the single viral core protein and surrounded by an envelope composed of the three viral envelope proteins. The largest of these envelope proteins contains an amino acid sequence that interacts with a cell-surface receptor present on hepatocytes (Yan *et al.*, 2014). The interaction between the HBV envelope protein and the receptor on the surface of hepatocytes is thought to be the major determinant of the species-specificity and narrow host-range of HBV;

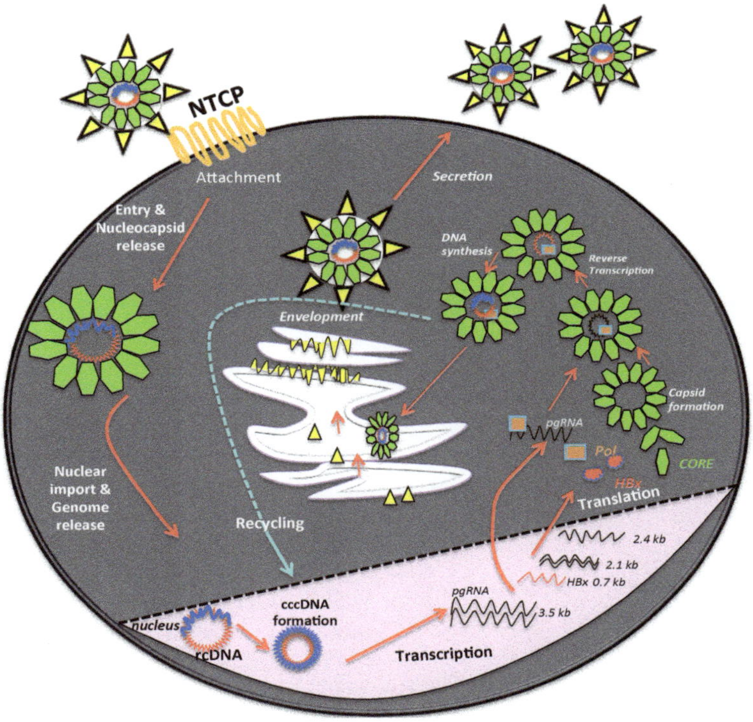

Figure 4.1. Schematic depiction of the HBV life cycle. See text for detailed description.

however, there may be additional factors that affect the host range and cell-type specificity of an HBV infection (Yan *et al.*, 2014). For example, hepatocyte-specific transcription factors that control the transcription of HBV RNAs may also contribute to the tropism of HBV (Block *et al.*, 2007; Bar-Yishay *et al.*, 2011; Seeger, 2013; Seeger and Mason, 2015).

The HBV life cycle is shown in Figure 4.1 and briefly described here. The infectious HBV virion binds to a cell-surface receptor on a hepatocyte and infects the hepatocyte. Recently, the human hepatocyte sodium taurocholate co-transporting polypeptide, hNTCP, was identified as a cell-surface entry factor for HBV (Yan *et al.*, 2012; Ni *et al.*, 2014). After binding to hNTCP, the HBV virion enters the hepatocyte, and the encapsidated genome is transported to the nucleus. The specific processes that facilitate cell entry of HBV and transport of its genome to the nucleus remain incompletely understood but are thought to involve clathrin-dependent endocytic mechanisms (Cooper and Shaul, 2006; Huang *et al.*, 2012). Once in the nucleus, the partially double-stranded HBV genome is repaired and completed by cellular and viral factors and is eventually converted into covalently closed circular DNA (cccDNA) (Seeger and Mason, 2000b; Glebe and Bremer, 2013; Seeger, 2013). The HBV cccDNA is the template for all HBV RNA transcripts. The viral transcripts are transported to the cytoplasm and translated into HBV proteins, and one of these transcripts, designated the pre-genomic RNA (pgRNA), is also encapsidated by viral core proteins and is the template for nascent HBV genomes. Genome replication occurs via reverse transcription of the pgRNA within these newly formed HBV core particles (Seeger and Mason, 2000b; Glebe and Bremer, 2013; Seeger, 2013). The cccDNA itself is not replicated, and its levels are instead maintained and controlled by cycling of DNA-containing core particles from the cytoplasm into the nucleus (Seeger *et al.*, 2007; Seeger, 2013). Consequently, cytoplasmic HBV core particles containing the replicating HBV genome have two fates: either to cycle back to the nucleus to maintain a pool of HBV cccDNA or to enter the endoplasmic reticulum (ER) where HBV envelope proteins are acquired, and new HBV virions are secreted from the cell (Seeger and Mason, 2000b; Glebe and Bremer, 2013; Seeger, 2013).

The sodium taurocholate co-transporting peptide: an HBV cell-entry factor

Although HBV was discovered over 50 years ago, it was only recently that the human sodium taurocholate co-transporting polypeptide (hNTCP) was identified as a specific entry factor for HBV infection of human hepatocytes (Yan *et al.*, 2012; Ni *et al.*, 2014). NTCP represents the founding member of the SLC10A family of membrane transporters. NTCP expression is tightly controlled, and NTCP is predominantly expressed on the basolateral surface of hepatocytes in the liver (Hagenbuch, 1997; Claro da Silva *et al.*, 2013; Anwer and Stieger, 2014). When expressed on the surface of hepatocytes, NTCP facilitates a major portion of the sodium-dependent transport of bile salts from the bloodstream into hepatocytes (Hagenbuch, 1997; Claro da Silva *et al.*, 2013; Anwer and Stieger, 2014). Membrane-expressed NTCP molecules have been reported to form multimeric structures; however, NTCP monomers have been reported to be sufficient for substrate transport (Bijsmans *et al.*, 2012; Anwer and Stieger, 2014). The C-terminus of NTCP has been shown to be important for mediating signals that influence the level of surface-expressed NTCP, and protein kinase C (PKC)-mediated phosphorylation of the C-terminus of NTCP can influence receptor internalization (Stross *et al.*, 2010; Stross *et al.*, 2013). Once internalized, the fate of NTCP has been tracked first to the early and then ultimately to the late endosomal compartments; NTCP can subsequently be degraded in lysosomes (Muhlfeld *et al.*, 2012; Stross *et al.*, 2013).

Consistent with the expression profile of hNTCP, and prior to the identification of hNTCP as an HBV entry factor, the HBV infection process was shown to be dependent on a protein present on the basolateral surface of hepatocytes (Meier *et al.*, 2013). Moreover, previous studies had also shown that susceptibility of human hepatocytes to an HBV infection rapidly diminished when primary hepatocytes were isolated and cultured, and general NTCP expression also rapidly decreases when primary hepatocytes are cultured (Yan *et al.*, 2012). Together, these early observations and more recent studies with hNTCP-expressing cells provide compelling evidence that hNTCP is a mediator of HBV infection of hepatocytes. hNTCP binds specifically to the HBV L-Ag (described in a later section: HBV envelope, precore, core and reverse-transcriptase proteins), and this interaction is thought to be critical in mediating HBV entry into hepatocytes as indicated by the observation that HBV L-Ag mimetic peptides can block HBV infection (Yan *et al.*, 2012; Volz *et al.*, 2013; Ni *et al.*, 2014). Whether hNTCP dimerization effects HBV entry or trafficking of HBV virions to the nucleus remains unknown. Although hNTCP can render non-susceptible human liver-derived cell lines susceptible to HBV infection, the infection process is highly inefficient, necessitating multiplicities of infection often as high as 1000 genome equivalents of HBV per cell and usually also requiring concomitant exposure of the cells to factors

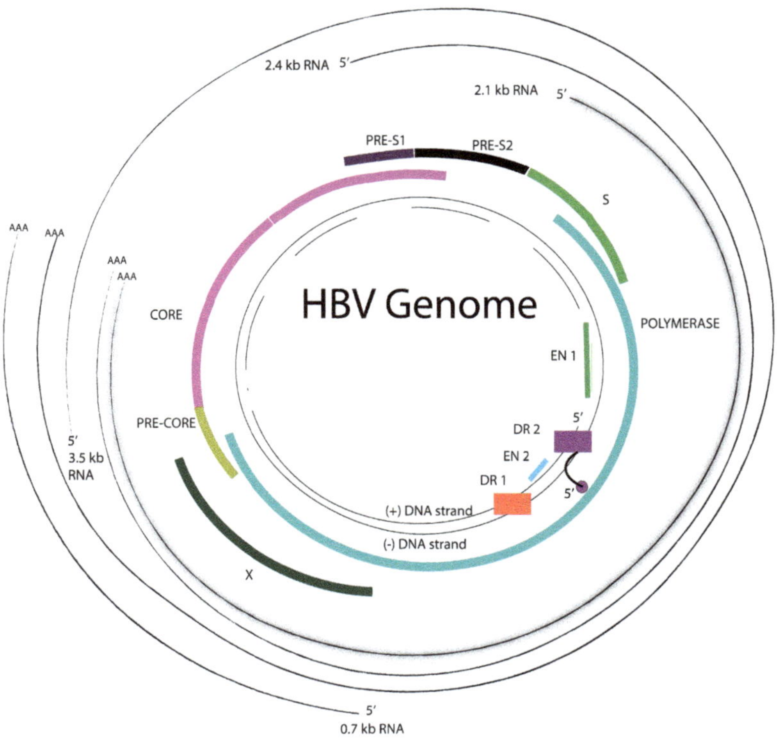

Figure 4.2. HBV genome organization. The center lines indicate the partially double stranded DNA genome of HBV. Genome regulatory elements and open reading frames (ORFs) are depicted by colored boxes. Outside of the ORF map is a depiction of the HBV transcripts. For further details and references see text.

such as polyethylene glycol to increase infection efficiency (Yan *et al.*, 2012; Ni *et al.*, 2014; Watashi *et al.*, 2014). Moreover, expression of hNTCP on rodent liver cells seems to only facilitate the initial steps of an HBV infection without subsequent delivery of the viral genome to the nucleus of the infected cell (Yan *et al.*, 2014). A recent study has highlighted the importance of hNTCP glycosylation in mediating hNTCP trafficking to the plasma membrane as well as HBV infection (Appelman *et al.*, 2017). Partial hNTCP glycosylation is sufficient for hNTCP to function as a bile acid transporter and as a cell-surface receptor for HBV, whereas mutation of all glycosylated amino acids in hNTCP results in loss of hNTCP-membrane localization and hNTCP degradation (Appelman *et al.*, 2017). Cumulatively, these observations support the notion that hNTCP is an entry factor for HBV but also suggest that additional cell-surface and cytosolic factors are essential for efficient HBV infection and replication in hepatocytes. Importantly, the discovery of hNTCP as a cell-surface receptor for HBV has facilitated generation of hNTCP-expressing cell lines that can serve as experimentally tractable models for studying the entire HBV infection process (Yan *et al.*, 2012; Iwamoto *et al.*, 2014; Ni *et al.*, 2014). However, the low efficiency of HBV infection in these models may limit their use in studies that require high infection efficiencies, such as those seeking to analyze subtle HBV affects on cellular signal-transduction pathways.

HBV transmission

Common modes of HBV transmission include exposure to HBV-contaminated blood, unsafe injection practices, sexual contact with an HBV-infected individual, and perinatal transmission from an HBV-infected mother to her child (WHO, 2008).

HBV infections are classified as acute, chronic, or occult (Seeger, 2013). In most people with acute or short-term HBV infections, the infection is resolved by immune-mediated viral clearance. The inflammatory response to the HBV infection causes many of the symptoms that have been associated with hepatitis; these symptoms include jaundice, fatigue, nausea, vomiting, and abdominal pain (Seeger *et al.*, 2007; WHO, 2008).

Because of their inability to mount a sufficient immune response that clears the virus, approximately 90% of infants infected with HBV develop a chronic HBV infection. In contrast, only about 5% of those infected as adults with HBV develop a chronic HBV infection (Seeger and Mason, 2000b; Seeger, 2013). During an HBV infection, the levels of HBV virions can reach up to 10^{10} particles/ml of blood (Seeger and Mason, 2000b; Ganem, 2001). Resolution of an acute infection is associated with mounting an effective antiviral response, including a cytotoxic T lymphocyte (CTL) response, which can lead to the destruction of infected hepatocytes to facilitate eventual clearance of the virus from the liver. In addition to the cytotoxic response, noncytolytic responses mediated by cytokines, including interferon-γ (IFN-γ), tumor necrosis factor-α (TNF-α), interleukin (IL)-12, and IL-6, also play important roles in terminating HBV replication and facilitating viral clearance (Guidotti *et al.*, 1996; Guidotti *et al.*, 1999; Chisari, 2000; Summers *et al.*, 2003; Thimme *et al.*, 2003b; Block *et al.*, 2007). The results of some studies suggest that the innate, cytokine-mediated, noncytolytic clearance of HBV may be responsible for clearing most HBV from infected hepatocytes and the liver prior to the induction of the adaptive, CTL-mediated clearance (Seeger, 2013; Xia and Protzer, 2017). The development of a chronic HBV infection has been linked to a poor immune response to the initial HBV infection and is characterized by the continued presence of detectable levels of the HBV small envelope protein in the blood of an HBV-infected individual for more than six months. Importantly, a chronic HBV infection is the major culprit in the development of severe liver diseases, including liver cirrhosis and HCC (Lok and McMahon, 2001). In contrast to acute and chronic HBV infections, occult HBV infection is defined by the presence of HBV DNA in the liver tissue and serum of an HBV-infected person in the absence of detectable HBsAg in the serum. Occult infections have also been linked to the development of liver diseases and HCC, and there is evidence that many previous cases of HCC that were of unknown etiology may have been caused by an occult HBV infection (Said, 2011; Nishikawa and Osaki, 2013).

HBV genome organization

The minus-strand of the partially double-stranded, relaxed circular HBV DNA genome is unit-length (Figure 4.2), is copied by reverse transcription of the pgRNA, and is typically fully completed in the infectious HBV virion. The designation of this strand as "minus" is derived from the mRNA role of the pgRNA (described below). The plus-strand of the HBV DNA genome is copied from the minus-strand DNA and is completed to varying lengths that are typically

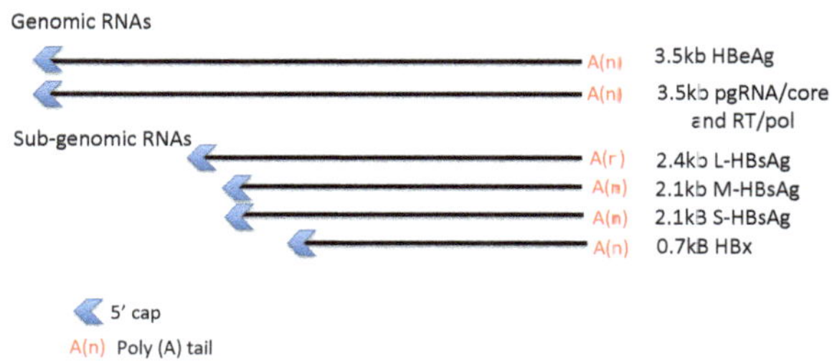

Figure 4.3. Viral transcripts. The HBV genome encodes two types of transcripts: genomic and subgenomic. For a detailed description see text.

less than the length of the full genome, generating the partially double-stranded circular HBV genome characteristic of infectious HBV virions. There are two short repetitive sequences within the genome: direct repeat (DR) 1 and DR2. DR1 serves as the initiation site for DNA synthesis of the minus strand, and DR2 serves as the initiation site for DNA synthesis of the plus strand (reviewed in: Seeger and Mason, 2000b). The HBV genome has two transcription enhancer elements (Enhancer I and II) and four transcription promoters: the preS1 promoter, the S promoter, the core promoter, and the HBx promoter (Figure 4.2) (Seeger and Mason, 2000a; Ganem, 2001; Seeger *et al.*, 2007; Glebe and Bremer, 2013).

All HBV RNA transcripts are generated from the nuclear-localized HBV cccDNA; these transcripts are capped, serve as mRNAs, and use a single polyadenylation signal that is present in the core gene of the HBV genome (Figures 4.2 and 4.3). Both subgenomic and genomic transcripts are synthesized from the cccDNA (Figure 4.3). The subgenomic RNAs are the 2.4kb, 2.1kb, 2.1kb, and 0.7kb transcripts, which encode the three envelope proteins and the HBx protein, respectively. The cccDNA also encodes two genomic RNAs, both of which are approximately 3.5kb in length. While the larger of the genomic transcripts acts as the mRNA for the HBV e antigen (HBeAg), which will be discussed below, the second shorter genomic RNA serves as the mRNA for both the HBV core protein and the reverse transcriptase/polymerase. The genomic transcript that encodes the core and polymerase is also the pgRNA, the template for synthesis of nascent viral DNA genomes which will be incorporated into newly formed virions. The genomic RNAs are both greater than unit length copies of the circular HBV genome and are therefore terminally redundant (Figures 4.2 and 4.3) (Ganem, 2001; Glebe and Bremer, 2013; Seeger, 2013).

HBV envelope, precore, core, and reverse-transcriptase proteins

HBV encodes three envelope proteins, also referred to as HBV surface antigens (HBsAg); these envelope proteins are the L (L-HBsAg), M (M-HBsAg), and S (S-HBsAg) surface antigens. mRNAs of the three surface antigens are transcribed from the preS-S (presurface-surface) region of the HBV genome (Figure 4.2) (De Meyer *et al.*, 1997; Seeger *et al.*, 2007). The three surface antigens are translated from three different-in-frame initiation codons within the same open reading frame (Ganem and Prince, 2004; Block *et al.*, 2007). Therefore, the L, M, and S surface antigens form what has been termed a "carboxy terminal nested set", where S is the smallest surface antigen, M is the next largest, and L is the largest (Seeger and Mason, 2015). The 24 kDa S-HBsAg, which is 226 amino acids, is the most abundant envelope protein present in the Dane and subviral particles. The 31 kDa M-HBsAg contains the S domain with a N-terminal extension of 55 amino acids encoded by the preS2 region of the genome. The function of the M-HBsAg remains unclear. The 42 kDa L-HBsAg contains S, preS2, and an additional N-terminal extension of 108-119 amino acids, which is referred to as preS1 (Coppola *et al.*, 2015). While the L-HBsAg comprises approximately 17% of the total HBsAgs on the Dane particles, it accounts for only 1-2% of the total HBsAgs on the subviral particles (Ganem, 2001). The L-HBsAg contains a region necessary for hNTCP binding and therefore plays a central role in binding of HBV to the host-cell receptor; L-HBsAg also interacts with cytosolic core particles containing replicating HBV genomes and is involved in assembly and secretion of infectious HBV virions (Heermann *et al.*, 1984; Neurath *et al.*, 1985; Ganem and Prince, 2004; Block *et al.*, 2007; Seeger and Mason, 2015).

The HBV precore protein, also referred to as E antigen (HBeAg), is produced from the largest HBV genomic RNA. E antigen is encoded in the same open reading frame as the HBV core protein, is an N-terminal extension of core, is targeted to the endoplasmic reticulum where additional processing events generate HBeAg, and is eventually secreted from the cell. Although the level of HBeAg in the serum of an HBV-infected individual has been commonly used as a marker of active HBV replication, for many years, the function of HBeAg in the course of HBV infections remained unclear, and studies have shown that once an HBV infection is established, the virus does not require HBeAg to maintain viral replication (Tong, 1991). Although some early studies suggested that HBeAg had a role in dampening an HBV immune response, this was only definitely demonstrated in a recent report by Ou and colleagues that clearly defined a function of HBeAg in promoting persistence of HBV infection in newborn offspring of HBV-infected mothers (Tian *et al.*, 2016). The results of these studies, which were conducted in HBV-transgenic mice, showed that HBeAg promotes priming of hepatic macrophages, such that the CD8+ T-cell response is dampened in the neonate, facilitating the development of a persistent HBV infection in the newborn (Tian *et al.*, 2016). This study has enhanced our understanding of the role of HBeAg in the establishment of an HBV infection and has defined another molecular mechanism that might underlie the high incidence of HBV persistence that can result from the transmission of HBV from an HBV-infected mother to her offspring.

The 21 kDa HBV core protein, or HBcAg, consists of 183 amino acids and forms the nucleocapsid of HBV. Capsids are composed of core protein dimers assembled into a 20-sided structure with icosahedral symmetry. HBcAg is translated from the 5′ end of the pregenomic RNA. The core protein can be divided into two functional domains: the assembly domain and the arginine-rich C-terminal domain. The first 144 amino acids form the assembly domain, which is sufficient for the assembly of HBV nucleocapsids. The remaining 39 amino acids of the core protein contain 16 arginine residues that are dispensable for nucleocapsid formation, but important for pgRNA encapsidation. Additionally, it has been reported that both phosphorylation and dephosphorylation of certain C-terminal serine residues of HBcAg are required for different stages of HBV replication (Birnbaum and Nassal, 1990; Machida *et al.*, 1991; Yeh and Ou, 1991; Nassal, 1992; Beames and Lanford, 1993; Liao and Ou, 1995; Lan *et al.*, 1999; Ganem and Prince, 2004; Seeger *et al.*, 2007).

The 90 kDa HBV reverse transcriptase/polymerase (RT/pol) is translated from the 3′ end of the pgRNA; the mechanisms that controls internal translation initiation of the RT/pol from the pgRNA remains incompletely understood but has been proposed to be regulated by ribosome shunting from the 5′ cap of the pgRNA to the internal translation start site of RT/pol (Fouillot, 1993). The RT/pol consists of 838 amino acids and has three functional domains: the terminal protein domain, the reverse-transcriptase domain, and the RNaseH domain. The 5′ end of the pgRNA contains an encapsidation signal, termed epsilon (ε). HBV RT/pol recognizes and binds ε to initiate packaging of the pgRNA in capsids. A specific tyrosine residue within the terminal-protein domain of RT/pol acts as the primer for reverse transcription. RT/pol reverse transcribes the pgRNA to form the viral DNA minus-strand. While the minus-strand is being synthesized, the RNaseH domain degrades the majority of the pgRNA. Approximately 18 nucleotides are left after the RNaseH activity of HBV RT/pol has degraded the pgRNA, and this serves as the 5′ primer for the synthesis of the plus-strand. The minus-strand serves as the template for the synthesis of the plus-strand (Lien *et al.*, 1986; Junker-Niepmann *et al.*, 1990; Loeb *et al.*, 1991; Wang and Seeger, 1992; Seeger *et al.*, 2007; Seeger *et al.*, 2013; Seeger and Mason, 2015).

Proposed mechanisms that may underlie development of HBV-associated HCC

Integration of the HBV genome into the host genome

HBV does not integrate into the host genome as a normal part of the virus replication cycle; however, integration of the HBV genome into the host genome is common and frequently observed in HBV-associated HCC samples (Shafritiz, 1981; Bonilla Guerrero and Roberts, 2005). Although this suggests that HBV integration may participate in promoting the development of HCC, whether integration of the HBV genome into the host genome actually contributes to HCC development remains unclear. Initial analyses of sites of HBV integration using more traditional DNA sequencing and polymerase chain reaction (PCR) technologies confirmed the frequent integration of the HBV genome into the host genome but did not identify common sites of integration that might help define how this integration could contribute to HCC development (Shafritiz, 1981). More recently, the use of deep-sequencing, high-throughput platforms has allowed identification of specific HBV integration sites in much larger numbers of HBV-associated HCCs and a comparison of these integration sites to HBV integration sites in non-tumor cells derived from the same HBV-infected liver. These studies identified a number of genes that were frequent integration sites for HBV, including genes that encode telomerase (*TERT*), cyclin E1 (*CCE1*), and Rho-associated kinase (*ROCK1*) (Sung *et al.*, 2012); however, whether these integrations of the HBV genome contributed to HCC development remains untested. A consistent observation from these whole-genome studies is that a common breakpoint in the integrated HBV genome is in the 3′ end of the HBx gene and 5′ end of the precore gene (see Figure 4.2 for genome structure) (Sung *et al.*, 2012). This observation is consistent with earlier studies with DHBV that suggested that a defective double-stranded linear replicative form of DHBV was preferentially integrated into the duck genome (Bill, 2004). Additional evidence that HBV integration into the host genome might contribute to HCC development was provided in a recent study by Yang and colleagues which showed that HBV integration events are more common in HBV-associated tumor tissue compared to adjacent non-tumor tissue. Interestingly, the number of common integration sites between HBV-associated tumor tissue and non-tumor tissue was low, resulting in less than 5% of shared sites (Yang *et al.*, 2017). Although this study supports the notion that integration events may play a specific role in the development of HCC, definitive proof that the integration sites unique to the HCCs were relevant for HCC development was not provided. A particularly interesting observation is that integration of the HBV genome into long interspersed nuclear elements (LINEs) can generate chimeric HBx-LINE transcripts that could function as long non-coding (lnc)RNAs. LncRNAs play important roles in regulation of gene expression, and some of the HBx-LINE lncRNAs could have important roles in regulating gene expression to influence HCC development (Yang *et al.*, 2017). Moreover, additional studies analyzing the integration events in HBV-associated HCCs have shown that HBV integration events are more common in male as compared to female HBV-associated HCCs (Zhao *et al.*, 2016). It is important to note, however, that while some studies have suggested that integration of HBV genome may have a preference for putative tumor-suppressor genes or fragile-chromosome sites, these types of integration events may have provided a growth advantage to cells and hence, integration itself may still be random while specific integrations appear preferred because the associated growth advantage facilitates a selective advantage. Consequently, whether integration of the HBV genome into the host chromosome targets specific sites in the chromosome or is a root cause of carcinogenesis remains unclear. It is also important to note that integration of the DHBV genome into the infected host genome is also common but not associated with HCC development in ducks (Cullen, 1990). Finally, although HBV genomes that are integrated into the host genome are not replication competent, these integrants often continue to generate transcripts for some viral proteins, such as envelope and HBx proteins, and these proteins may also influence HCC development when continually expressed from the integrated HBV genome (Quetier *et al.*, 2015; Lee *et al.*, 2016; Ng, 2016).

HBV-associated inflammation

The immune response to an HBV infection is complex and occurs in distinct phases. The initial immune responses is characterized by low levels of inflammation, followed by an inflammatory phase, and ultimately a low-replicative, or HBV-replicative

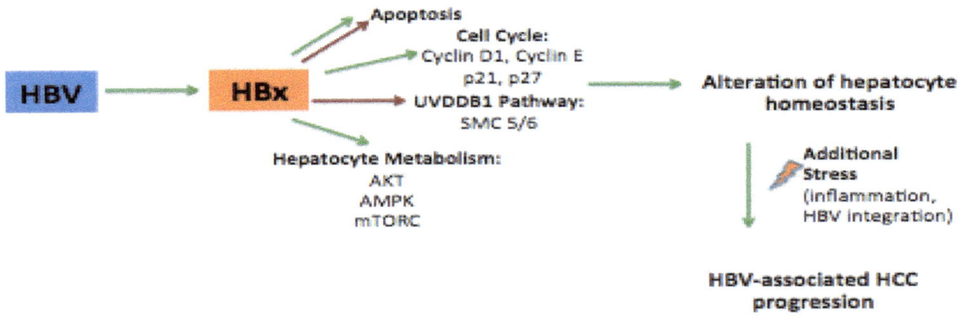

Figure 4.4. Potential HBx-mediated changes in hepatocyte physiology. HBx-mediated effects thought to be positive are indicated by green arrows, while repressive effects are indicated by red arrows. For specific details and references see section on *HBx-mediated effects on hepatocyte physiology and HBV replication.*

"inactive", phase (Fourati, 2016; Shin *et al.*, 2016). Immunological features of HBV infections include an initial rise in HBV DNA that ultimately subsides after induction of specific anti-HBV antibodies; this however, is associated with increased immune-mediated destruction of HBV-infected hepatocytes, elevation of liver enzyme levels in the serum of the HBV-infected patients, and typical sequelae, such as hepatitis, that can be associated with an HBV infection (Shin *et al.*, 2016).

Most adults who are infected by HBV will clear the infection rapidly; these adults typically display a strong anti-HBV CTL response. Specifically, CD8+ T-cells are the crucial T cell type in clearing HBV infections, as shown in a chimpanzee model for HBV infection (Thimme *et al.*, 2003a). The actions of CD8+ T-cells are also thought to mediate liver injury in the process of clearing HBV-infected hepatocytes (Shin *et al.*, 2016), and inflammation in the liver and associated pathologies of liver disease, such as cirrhosis, coincide with the appearance of HBV-specific CD8+ T-cells in the liver (reviewed in: Guidotti, 2006). The activities of HBV-specific immune cells may also recruit non-specific immune cells to the liver, exacerbating inflammation and potentially increasing the severity of the liver disease caused by an HBV infection (reviewed in: Guidotti, 2006). The host inflammatory response resulting from an HBV infection has been considered key for the development of liver pathologies, and symptoms associated with an HBV infection are believed to be caused by the inflammatory response because the virus itself in non-cytopathic. Liver inflammation caused by an HBV infection, and the resultant, repeated cycles of destruction of hepatocytes followed by regeneration, have been hypothesized to be key factors in the development of HBV-associated HCC; however, recent evidence in HBV-transgenic mice, where an anti-HBV inflammatory response is not present due to the recognition of HBV antigens as self, suggests that HCC can still arise in an HBV-infected liver in the absence of inflammation. Interestingly, although HBV-transgenic mice developed HCC at a significantly higher rate than wild-type littermates, this occurred only when the mice were 2 or more years old, suggesting that age may also have a role in HCC development in these mice (Na *et al.*, 2011).

HBx-mediated effects on hepatocyte physiology and HBV replication

The HBV HBx protein is encoded by the smallest open reading frame in the HBV genome (Figure 4.2). HBx is a 154 amino acid protein that plays a variety of roles in modulation of HBV transcription and HBV genome replication. HBx can interact with numerous cellular proteins across a variety of cellular signaling pathways (Bouchard, 2001; Melegari *et al.*, 2005; Clippinger and Bouchard, 2008; Gearhart and Bouchard, 2010b; Rawat *et al.*, 2012; Yang and Bouchard, 2012; Rawat and Bouchard, 2014; Decorsiere *et al.*, 2016; Murphy *et al.*, 2016; Casciano *et al.*, 2017) (Figure 4.4). In the context of HBV replication, HBx is typically expressed at low levels; this is an important aspect of HBx biology that is important to consider when attempting to evaluate the many, sometimes discrepant HBx activities that have been reported. For example, many studies that have analyzed HBx activities were conducted in systems where HBx was expressed independent of the HBV genome and at higher levels than apparent during an HBV infection; many studies were also conducted in non-liver-derived cell-lines (Moon, 2003; Lim *et al.*, 2013; Zhang, 2016). Although these studies have been important for identifying putative HBx activities in specific cellular contexts, the absence of confirmatory studies in the context of HBx expression during HBV replication in hepatocytes has lead to some confusion as to which identified HBx activities and interacting partners are relevant to an HBV infection.

In this section, we will summarize HBx-mediated changes in signaling pathways that are thought to play roles in promoting hepatocarcinogenesis. Rather than review decades of HBx studies, we will provide an overview of these studies and an update of some recent studies. We will specifically focus on four major areas of HBx-mediated effects on cellular physiology that could promote HCC development. More extensive descriptions of HBx activities, including those that might also contribute to the development of HBV-associated HCC, can be found in a number of recent reviews (Bouchard and Navas-Martin, 2011; Rawat and Bouchard, 2014; Lamontagne *et al.*, 2016).

HBx and apoptosis

Altered apoptotic signals and pathways have been implicated as important characteristics of tumor cells and are among the most frequently identified alterations that can distinguish neoplastic tissue from normal tissue (Hanahan and Weinberg, 2011). Depending on the cell-model system, and possibly the level of HBx expression, HBx has been reported to activate, inhibit, have no effect on, or sensitize cells to apoptotic signals (reviewed in: Rawat *et al.*, 2012). Although contradictory reports of HBx regulation of apoptotic pathways initially caused considerable controversy in the field of HBx studies, it has now become apparent that HBx-mediated effects on apoptotic pathways are context specific, necessitating the study of HBx effects in biologically relevant systems such as primary hepatocyte culture systems or *in vivo* models where HBx is expressed at physiologically relevant levels or in the context of replicating HBV (reviewed in: Rawat *et al.*, 2012).

Recent studies in cultured primary rat hepatocytes, a surrogate model of human hepatocytes that can be used to study HBV biology, have suggested that HBx plays a multifaceted role in modulation of apoptosis (Clippinger *et al.*, 2009; Rawat and Bouchard, 2014). For example, HBx, expressed independently and in the context of HBV replication, was found to be anti-apoptotic via an NFκB-dependent mechanism (Clippinger *et al.*, 2009). NFκB acts as an important inflammation and stress-induced transcription factor that is thought to mediate progression of liver disease (reviewed in: Luedde and Schwabe, 2011). In contrast, when NFκB was inhibited in cultured primary rat hepatocytes, HBx became pro-apoptotic both on its own and in the context of HBV replication (Clippinger *et al.*, 2009). These observations, in fact, support the notion that HBx has both pro- and anti-apoptotic activities that may be affected and regulated by other cellular factors; these observations could help explain the seemingly discrepant HBx apoptotic effects identified in various models systems. Finally, these HBx effects are reminiscent of effects of cytokines such as tumor necrosis factor (TNF) α, which can be either pro- or anti-apoptotic depending on the status of NF-κB activation (Su *et al.*, 2001). Additional studies in the primary rat hepatocyte model showed that HBx can also activate the anti-apoptotic factor AKT and that HBx activation of AKT promoted hepatocyte survival at the expense of higher rates of HBV genome replication: inhibition of AKT stimulated HBV replication but also led to higher levels of apoptosis in HBV-infected hepatocytes than in control hepatocytes (Rawat and Bouchard, 2014). Similar AKT-mediated effects on HBV replication have also been observed in liver-derived cell-culture models (Guo *et al.*, 2007).

Numerous studies support the notion that HBx expression might not directly affect apoptosis in the absence of other factors, hence sensitizing cells to apoptotic signals (Su *et al.*, 2001; Lee *et al.*, 2004; Wang *et al.*, 2004; Liang *et al.*, 2006; Zhang *et al.*, 2015). These observations might also be consistent with those described above in primary-hepatocyte systems; HBx anti-apoptotic effects are masked by activation of NF-κB but become apparent in the context of additional pro-apoptotic signals. Studies in HBV- and HBx-transgenic mouse models, and in some liver-derived cell-line models, have also suggested that HBV, probably through activities of HBx, and HBx itself, can cause expression of higher levels of the pro-apoptotic BH3-only protein BAX and lower levels of anti-apoptotic Bcl-xL that sensitize hepatocytes to pro-apoptotic signals (Koike, 1998; Liang *et al.*, 2006; Miao *et al.*, 2006). Moreover, this observation would be consistent with the generally noncytopathic effect of HBV replication that could be altered in response to a cytokine response to stimulate apoptosis.

HBx-induced changes in mitochondrial physiology, and the subsequent consequences of these subtle alterations, might also affect apoptotic signals (Lee *et al.*, 2004; Clippinger and Bouchard, 2008; Yang and Bouchard, 2012; Kim *et al.*, 2013; Zou *et al.*, 2015; Casciano *et al.*, 2017). For example, the results of one study demonstrated that HBx-expressing cells had higher levels of reactive oxygen species (ROS) and decreased levels of mitochondrial oxidative phosphorylation enzymes; this was linked to pro-apoptotic effects, but only when additional pro-apoptotic factors were added (Lee *et al.*, 2004). Finally, a more recent study has shown that in both HBV- and HBx-expressing liver-derived cells, mitophagy is induced by HBV/HBx, and that

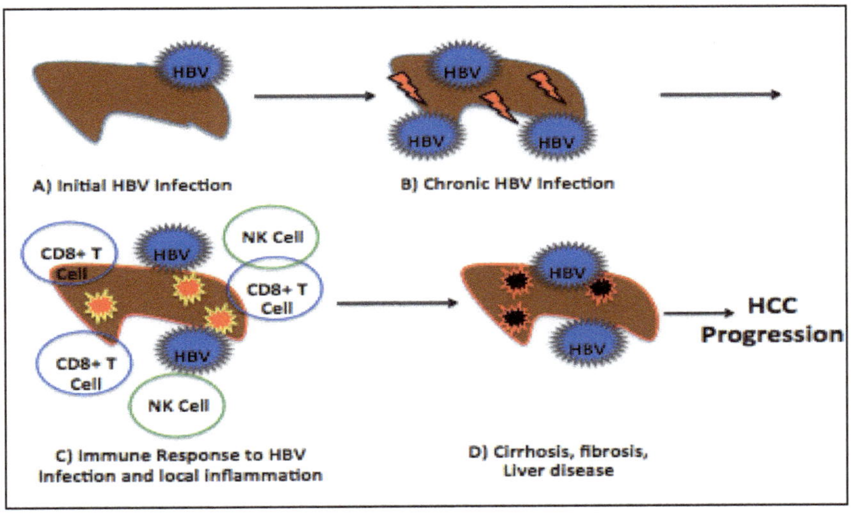

Figure 4.5. Possible mechanism of chronic HBV infection resulting in HCC Progression. **A)** Initial HBV infection occurs in the liver. **B)** In some individuals, a chronic HBV infection is established, resulting in continual HBV replication and expression of HBV proteins. **C)** Chronic HBV infection causes persistent immune cell recruitment and immune mediated destruction of hepatocytes. **D)** Chronic HBV infection and associated inflammation, integration of the HBV genome into the host genome, and activities of HBV proteins such as HBx could individually, or in combination, cause liver diseases such as fibrosis and cirrhosis that can eventually contributing to HCC development. See text for more details.

inhibition of HBV/HBx-mediated mitophagy resulted in induction of apoptosis (Kim *et al.*, 2013). The authors of this study postulate that HBV/HBx induces mitophagy as a protective mechanism in HBV-expressing cells; in the context of HBV expression, inhibition of mitophagy results in induction of apopotosis (Kim *et al.*, 2013). The results from this study suggest that HBV, via the actions of HBx, induces mitophagy to prevent apoptosis in HBV-expressing cells (Kim *et al.*, 2013).

A few recent studies have analyzed the contribution of HBx to apoptosis that is regulated by micro (mi)RNAs such as miR-181a and miR-125a (Zhang *et al.*, 2015; Tian *et al.*, 2017). The results of one study showed that miR-181a levels are elevated in HBV-infected cells, specifically via HBx expression, and that this increased level of miR-181a was responsible for decreased expression of the tumor suppressor phosphatase and tensin homolog (PTEN) and inhibition of apoptosis (Tian *et al.*, 2017). The results of another study demonstrated that HBx expression increased the level of miR-125a, a negative regulator of expression of A20, which is an E3 ubiquitin ligase critical for suppressing apoptosis (Zhang *et al.*, 2015). In this model, re-establishing A20 expression limited pro-apoptotic effects of HBx (Zhang *et al.*, 2015).

Cumulatively, studies in cell-lines, *in vivo* models, and in cultured primary hepatocytes that we have described here and that have been more extensively covered in previous reviews suggest that HBx maintains a balance between pro- and anti-apoptotic signals that could help promote persistent HBV replication. Moreover any disruption of the pro- and anti-apoptotic balance mediated by HBx, whether to favor pro- or anti-apoptotic effects, could contribute to the development of HBV-associated HCC (Rawat *et al.*, 2012; Lamontagne *et al.*, 2016). The model system used to study HBx pro- or anti-apoptotic effects is of critical importance, and HBx apoptotic effects are likely context specific (Slagle *et al.*, 2015). Most studies in biologically relevant systems suggest that HBV, via HBx, is not directly pro-apoptotic and may be anti-apoptotic but can still sensitize HBV-infected cells to pro-apoptotic factors. The subtle, yet volatile combination of HBV and additional stresses in the context of inflammatory signals brought forth by an HBV-specific immune response could promote an environment that is potentially oncogenic, and HBx modulation of apoptotic signals is likely to be an important, although incompletely understood, contributor to the development of HBV-associated HCC.

HBx and the cell cycle

Regulated growth and division of normal cells ensures that genome integrity is maintained and conditions are right for cell proliferation. In contrast, cancer cells have accumulated genetic or epigenetic changes that alter the activity or expression of oncogenes or tumor-suppressor genes that regulate cell growth and proliferation, allowing cell proliferation even in conditions that might not normally favor cell growth and replication (reviewed in: Hanahan and Weinberg, 2011). Dis-regulation of oncogenes or tumor suppressors can be associated with hyper-proliferation or a lost ability to respond to cell-division-repressive signals. Numerous studies have demonstrated that HBx can alter cell-proliferation and growth pathways, but because these studies have been conducted in a variety of cell-culture models and under various cell-culture and HBx-expression conditions, discrepant HBx effects have been reported. Consequently, HBx has been reported to cause cells to progress more rapidly through the cell cycle, stall cells at specific stages of the cell cycle, or have no effect on the cell cycle (Lee *et al.*, 2002; Wu *et al.*, 2006; Hodgson *et al.*, 2008; Gearhart and Bouchard, 2010a, 2010b). Despite the differences in model systems, most studies have demonstrated that HBx can affect the cell cycle, and it has been postulated that these HBx effects could contribute to the development of HBV-associated HCC (Benn, 1995; Ahn, 2002; Wu *et al.*, 2006; Gearhart and Bouchard, 2010a). Alteration in normal cell growth and proliferation pathways is not a unique characteristic of an HBV infection, and many viruses alter normal cell cycle control mechanisms to promote an environment that is more favorable for their own replication (reviewed in: Bagga and Bouchard, 2014). This is especially true of many DNA viruses, such as HBV, which require nucleotides for replication but infect quiescent, differentiated cells that may not present an ideal environment for viral genome replication. Unfortunately, altered cell proliferation may be detrimental to the host cell, necessitating adaptive responses that could contribute to viral-associated diseases (reviewed in: Bagga and Bouchard, 2014). A summary of HBV- and HBx-mediated changes to normal cell-cycle-control mechanisms will be discussed in this section; more extensive descriptions of HBV- and HBx-induced changes in the cell cycle can found in a number of excellent reviews (Bouchard and Navas-Martin, 2011; Bagga and Bouchard, 2014; Lamontagne *et al.*, 2016).

Early studies that characterized HBx effects on the cell cycle demonstrated that HBx expression accelerated entry and progression through the cell cycle; in some studies, this was shown to be dependent upon activation of the RAS kinase pathways (Benn, 1995). In contrast, in other studies, HBx expression was shown to increase p21 protein levels, which was linked to stalled cell proliferation (Koike *et al.*, 1994; Ahn, 2002). A possible reason for the differences in these reported HBx effects is that they were performed in different non-liver cell lines. Moreover, these results highlight the context-specific effect of HBx activities. In support of the notion that HBx activities can be context specific, a study performed by Lee and colleagues clearly demonstrated that HBx can have different effects on cell-cycle progression depending on the model used (Lee *et al.*, 2002). This study analyzed HBx-mediated changes in two different cell lines derived from AML-12 cells, a liver-derived cell line (Wu, 1994). The first of these cell lines was more dedifferentiated, while the other had characteristics more similar to differentiated hepatocytes. HBx stalled cell-cycle progression shortly after transition into S-phase in the dedifferentiated AML12 cell line, whereas HBx induced a rapid progression through the cell-cycle in the more hepatocyte-like AML12 cell line and maintained aberrantly elevated levels of G1 phase cyclins (Lee *et al.*, 2002).

Studies in cultured primary rat hepatocytes have shown that HBx, both when expressed independently or in the context of replicating HBV, elevated the protein levels of p21 and p27 and the G1 phase cyclins D1 and E, without changing the level of S phase-specific cyclin A. In this system, HBx caused quiescent (G0) hepatocytes to exit quiescence, enter G1 but stall prior to entering into S phase. This HBx effect was also linked to HBx-mediated control of calcium signaling and was required for efficient HBV replication (Gearhart and Bouchard, 2010b). Similar HBx effects on cell cycle progression have been observed in hepatocytes of HBx-transgenic mice; hepatocytes in these mice have stalled cell-cycle progression that results in altered rates of DNA synthesis, which may also be linked to HCC development in these mice (Wu *et al.*, 2006). Additional work in HBx-transgenic mouse models confirmed that hepatocyte proliferation is changed by the presence of HBx (Tralhao *et al.*, 2002; Wu *et al.*, 2006; Hodgson *et al.*, 2008). In two independent HBx-transgenic mouse models, hepatocytes in these mice displayed impaired proliferative capacity compared to hepatocytes in wild-type littermates following a partial hepatectomy (Tralhao *et al.*, 2002; Wu *et al.*, 2006). One of these models showed impaired proliferative capacity of hepatocytes following partial hepatectomy both when hepatocytes were derived from HBx-transgenic mice or in a separate mouse model that had been implanted with a subset of HBx-expressing hepatocytes derived from HBx-transgenic mice (Tralhao *et al.*, 2002). However, in a separate HBx-transgenic mouse model, a subset of hepatocytes was shown to undergo early cell-cycle entry following partial hepatectomy (Hodgson *et al.*, 2008). This study also showed that HBx-expressing hepatocytes exhibit early DNA synthesis following partial hepatectomy; however, total regeneration following hepatectomy was not affected by HBx expression. Differences in the levels of HBx expression in the various HBx-transgenic mice that were used in these studies make it difficult to directly compare HBx-mediated effects in these mice; however, all three studies support the conclusion that HBx can affect hepatocyte proliferation pathways. Overall, these studies suggest that HBV-associated changes that are mediated by HBx alter control of the cell cycle via modulation of cell-cycle-control factors such as p21, p 27, Cyclin D1, Cyclin E, and Cyclin A, setting off a chain of events that is potentially more oncogenic and could play a role in development of HBV-associated HCCs (Bouchard *et al.*, 2001; Ahn, 2002; Tralhao *et al.*, 2002; Mukherji *et al.*, 2007; Hodgson *et al.*, 2008; Gearhart and Bouchard, 2010b). For example, the liver might respond to delayed hepatocyte proliferation by selecting for mutant hepatocytes that could overcome these HBx-mediated effects, ultimately selecting for hepatocytes that no longer respond to normal signals that regulate cell proliferation.

HBx expression and an HBV infection may also utilize regulation of cell-cycle factors to control expression of non-cell-cycle proteins. For example, expression of hNTCP, the HBV cell-surface receptor, is regulated during the cell cycle, and this could have clinical significance in HCC prognosis (Kang, 2016). In fact, Kang *et al.* reported that ectopic hNTCP expression inhibits proliferation of liver cancer-derived cell lines; interestingly, NTCP is often a factor that is lost upon hepatocyte de-differentiation (Yan *et al.*, 2012; Ni *et al.*, 2014; Kang, 2016). Cyclin D1 negatively regulates transcription of the NTCP gene, and low NTCP expression levels were also correlated with poorer HCC survival rates (Kang, 2016). Interesting HBV and HBx-expressing cells have elevated levels of cyclin D1 (Gearhart and Bouchard, 2010a, b). This suggests a possible scenario wherein HBV-infected cells stall the cell cycle to promote HBV replication, while simultaneously decreasing expression of hNTCP. The specific consequence of this for the development of HBV-associated HCC remains to be investigated.

HBx might also effect responses to chemotherapeutic agents by affecting the cell cycle. Chen *et al.* have shown that in response to treatment in cell culture with the chemotherapeutic agent doxorubicin, HBV transcript levels, genome replication, and core particle levels were regulated by p21-CCAAT/enhancer binding protein (C/EBP) in a doxorubicin-dependent manner (Chen *et al.*, 2015). C/EBP binding to the HBV genome was linked to doxorubicin-induced increases in p21 expression (Chen *et al.*, 2015). The findings of this study linked HBV replication and cell-cycle components that modulate the levels of replicating HBV in infected cells. Another important observation of this study was that certain chemotherapeutic agents, especially ones that limit cell-cycle progression in a similar manner as HBV/HBx may not be ideal to use as a therapy for HBV-associated HCCs, might contribute to HBV-reactivation in HBV-associated HCCs, or could enhance HBV replication in individuals with chronic HBV infections at risk for HCC development.

HBV- and HBx-mediated changes in cellular metabolism

The liver is the major metabolic internal organ of the body and serves as a metabolic hub that is critical in fed and starvation conditions for regulation of the levels of blood glucose, an energy source of multiple organs in the body (reviewed in: Nordlie, 1999; Berg, 2002). There is emerging evidence that HBV can affect expression and activities of metabolic factors in hepatocytes and that metabolic conditions within hepatocytes during HBV infection are altered to favor viral replication (reviewed in: Shlomai and Shaul, 2008). The next section will discuss recent studies of HBV/HBx-related changes in cellular metabolic pathways.

HBx-mediated changes in metabolic factors have been analyzed in HBx-transgenic mice and in HBx-expressing, liver-derived cell lines (Shin *et al.*, 2011). The results of one study demonstrated that HBx elevated the expression of both mRNAs and proteins of hepatic gluconeogenic genes, including peroxisome proliferator-activated receptor gamma coactivator-1α (PGC1α) and phosphoenolpyruvate carboxykinase (PEPCK), via a mechanism involving nitrogen oxide synthase (NOS) signaling pathways (Shin *et al.*, 2011). Additional studies have analyzed HBx-mediated changes in major metabolic signaling pathways including PI3K/AKT, the mammalian target of rapamycin complex (mTORC), and the AMP-activated protein kinase (AMPK) pathways in both liver-derived cell lines and primary rat hepatocytes (Guo *et al.*, 2007; Liu *et al.*, 2011; Rawat and Bouchard, 2014; Bagga *et al.*, 2016); these pathways, and cross communication between these pathways, have been implicated in the development of HCC (Bupathi *et al.*, 2015). Importantly, these studies have linked control of HBV transcription and genome replication directly to the PI3K-AKT-mTORC pathway in liver-cancer cell lines and primary hepatocytes (Guo *et al.*, 2007; Rawat and Bouchard, 2014; Bagga *et al.*, 2016). HBx has been shown to activate AKT, and this activation is inhibitory to both HBV transcription and genome replication in liver-cancer cell lines and cultured primary rat hepatocytes (discussed above in the HBx and apoptosis section) (Guo *et al.*, 2007; Rawat and Bouchard, 2014).

Additional studies have explored the role of the mTORC1- and AMPK-signaling pathways in HBV-expressing primary rat hepatocytes (Bagga *et al.*, 2016). HBx was found to simultaneously activate the opposing mTORC1 and AMPK signaling pathways when HBx was expressed independently and in the context of replicating HBV. Activation of mTORC1 inhibited HBV replication, whereas AMPK activation enhanced HBV replication (Bagga *et al.*, 2016). The results of a study that analyzed PI3K/AKT and mTORC signaling in HBV-positive patient liver samples demonstrated that mTORC1 and AKT levels were elevated; this study also tied activation of the mTORC and PI3K/AKT pathway to upregulation of alpha-fetoprotein protein (AFP) expression via an HBx-dependent mechanism. The authors of this study speculated that HBx-mediated AFP expression is a potential initiating transformative event that could promote HBV-associated HCC development and progression (Zhu, 2015).

The cumulative consequences of HBx-induced alterations of metabolic signaling pathways could be the creation of an environment that may be oncogenic, especially when considered in the context of the inflammatory response associated with an HBV infection. mTORC pathway activation has been showed to occur in up to 50% of HCC cases and is thought to mediate roles in HCC progression (Sahin, 2004). Recent histological analysis suggests that HBV-associated HCCs have more significant changes in AKT activation than mTORC activation, which is altered more often in nonviral HCC patients (Golob-Schwarzl *et al.*, 2017). This suggests that HBV-related cancer progression may be driven more by alterations in the PI3K/ATK pathway than the mTORC pathway. Additionally, therapeutic interventions targeting the AKT pathway may be beneficial for patients with virally-associated HCCs. While many of the HBV- or HBx-related changes in metabolism are still being investigated, the results of published studies suggest that HBV infection may be altering hepatocyte metabolism such that a more permissive environment for viral replication is maintained and that these metabolic changes could promote or contribute to progression of HBV-associated carcinogenesis.

HBx regulation of UV-DDB1

HBx can directly interact with Ultra Violet light-DNA Damage Binding protein 1 (UV-DDB1/DDB1) (Lee, 1995; Sitterlin, 1997). DDB1 affects DNA repair pathways and is a component of the CRL4 E3 ubiquitin ligase complex (reviewed in: Iovine *et al.*, 2011). The HBx-DDB1 interaction has been shown to interfere with cellular DNA repair pathways (Becker, 1998). The interaction of HBx and DDB1 has also been shown to be important for HBx stimulation of HBV replication, and depletion of DDB1 has a negative effect on HBV replication (Leupin *et al.*, 2005). Two very recent reports have indicated a mechanistic link between the HBx-DDB1 interaction and activation of HBV transcription by HBx (Decorsiere *et al.*, 2016; Murphy *et al.*, 2016). Both of these studies showed that HBx-DDB1 complexes interact with structural maintenance of chromosome complex (SMC) members 5/6 (Decorsiere *et al.*, 2016; Murphy *et al.*, 2016). The SMC5/6 complex is involved in DNA repair and replication and largely functions in association with the homologous recombination (HR) DNA-repair pathways (reviewed in: Murray, 2008; Kegel, 2010). In the presence of HBx, SMC5/6 was shown to be degraded by CRL4 E3 ubiquitin ligase complexes containing DDB1 and HBx (Decorsiere *et al.*, 2016; Murphy *et al.*, 2016). Importantly, in the context of expression of an HBV mutant virus that did not express HBx, depletion of SMC5/6 was able to rescue HBV gene transcription, indicating that at least one transcriptional function of HBx is to promote degradation of SMC5/6 (Decorsiere *et al.*, 2016; Murphy *et al.*, 2016). The targeting of SMC complexes by HBx was found to be specific to the SMC5/6 complex, and other SMC complexes were not affected (Murphy *et al.*, 2016). Loss of SMC5/6 was also shown to occur in humanized-mouse livers that were infected with HBV, indicating that this HBV/HBx-mediated effect occurs *in vivo* and is important for regulation of HBV transcription (Decorsiere *et al.*, 2016). Finally, studies of HBx-DDB1 interactions and their effect on the stability of the SMC5/6 complex have highlighted an important role of HBx in promoting HBV gene transcription. Although it has been suggested that HBx regulation of SMC5/6 stability, and resultant effects on transcription of HBV mRNAs, represents the essential role of HBx in the HBV life cycle, it is important to note that HBx has other clearly defined and distinct functions that are also essential for HBV replication. For example, specific phosphorylation events on the HBV core protein have been shown to be stimulated by HBx, and these promote HBV genome replication and maturation of HBV core particles (Lan, 1999; Melegari *et al.*, 2005). Moreover, there is also evidence that HBx modulation of cellular calcium signaling is essential for promoting HBV genome replication (Bouchard, 2001; Yang and Bouchard, 2012; Casciano *et al.*, 2017). Overall, these many studies demonstrate that HBx affects the HBV lifecycle at distinct points, both transcriptionally and post-transcriptionally.

Moreover the many HBx activities could also independently, or in combination, impact signal-transduction pathways that influence hepatocyte transformation and HCC development. For example, HBx-mediated changes in the cellular levels of SMC5/6, a critical component of DNA repair pathways, could affect hepatocyte physiology to promote genetic instability and influence the development of HCC.

HBV vaccination

The best strategy for decreasing the global incidence of HBV infection and associated diseases is effective implementation of HBV vaccination strategies. Highly effective vaccines are available, with the most widely used vaccine consisting of recombinant HBV sAg that is produced in yeast and administered as three injections over a six-month period. Approximately 95% of the individuals that receive the HBV vaccine develop a sufficient immune response resulting in the production of protective anti-HBsAg antibodies (Block *et al.*, 2007; WHO, 2008). Results of a recent study by Pinto and colleagues demonstrated that challenge after HBV vaccination results in a protective response in those more recently vaccinated. Moreover, the anti-HBV response can decline significantly in some individuals, and these individuals may require booster doses of the HBV vaccine to maintain protective immunity (Pinto *et al.*, 2017). Another recent study analyzed the prevalence of HBV infection in response to the introduction of the HBV vaccine in the United States-controlled pacific islands region, an HBV endemic area. The results of this study showed that the introduction of the HBV vaccine was protective and resulted in a decrease in the number of HBV infections; the study cohort had a drop in HBV infections from 10% of total newborns being infected to less than 1% after two decades of implementation of strict HBV vaccination strategies for newborn children, with infants being vaccinated at a time after birth (Abara *et al.*, 2017). Unfortunately, the cost of the vaccine, as well as the requirement for three injections, can lead to compliance issues as well as difficulties in vaccinating populations in some less-developed countries. However, recent global vaccination initiatives sponsored by a variety of public and private agencies have begun to address this issue and are likely to continue to help decrease the global incidence of HBV infection and hopefully, by association, the global incidence of HCC (PATH; National Expanded Program on Immuniazation, 2007; PATH, 2014). Please refer the accompanying chapter on HPV and HPV vaccines for more detail of the HBV vaccine.

Treatment strategies: PEG-Interferon-α (IFNα) and nucleos(t)side analogs (NUCs)

Although current anti-HBV therapies can decrease the levels of circulating HBV in the blood and replicating HBV in hepatocytes of an HBV-infected individual, no available therapies are curative. The inability of current therapies to completely cure an HBV infection is linked to the persistence of HBV cccDNA, a stable replicative intermediate of HBV that is localized to the nucleus of HBV-infected hepatocytes and is not directly targeted by currently available therapies. Hence, persistence of cccDNA remains a major barrier for the treatment of chronic HBV infections (Zoulim, 2004, 2007; Ghany and Doo, 2009; Cheng *et al.*, 2011; Yang and Kao, 2014; Kim *et al.*, 2016). Approved therapeutic options for the treatment of a chronic HBV infection are limited but include interferon-α and nucleoside analogs. Antiviral drug resistance, which develops due to the emergence of HBV mutants, is another major limitation of long-term nucleoside analog therapy. Therefore, treatment of chronic HBV infections remains a clinical challenge, and there is considerable interest in understanding the mechanisms that regulate HBV replication and identifying potential therapeutic targets for inhibiting viral replication and clearing HBV cccDNA from HBV-infected hepatocytes (reviewed in: Zoulim and Locarnini, 2009; Thomas *et al.*, 2015; Revill and Locarnini, 2016; Emery and Feld, 2017; Xia and Protzer, 2017).

Current anti-HBV therapies aim to control HBV replication and slow the occurrence of HBV-associated liver disease. Approved therapies for HBV infection include pegylated interferon alpha (here on referred to as: IFNα) and nucleos(t)ide analogs (NUCs). IFNα acts to increase the anti-HBV inflammatory response, although in a non-specific manner (Sonneveld and Janssen, 2010). NUCs target the HBV RT/pol and specifically decrease HBV genome replication, although they do not necessarily affect expression of HBV RNAs from the HBV cccDNA or subsequent expression of HBV proteins (Zoulim *et al.*, 2016). Treatments using IFNα are typically administered for a limited time and are often associated with adverse effects on a patient that may limit use and can be associated with noncompliance. Due to the persistence of HBV cccDNA, NUCs treatment is lifelong, although the rise of NUC-resistant mutants may eventually limit their use in specific patients. Both IFNα and NUC treatments aim to reduce HBV DNA levels below a threshold level, while stabilizing patient liver-enzyme levels (Kao, 2014). NUCs are generally better tolerated by treated individuals, but can also have side effects, albeit less severe than the side effects

Table 4.1. Approved Therapies in USA for chronic HBV infection. A list of approved anti-HBV therapies in the United States of America. See text for more details.

Drug Name	Mechanism of Action/Target
Intron A (Interferon α)	Immune System Agent
Peg-Interferon	Immune System Agent
Lamivudine	Nucleos(t)ide Analog/HBV Polymerase
Adefovir	Nucleos(t)ide Analog/HBV Polymerase
Entecavir	Nucleos(t)ide Analog/HBV Polymerase
Telbivudine	Nucleos(t)ide Analog/HBV Polymerase
Tenofovir	Nucleos(t)ide Analog/HBV Polymerase

Modified from Hepatitis B Foundation: Drug Watch

seen with IFNα treatment (Sonneveld and Janssen, 2010; Buti, 2014; Kao, 2014; Zoulim *et al.*, 2016).

There are multiple NUCs approved for HBV treatment; these include lamivudine, telbivudine, adefovir, entecavir, and tenofovir (Table 4.1) (Sonneveld and Janssen, 2010; Zoulim *et al.*, 2016). The highest occurrence of HBV mutants resistant to treatment has been associated with lamivudine treatment. In contrast, entecavir treatment has been associated with the lowest appearance of NUC-resistant HBV (Zoulim and Locarnini, 2009), although resistance to entecavir has been reported to be more common in patients who have used lamivudine or other NUC treatment prior to starting entecavir treatment as compared to those only treated with entacavir (Sun *et al.*, 2016). Treatment groups of chronic HBV-infected individuals are traditionally stratified into HBeAg positive or HBeAg negative groups (Buti, 2014; Kao, 2014). The results of some studies suggest that HBeAg-negative patients respond more readily to NUC treatment than IFNα treatment (Sonneveld and Janssen, 2010; Zoulim *et al.*, 2016); however, a recent meta-analysis study performed by Kim and colleagues suggested that the better treatment option is a combination IFNα and NUC therapy for both HBeAg-positive and negative patients (Kim *et al.*, 2016). Despite both IFNα and NUC therapy being available for treatment of chronic HBV infections, both of these therapies only act to manage chronic HBV infection and are typically not curative because HBV cccDNA is not targeted by these therapies (Revill and Locarnini, 2016). Although a goal of anti-HBV therapy is to lower replication of HBV and prevent the development of HCC, recent studies suggest that treated individuals still have a higher incidence of HCC than non-HBV-infected controls. While this incidence of HBV-associated HCC is significantly lower than in untreated, HBV-infected individuals, the continued expression of HBV proteins, such as HBx and the HBV envelope proteins, in the context of NUC treatment, may be enough to increase the risk for HCC development in these individuals, and there remains a need to develop more effective anti-HBV therapies that either eliminate the cccDNA or target activities of HBV proteins other than just the RT/pol (reviewed in: Honda *et al.*, 2015; Zoulim *et al.*, 2016).

Emerging treatments and new therapies for HBV infection

Targeting hNTCP

With the discovery of hNTCP as a receptor for HBV entry, there is significant interest in exploiting hNTCP as a new therapeutic target for preventing HBV infection of hepatocytes. Although this would not necessarily affect the initial infection unless delivered as a prophylactic therapy, anti-hNTCP therapies, such as those blocking interaction of HBV with hNTCP, could prevent spread of the virus in an already HBV-infected liver. A number of recent studies have analyzed the effects of compounds targeting hNTCP and their impact on HBV infections. Early reports indicate that cyclosporin A (CsA) and some related compounds can decrease HBV infections in cultured primary human hepatocytes and in HepaRG cells, a liver-derived cell line that can express hNTCP; the tested compounds interacted with hNTCP and competed for binding with HBV (Nkongolo *et al.*, 2014; Watashi *et al.*, 2014). CsA impairs normal NTCP transport of bile salts and only exerted an anti-HBV effect when administered prior to or during an infection (Nkongolo *et al.*, 2014).

Additional screens in hNTCP-expressing HepG2 cells have tested a variety of potential HBV-entry inhibitors, hNTCP inhibitors, anti-HBV antibodies, and other compounds (Iwamoto *et al.*, 2014). Oxysterol compounds, CsA derivatives, and bile salts all antagonized HBV infection in this model (Iwamoto *et al.*, 2014). Additional studies have analyzed pathways related to hNTCP signaling and consequential effects on HBV infection. One study showed that

Table 4.2. Emerging therapies for treatment of chronic HBV infection.

Drug Name	Mechanism of Action/Target
ARB-1467	siRNA/HBV RNA
ARB-1740	siRNA/HBV RNA
RG6004 (HBV LNA)	Locked Nucleic Acid/HBV RNA
Myrcludex B	HBV Entry Inhibitor/L-HBsAg Mimetic
Morphothiadin (GLS4)	Capsid Inhibitor/HBcAg
REP 2139	sAg Inhibitor/HBsAg
IONIS-HBVRx	mRNA inhibitor/HBV mRNA

Modified from Hepatitis B Foundation: Drug Watch

antagonizing the activity of the retinoic-acid receptor decreased HBV infection efficiency by lowering hNTCP-expression levels (Tsukuda *et al.*, 2015). Other studies have demonstrated that epigallocatechin-3-gallate (EGCG) promotes hNTCP internalization and prevents HBV infection; however, EGCG did not affect HBV replication if the viral genome was already present in EGCG-treated cells (Huang *et al.*, 2014). Additional compounds, such as irbesartan, have also been shown to antagonize HBV infection by targeting hNTCP, but the specific mechanism that underlies this effect on hNTCP and HBV replication remains unknown (Wang *et al.*, 2015b).

Finally, preS1-derived (HB-LAg) peptides have been shown to be targeted to the liver and effectively block HBV infections (Glebe *et al.*, 2005; Gripon *et al.*, 2005). The compound Myrcludex B (Myr B) is an HBV entry inhibitor modeled after the presS1 region of HBV that was previously shown to inhibit both HBV entry and subsequent infection in a humanized hepatocyte mouse model (Lutgehetmann *et al.*, 2012; Volz *et al.*, 2013). More recently, Myr B was shown to effectively block HBV infection of hNTCP-expressing HepG2 cells (Iwamoto *et al.*, 2014; Ni *et al.*, 2014; Nkongolo *et al.*, 2014). Myr B also inhibits bile acid uptake, but at a much higher concentration than is required to inhibit HBV infection, suggesting a potentially safe therapeutic profile for use in patients (Ni *et al.*, 2014). Myr B is currently being tested in early-phase-human-clinical trials for use as an anti-HBV therapy (Table 4.2) (Blank *et al.*, 2016).

Targeting HBV cccDNA

APOBEC3A and APOBEC3B targeting of HBV cccDNA

Lucifora and colleagues have demonstrated that treatment of HBV-infected cells with high doses of IFNα or antagonists of the lymphotoxin-β-receptor resulted in degradation of HBV cccDNA (Lucifora, 2014). In this study, treatment with IFNα or stimulation of the lymphotoxin-β-receptor pathway resulted in up-regulation of the cytosine deaminases APOBEC3A and APOBEC3B, which were shown to be associated with the HBcAg protein and recruited to the HBV cccDNA in the nuclei of infected cells; this interaction promoted the modification and subsequent degradation of HBV cccDNA (Lucifora, 2014). In an extension of this work, additional studies have shown that in acute HBV-infected patients, IFNγ and TNFα levels are elevated in the sera and that elevation of these immunomodulatory agents results in activation of the same cytosine deaminase enzymes which are targeted to HBV cccDNA to promote cccDNA degradation (Xia *et al.*, 2016). These studies have identified a specific pathway that targets cccDNA in HBV-infected cells, and this pathway could be exploited as a new therapeutic alternative for elimination of HBV in chronically infected individuals.

CRISPR/CAS-mediated targeting of HBV and HBV cccDNA

The advent of clustered regularly interspaced short palindromic repeats (CRISPR)/CRISPR-associated protein (CAS) (CRISPR/CAS) technology has created a variety of highly specific gene-targeting platforms (Donohoue *et al.*, 2017). Application of this technology could be beneficial for treating an HBV infection as this would present a specific method for targeting HBV cccDNA. Initially identified as a primitive anti-bacteriophage immune system in prokaryotes (Barrangou, 2007), the CRISPR/CAS system relies on the expression of guide RNAs (gRNAs), which direct site-specific nucleases to a location complementary to the gRNA; once gRNAs bind to a target locus, the CAS nuclease cleaves the target site, typically resulting in nucleotide deletions that can silence the target gene (reviewed in: Tsai and Joung, 2016; Wang *et al.*, 2016). Lin and colleagues demonstrated specific targeting of HBV in HBV-transfected cell lines; they also showed that CRISPR/CAS targeting of HBV is possible in a hydrodynamic injection mouse model where a plasmid containing the HBV genome was co-injected with CRISPR/CAS expression plasmids into the tail vein of mice (Lin *et al.*, 2014). The hydrodynamic injection model involves uptake of DNA-expression vectors to the liver after tail-vein injection into a rodent model of choice, often mice (reviewed by:Yokoo *et al.*, 2016). CRISPR/CAS-mediated targeting of HBV when co-injected with HBV during hydrodynamic injection caused a drop in HBsAg from the sera and a drop in HBV replicative DNA intermediates in the livers of mice that were hydrodynamically injected with HBV and CRISPR/CAS-targeting vectors, compared to the control group of HBV- and mock-targeting CRISPR/CAS vectors (Lin *et al.*, 2014). This study showed that targeting of HBV by CRISPR/CAS technology was possible, but did not specifically demonstrate targeting of HBV cccDNA, as the hydrodynamic injection model does not support cccDNA formation in mouse livers (Yang *et al.*, 2002). Similar types of CRISPR/CAS-targeting systems have been applied and were successful at targeting multiple HBV genotypes (Liu *et al.*, 2015; Wang *et al.*, 2015a). Seeger and colleagues also used hNTCP-expressing HepG2 cells to demonstrate that cccDNA can be targeted in HBV-infected cells (Seeger and Sohn, 2014). Analysis of cccDNA from CRISPR/CAS-targeted, HBV-infected, hNTCP-expressing HepG2 cells showed that mutations and deletions of various sizes were recovered from cccDNA in infected cells (Seeger and Sohn, 2014). An important aspect of this study was that the cccDNA was derived from an authentic HBV infection, demonstrating that this cccDNA could be targeted by the CRISPR/CAS system. The targeting of cccDNA by CRISPR/CAS is an important step forward for the development of more specific anti-HBV therapeutics, but this strategy still needs to be developed further such that it can be used as an actual therapy in patients chronically infected with HBV. Certain concerns currently exist when considering the use of a CRISPR/CAS targeting system as an anti-HBV therapy. Specifically, possible integration of HBV cccDNA into the host chromosome as a result of CRISPR/CAS targeting is a potential concern for random integration events that could still promote genomic instability in treated individuals. These concerns will need to be explored prior to use of CRISPR/CAS targeting as a patient therapy (reviewed in: Soppe and Lebbink, 2017; Yang and Chen, 2017).

HBV capsid-assembly inhibitors

Recently identified HBV capsid-assembly inhibitors have become an exciting new area of investigation as novel compounds for treating a chronic HBV infection (Cho *et al.*, 2014; Wang *et al.*, 2015c; Yang *et al.*, 2016). These compounds act to prevent capsid assembly, leading to reported formation of genome-free capsids in cell-culture models and subsequent decreases in intracellular HBV DNA levels (Cho *et al.*, 2014; Wang *et al.*, 2015c; Yang *et al.*, 2016). At least one of the capsid assembly inhibitors was tested for combinatorial effects with NUCs and had a positive synergistic effect (Cho *et al.*, 2014). Additional work will likely be required for optimization and testing of the capsid assembly inhibitors such that effective and safe pharmacological compounds can be developed, but capsid inhibitors present another new opportunity to target HBV more specifically than some of the currently available therapies (Table 4.2).

Summary and future perspectives

Chronic infection by HBV has been linked to the development of liver cancer; however, the specific mechanisms that influence the progression from a chronic HBV infection to the development of HBV-associated liver cancer are still not completely understood. It is not likely that a single HBV-related change in the cell-cycle, hepatocyte metabolism, or apoptotic pathways is causative, and it is more likely that development of HBV-associated disease or cancer is a complex series of events that results from decades of subtle HBV-induced changes to infected hepatocytes in an environment of inflammation and potential exposure to other carcinogenic factors. HBV-mediated changes to hepatocytes could include consequences of insertion of the HBV genome into host chromosomes, chronic inflammation, and actions of HBV proteins, such as HBx. Cumulatively, HBV and HBx-mediated changes to hepatocyte physiology could create an environment that is favorable to HBV replication but could also have long-term negative consequences for hepatocytes and the liver, thus generating a potentially oncogenic environment.

Despite the availability of a protective vaccine, HBV infection and associated diseases remain a global health concern. Although some effective therapies exist that can control HBV replication, such as NUCs and INFα, none are curative, and ongoing research centered around the changes in cellular signaling pathways induced by HBV infections are critical for developing novel anti-HBV therapies that are needed to eradicate chronic HBV infection as a global health concern. Advances in new molecular technologies, such as CRISPR/CAS system, present a targeting platform specific to HBV cccDNA, a long standing and previously intractable therapeutic target. The recent discovery of a HBV cell surface receptor, hNTCP, also presents a new opportunity to develop drugs that target hNTCP and control or prevent HBV infection of hepatocytes. Finally, the recent identification of new HBV capsid-assembly inhibitors provides another promising avenue for developing treatments for HBV infections. The outlook for the next generation of specific anti-HBV therapeutics is promising, and with continued research to identify novel therapeutic strategies to clear an HBV infection, there is hope that HBV infection and associated diseases such as HCC will fade as a global health concern.

References

Abara, W.E., Collier, M.G., and Teshale, E.H. (2017). Impact of universal infant hepatitis B vaccination in the US-affiliated Pacific Islands, 1985-2015. Vaccine *35*, 997-1000.

Ahn, J.-Y., Jung, E-Y., Kwun, H-J., Lee, C-W., Sung, Y-C., and Jang, K. (2002). Dual effects of hepatitis B virus X protein on the regulation of cell-cycle control depdending on the status of cellular p53. Journal of General Virology *83*, 2765-2772.

Anwer, M.S., and Stieger, B. (2014). Sodium-dependent bile salt transporters of the SLC10A transporter family: more than solute transporters. Pflugers Archiv : European journal of physiology *466*, 77-89.

Appelman, M.D., Chakraborty, A., Protzer, U., McKeating, J.A., and van de Graaf, S.F. (2017). N-Glycosylation of the Na+-Taurocholate Cotransporting Polypeptide (NTCP) Determines Its Trafficking and Stability and Is Required for Hepatitis B Virus Infection. PloS one *12*, e0170419.

Bagga, S., and Bouchard, M.J. (2014). Cell cycle regulation during viral infection. Methods Mol Biol *1170*, 165-227.

Bagga, S., Rawat, S., Ajenjo, M., and Bouchard, M.J. (2016). Hepatitis B virus (HBV) X protein-mediated regulation of hepatocyte metabolic pathways affects viral replication. Virology *498*, 9-22.

Bancroft, W.H., Mundon, F.K., and Russell, P.K. (1972). Detection of additional antigenic determinants of hepatitis B antigen. Journal of immunology *109*, 842-848.

Bar-Yishay, I., Shaul, Y., and Shlomai, A. (2011). Hepatocyte metabolic signalling pathways and regulation of hepatitis B virus expression. Liver international : official journal of the International Association for the Study of the Liver *31*, 282-290.

Barrangou, R., Fremaux, C., Deveau, H., Richards, M., Boyaval, P., Moineau, S., Romero, D., and Horvath, P. (2007). CRISPR Provides Acquired Resistance Against Viruses in Prokaryotes Science *315*.

Beames, B., and Lanford, R.E. (1993). Carboxy-terminal truncations of the HBV core protein affect capsid formation and the apparent size of encapsidated HBV RNA. Virology *194*, 597-607.

Becker, S., Lee, T-H., Butel, J. and Slagle, B. (1998). Hepatitis B Virus X Protein Interferes with Cellular DNA Repair. Journal of virology, 1.

Benn, J., Schneider, R. (1995). Hepatitis B virus HBx protein deregulates cell cycle checkpoint controls Proceedings of the National Academy of Sciences of the United States of America *92*, 11215-11219.

Berg, J., Tymoczko, JL, Stryer, L (2002). Chapter 30: The Integration of Metabolism. In Biochemistry (Michelle Julet), pp. 851-854.

Bijsmans, I.T., Bouwmeester, R.A., Geyer, J., Faber, K.N., and van de Graaf, S.F. (2012). Homo- and hetero-dimeric architecture of the human liver Na(+)-dependent taurocholate co-transporting protein. The Biochemical journal *441*, 1007-1015.

Bill, C., Summers, J (2004). Genomic DNA double-strand breaks are targets for hepadnaviral DNA integration. Proceedings of the National Academy of Sciences of the United States of America *101*, 11135-11140.

Birnbaum, F., and Nassal, M. (1990). Hepatitis B virus nucleocapsid assembly: primary structure requirements in the core protein. Journal of virology *64*, 3319-3330.

Blank, A., Markert, C., Hohmann, N., Carls, A., Mikus, G., Lehr, T., Alexandrov, A., Haag, M., Schwab, M., Urban, S., *et al.* (2016). First-in-human application of the novel hepatitis B and hepatitis D virus entry inhibitor myrcludex B. Journal of hepatology *65*, 483-489.

Block, T.M., Guo, H., and Guo, J.T. (2007). Molecular virology of hepatitis B virus for clinicians. Clinics in liver disease *11*, 685-706, vii.

Bonilla Guerrero, R., and Roberts, L.R. (2005). The role of hepatitis B virus integrations in the pathogenesis of human hepatocellular carcinoma. Journal of hepatology *42*, 760-777.

Bouchard, M., Giannakopoulos, S., Wang, E.H., Tanese, N., and Schneider, R.J. (2001). Hepatitis B virus HBx protein activation of cyclin A-cyclin-dependent kinase 2 complexes and G1 transit via a Src kinase pathway. Journal of virology *75*, 4247-4257.

Bouchard, M., Wang, Li-Hua, Schneider, R. (2001). Calcium Signaling by HBx Protein in Hepatitis B Virus DNA Replication. Science *294*.

Bouchard, M.J., and Navas-Martin, S. (2011). Hepatitis B and C virus hepatocarcinogenesis: lessons learned and future challenges. Cancer letters *305*, 123-143.

Bupathi, M., Kaseb, A., Meric-Bernstam, F., and Naing, A. (2015). Hepatocellular carcinoma: Where there is unmet need. Mol Oncol *9*, 1501-1509.

Buti, M. (2014). HBeAg-positive chronic hepatitis B: why do i treat my patients with Nucleos(t)ide analogs? Liver international : official journal of the International Association for the Study of the Liver *34 Suppl 1*, 108-111.

Carman, W. (1997). The clinical significance of surface antigen variants of hepatitis B virus. Journal of Viral Hepatitis *4*, 11-20.

Casciano, J.C., Duchemin, N.J., Lamontagne, R.J., Steel, L.F., and Bouchard, M.J. (2017). Hepatitis B virus modulates store-operated calcium entry to enhance viral replication in primary hepatocytes. PloS one *12*, e0168328.

Chen, Y.F., Chong, C.L., Wu, Y.C., Wang, Y.L., Tsai, K.N., Kuo, T.M., Hong, M.H., Hu, C.P., Chen, M.L., Chou, Y.C., *et al.* (2015). Doxorubicin Activates Hepatitis B Virus Replication by Elevation of p21

(Waf1/Cip1) and C/EBPalpha Expression. PloS one *10*, e0131743.

Cheng, P.N., Liu, W.C., Tsai, H.W., Wu, I.C., Chang, T.T., and Young, K.C. (2011). Association of intrahepatic cccDNA reduction with the improvement of liver histology in chronic hepatitis B patients receiving oral antiviral agents. Journal of medical virology *83*, 602-607.

Chisari, F.V. (2000). Rous-Whipple Award Lecture. Viruses, immunity, and cancer: lessons from hepatitis B. The American journal of pathology *156*, 1117-1132.

Cho, M.H., Jeong, H., Kim, Y.S., Kim, J.W., and Jung, G. (2014). 2-Amino-N-(2,6-dichloropyridin-3-yl)acetamide derivatives as a novel class of HBV capsid assembly inhibitor. Journal of Viral Hepatitis *21*, 843-852.

Claro da Silva, T., Polli, J.E., and Swaan, P.W. (2013). The solute carrier family 10 (SLC10): beyond bile acid transport. Molecular aspects of medicine *34*, 252-269.

Clippinger, A.J., and Bouchard, M.J. (2008). Hepatitis B virus HBx protein localizes to mitochondria in primary rat hepatocytes and modulates mitochondrial membrane potential. Journal of virology *82*, 6798-6811.

Clippinger, A.J., Gearhart, T.L., and Bouchard, M.J. (2009). Hepatitis B virus X protein modulates apoptosis in primary rat hepatocytes by regulating both NF-kappaB and the mitochondrial permeability transition pore. Journal of virology *83*, 4718-4731.

Cooper, A., and Shaul, Y. (2006). Clathrin-mediated endocytosis and lysosomal cleavage of hepatitis B virus capsid-like core particles. The Journal of biological chemistry *281*, 16563-16569.

Coppola, N., Onorato, L., Minichini, C., Di Caprio, G., Starace, M., Sagnelli, C., and Sagnelli, E. (2015). Clinical significance of hepatitis B surface antigen mutants. World J Hepatol *7*, 2729-2739.

Cullen, J., Marion, P., Sherman, G., Hong, X., and Newbold, J. (1990). Hepatic Neoplasm in Aflatoxin-B1-treated, Congential Duck Hepatitis B Virus-infected, and Virus-free Pekin Ducks Cancer Res *50*, 4072-4080.

Dane, D., Cameron, CH., Briggs, M (1970). Virus-like paritcles in serum of patients with Australia-antigen-associated hepatitis Lancet *1*.

De Meyer, S., Gong, Z.J., Suwandhi, W., van Pelt, J., Soumillion, A., and Yap, S.H. (1997). Organ and species specificity of hepatitis B virus (HBV) infection: a review of literature with a special reference to preferential attachment of HBV to human hepatocytes. Journal of viral hepatitis *4*, 145-153.

Decorsiere, A., Mueller, H., van Breugel, P.C., Abdul, F., Gerossier, L., Beran, R.K., Livingston, C.M., Niu, C., Fletcher, S.P., Hantz, O., *et al.* (2016). Hepatitis B virus X protein identifies the Smc5/6 complex as a host restriction factor. Nature *531*, 386-389.

Donohoue, P.D., Barrangou, R., and May, A.P. (2017). Advances in Industrial Biotechnology Using CRISPR-Cas Systems. Trends Biotechnol.

El-Serag, H.B. (2012). Epidemiology of viral hepatitis and hepatocellular carcinoma. Gastroenterology *142*, 1264-1273 e1261.

El-Serag, H.B., and Rudolph, K.L. (2007). Hepatocellular carcinoma: epidemiology and molecular carcinogenesis. Gastroenterology *132*, 2557-2576.

Emery, J.S., and Feld, J.J. (2017). Treatment of hepatitis B virus with combination therapy now and in the future. Best Pract Res Clin Gastroenterol *31*, 347-355.

Fouillot, N., Tlouzeau, S., Rossignol, J-M., and Jean-Jean, O. (1993). Translation of the Hepatitis B Virus P Gene by Ribosomal Scanning as an Alternative to Internal Initiation Journal of virology *67*, 4886-4895.

Fourati, S., Pawlotsky, Jean-Michel (2016). Recent Advances in understanding and diagnosing hepatitis B virus infection F1000 Research.

Ganem, D., and Prince, A.M. (2004). Hepatitis B virus infection--natural history and clinical consequences. The New England journal of medicine *350*, 1118-1129.

Ganem, D., and Schneider, R. (2001). Hepadnaviridae: The Viruses and Their Replication, Vol 2, 4th edn (Fields Virology Lippincott Williams & Wilkins).

Gearhart, T.L., and Bouchard, M.J. (2010a). The hepatitis B virus X protein modulates hepatocyte proliferation pathways to stimulate viral replication. Journal of virology *84*, 2675-2686.

Gearhart, T.L., and Bouchard, M.J. (2010b). Replication of the hepatitis B virus requires a calcium-dependent HBx-induced G1 phase arrest of hepatocytes. Virology *407*, 14-25.

Ghany, M.G., and Doo, E.C. (2009). Antiviral resistance and hepatitis B therapy. Hepatology *49*, S174-184.

Glebe, D., and Bremer, C.M. (2013). The molecular virology of hepatitis B virus. Seminars in liver disease *33*, 103-112.

Glebe, D., Urban, S., Knoop, E.V., Çağ, N., Krass, P., Grün, S., Bulavaite, A., Sasnauskas, K., and Gerlich, W.H. (2005). Mapping of the Hepatitis B Virus Attachment Site by Use of Infection-Inhibiting preS1 Lipopeptides and Tupaia Hepatocytes. Gastroenterology *129*, 234-245.

Golob-Schwarzl, N., Krassnig, S., Toeglhofer, A.M., Park, Y.N., Gogg-Kamerer, M., Vierlinger, K., Schroder, F., Rhee, H., Schicho, R., Fickert, P., *et al.* (2017). New liver cancer biomarkers: PI3K/AKT/mTOR pathway members and eukaryotic translation initiation factors. Eur J Cancer *83*, 56-70.

Gripon, P., Cannie, I., and Urban, S. (2005). Efficient inhibition of hepatitis B virus infection by acylated peptides derived from the large viral surface protein. Journal of virology *79*, 1613-1622.

Guidotti, C. (2006). Immunobiology and Pathogenesis of Viral Hepatitis Annu Rev Pathol Mech Dis, 23-61.

Guidotti, L.G., Ishikawa, T., Hobbs, M.V., Matzke, B., Schreiber, R., and Chisari, F.V. (1996). Intracellular inactivation of the hepatitis B virus by cytotoxic T lymphocytes. Immunity *4*, 25-36.

Guidotti, L.G., Rochford, R., Chung, J., Shapiro, M., Purcell, R., and Chisari, F.V. (1999). Viral clearance without destruction of infected cells during acute HBV infection. Science *284*, 825-829.

Guo, H., Zhou, T., Jiang, D., Cuconati, A., Xiao, G.H., Block, T.M., and Guo, J.T. (2007). Regulation of hepatitis B virus replication by the phosphatidylinositol

3-kinase-akt signal transduction pathway. Journal of virology *81*, 10072-10080.

Hagenbuch, B. (1997). Molecular Properties of hepatic uptake systems for bile acids and organic anions. Journal of Membrane Biology *160*, 1-8.

Hanahan, D., and Weinberg, R.A. (2011). Hallmarks of cancer: the next generation. Cell *144*, 646-674.

Heermann, K.H., Goldmann, U., Schwartz, W., Seyffarth, T., Baumgarten, H., and Gerlich, W.H. (1984). Large surface proteins of hepatitis B virus containing the pre-s sequence. Journal of virology *52*, 396-402.

Hodgson, A.J., Keasler, V.V., and Slagle, B.L. (2008). Premature cell cycle entry induced by hepatitis B virus regulatory HBx protein during compensatory liver regeneration. Cancer Res *68*, 10341-10348.

Honda, K., Seike, M., and Murakami, K. (2015). Benefits of nucleos(t)ide analog treatments for hepatitis B virus-related cirrhosis. World J Hepatol *7*, 2404-2410.

Huang, H.C., Chen, C.C., Chang, W.C., Tao, M.H., and Huang, C. (2012). Entry of hepatitis B virus into immortalized human primary hepatocytes by clathrin-dependent endocytosis. Journal of virology *86*, 9443-9453.

Huang, H.C., Tao, M.H., Hung, T.M., Chen, J.C., Lin, Z.J., and Huang, C. (2014). (-)-Epigallocatechin-3-gallate inhibits entry of hepatitis B virus into hepatocytes. Antiviral research.

Iovine, B., Iannella, M.L., and Bevilacqua, M.A. (2011). Damage-specific DNA binding protein 1 (DDB1): a protein with a wide range of functions. Int J Biochem Cell Biol *43*, 1664-1667.

Iwamoto, M., Watashi, K., Tsukuda, S., Aly, H.H., Fukasawa, M., Fujimoto, A., Suzuki, R., Aizaki, H., Ito, T., Koiwai, O., *et al.* (2014). Evaluation and identification of hepatitis B virus entry inhibitors using HepG2 cells overexpressing a membrane transporter NTCP. Biochemical and biophysical research communications *443*, 808-813.

Jeang, N., Wu T-C (EDs) (2014). Human Cancer Viruses: Principles of Transformation and Pathogenesis Translation Rese Biomed *1*, 108-136.

Junker-Niepmann, M., Bartenschlager, R., and Schaller, H. (1990). A short cis-acting sequence is required for hepatitis B virus pregenome encapsidation and sufficient for packaging of foreign RNA. The EMBO journal *9*, 3389-3396.

Kang, J., Wang, J., Cheng, J., Cao, Z., Chen, R., Li, H., Liu, S., Chen, X., Sui, J., Lu, F. (2016). Down-regulation of NTCP expression by cyclin D1 in hepatitis B virus-related hepatocellular carcinoma has clinical significance Oncotarget.

Kao, J.H. (2014). HBeAg-positive chronic hepatitis B: why do I treat my patients with pegylated interferon? Liver international : official journal of the International Association for the Study of the Liver *34 Suppl 1*, 112-119.

Kay, A., and Zoulim, F. (2007). Hepatitis B virus genetic variability and evolution. Virus Res *127*, 164-176.

Kegel, A., Sjögren, C. (2010). The SMC5/6 Complex: More than Repair? Cold Spring Harb Symposia on Quantitative Biology *75*.

Kim, S.J., Khan, M., Quan, J., Till, A., Subramani, S., and Siddiqui, A. (2013). Hepatitis B virus disrupts mitochondrial dynamics: induces fission and mitophagy to attenuate apoptosis. PLoS pathogens *9*, e1003722.

Kim, V., Abreu, R.M., Nakagawa, D.M., Baldassare, R.M., Carrilho, F.J., and Ono, S.K. (2016). Pegylated interferon alfa for chronic hepatitis B: systematic review and meta-analysis. J Viral Hepat *23*, 154-169.

Koike, K., Moriya, K., Yotsuyanagi, H., Iino, S., and Kurokawa, K. (1994). Induction of cell cycle progression by hepatitis B virus HBx gene expression in quiescent mouse fibroblasts. J Clin Invest *94*, 44-49.

Koike, K.M., K. Yotsuyanagi, H., Shintani, Y., Fujie, H., Tsutsumi, T., Kimura, S. (1998). Compensatory apoptosis in preneoplastic liver of a transgenic mouse model for viral hepatocarcinogenesis Cancer letters *134*, 181-186.

Kurbanov, F., Tanaka, Y., Kramvis, A., Simmonds, P., and Mizokami, M. (2008). When should "I" consider a new hepatitis B virus genotype? Journal of virology *82*, 8241-8242.

Lamontagne, R.J., Bagga, S., and Bouchard, M.J. (2016). Hepatitis B virus molecular biology and pathogenesis. Hepatoma Res *2*, 163-186.

Lan, Y., Li, J., Liao, W-Y., Ou, J-H. (1999). Roles of the Three Major Phosphorlation Sites of Hepatitis B Virus Core Protein in Viral Replication. Virology *259*.

Lan, Y.T., Li, J., Liao, W., and Ou, J. (1999). Roles of the three major phosphorylation sites of hepatitis B virus core protein in viral replication. Virology *259*, 342-348.

Lee, S., Lee, M.J., Zhang, J., Yu, G.R., and Kim, D.G. (2016). C-terminal-truncated HBV X promotes hepato-oncogenesis through inhibition of tumor-suppressive beta-catenin/BAMBI signaling. Exp Mol Med *48*, e275.

Lee, S., Tarn, C., Wang, W.H., Chen, S., Hullinger, R.L., and Andrisani, O.M. (2002). Hepatitis B virus X protein differentially regulates cell cycle progression in X-transforming versus nontransforming hepatocyte (AML12) cell lines. The Journal of biological chemistry *277*, 8730-8740.

Lee, T.-H., Elledge, S., and Butel, J. (1995). Hepatitis B Virus X Protein Interactrs with a Probably Cellular DNA Repair Protein. Journal of virology *69*.

Lee, Y.I., Hwang, J.M., Im, J.H., Lee, Y.I., Kim, N.S., Kim, D.G., Yu, D.Y., Moon, H.B., and Park, S.K. (2004). Human hepatitis B virus-X protein alters mitochondrial function and physiology in human liver cells. The Journal of biological chemistry *279*, 15460-15471.

Leupin, O., Bontron, S., Schaeffer, C., and Strubin, M. (2005). Hepatitis B virus X protein stimulates viral genome replication via a DDB1-dependent pathway distinct from that leading to cell death. Journal of virology *79*, 4238-4245.

Liang, X., Liu, Y., Zhang, Q., Gao, L., Han, L., Ma, C., Zhang, L., Chen, Y.H., and Sun, W. (2006). Hepatitis B Virus Sensitizes Hepatocytes to TRAIL-Induced Apoptosis through Bax. The Journal of Immunology *178*, 503-510.

Liao, W., and Ou, J.H. (1995). Phosphorylation and nuclear localization of the hepatitis B virus core protein: significance of serine in the three repeated SPRRR motifs. Journal of virology *69*, 1025-1029.

Lien, J.M., Aldrich, C.E., and Mason, W.S. (1986). Evidence that a capped oligoribonucleotide is the primer for duck hepatitis B virus plus-strand DNA synthesis. Journal of virology *57*, 229-236.

Lim, K.H., Choi, H.S., Park, Y.K., Park, E.S., Shin, G.C., Kim, D.H., Ahn, S.H., and Kim, K.H. (2013). HBx-induced NF-kappaB signaling in liver cells is potentially mediated by the ternary complex of HBx with p22-FLIP and NEMO. PloS one *8*, e57331.

Lin, S.R., Yang, H.C., Kuo, Y.T., Liu, C.J., Yang, T.Y., Sung, K.C., Lin, Y.Y., Wang, H.Y., Wang, C.C., Shen, Y.C., *et al.* (2014). The CRISPR/Cas9 System Facilitates Clearance of the Intrahepatic HBV Templates In Vivo. Molecular therapy Nucleic acids *3*, e186.

Liu, H., Xu, J., Zhou, L., Yun, X., Chen, L., Wang, S., Sun, L., Wen, Y., and Gu, J. (2011). Hepatitis B virus large surface antigen promotes liver carcinogenesis by activating the Src/PI3K/Akt pathway. Cancer Res *71*, 7547-7557.

Liu, X., Hao, R., Chen, S., Guo, D., and Chen, Y. (2015). Inhibition of hepatitis B virus by the CRISPR/Cas9 system via targeting the conserved regions of the viral genome. J Gen Virol *96*, 2252-2261.

Locarnini, S., Littlejohn, M., Aziz, M.N., and Yuen, L. (2013). Possible origins and evolution of the hepatitis B virus (HBV). Seminars in cancer biology *23*, 561-575.

Loeb, D.D., Hirsch, R.C., and Ganem, D. (1991). Sequence-independent RNA cleavages generate the primers for plus strand DNA synthesis in hepatitis B viruses: implications for other reverse transcribing elements. The EMBO journal *10*, 3533-3540.

Lok, A., and McMahon, B. (2001). Chronic hepatitis B. Hepatology *34*, 1225-1241.

Lozano, R., et al. (2012). Global and regional mortality from 235 causes of death for 20 age groups in 1990 and 2010: a systematic analysis for the Global Burden of Disease Study 2010. The Lancet *380*.

Lucifora, J., Xia, Y., Reisinger, F., Zhang, K., Stadler, D., Cheng, X., Sprinzl, M., Koppensteiner, H., Makowska, Z., Volz, T., Remouchamps, C., Chou, Wen-Min, Thasler, W., Browning, J., Dejardin, E., Dandri, M., Schindler, M., Heikenwalder, M., Protzer, U. (2014). Specific and Nonhepatotoxic degradation of nuclear hepatitis b virus cccDNA. Science *343*.

Luedde, T., and Schwabe, R.F. (2011). NF-kappaB in the liver--linking injury, fibrosis and hepatocellular carcinoma. Nat Rev Gastroenterol Hepatol *8*, 108-118.

Lutgehetmann, M., Mancke, L.V., Volz, T., Helbig, M., Allweiss, L., Bornscheuer, T., Pollok, J.M., Lohse, A.W., Petersen, J., Urban, S., *et al.* (2012). Humanized chimeric uPA mouse model for the study of hepatitis B and D virus interactions and preclinical drug evaluation. Hepatology *55*, 685-694.

Machida, A., Ohnuma, H., Tsuda, F., Yoshikawa, A., Hoshi, Y., Tanaka, T., Kishimoto, S., Akahane, Y., Miyakawa, Y., and Mayumi, M. (1991). Phosphorylation in the carboxyl-terminal domain of the capsid protein of hepatitis B virus: evaluation with a monoclonal antibody. Journal of virology *65*, 6024-6030.

Meier, A., Mehrle, S., Weiss, T.S., Mier, W., and Urban, S. (2013). Myristoylated PreS1-domain of the hepatitis B virus L-protein mediates specific binding to differentiated hepatocytes. Hepatology *58*, 31-42.

Melegari, M., Wolf, S.K., and Schneider, R.J. (2005). Hepatitis B virus DNA replication is coordinated by core protein serine phosphorylation and HBx expression. Journal of virology *79*, 9810-9820.

Miao, J., Chen, G.G., Chun, S.Y., and Lai, P.P. (2006). Hepatitis B virus X protein induces apoptosis in hepatoma cells through inhibiting Bcl-xL expression. Cancer letters *236*, 115-124.

Mina, T., Amini Bavil Olyaee, S., Tacke, F., Maes, P., Van Ranst, M., and Pourkarim, M.R. (2015). Genomic Diversity of Hepatitis B Virus Infection Associated With Fulminant Hepatitis B Development. Hepat Mon *15*, e29477.

Moon, E.-J., Jeong, Chul-Ho, Jeong, Joo-Won, Rok Kim, Kwang, Yu, Dae-Yeul, Murakami, S., Kim, Chul Woo, Kim, Kyu-Won (2003). Hepatitis B virus X protein induced angiogenesis by stabililzing hypoxia-inducible factor 1 alpha. FASEB Journal.

Muhlfeld, S., Domanova, O., Berlage, T., Stross, C., Helmer, A., Keitel, V., Haussinger, D., and Kubitz, R. (2012). Short-term feedback regulation of bile salt uptake by bile salts in rodent liver. Hepatology *56*, 2387-2397.

Mukherji, A., Janbandhu, V.C., and Kumar, V. (2007). HBx-dependent cell cycle deregulation involves interaction with cyclin E/A-cdk2 complex and destabilization of p27Kip1. The Biochemical journal *401*, 247-256.

Murphy, C.M., Xu, Y., Li, F., Nio, K., Reszka-Blanco, N., Li, X., Wu, Y., Yu, Y., Xiong, Y., and Su, L. (2016). Hepatitis B Virus X Protein Promotes Degradation of SMC5/6 to Enhance HBV Replication. Cell Rep *16*, 2846-2854.

Murray, J., Carr, A. (2008). SMC5/6: a link between DNA repair and unidirectional replication. Nature reviews Molecular cell biology *9*.

Na, B., Huang, Z., Wang, Q., Qi, Z., Tian, Y., Lu, C.C., Yu, J., Hanes, M.A., Kakar, S., Huang, E.J., *et al.* (2011). Transgenic expression of entire hepatitis B virus in mice induces hepatocarcinogenesis independent of chronic liver injury. PloS one *6*, e26240.

Nassal, M. (1992). The Arginine-Rich Domain of the Hepatitis B Virus Core Protein is Required for Pregenome Encapsidation and Productive Viral Positive-Strand DNA Synthesis but not for Virus Assembly Journal of virology *66*, 4107-4116.

Neurath, A.R., Kent, S.B., Strick, N., Taylor, P., and Stevens, C.E. (1985). Hepatitis B virus contains pre-S gene-encoded domains. Nature *315*, 154-156.

Ng, K.-Y., Chai, S., Tong, M., Guan, X-Y., Lin, C-H., Ching, Y-P., Xie, D., Sze-Lok Cheng, A., Ma, S. (2016). C-terminal truncated hepatitis B virus X protein promotes hepatocellular carcinogenesis through induction of cancer and stem cell-like properties. Oncotarget *7*, 24005-24017.

Ni, Y., Lempp, F.A., Mehrle, S., Nkongolo, S., Kaufman, C., Falth, M., Stindt, J., Koniger, C., Nassal, M., Kubitz, R., *et al.* (2014). Hepatitis B and D viruses exploit sodium taurocholate co-transporting polypeptide for species-specific entry into hepatocytes. Gastroenterology *146*, 1070-1083.

Nishikawa, H., and Osaki, Y. (2013). Clinical significance of occult hepatitis B infection in progression of liver disease and carcinogenesis. Journal of Cancer *4*, 473-480.

Nkongolo, S., Ni, Y., Lempp, F.A., Kaufman, C., Lindner, T., Esser-Nobis, K., Lohmann, V., Mier, W., Mehrle, S., and Urban, S. (2014). Cyclosporin A inhibits hepatitis B and hepatitis D virus entry by cyclophilin-independent interference with the NTCP receptor. Journal of hepatology *60*, 723-731.

Nordlie, R.a.F., J. (1999). Regulation of glucose production by the liver Annu Rev Nutr *19*, 379-406.

PATH (2014). Protecting newborns from hepatitis B in Vietnam. PATH: Vaccines and Immunization

PATH; National Expanded Program on Immuniazation, V. (2007). Increasing Hepatitis B Vaccine Coverage Rates by Implementing an Out-of-the-Cold-Chain Strategy. PATH.

Pinto, M., Dawar, M., Krajden, M., Naus, M., and Scheifele, D.W. (2017). Will Infant Hepatitis B Vaccination Protect into Adulthood? Extended Canadian Experience After a 2, 4, 6 Month Immunization Schedule. Pediatr Infect Dis J.

Pourkarim, M.R., Amini-Bavil-Olyaee, S., Kurbanov, F., Van Ranst, M., and Tacke, F. (2014). Molecular identification of hepatitis B virus genotypes/subgenotypes: revised classification hurdles and updated resolutions. World journal of gastroenterology : WJG *20*, 7152-7168.

Quetier, I., Brezillon, N., Revaud, J., Ahodantin, J., DaSilva, L., Soussan, P., and Kremsdorf, D. (2015). C-terminal-truncated hepatitis B virus X protein enhances the development of diethylnitrosamine-induced hepatocellular carcinogenesis. J Gen Virol *96*, 614-625.

Rawat, S., and Bouchard, M. (2014). The Hepatitis B Virus HBx protein activates AKT to simultaneously regulate HBV replication and hepatocyte survival. Journal of virology.

Rawat, S., Clippinger, A.J., and Bouchard, M.J. (2012). Modulation of apoptotic signaling by the hepatitis B virus X protein. Viruses *4*, 2945-2972.

Revill, P., and Locarnini, S. (2016). Antiviral strategies to eliminate hepatitis B virus covalently closed circular DNA (cccDNA). Curr Opin Pharmacol *30*, 144-150.

Sahin, F., Kannangai, R., Adegbola, O., Wang, J., Su, G., and Torbenson, M. (2004). mTOR and P70 S6 Kinase Expression in Primary Liver Neoplasms Clinical Cancer Research *10*, 8421-8425.

Said, Z.N. (2011). An overview of occult hepatitis B virus infection. World journal of gastroenterology : WJG *17*, 1927-1938.

Schaefer, S. (2007). Hepatitis B virus taxonomy and hepatitis B virus genotypes. World J Gastroenterol *13*, 14-21.

Seeger, C., and Mason, W.S. (2000a). Hepatitis B virus biology. Microbiology and molecular biology reviews : MMBR *64*, 51-68.

Seeger, C., and Mason, W.S. (2000b). Hepatitis B Virus Biology. Microbiology and Molecular Biology Reviews *64*, 51-68.

Seeger, C., and Mason, W.S. (2015). Molecular biology of hepatitis B virus infection. Virology *479-480*, 672-686.

Seeger, C., and Sohn, J.A. (2014). Targeting Hepatitis B Virus With CRISPR/Cas9. Molecular therapy Nucleic acids *3*, e216.

Seeger, C., Mason, W., and Zoulim, F. (2013). Hepadnaviruses. In Field's virology, H.P. Knipe DM, ed. (Philadelphia, PA: Lippincott Williams & Wilkins), pp. 3376-3436.

Seeger, C., Mason, W.S., and Zoulim, F. (2007). Hepadnaviruses. In Fields virology, D.M. Knipe, and P.M. Howley, eds. (Philadelphia, PA: Lippincot Williams and Wilkins), pp. 2977-3029.

Seeger, C., Zoulim, F., and Mason, W. (2013). Hepadnaviruses In Fields Virology (eds Knipe, DM & Howley, PM), D.a.H. Knipe, P., ed. (Lippincott Williams & Wilkins), pp. 2185-2221.

Shafritiz, D., Kew, M (1981). Identificaiton of Integrated Hepatitis B Virus DNA Sequences in Human Hepatocellular Carcinomas Hepatology *1*.

Shin, E.C., Sung, P.S., and Park, S.H. (2016). Immune responses and immunopathology in acute and chronic viral hepatitis. Nat Rev Immunol *16*, 509-523.

Shin, H.J., Park, Y.H., Kim, S.U., Moon, H.B., Park, D.S., Han, Y.H., Lee, C.H., Lee, D.S., Song, I.S., Lee, D.H., *et al.* (2011). Hepatitis B virus X protein regulates hepatic glucose homeostasis via activation of inducible nitric oxide synthase. The Journal of biological chemistry *286*, 29872-29881.

Shlomai, A., and Shaul, Y. (2008). The "metabolovirus" model of hepatitis B virus suggests nutritional therapy as an effective anti-viral weapon. Med Hypotheses *71*, 53-57.

Sitterlin, D., Lee, T-H., Prigent, S., Tiollais, P.,Butel, J., and Transy, C. (1997). Interaction of the UV-Damged DNA-Binding Protein with Hepatitis B Virus X Protein Is Conversved among Mammalian Hepadnaviruses and Restricted to Transactivation-Proficient X-insertion Mutants Journal of virology *71*.

Slagle, B.L., Andrisani, O.M., Bouchard, M.J., Lee, C.G., Ou, J.H., and Siddiqui, A. (2015). Technical standards for hepatitis B virus X protein (HBx) research. Hepatology *61*, 1416-1424.

Sonneveld, M.J., and Janssen, H.L. (2010). Pros and Cons of Peginterferon Versus Nucleos(t)ide Analogues for Treatment of Chronic Hepatitis B. Curr Hepat Rep *9*, 91-98.

Soppe, J.A., and Lebbink, R.J. (2017). Antiviral Goes Viral: Harnessing CRISPR/Cas9 to Combat Viruses in Humans. Trends in microbiology.

Stross, C., Helmer, A., Weissenberger, K., Gorg, B., Keitel, V., Haussinger, D., and Kubitz, R. (2010). Protein kinase C induces endocytosis of the sodium taurocholate cotransporting polypeptide. American journal of physiology Gastrointestinal and liver physiology *299*, G320-328.

Stross, C., Kluge, S., Weissenberger, K., Winands, E., Haussinger, D., and Kubitz, R. (2013). A dileucine motif is involved in plasma membrane expression and endocytosis of rat sodium taurocholate cotransporting polypeptide (Ntcp). American journal of physiology Gastrointestinal and liver physiology *305*, G722-730.

Su, F., Theodosis, C.N., and Schneider, R.J. (2001). Role of NF-kappaB and myc proteins in apoptosis induced by hepatitis B virus HBx protein. Journal of virology *75*, 215-225.

Summers, J., Jilbert, A.R., Yang, W., Aldrich, C.E., Saputelli, J., Litwin, S., Toll, E., and Mason, W.S. (2003). Hepatocyte turnover during resolution of a transient hepadnaviral infection. Proceedings of the National Academy of Sciences of the United States of America *100*, 11652-11659.

Sun, M., Tan, G., Song, J., Wang, J., and Wu, X. (2016). Profile of HBV polymerase gene mutations during entecavir treatment in patients with chronic hepatitis B. Clin Res Hepatol Gastroenterol *40*, 590-596.

Sunbul, M. (2014). Hepatitis B virus genotypes: global distribution and clinical importance. World J Gastroenterol *20*, 5427-5434.

Sung, W.K., Zheng, H., Li, S., Chen, R., Liu, X., Li, Y., Lee, N.P., Lee, W.H., Ariyaratne, P.N., Tennakoon, C., *et al.* (2012). Genome-wide survey of recurrent HBV integration in hepatocellular carcinoma. Nat Genet *44*, 765-769.

Tatematsu, K., Tanaka, Y., Kurbanov, F., Sugauchi, F., Mano, S., Maeshiro, T., Nakayoshi, T., Wakuta, M., Miyakawa, Y., and Mizokami, M. (2009). A genetic variant of hepatitis B virus divergent from known human and ape genotypes isolated from a Japanese patient and provisionally assigned to new genotype J. Journal of virology *83*, 10538-10547.

Thimme, R., Wieland, S., Steiger, C., Ghrayeb, J., Reimann, K.A., Purcell, R.H., and Chisari, F.V. (2003a). CD8+ T Cells Mediate Viral Clearance and Disease Pathogenesis during Acute Hepatitis B Virus Infection. Journal of virology *77*, 68-76.

Thimme, R., Wieland, S., Steiger, C., Ghrayeb, J., Reimann, K.A., Purcell, R.H., and Chisari, F.V. (2003b). CD8(+) T cells mediate viral clearance and disease pathogenesis during acute hepatitis B virus infection. Journal of virology *77*, 68-76.

Thomas, E., Yoneda, M., and Schiff, E.R. (2015). Viral Hepatitis: Past and Future of HBV and HDV. Cold Spring Harbor perspectives in medicine *5*.

Tian, Y., Kuo, C.F., Akbari, O., and Ou, J.H. (2016). Maternal-Derived Hepatitis B Virus e Antigen Alters Macrophage Function in Offspring to Drive Viral Persistence after Vertical Transmission. Immunity *44*, 1204-1214.

Tian, Y., Xiao, X., Gong, X., Peng, F., Xu, Y., Jiang, Y., and Gong, G. (2017). HBx promotes cell proliferation by disturbing the cross-talk between miR-181a and PTEN. Sci Rep *7*, 40089.

Tong, S., Diot, C., Gripon, P., Li, J., Vitvitski, L., Trépo, C., Guguen-Guillouzo, C (1991). In vitro replication competence of a cloned hepatitis B virus variant with a nonsense mutation in the distal pre-C region Virology *181*, 733-737.

Tralhao, J.G., Roudier, J., Morosan, S., Giannini, C., Tu, H., Goulenok, C., Carnot, F., Zavala, F., Joulin, V., Kremsdorf, D., *et al.* (2002). Paracrine in vivo inhibitory effects of hepatitis B virus X protein (HBx) on liver cell proliferation: an alternative mechanism of HBx-related pathogenesis. Proceedings of the National Academy of Sciences of the United States of America *99*, 6991-6996.

Tran, T.T., Trinh, T.N., and Abe, K. (2008). New complex recombinant genotype of hepatitis B virus identified in Vietnam. Journal of virology *82*, 5657-5663.

Trepo, C., Chan, H., Lok, A. (2014). Hepatits B virus infection The Lancet *384*.

Tsai, S.Q., and Joung, J.K. (2016). Defining and improving the genome-wide specificities of CRISPR-Cas9 nucleases. Nat Rev Genet *17*, 300-312.

Tsukuda, S., Watashi, K., Iwamoto, M., Suzuki, R., Aizaki, H., Okada, M., Sugiyama, M., Kojima, S., Tanaka, Y., Mizokami, M., *et al.* (2015). Dysregulation of retinoic acid receptor diminishes hepatocyte permissiveness to hepatitis B virus infection through modulation of sodium taurocholate cotransporting polypeptide (NTCP) expression. The Journal of biological chemistry *290*, 5673-5684.

Volz, T., Allweiss, L., Ben, M.M., Warlich, M., Lohse, A.W., Pollok, J.M., Alexandrov, A., Urban, S., Petersen, J., Lutgehetmann, M., *et al.* (2013). The entry inhibitor Myrcludex-B efficiently blocks intrahepatic virus spreading in humanized mice previously infected with hepatitis B virus. Journal of hepatology *58*, 861-867.

Wang, G.H., and Seeger, C. (1992). The reverse transcriptase of hepatitis B virus acts as a protein primer for viral DNA synthesis. Cell *71*, 663-670.

Wang, J., Xu, Z.W., Liu, S., Zhang, R.Y., Ding, S.L., Xie, X.M., Long, L., Chen, X.M., Zhuang, H., and Lu, F.M. (2015a). Dual gRNAs guided CRISPR/Cas9 system inhibits hepatitis B virus replication. World J Gastroenterol *21*, 9554-9565.

Wang, W.H., Gregori, G., Hullinger, R.L., and Andrisani, O.M. (2004). Sustained activation of p38 mitogen-activated protein kinase and c-Jun N-terminal kinase pathways by hepatitis B virus X protein mediates apoptosis via induction of Fas/FasL and tumor necrosis factor (TNF) receptor 1/TNF-alpha expression. Mol Cell Biol *24*, 10352-10365.

Wang, X., Huang, X., Fang, X., Zhang, Y., and Wang, W. (2016). CRISPR-Cas9 System as a Versatile Tool for Genome Engineering in Human Cells. Molecular Therapy - Nucleic Acids *5*, e388.

Wang, X.J., Hu, W., Zhang, T.Y., Mao, Y.Y., Liu, N.N., and Wang, S.Q. (2015b). Irbesartan, an FDA approved drug for hypertension and diabetic nephropathy, is a potent inhibitor for hepatitis B virus entry by disturbing Na(+)-dependent taurocholate cotransporting polypeptide activity. Antiviral research *120*, 140-146.

Wang, Y.J., Lu, D., Xu, Y.B., Xing, W.Q., Tong, X.K., Wang, G.F., Feng, C.L., He, P.L., Yang, L., Tang, W., *et al.* (2015c). A novel pyridazinone derivative inhibits hepatitis B virus replication by inducing genome-free capsid formation. Antimicrobial agents and chemotherapy *59*, 7061-7072.

Watashi, K., Sluder, A., Daito, T., Matsunaga, S., Ryo, A., Nagamori, S., Iwamoto, M., Nakajima, S., Tsukuda, S., Borroto-Esoda, K., *et al.* (2014). Cyclosporin A and its analogs inhibit hepatitis B virus entry into cultured hepatocytes through targeting a membrane transporter, sodium taurocholate cotransporting polypeptide (NTCP). Hepatology *59*, 1726-1737.

WHO (2008). Hepatitis B Fact Sheet.

W.H.O. (2016). Hepatitis B Fact Sheet.

WHO, W.H.O. (2017). Hepatitis B: Fact Sheet.

Wu, B.K., Li, C.C., Chen, H.J., Chang, J.L., Jeng, K.S., Chou, C.K., Hsu, M.T., and Tsai, T.F. (2006). Blocking of G1/S transition and cell death in the regenerating liver of Hepatitis B virus X protein transgenic mice. Biochemical and biophysical research communications *340*, 916-928.

Wu, J., Merlino, G., and Fausto, N. (1994). Establishment and characterization of differentiated nontransformed hepatocyte cell lines derived from mice transgenic for transforming growth factor alpha Proceedings of the National Academy of Sciences of the United States of America *91*, 674-678.

Xia, Y., and Protzer, U. (2017). Control of Hepatitis B Virus by Cytokines. Viruses *9*.

Xia, Y., Stadler, D., Lucifora, J., Reisinger, F., Webb, D., Hösel, M., Michler, T., Wisskirchen, K., Cheng, X., Zhang, K., *et al.* (2016). Interferon-γ and Tumor Necrosis Factor-α Produced by T Cells Reduce the HBV Persistence Form, cccDNA, Without Cytolysis. Gastroenterology *150*, 194-205.

Yan, H., Peng, B., Liu, Y., Xu, G., He, W., Ren, B., Jing, Z., Sui, J., and Li, W. (2014). Viral entry of hepatitis B and D viruses and bile salts transportation share common molecular determinants on sodium taurocholate cotransporting polypeptide. Journal of virology *88*, 3273-3284.

Yan, H., Zhong, G., Xu, G., He, W., Jing, Z., Gao, Z., Huang, Y., Qi, Y., Peng, B., Wang, H., *et al.* (2012). Sodium taurocholate cotransporting polypeptide is a functional receptor for human hepatitis B and D virus. eLife *1*, e00049.

Yang, B., and Bouchard, M.J. (2012). The hepatitis B virus X protein elevates cytosolic calcium signals by modulating mitochondrial calcium uptake. Journal of virology *86*, 313-327.

Yang, H.C., and Chen, P.J. (2017). The potential and challenges of CRISPR-Cas in eradication of hepatitis B virus covalently closed circular DNA. Virus Res.

Yang, H.C., and Kao, J.H. (2014). Persistence of hepatitis B virus covalently closed circular DNA in hepatocytes: molecular mechanisms and clinical significance. Emerging microbes & infections *3*, e64.

Yang, L., Wang, Y.J., Chen, H.J., Shi, L.P., Tong, X.K., Zhang, Y.M., Wang, G.F., Wang, W.L., Feng, C.L., He, P.L., *et al.* (2016). Effect of a hepatitis B virus inhibitor, NZ-4, on capsid formation. Antiviral research *125*, 25-33.

Yang, P.L., Althage, A., Chung, J., and Chisari, F.V. (2002). Hydrodynamic injection of viral DNA: a mouse model of acute hepatitis B virus infection. Proceedings of the National Academy of Sciences of the United States of America *99*, 13825-13830.

Yang, X., Wu, L., Lin, J., Wang, A., Wan, X., Wu, Y., Robson, S.C., Sang, X., and Zhao, H. (2017). Distinct hepatitis B virus integration patterns in hepatocellular carcinoma and adjacent normal liver tissue. Int J Cancer *140*, 1324-1330.

Yeh, C.T., and Ou, J.H. (1991). Phosphorylation of hepatitis B virus precore and core proteins. Journal of virology *65*, 2327-2331.

Yokoo, T., Kamimura, K., Abe, H., Kobayashi, Y., Kanefuji, T., Ogawa, K., Goto, R., Oda, M., Suda, T., and Terai, S. (2016). Liver-targeted hydrodynamic gene therapy: Recent advances in the technique. World J Gastroenterol *22*, 8862-8868.

Zhang, H., Huang, C., Wang, Y., Lu, Z., Zhuang, N., Zhao, D., He, J., and Shi, L. (2015). Hepatitis B Virus X Protein Sensitizes TRAIL-Induced Hepatocyte Apoptosis by Inhibiting the E3 Ubiquitin Ligase A20. PloS one *10*, e0127329.

Zhang, Y., Chengs, J., Liu, C. (2016). The Unique Role of the Hepatitis virus B X Protein on HEK293 cell morphology and cellular change. Arch Virol *161*, 1347-1352.

Zhao, L.H., Liu, X., Yan, H.X., Li, W.Y., Zeng, X., Yang, Y., Zhao, J., Liu, S.P., Zhuang, X.H., Lin, C., *et al.* (2016). Genomic and oncogenic preference of HBV integration in hepatocellular carcinoma. Nat Commun *7*, 12992.

Zhu, M., Guo, J., Li, W., Lu, Y., Fu, S., Xie, X., Xia, H., Dong, X., Chen, Y., Quan, M., Zheng, S., Xie, K., Li, M. (2015). Hepatitis B Virus X protein induces expression of alpha-fetoprotein and activates PI3K/mTOR signaling pathway in liver cells Oncotarget *6* 12196-12208.

Zou, L.Y., Zheng, B.Y., Fang, X.F., Li, D., Huang, Y.H., Chen, Z.X., Zhou, L.Y., and Wang, X.Z. (2015). HBx co-localizes with COXIII in HL-7702 cells to upregulate mitochondrial function and ROS generation. Oncol Rep *33*, 2461-2467.

Zoulim, F. (2004). Mechanism of viral persistence and resistance to nucleoside and nucleotide analogs in chronic hepatitis B virus infection. Antiviral research *64*, 1-15.

Zoulim, F. (2007). Emerging drugs for hepatitis B. Expert opinion on emerging drugs *12*, 199-217.

Zoulim, F., and Locarnini, S. (2009). Hepatitis B virus resistance to nucleos(t)ide analogues. Gastroenterology *137*, 1593-1608 e1591-1592.

Zoulim, F., Lebosse, F., and Levrero, M. (2016). Current treatments for chronic hepatitis B virus infections. Curr Opin Virol *18*, 109-116.

5

Controlling Human Papillomavirus and Hepatitis B Virus Infections: Prophylatic Vaccines and Future Prospects

Margaret Stanley

Department of Pathology, Cambridge University, Cambridge, UK

Corresponding author: mas1001@cam.ac.uk

Abstract

It is estimated that, globally, 15% of cancers can be attributed to viral infections. Hepatitis B virus (HBV) causes more than 300,000 cases of hepatocellular carcinoma per year and a subset of 13 human papillomaviruses (HPV) particularly HPV type 16 cause more than 530,000 cases of cervical cancer and 100,000 other anogenital and oropharyngeal cancers per year.

HBV infection has been preventable by vaccination since 1982 and vaccination of neonates and infants is highly effective resulting already in decreased rates of new infections, chronic liver disease and HCC. Nonetheless HBV remains a global public health problem with high rates of vertical transmission from mother to child in some regions.

Two prophylactic human papillomavirus (HPV) vaccines composed of virus like particles (VLPs) of the L1 capsid protein have been licensed since 2006/7. Both target infection by the oncogenic HPV's 16 and 18 (the cause of 70% of cervical cancers) and one also targets the low risk HPV types 6 and 11 that cause genital and laryngeal warts. The vaccines are now included in the national immunisation programmes in many countries, with young adolescent peri-pubertal girls the usual cohort for immunisation. Population effectiveness in women is now being demonstrated in countries with high vaccine coverage with significant reductions in cervical intra epithelial neoplasia, genital warts, vaccine HPV type genoprevalence and herd effects in young heterosexual men and older women. A third VLP vaccine targeting 7 oncogenic HPVs, 16, 18, 31, 33, 45, 52 and 58 plus HPV6 and 11 was licensed in the USA and Europe in 2014/2015 and it is predicted could reduce the incidence of cervical cancer by >85%.

Introduction

It is estimated that, globally, 15% of cancers can be attributed to viral infections with the greatest burden falling in the less economically developed countries (de Martel *et al.*, 2012). The evidence that some cancers are caused by infection with a specific virus implies that intervention against these infections, by prevention or treatment, should be able to prevent the vast majority of such cancers worldwide and have a major public health impact. With respect to HBV (hepatitis B virus) and HPV (human papillomavirus) associated cancers this presumption is now underpinned by solid evidence with the development and implementation over the past 3-4 decades of effective prophylactic vaccines against these viral infections.

Hepatitis B

Burden of disease and transmission

Infection with the hepatitis B virus (HBV) is a major global health problem. More than 2 billion individuals are exposed, 250 million are chronically infected (hepatitis B surface antigen HBsAg +ve) and about 25% of these develop serious liver disease including acute and chronic hepatitis, cirrhosis and hepatocellular carcinoma (Dienstag and Delemos, 2014). It is estimated that each year 500,000 new cases of HCC (hepatocellular carcinoma) are attributable to chronic HBV infection with an annual mortality of 300,000 cases. HBV is species specific, only infecting humans and tissue tropic, only undergoing productive infection in the liver (Liaw *et al.*, 2010). The virus is highly infectious, abundantly found in blood and other body fluids of infected persons many of whom are asymptomatic and unaware of their disease. HBV is transmitted via the parenteral route, infusion of infected blood or blood products, needle sharing, penetration of microwounds on skin or mucosae etc (Orlando *et al.*, 2015). In low and middle income countries the vast majority of infections are vertically transmitted from infected mother to child during parturition. Ninety percent of those infected at birth and 30-50% infected between 1 and 4 years of age develop chronic hepatitis (hepatitis B surface antigen, HBsAg +ve) and have a 25% risk of dying prematurely from cirrhosis and/or HCC (Centers for Disease and Prevention, 2011). Current strategies for controlling HBV infection include active immunisation, passive immunisation and chemoprophylaxis with antiviral drugs (Orlando *et al.*, 2015).

HBV prophylactic vaccines

Two types of vaccine, plasma derived and recombinant are available for active immunisation against HBV. HBV cannot be grown efficiently in tissue culture and the first generation vaccines were inactivated and purified HBsAg particles derived from the plasma of chronic carriers. Subsequently this was replaced in most countries by second generation vaccines generated by expression of the gene sequence encoding the HBsAg in the yeast Saccharomyces cereviseae, a molecular biological triumph of its time (Harford *et al.*, 1983; McAleer *et al.*, 1984; Valenzuela *et al.*, 1982). The HBsAg recombinant protein product could be produced in large amounts, was highly immunogenic and immunised chimpanzees were protected from viral challenge (McAleer *et al.*, 1984). Field studies immunising neonates born to highly infectious mothers showed high efficacy and safety (Stevens *et al.*, 1987) paving the way for universal neonatal and worldwide vaccination. Plasma derived vaccine is no longer available in North America and Europe but is still widely used in parts of Asia and India. Both vaccines are comparable in terms of efficacy and durability, contain the HBsAg and elicit the development of specific antibodies, anti-HBs which alone confer protective immunity (Batdelger *et al.*, 2007). The protective titre (immune correlate) is accepted as >10 IU/ml of anti-HBs antibody (Dienstag and Delemos, 2014).

The recommended immunisation schedule is a 3 dose prime, prime, boost of 0, 1 or 2 and 6 months delivered intra muscularly (Centers for Disease and Prevention, 2011). Anti HBs antibody persists in most subjects for many years (≥ 20 years) after immunisation but even when titres fall below 10 IU/ml most subjects continue be protected and show strong recall responses to a booster dose of vaccine. Booster doses are not recommended for healthy immunocompetent individuals who underwent routine immunisation, but booster doses are recommended for immuno-compromised e.g. HIV infected and those at risk of HBV exposure such as health care workers.

Initially the strategy for HBV vaccination was to immunise at risk groups such as babies born to infected mothers, intra venous drug users, sex workers and health care professionals. This strategy was not successful and it was recognised that to reduce virus prevalence and block transmission the infant reservoir of virus had to be targeted by vaccinating babies soon after birth to prevent infection and carriage in early life. This strategy was validated by studies in high prevalence locations where significant decreases after neonatal immunisation in the incidence of acute hepatitis and chronic carriage were shown (Chang, 2014; Wichajarn *et al.*, 2008). In 1992 the World Health Organisation (WHO) advised all countries to include HBV in all universal vaccination programmes and the implementation of this particularly in low and middle income countries with high virus prevalence has resulted in a world wide decrease in carrier rate, disease burden and the preliminary evidence of impact on HCC (Chang *et al.*, 2009).

HBV vaccine impact

Universal hepatitis B immunisation in infancy can effectively reduce the incidence of HCC in children and adolescents (Centers for Disease and Prevention, 2011; Cooper and Mackie, 2011; Gerlich, 2015). A central question however is whether HBV immunisation at birth provides protection against HCC in adults. The incidence rate of liver cancer increases with age, HCC incidence is much higher in adults than children and HBV antigen levels after infant immunisation decline over time (Valenzuela *et al.*, 1982). Taiwan is endemic for chronic HBV infection and HCC, it has the world's first universal HBV immunisation programme for infants starting in July 1986 and in a recent study the long term effects of infant immunisation in preventing HCC were reported (Chang *et al.*, 2016). In this analysis 1509 patients 6-26 years of age diagnosed with HCC between 1983 and 2011 were identified, their HBV vaccination history and prenatal maternal levels of HBVsAg collated and the relative risks (RRs) of vaccinated versus unvaccinated calculated. The RRs for different birth cohorts 6-9, 10-14, 15-19 and 20-26 years were respectively 0.26 [CI 0.17-0.40], 0.34 [CI 0.25-0.48], 0.37 [CI 0.25-0.51] and 0.42 [0.32-0.36]. The incidence of HCC/105 person years in the non-vaccinated cohort was 0.92 compared with 0.23 in the vaccinated cohort leading the authors to conclude that "universal HBV immunisation in infants has successfully prevented liver cancer both in children and young adults".

HBV vaccines what next?

Routine neonatal and infant vaccination programmes have significantly changed the epidemiology of HBV infection (Bonanni *et al.*, 2003; Huang and Lin, 2000) and can be anticipated now to reduce the disease burden of HCC globally (Chang, 2014). Nonetheless about 5% of immunocompetent individuals are non responders and fail to seroconvert after the regular 3 dose regimen or a fourth booster dose. Male gender, old age, individuals with chronic disease states such as kidney, liver, type I diabetes and coeliac disease are major factors in non-responses aside from frank immunodeficiency inherited or acquired particularly HIV infection (Walayat *et al.*, 2015). Of most concern for the long term goal of prevention of HBV caused HCC is the residual rate of mother to child transmission (MTCT). This remains at 10-30% of immunised newborns born to HBsAg+ve mothers despite the combination of passive (with specific HBV immunoglobulins, HBIG) and an accelerated schedule of active immunisation (Lee *et al.*, 2006; Wichajarn *et al.*, 2008). Poor compliance with HBV immunisation either no HBV or HBIG are the two most important causes of HCC prevention failure in Taiwan (Chang *et al.*, 2016) but newer more immunogenic vaccines could enhance HCC prevention.

Various strategies have been, or are under development, to develop vaccines that elicit a more potent immune response. The current licensed yeast recombinant second generation vaccines contain partially misfolded HBsAg and lack the preS1 antigen that carries the major HBV attachment site and neutralising epitopes. Third generation vaccines produced in mammalian cells contain correctly folded HBsAg and the neutralising epitopes of the preS antigens, [reviewed in (Gerlich, 2015)]. These vaccines increase sero-conversion amongst non-responders (Bertino *et al.*, 1997; Rendi-Wagner *et al.*, 2006) and may provide better protection for newborns born to highly infectious mothers. Other strategies to enhance immunogenicity focus on novel adjuvants (Cooper and Mackie, 2011) such as MPL (3-deacylatedmonophosphoryl lipid A) an agonist of toll like receptor 4, which has been shown to induce immune responses in non-responders (Jacques *et al.*, 2002) and the polysaccharide adjuvant delta inulin (Saade *et al.*, 2013).

Human papillomaviruses: Burden of disease

HPVs are a very large branch of the papillomavirus family with the DNA of more than 170 HPV types, many of which have been fully sequenced, having been isolated from tissue biopsies (de Villiers, 2013). Despite the daunting number of HPVs they fall basically into two groups: those that infect skin, or cutaneous surfaces, and those that infect the internal wet squamous mucosal surfaces, particularly the genital tract. Within these groups there are low-risk types (LRHPV), which generate benign lesions, in other words warts, and high-risk or oncogenic types (HRHPV), that are associated with cancers and their precursor lesions (Forman *et al.*, 2012).

Approximately 40 HPV types regularly or sporadically infect the mucosal epithelial surfaces of the lower genital tract: LRHPV types, HPV6 and 11 cause more than 90% of external genital warts with minor types (HPV42, 44) and some HRHPVs contributing to the remaining 10% (Ball *et al.*, 2011). Genital warts are the most common viral sexually transmitted disease, they are highly infectious, refractory to therapy, result in significant morbidity and are a health economic burden (Lacey *et al.*, 2006). A maternal history of genital warts is associated with a 231-fold risk for recurrent respiratory papillomatosis (RRP), an uncommon but potentially devastating disease, characterized by the growth of wart-like benign neoplasms throughout the aerodigestive tract that often requires repeated surgeries (Larson and Derkay, 2010)
HPV associated cancer in the genital tract is dominated by HPV16 and HPV18 which, with their close relatives, 31, 33, 35, 39, 45, 51, 52, 56, 58, and 59, are the cause of virtually all cervical cancers and the majority of the high grade cervical cancer precursor lesions CIN2/3 (cervical intra-epithelial neoplasia grades 2 and 3) (Serrano *et al.*, 2015). Thus 99% or more of biopsies of invasive cervical cancer

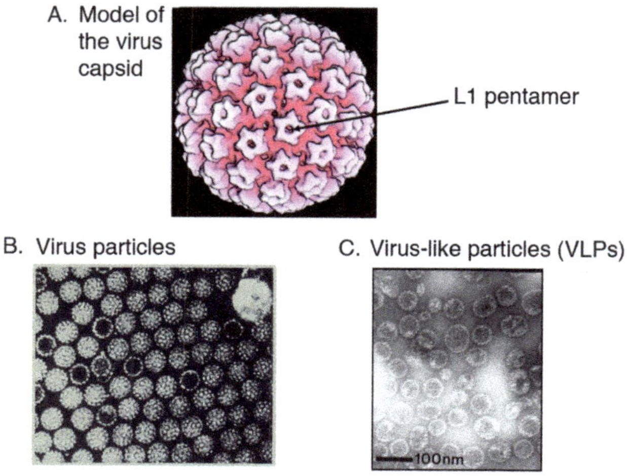

Figure 5.1. Panel A. A model of the papillomavirus coat or capsid. There are 2 coat proteins L1 and L2. The rosette like surface structures (arrowed) are pentamers each consisting of 5 molecules of L1, one molecule of L2 fits into the central dimple of each pentamer. **Panel B.** Papillomavirus particles, both full (contain DNA) and empty particles can be seen. **Panel C.** HPV16 L1 virus-like particles, VLPs, made by expressing the HPV16 *L1* gene in insect cells using a recombinant baculovirus. The L1 protein so expressed, spontaneously assembles into empty capsids or VLPs that are morphologically similar to the empty virus particles seen in panel B. From Stanley *et al.* (2006) with permission.

worldwide, and approximately 80% of CIN 2/3 contain HRHPV DNA sequences (Guan *et al.*, 2012). HPV16 dominates, and is present in at least 50% of cancers irrespective of geographical location, followed by HPV18, 7-20%. However invasive cervical cancer is not the only malignant disease associated with HRHPV infection, HPV DNA sequences are found in most anal and tonsil carcinomas, and a proportion of vulvar, vaginal, penile and head and neck cancers (de Martel *et al.*, 2012). HPV16 again is the dominant oncogenic type (Alemany *et al.*, 2016; Alemany *et al.*, 2015; Serrano *et al.*, 2015). Overall, the global malignant burden attributable to HPV infection is calculated to be 5.2% of all cancers.

Human papillomaviruses: Prophylactic vaccines

The currently licensed HPV vaccines are comprised of virus-like particles (VLPs) formed by the major capsid protein L1. HPV VLPs are made using sophisticated recombinant technologies in which the L1 gene of specific HPV types is recombined into the host genome of the yeast *Saccharomyces cerevisiae* or the insect virus baculovirus and the L1 protein expressed via these recombinant vectors. The chemistry of the expressed protein is such that it spontaneously assembles into VLPs that are morphologically and antigenically similar to the wild-type virus particle (Figure 5.1). However, VLPs lack DNA and are non-infectious and non-oncogenic (Stanley *et al.*, 2006). There are three licensed prophylactic HPV L1 VLP vaccines. These are Cervarix®, a 2 valent HPV (2HPV) 16/18 product from GlaxoSmithKline Biologicals Rixensart, Belgium, Gardasil® (also known as Silgard), a 4 valent HPV (4HPV,) 6, 11, 16, 18 product and Gardasil 9 a 9 valent (9HPV) 6, 11, 16, 18, 31, 33, 45, 52, 58 product, both from MSD, Whitehouse Station, New Jersey, USA (Table 5.1).

All vaccines have undergone large, randomised, placebo controlled, double blind trials (RCTs) in young women, 15-26 years old and have demonstrated >90% efficacy against HPV vaccine type associated ano-genital disease and persistent infection in individuals naïve for vaccine HPV types at trial entry, and at the completion of the 3 dose immunisation (Joura *et al.*, 2015; Schiller *et al.*, 2012). Trials have been undertaken in older women >25 years (2HPV) and >26years (Skinner *et al.*, 2014) but < 45 years (4HPV) (Munoz *et al.*, 2009), and for 4HPV in heterosexual (Giuliano *et al.*, 2011) and homosexual men (Palefsky *et al.*, 2011). In all trials >90% efficacy against high grade intra-epithelial disease and persistent infection with vaccine HPV types was shown in individuals who were naïve for vaccine HPV types at trial entry and throughout the 6 month immunisation period.

HPV Vaccines: Implementation and Impact

The 4HPV vaccine was licensed in 2006 in the USA and the 2HPV vaccine in Europe in 2007. In the 10 years that have since elapsed, these vaccines have been introduced into the National Immunisation programmes of >62 countries and more than 200 million doses have been distributed worldwide. In general, HPV immunisation programmes target adolescent girls 9-15 years of age, with or without catch up for older adolescents and young women. Anogenital HPV infections are sexually transmitted with first infection occurring most frequently within the months following the start of sexual activity (Winer *et al.*, 2003). To achieve optimal vaccine effectiveness, immunisation ideally should be

Table 5.1. Prophylactic HPV VLP vaccines.

	Cervarix Bivalent vaccine (2HPV)	Gardasil Quadrivalent vaccine (4HPV)	Gardasil9 Nonovalent vaccine (9HPV)
Manufacturer	Glaxo Smith Kline	Merck (MSD)	Merck (MSD)
Volume	Per dose 0.5ml	Per dose 0.5ml	Per dose 0.5ml
Adjuvant	ASO4: $Al(OH)_3$ 500mg MPL® 50mg	Amorphous Aluminium Hydroxyphosphate sulphate® 225mg	Amorphous Aluminium Hydroxyphosphate sulphate® 500mg
Antigens	L1 HPV16 20μg L1 HPV18 20μg	L1 HPV6 20μg L1 HPV11 40μg L1 HPV16 40μg L1 HPV18 20μg	L1 HPV6 30μg L1 HPV11 40μg L1 HPV16 60μg L1 HPV18 40μg L1 HPV31 20μg L1 HPV33 20μg L1 HPV45 20μg L1 HPV52 20μg L1 HPV58 20μg
Expression system	Hi-5 Baculovirus	Yeast: *Saccharomyces cereviseae*	Yeast: *Saccharomyces cereviseae*
Immunisation Schedule	Intra muscular	Intra muscular	Intra muscular
>15 years age	0, 1, 6 months	0, 2, 6 months	0, 2, 6 months
<15 years age	0, 6 or 12 months	0, 6 or 12 months	0, 6 or 12 months
FDA licensed	2009	2006	2014
EMA licensed	2007	2007	2015

completed before the start of sexual activity and this is why the target group for immunisation are young adolescents. Cervical cancer in women is the most prevalent HPV associated cancer and this plus the cost of cervical cancer screening programmes made the cost effectiveness case initially for immunising perimenarchal girls (Brisson *et al.*, 2009).

Two recent reviews have assessed the impact, at the population level, on the incidence of vaccine HPV type infection and disease of a 3 dose female only HPV vaccination (Drolet *et al.*, 2015; Garland *et al.*, 2016). Not unexpectedly the key determinants were vaccine coverage, age of birth cohorts targeted for vaccination and extent and duration of catch up programmes. In countries achieving >50% vaccine coverage, HPV 16 and 18 infections decreased by 68%; significant reductions in infections with the non-vaccine HPV types 31, 45, 33 were also recorded suggesting some cross protection. In populations immunised with the 4HPV vaccine ano-genital warts fell by 61% in girls 13-19 years of age and reductions in warts were reported in men <20years of age and women 20-39 years implying herd effects. In countries with <50% vaccine coverage significant reductions in 16/18 infection and anogenital warts occurred in women <20 years but no cross protection or herd effects were demonstrated.

Reductions in cervical cancer will only be seen in the long term – decades after vaccination - but a reduction in high grade cervical intra epithelial lesions CIN2/CIN3 caused by HPV vaccine types in vaccinated cohorts should be a surrogate for the long term effects on cervical cancer incidence. The Australian National Vaccination Programme, a school based programme targeted to girls aged 12-13 years with a catch up over 2 years for 13-26 year old young women was introduced in 2007 and achieved ≥ 70% coverage in the school cohort. Four years after the implementation of vaccination with the 4HPV vaccine, high grade cervical abnormalities (CIN2/3, adenocarcinoma *in situ* (AIS)) in the vaccinated cohorts (age 12-26 years) receiving 3 doses of vaccine, decreased by about 48% (Crowe *et al.*, 2014). Comparable data have been reported from several countries using the 4HPV vaccine (for review see (Garland *et al.*, 2016) and from Scotland with the 2HPV vaccine where reductions of 50% for CIN2 and 55% for CIN3 in 20/21 year old females were observed in the catch up cohort (15-18 year olds), mean vaccine coverage 66% (Pollock *et al.*, 2014). In all studies the largest declines in CIN 2, CIN3 and AIS were shown in the cohorts vaccinated at the youngest age. As an example, in a nationwide analysis in Sweden over a 2 and 5 year period, declines of CIN2+ and CIN3+ among fully vaccinated females <17 years of age at vaccination were 75% and 84%, respectively, compared with unvaccinated and partially vaccinated females. In contrast, among those vaccinated between 20 and 29 years of age, declines in CIN2+ and CIN3+ in vaccinated compared to unvaccinated were 22% and 25% [quoted in (Garland *et al.*, 2016)].

Vaccine induced immune responses – mechanism of protection

Humoral immunity

The current assumption is that neutralising antibody is the mechanism of protection afforded by HPV VLP vaccines (Stanley *et al.*, 2012). In contrast to natural infections, in which the humoral immune response is slow and weak and not all individuals sero convert, systemic immunisation with L1 VLP vaccines generates high serum neutralising antibody concentrations at least 50-1000 times greater than those measured in natural infections and virtually all vaccinees sero convert (Scherpenisse *et al.*, 2013; Villa *et al.*, 2006). Following the three dose immunisation schedule geometric mean titres (GMTs) for antibodies to the vaccine HPV types peak at month 7; GMTs then wane until 18-24 months at which there is a plateau level at about 10 times natural infection and this level persists for at least 10-12 years after the primary immunisation (Harper *et al.*, 2004). Antigen challenge at 60 months post dose 1, with either vaccine, results in a rapid, and robust, anamnestic response with antibody concentrations rising within a week to levels greater than that achieved at peak (one month post dose 3) in the initial immunisation schedule, demonstrating the presence of reactive memory B cells (Moscicki *et al.*, 2012; Olsson *et al.*, 2007; Scherer *et al.*, 2014). Mathematical modelling predicts slow decay of antibody over a 30-50 year period and potentially, therefore, protection over that time (David *et al.*, 2009; Fraser *et al.*, 2007)

The current evidence from experimental murine challenge systems (Longet *et al.*, 2011) is that very low serum neutralising antibody concentrations are protective but the importance of antibody affinity and avidity in relation to this in humans is not known. The classical prime, prime, boost schedule with the extended interval of several months between prime and boost for protein vaccines is designed to generate high affinity, high avidity antibodies and a large B memory pool (McHeyzer-Williams *et al.*, 2012). The importance of avidity to antibody mediated protection to viruses has been demonstrated in experimental systems (Bachmann *et al.*, 1997) but there is limited data on avidity in either natural HPV infections (Namujju *et al.*, 2011; Scherpenisse *et al.*, 2013) or after vaccination (Kemp *et al.*, 2012; Petras *et al.*, 2012; Sankaranarayanan *et al.*, 2016). The available serological assays provide only a partial characteris-ation of the immune status in vaccinated individuals and current understanding of this response is superficial. The kinetics of the response and specific details – avidity, affinity, epitope specificity are not well documented. Both type specific and cross neutralising antibodies are generated by the 2HPV and 4HPV vaccines. The 2HPV vaccine generates higher GMTs than the 4HPV for both type specific and cross neutralising species (Einstein *et al.*, 2014) although cross neutralising GMTs are on average 10-100 times lower than type specific. HPV VLP vaccines are highly efficacious and immunogenic, no breakthrough cases of disease have been reported from the follow up studies of the RCTs which now extend for more than 10 years, but there is no immune correlate of protection against infection or disease; the minimum level of antibody, the affinity/avidity of such antibody needed for protection and the role of B cell memory if antibody wanes have yet to be established.

Vaccine elicited T cell responses

L1 HPV VLP vaccines elicit robust antigen specific and cross reactive CD4+ve and CD8+ve T cell responses that are functional in *in vitro* assays (Evans *et al.*, 2001; Herrin *et al.*, 2014; Pinto *et al.*, 2003; Pinto *et al.*, 2006; Williams *et al.*, 2002). The contribution of these responses to vaccine efficacy against infection and/or disease is likely to be negligible. HPV infects basal epithelial cells but L1 protein expression is restricted to the upper terminally differentiated keratinocytes at the surface of squamous epithelium and is both temporally and spatially separated from the infected basal cell (Doorbar *et al.*, 2012). Furthermore the absence of a therapeutic effect by the licensed vaccines implies that the protection elicited by the vaccines is not mediated by L1 antigen specific T cells.

Immunisation schedules

In the RCTs that led to licensure for the current products an immunisation schedule of 3 doses at 0, 1-2 and 6 months was used. Since the intended cohorts for immunisation were young adolescents immunobridging studies were undertaken in 9-15 year olds and showed that GMTs in these cohorts were twice those achieved in the 16-26 year old women in whom efficacy had been demonstrated in the RCTs (Block *et al.*, 2006; Schwarz *et al.*, 2014). Studies have investigated the impact on immunogenicity in the young adolescent cohort (9-14 year old girls and boys), of changing from the 3 dose 'prime, prime, boost' to a 2 dose 'prime, boost' at 0 and 6 months. For all three licensed vaccines these studies showed that 2 doses of vaccine at 0 and 6 months resulted in antibody responses (GMTs) that are non-inferior over a 3 or 4 year period to those achieved after 3 doses in 16-26 year old women. Since antibody concentration and quality for 2 doses given at 0 and 6 months in adolescents are as good or better than from 3 doses given at 0, 1/2 and 6 months in older women then vaccine mediated protection and duration of protection should be comparable. The Strategic Advisory Group of Experts on Immunization (SAGE) of WHO in April 2014 reviewed the evidence and "recommended a 2-dose schedule for girls, if vaccination is initiated prior to 15 years of age. A 3-dose schedule remains necessary if immunization is initiated after the girls' 15th birthday. The recommended minimal interval between the 2 doses is 6 months. This interval may be extended to 12 months if this facilitates administration. A 3-dose schedule (i.e. at 0, 1-2, and 6 months) remains recommended for immunocompromised individuals, including those known to be HIV-infected."

The immunological evidence is robust in 9-14 year old girls for a 2 dose schedule with a minimum interval of 6 months between doses but at the present there are no published data on vaccine effectiveness either for infection or disease in the adolescent 2 and 3 dose cohorts in any randomised control trial. These cohorts are being prospectively followed. A post hoc analysis of data from two trials using the 2HPV vaccine show that 4 years after vaccination of women aged 15-25 years (Kreimer *et al.*, 2011) or 18-25 years (Kreimer *et al.*, 2015), one and two doses of vaccine protected against persistent infection with HPV 16 and 18 (a pre-requisite for the development of high grade CIN) raising the provocative suggestion that one dose of vaccine might be enough for protection, a notion supported by data with the 4HPV vaccine from an observational cohort study sponsored by IARC/WHO in India (Sankaranarayanan *et al.*, 2016).

A central issue for the 1 dose scenario is duration of protection. Protection will have to be maintained in adolescents who are immunised at 12/13years of age for the following 2-4 decades. In all studies reported to date, GMTs after 2 doses 0, 6 or 0, 12 months are non-inferior to the standard 3 dose schedule and the expectation is that, in view of this, protection will be equivalent to the standard 3 dose schedule. However after 1 dose GMTs are inferior to 2 or 3 doses although the kinetics of antibody persistence do not differ, with antibody concentrations in all dosage regimens achieving a plateau after 18-24 months that remains stable for up to 48 months (Safaeian *et al.*, 2013). Importantly in the WHO trial all the geometric mean avidity indices after fewer than three doses in any group were non-inferior to those after three doses of vaccine suggesting that antibody quality was as good as that achieved after 2 or 3 doses (Sankaranarayanan *et al.*, 2016). Furthermore in a recent study (Scherer *et al.*, 2016) a single dose of the 4HPV vaccine elicited high titres of neutralising antibody and potent B memory cells confirming the remarkable immunogenicity of the VLP. This, it is suggested (Schiller and Lowy, 2015) is due to the combination in the VLP, of particle size and geometry with the dense repetitive display of epitopes on the L1 pentamers activating B cells (Bachmann and Zinkernagel, 1996) and other key immunocytes via pattern recognition receptors generating potent immune responses. All these findings are very encouraging but the issue of whether 1 dose of vaccine will be enough for protection against infection and disease can only be addressed by randomised controlled trials.

Gender neutral vaccination

The evidence from "real life" data from countries that have implemented a female only vaccine programme with high coverage targeting adolescents is that this results in reduction in disease (Drolet *et al.*, 2015) and virus prevalence. HPV vaccines in common with most vaccines, depend for their impact at the population level upon the indirect effects of reducing transmission and carriage by herd immunity. The rationale for immunising only one gender (females) against a sexually transmitted infection is that where immunisation coverage is high enough this generates herd protection by blocking transmission to effectively protect the sexual partners (Garnett, 2005); immunising boys in such a scenario is not cost effective (Jit *et al.*, 2008). Female only vaccination by definition cannot achieve herd immunity since heterosexual men are not immune but protected if their sex partners are immune or uninfected but remain susceptible to infection if this scenario changes. Men who have sex with men (MSM) are left entirely unprotected by the female only approach (Ali *et al.*, 2013), HPV geno-prevalence remains high in this group and they represent a reservoir of infection (Poynten *et al.*, 2016). Male immunisation is a controversial issue but men have a significant and rising burden of HPV associated cancer (Stanley, 2012), HPV vaccines are efficacious in men (Giuliano *et al.*, 2011; Palefsky *et al.*, 2011) and modelling studies show that gender neutral vaccination gives programme resilience if vaccine coverage in females falls (Elfstrom *et al.*, 2016). If transmission efficiency of vaccine HPV genotypes is to be reduced to almost undetectable levels at the population level then immunisation of boys is required: this provides immunity in men irrespective of their sexual preference and real herd immunity.

HPV Vaccines: Safety

The safety profile of the licensed vaccines was assessed extensively in the RCTs (Vichnin *et al.*, 2015)

(Moreira *et al.*, 2016). In the 10 years since the 2HPV and 4HPV vaccines have been licensed the safety profile has been intensively monitored by robust pharmacovigilance in the post licensure setting using both passive and active vaccine surveillance (Angelo *et al.*, 2014; Vichnin *et al.*, 2015). The key challenge faced in pharmacovigilance is to distinguish real adverse events (AE's) from background conditions that would occur regardless of vaccination. Studies providing population based data on incidence of potential adverse events prior to vaccination allow analysis of observed/expected rates in vaccinated populations (Callreus *et al.*, 2009; Siegrist *et al.*, 2007) The most commonly reported HPV vaccine related adverse events (AE's) are injection site reactions including pain, swelling, erythema, these are usually of short duration and resolve spontaneously. Systemic AE's, such as myalgia, fatigue, have been mild and self-limited. Serious vaccine related AE's such as anaphylaxis are very rare. No associations with new onset chronic conditions such as auto-immune or neurological disease have been identified in large well conducted population based studies (Scheller *et al.*, 2015). HPV vaccines are not recommended for administration during pregnancy but surveillance of pregnancy outcomes following inadvertent vaccination has revealed no adverse outcomes such as miscarriage, congenital abnormalities or premature labour beyond background rates (Goss *et al.*, 2015; Panagiotou *et al.*, 2015).

L2 Vaccines – next generation prophylaxis?

The L2 capsid protein is a key player in the process of viral entry and potentially a vaccine target. Antibodies to L2 cannot be detected in natural infections in animals or man and are not induced after immunisation with L1/L2 VLPs. However it was shown more than 2 decades ago that prophylactic vaccination of cattle with a peptide comprising the N' terminal 200 amino acids (aa) of BPV-4 completely prevented BPV-4 infection of the alimentary tract and that this was attributable to the generation of L2 neutralising antibody (Chandrachud *et al.*, 1995). Importantly, it turns out, that cross reactive, sub-dominant linear epitopes are present in L2 and immunisation of animals with peptides of the N' terminus of HPV-16 L2 induce antibodies that neutralise a broad range of both cutaneous and mucosal HPV types, suggesting that L2 polypeptide vaccines could provide broad spectrum protection (for review see (Jiang *et al.*, 2016). However non-particulate protein vaccines such as L2 are poorly immunogenic compared to L1 VLP vaccines and because of this, a key challenge in L2-based vaccine development is effective display of these broadly cross-neutralizing L2 epitopes to the immune system to trigger high titer, high avidity and durable neutralizing antibody responses. One of the most promising approaches to this is to use the L1VLP as a platform to display the peptide encompassing the broadly cross neutralising L2 epitope by inserting the peptide into L1 thus generating a chimeric VLP (Boxus *et al.*, 2016; Schellenbacher *et al.*, 2009). Such a structure results in a 360 fold display of the peptide, the latter by itself is not immunogenic but the VLP with its repetitive crystalline array acts as a hapten. The chimeric particle is highly immunogenic generating antibodies that neutralise a broad spectrum of high and low risk mucosal and cutaneous HPVs (Schellenbacher *et al.*, 2009) which protect against *in vivo* challenge with the relevant pseudovirions in rodents (Schellenbacher *et al.*, 2013).

The L2 vaccines appropriately adjuvanted show great promise in meeting the objectives of broad protection, low cost, thermostability and non-invasive delivery but are not yet in Phase II/III trials. These are "new vaccines" in regulatory terms and it is not clear what efficacy endpoints they would be required to meet in the large clinical trials that will be required. These trials will not have a placebo group but a "control group" vaccinated with one of the current HPV VLP vaccines. This will substantially reduce the incidence of anogenital high grade lesions in the reference group, increasing the size and duration of the trials. Virological endpoints (persistent infection, antibody titres and concentrations over time) have not been accepted by regulatory authorities to date although persistent infection is a surrogate for high grade cervical disease (Radley *et al.*, 2015).

Conclusions

HBV infection has been preventable by vaccination since 1982; it is highly effective and has resulted already in decreased rates of new infections, chronic liver disease and HCC. Nonetheless HBV remains a global public health problem. Passive immunisation and chemoprophylaxis can contribute to control of HBV infection but the only preventive strategy remains global neonatal vaccination. Implementation of this with optimal vaccines and high vaccination coverage could control HBV infection and make HBV caused HCC a rare disease.

Cancer resulting from HPV infection, particularly HPV16, is a global public health problem which is likely, if there is no effective intervention, to increase over the short to medium term because of the population demographics in economically developing countries which bear the major disease burden (Bruni *et al.*, 2016). The current HPV vaccines are highly effective, have a good safety profile and with high vaccine coverage can reduce the circulation of vaccine HPV types in the population to virtually zero (Chow *et al.*, 2015; Chow *et al.*, 2016) and reduce the incidence of CIN3 the obligate precursor to cervical cancer by 80% or more (Garland *et al.*, 2016). The major barriers to uptake of these vaccines in countries with the greatest burden of disease is cost – the cost per dose of vaccine and the costs of delivery. In the best case scenario reducing the number of doses, innovative manufacturing processes and manufacturing in developing countries could reduce costs allowing the delivery of these anti cancer vaccines to those with greatest need.

References

Hepatitis B. Fact sheet Nr. 204.http://www.who.int/mediacentre/factsheets/fs204/en/ (Accessed: 29 September 2014).

Alemany, L., Cubilla, A., Halec, G., Kasamatsu, E., Quiros, B., Masferrer, E., Tous, S., Lloveras, B., Hernandez-Suarez, G., Lonsdale, R., *et al.* (2016). Role of Human Papillomavirus in Penile Carcinomas Worldwide. European urology. http://dx.doi.org/10.1000/10.1016/j.eururo.2015.12.007.

Alemany, L., Saunier, M., Alvarado-Cabrero, I., Quiros, B., Salmeron, J., Shin, H.R., Pirog, E.C., Guimera, N., Hernandez-Suarez, G., Felix, A., *et al.* (2015). Human papillomavirus DNA prevalence and type distribution in anal carcinomas worldwide. Int J Cancer 136, 98-107. http://dx.doi.org/10.1000/10.1002/ijc.28963 [doi].

Ali, H., Donovan, B., Wand, H., Read, T.R., Regan, D.G., Grulich, A.E., Fairley, C.K., and Guy, R.J. (2013). Genital warts in young Australians five years into national human papillomavirus vaccination programme: national surveillance data. Bmj 346, f2032.

Angelo, M.G., Zima, J., Tavares Da Silva, F., Baril, L., and Arellano, F. (2014). Post-licensure safety surveillance for human papillomavirus-16/18-AS04-adjuvanted vaccine: more than 4 years of experience. Pharmacoepidemiol Drug Saf 23, 456-465. http://dx.doi.org/10.1000/10.1002/pds.3593.

Bachmann, M.F., Kalinke, U., Althage, A., Freer, G., Burkhart, C., Roost, H., Aguet, M., Hengartner, H., and Zinkernagel, R.M. (1997). The role of antibody concentration and avidity in antiviral protection. Science 276, 2024-2027.

Bachmann, M.F., and Zinkernagel, R.M. (1996). The influence of virus structure on antibody responses and virus serotype formation. Immunology today 17, 553-558.

Ball, S.L., Winder, D.M., Vaughan, K., Hanna, N., Levy, J., Sterling, J.C., Stanley, M.A., and Goon, P.K. (2011). Analyses of human papillomavirus genotypes and viral loads in anogenital warts. J Med Virol 83, 1345-1350. http://dx.doi.org/10.1000/10.1002/jmv.22111 [doi].

Batdelger, D., Dandii, D., Jirathitikal, V., and Bourinbaiar, A.S. (2007). Open label trial of therapeutic hepatitis B vaccine V-5 Immunitor (V5) delivered by oral route. Letters in Drug Design and Discovery 4, 540-544.

Bertino, J.S., Jr., Tirrell, P., Greenberg, R.N., Keyserling, H.L., Poland, G.A., Gump, D., Kumar, M.L., and Ramsey, K. (1997). A comparative trial of standard or high-dose S subunit recombinant hepatitis B vaccine versus a vaccine containing S subunit, pre-S1, and pre-S2 particles for revaccination of healthy adult nonresponders. J Infect Dis 175, 678-681.

Block, S.L., Nolan, T., Sattler, C., Barr, E., Giacoletti, K.E., Marchant, C.D., Castellsague, X., Rusche, S.A., Lukac, S., Bryan, J.T., *et al.* (2006). Comparison of the immunogenicity and reactogenicity of a prophylactic quadrivalent human papillomavirus (types 6, 11, 16, and 18) L1 virus-like particle vaccine in male and female adolescents and young adult women. Pediatrics 118, 2135-2145.

Bonanni, P., Pesavento, G., Bechini, A., Tiscione, E., Mannelli, F., Benucci, C., and Nostro, A.L. (2003). Impact of universal vaccination programmes on the epidemiology of hepatitis B: 10 years of experience in Italy. Vaccine 21, 685-691.

Boxus, M., Fochesato, M., Miseur, A., Mertens, E., Dendouga, N., Brendle, S., Balogh, K.K., Christensen, N.D., and Giannini, S.L. (2016). Broad cross protection is induced by a human papillomavirus vaccine composed of L1/L2 chimeric virus-like particles in preclinical models. J Virol. http://dx.doi.org/10.1000/10.1128/jvi.00449-16.

Brisson, M., Van de Velde, N., and Boily, M.C. (2009). Economic evaluation of human papillomavirus vaccination in developed countries. Public Health Genomics 12, 343-351.

Bruni, L., Diaz, M., Barrionuevo-Rosas, L., Herrero, R., Bray, F., Bosch, F.X., de Sanjose, S., and Castellsague, X. (2016). Global estimates of human papillomavirus vaccination coverage by region and income level: a pooled analysis. Lancet Glob Health 4, e453-463. http://dx.doi.org/10.1000/10.1016/s2214-109x(16)30099-7.

Callreus, T., Svanstrom, H., Nielsen, N.M., Poulsen, S., Valentiner-Branth, P., and Hviid, A. (2009). Human papillomavirus immunisation of adolescent girls and anticipated reporting of immune-mediated adverse events. Vaccine 27, 2954-2958.

Centers for Disease, C., and Prevention (2011). Assessing completeness of perinatal hepatitis B virus infection reporting through comparison of immunization program and surveillance data--United States. MMWR Morb Mortal Wkly Rep 60, 410-413.

Chandrachud, L.M., Grindlay, G.J., McGarvie, G.M., O'Neil, B.W., Wagner, E.R., Jarrett, W.F., and Campo, M.S. (1995). Vaccination of cattle with the N-terminus of L2 is necessary and sufficient for preventing infection by bovine papillomavirus-4. Virology 211, 204-208.

Chang, M.H. (2014). Prevention of hepatitis B virus infection and liver cancer. Recent results in cancer research Fortschritte der Krebsforschung Progres dans les recherches sur le cancer 193, 75-95.

Chang, M.H., You, S.L., Chen, C.J., Liu, C.J., Lai, M.W., Wu, T.C., Wu, S.F., Lee, C.M., Yang, S.S., Chu, H.C., et al. (2016). Long-term Effects of Hepatitis B Immunization of Infants in Preventing Liver Cancer. Gastroenterology 151, 472-480 e471. http://dx.doi.org/10.1000/10.1053/j.gastro.2016.05.048.

Chang, M.H., You, S.L., Chen, C.J., Liu, C.J., Lee, C.M., Lin, S.M., Chu, H.C., Wu, T.C., Yang, S.S., Kuo, H.S., *et al.* (2009). Decreased incidence of hepatocellular carcinoma in hepatitis B vaccinees: a 20-year follow-up study. J Natl Cancer Inst 101, 1348-1355. http://dx.doi.org/10.1000/10.1093/jnci/djp288.

Chow, E.P., Danielewski, J.A., Fehler, G., Tabrizi, S.N., Law, M.G., Bradshaw, C.S., Garland, S.M., Chen, M.Y., and Fairley, C.K. (2015). Human papillomavirus in young women with Chlamydia trachomatis infection 7 years after the Australian human papillomavirus vaccination programme: a cross-sectional study. Lancet Infect Dis 15, 1314-1323. http://dx.doi.org/10.1000/10.1016/s1473-3099(15)00055-9.

Chow, E.P., Machalek, D.A., Tabrizi, S.N., Danielewski, J.A., Fehler, G., Bradshaw, C.S., Garland, S.M., Chen, M.Y., and Fairley, C.K. (2016). Quadrivalent vaccine-targeted human papillomavirus genotypes in heterosexual men after the Australian female human papillomavirus vaccination programme: a retrospective observational study. Lancet Infect Dis. http://dx.doi.org/10.1000/10.1016/s1473-3099(16)30116-5.

Cooper, C., and Mackie, D. (2011). Hepatitis B surface antigen-1018 ISS adjuvant-containing vaccine: a review of HEPLISAV safety and efficacy. Expert Rev Vaccines 10, 417-427. http://dx.doi.org/10.1000/10.1586/erv.10.162.

Crowe, E., Pandeya, N., Brotherton, J.M., Dobson, A.J., Kisely, S., Lambert, S.B., and Whiteman, D.C. (2014). Effectiveness of quadrivalent human papillomavirus vaccine for the prevention of cervical abnormalities: case-control study nested within a population based screening programme in Australia. Bmj 348, g1458.

David, M.P., Van Herck, K., Hardt, K., Tibaldi, F., Dubin, G., Descamps, D., and Van Damme, P. (2009). Long-term persistence of anti-HPV-16 and -18 antibodies induced by vaccination with the AS04-adjuvanted cervical cancer vaccine: modeling of sustained antibody responses. Gynecol Oncol 115, S1-6. http://dx.doi.org/10.1000/10.1016/j.ygyno.2009.01.011.

de Martel, C., Ferlay, J., Franceschi, S., Vignat, J., Bray, F., Forman, D., and Plummer, M. (2012). Global burden of cancers attributable to infections in 2008: a review and synthetic analysis. Lancet Oncol 13, 607-615. http://dx.doi.org/10.1000/10.1016/S1470-2045(12)70137-7.

de Villiers, E.M. (2013). Cross-roads in the classification of papillomaviruses. Virology 445, 2-10. http://dx.doi.org/10.1000/10.1016/j.virol.2013.04.023.

Dienstag, J.L., and Delemos, A.S. (2014). Viral hepatitis. In Priciples and practice of infectious diseases, 8th edition, G.L. Mandell, R.G. Douglas, and J.E. Bennett, eds. (Philadelphia: Churchill Livingstone Elsevier), pp. 1439-1468.

Doorbar, J., Quint, W., Banks, L., Bravo, I.G., Stoler, M., Broker, T.R., and Stanley, M.A. (2012). The biology and life-cycle of human papillomaviruses. Vaccine 30 Suppl 5, F55-70. http://dx.doi.org/10.1000/10.1016/j.vaccine.2012.06.083.

Drolet, M., Benard, E., Boily, M.C., Ali, H., Baandrup, L., Bauer, H., Beddows, S., Brisson, J., Brotherton, J.M., Cummings, T., *et al.* (2015). Population-level impact and herd effects following human papillomavirus vaccination programmes: a systematic review and meta-analysis. Lancet Infect Dis 15, 565-580. http://dx.doi.org/10.1000/10.1016/S1473-3099(14)71073-4.

Einstein, M.H., Takacs, P., Chatterjee, A., Sperling, R.S., Chakhtoura, N., Blatter, M.M., Lalezari, J., David, M.P., Lin, L., Struyf, F., *et al.* (2014). Comparison of long-term immunogenicity and safety of human papillomavirus (HPV)-16/18 AS04-adjuvanted vaccine and HPV-6/11/16/18 vaccine in healthy women aged 18-45 years: end-of-study analysis of a Phase III randomized trial. Human vaccines and immunotherapeutics 10, 3435-3445. http://dx.doi.org/10.1000/10.4161/hv.36121.

Elfstrom, K.M., Lazzarato, F., Franceschi, S., Dillner, J., and Baussano, I. (2016). Human Papillomavirus Vaccination of Boys and Extended Catch-up Vaccination: Effects on the Resilience of Programs. J Infect Dis 213, 199-205. http://dx.doi.org/10.1000/10.1093/infdis/jiv368.

Evans, T.G., Bonnez, W., Rose, R.C., Koenig, S., Demeter, L., Suzich, J.A., O'Brien, D., Campbell, M., White, W.I., Balsley, J., *et al.* (2001). A Phase 1 study of a recombinant viruslike particle vaccine against human papillomavirus type 11 in healthy adult volunteers. J Infect Dis 183, 1485-1493.

Forman, D., de Martel, C., Lacey, C.J., Soerjomataram, I., Lortet-Tieulent, J., Bruni, L., Vignat, J., Ferlay, J., Bray, F., Plummer, M., *et al.* (2012). Global burden of human papillomavirus and related diseases. Vaccine 30 Suppl 5, F12-23. http://dx.doi.org/10.1000/10.1016/j.vaccine.2012.07.055.

Fraser, C., Tomassini, J.E., Xi, L., Golm, G., Watson, M., Giuliano, A.R., Barr, E., and Ault, K.A. (2007). Modeling the long-term antibody response of a human papillomavirus (HPV) virus-like particle (VLP) type 16 prophylactic vaccine. Vaccine 25, 4324-4333.

Garland, S.M., Kjaer, S.K., Munoz, N., Block, S.L., Brown, D.R., DiNubile, M.J., Lindsay, B.R., Kuter, B.J., Perez, G., Dominiak-Felden, G., *et al.* (2016). Impact and Effectiveness Of the Quadrivalent Human Papillomavirus Vaccine: A Systematic Review of Ten Years of Real-World Experience. Clin Infect Dis. http://dx.doi.org/10.1000/10.1093/cid/ciw354.

Garnett, G.P. (2005). Role of herd immunity in determining the effect of vaccines against sexually transmitted disease. J Infect Dis 191 Suppl 1, S97-106. http://dx.doi.org/10.1000/10.1086/425271.

Gerlich, W.H. (2015). Prophylactic vaccination against hepatitis B: achievements, challenges and perspectives. Med Microbiol Immunol 204, 39-55. http://dx.doi.org/10.1000/10.1007/s00430-014-0373-y.

Giuliano, A.R., Palefsky, J.M., Goldstone, S., Moreira, E.D., Penny, M.E., Aranda, C., Vardas, E., Moi, H., Jessen, H., Hillman, R., *et al.* (2011). Efficacy of Quadrivalent HPV Vaccine against HPV Infection and Disease in Males. New Engl. J. Med. 364, 401-411.

Goss, M.A., Lievano, F., Buchanan, K.M., Seminack, M.M., Cunningham, M.L., and Dana, A. (2015). Final report on exposure during pregnancy from a pregnancy registry for quadrivalent human papillomavirus vaccine. Vaccine 33, 3422-3428. http://dx.doi.org/10.1000/10.1016/j.vaccine.2015.04.014.

Guan, P., Howell-Jones, R., Li, N., Bruni, L., de Sanjose, S., Franceschi, S., and Clifford, G.M. (2012). Human papillomavirus types in 115,789 HPV-positive women: a meta-analysis from cervical infection to cancer. Int J Cancer 131, 2349-2359. http://dx.doi.org/10.1000/10.1002/ijc.27485.

Harford, N., Cabezon, T., Crabeel, M., Simoen, E., Rutgers, A., and De Wilde, M. (1983). Expression of hepatitis B surface antigen in yeast. DEv. Biol. Standard. 54, 125-130.

Harper, D.M., Franco, E.L., Wheeler, C., Ferris, D.G., Jenkins, D., Schuind, A., Zahaf, T., Innis, B., Naud, P., De Carvalho, N.S., *et al*. (2004). Efficacy of a bivalent L1 virus-like particle vaccine in prevention of infection with human papillomavirus types 16 and 18 in young women: a randomised controlled trial. Lancet 364, 1757-1765.

Herrin, D.M., Coates, E.E., Costner, P.J., Kemp, T.J., Nason, M.C., Saharia, K.K., Pan, Y., Sarwar, U.N., Holman, L., Yamshchikov, G., *et al*. (2014). Comparison of adaptive and innate immune responses induced by licensed vaccines for Human Papillomavirus. Human vaccines and immunotherapeutics 10, 3446-3454. http://dx.doi.org/10.1000/10.4161/hv.34408.

Huang, K., and Lin, S. (2000). Nationwide vaccination: a success story in Taiwan. Vaccine 18 Suppl 1, S35-38.

Jacques, P., Moens, G., Desombere, I., Dewijngaert, J., Leroux-Roels, G., Wettendorff, M., and Thoelen, S. (2002). The immunogenicity and reactogenicity profile of a candidate hepatitis B vaccine in an adult vaccine non-responder population. Vaccine 20, 3644-3649.

Jiang, R.T., Schellenbacher, C., Chackerian, B., and Roden, R.B. (2016). Progress and prospects for L2-based human papillomavirus vaccines. Expert Rev Vaccines. http://dx.doi.org/10.1000/ 10.1586/14760584.2016.1157479.

Jit, M., Choi, Y.H., and Edmunds, W.J. (2008). Economic evaluation of human papillomavirus vaccination in the United Kingdom. Bmj 337, a769.

Joura, E.A., Giuliano, A.R., Iversen, O.E., Bouchard, C., Mao, C., Mehlsen, J., Moreira, E.D., Jr., Ngan, Y., Petersen, L.K., Lazcano-Ponce, E., *et al*. (2015). A 9-valent HPV vaccine against infection and intraepithelial neoplasia in women. New Engl. J. Med. 372, 711-723. http://dx.doi.org/10.1000/10.1056/NEJMoa1405044 [doi].

Kemp, T.J., Safaeian, M., Hildesheim, A., Pan, Y., Penrose, K.J., Porras, C., Schiller, J.T., Lowy, D.R., Herrero, R., and Pinto, L.A. (2012). Kinetic and HPV infection effects on cross-type neutralizing antibody and avidity responses induced by Cervarix((R)). Vaccine 31, 165-170. http://dx.doi.org/10.1000/10.1016/j.vaccine.2012.10.067.

Kreimer, A.R., Rodriguez, A.C., Hildesheim, A., Herrero, R., Porras, C., Schiffman, M., Gonzalez, P., Solomon, D., Jimenez, S., Schiller, J.T., *et al*. (2011). Proof-of-Principle Evaluation of the Efficacy of Fewer Than Three Doses of a Bivalent HPV16/18 Vaccine. J Natl Cancer Inst. http://dx.doi.org/10.1000/djr319 [pii] 10.1093/jnci/djr319 [doi].

Kreimer, A.R., Struyf, F., Del Rosario-Raymundo, M.R., Hildesheim, A., Skinner, S.R., Wacholder, S., Garland, S.M., Herrero, R., David, M.P., and Wheeler, C.M. (2015). Efficacy of fewer than three doses of an HPV-16/18 AS04-adjuvanted vaccine: combined analysis of data from the Costa Rica Vaccine and PATRICIA trials. Lancet Oncol 16, 775-786. http://dx.doi.org/10.1000/10.1016/s1470-2045(15)00047-9.

Lacey, C.J., Lowndes, C.M., and Shah, K.V. (2006). Chapter 4: Burden and management of non-cancerous HPV-related conditions: HPV-6/11 disease. Vaccine 24 Suppl 3, S3/35-41.

Larson, D.A., and Derkay, C.S. (2010). Epidemiology of recurrent respiratory papillomatosis. APMIS 118, 450-454.

Lee, C., Gong, Y., Brok, J., Boxall, E.H., and Gluud, C. (2006). Effect of hepatitis B immunisation in newborn infants of mothers positive for hepatitis B surface antigen: systematic review and meta-analysis. Bmj 332, 328-336. http://dx.doi.org/10.1000/10.1136/bmj.38719.435833.7C.

Liaw, Y.F., Brunetto, M.R., and Hadziyannis, S. (2010). The natural history of chronic HBV infection and geographical differences. Antivir Ther 15 Suppl 3, 25-33. http://dx.doi.org/10.1000/10.3851/IMP1621.

Longet, S., Schiller, J.T., Bobst, M., Jichlinski, P., and Nardelli-Haefliger, D. (2011). A murine genital-challenge model is a sensitive measure of protective antibodies against human papillomavirus infection. J Virol 85, 13253-13259. http://dx.doi.org/10.1000/10.1128/JVI.06093-11.

McAleer, W.J., Buynak, E.B., Maigetter, R.Z., Wampler, D.E., Miller, W.J., and Hilleman, M.R. (1984). Human hepatitis B vaccine from recombinant yeast. Nature 307, 178-180.

McHeyzer-Williams, M., Okitsu, S., Wang, N., and McHeyzer-Williams, L. (2012). Molecular programming of B cell memory. Nat Rev Immunol 12, 24-34. http://dx.doi.org/10.1000/10.1038/nri3128.

Moreira, E.D., Jr., Block, S.L., Ferris, D., Giuliano, A.R., Iversen, O.E., Joura, E.A., Kosalaraksa, P., Schilling, A., Van Damme, P., Bornstein, J., *et al*. (2016). Safety Profile of the 9-Valent HPV Vaccine: A Combined Analysis of 7 Phase III Clinical Trials. Pediatrics. http://dx.doi.org/10.1000/10.1542/peds.2015-4387.

Moscicki, A.B., Wheeler, C.M., Romanowski, B., Hedrick, J., Gall, S., Ferris, D., Poncelet, S., Zahaf, T., Moris, P., Geeraerts, B., *et al*. (2012). Immune responses elicited by a fourth dose of the HPV-16/18 AS04-adjuvanted vaccine in previously vaccinated adult women. Vaccine 31, 234-241. http://dx.doi.org/10.1000/10.1016/j.vaccine.2012.09.037.

Munoz, N., Manalastas, R., Jr., Pitisuttithum, P., Tresukosol, D., Monsonego, J., Ault, K., Clavel, C., Luna, J., Myers, E., Hood, S., *et al*. (2009). Safety, immunogenicity, and efficacy of quadrivalent human papillomavirus (types 6, 11, 16, 18) recombinant vaccine in women aged 24-45 years: a randomised, double-blind trial. Lancet 373, 1949-1957. http://dx.doi.org/10.1000/10.1016/S0140-6736(09)60691-7.

Namujju, P.B., Hedman, L., Hedman, K., Banura, C., Mbidde, E.K., Kizito, D., Byaruhanga, R.N., Muwanga, M., Kirnbauer, R., Surcel, H.M., *et al.* (2011). Low avidity of human papillomavirus (HPV) type 16 antibodies is associated with increased risk of low-risk but not high-risk HPV type prevalence. BMC Res Notes 4, 170. http://dx.doi.org/10.1000/1756-0500-4-170 [pii] 10.1186/1756-0500-4-170 [doi].

Olsson, S.E., Villa, L.L., Costa, R.L., Petta, C.A., Andrade, R.P., Malm, C., Iversen, O.E., Hoye, J., Steinwall, M., Riis-Johannessen, G., *et al.* (2007). Induction of immune memory following administration of a prophylactic quadrivalent human papillomavirus (HPV) types 6/11/16/18 L1 virus-like particle (VLP) vaccine. Vaccine 25, 4931-4939.

Orlando, R., Foggia, M., Maraolo, A.E., Mascolo, S., Palmiero, G., Tambaro, O., and Tosone, G. (2015). Prevention of hepatitis B virus infection: from the past to the future. Eur J Clin Microbiol Infect Dis 34, 1059-1070. http://dx.doi.org/10.1000/10.1007/s10096-015-2341-x.

Palefsky, J.M., Giuliano, A.R., Goldstone, S., Moreira, E.D., Jr., Aranda, C., Jessen, H., Hillman, R., Ferris, D., Coutlee, F., Stoler, M.H., *et al.* (2011). HPV vaccine against anal HPV infection and anal intraepithelial neoplasia. New Engl. J. Med. 365, 1576-1585. http://dx.doi.org/10.1000/10.1056/NEJMoa1010971 [doi].

Panagiotou, O.A., Befano, B.L., Gonzalez, P., Rodriguez, A.C., Herrero, R., Schiller, J.T., Kreimer, A.R., Schiffman, M., Hildesheim, A., Wilcox, A.J., *et al.* (2015). Effect of bivalent human papillomavirus vaccination on pregnancy outcomes: long term observational follow-up in the Costa Rica HPV Vaccine Trial. Bmj 351, h4358. http://dx.doi.org/10.1000/10.1136/bmj.h4358.

Petras, M., Sykora, T., Andrys, C., and Drahosova, M. (2012). Post-vaccination anti-human papillomavirus antibody seroprevalence among Czech teenaged girls and women. Int J Gynaecol Obstet 119, 178-181. http://dx.doi.org/10.1000/10.1016/j.ijgo.2012.06.007.

Pinto, L.A., Edwards, J., Castle, P.E., Harro, C.D., Lowy, D.R., Schiller, J.T., Wallace, D., Kopp, W., Adelsberger, J.W., Baseler, M.W., *et al.* (2003). Cellular immune responses to human papillomavirus (HPV)-16 L1 in healthy volunteers immunized with recombinant HPV-16 L1 virus-like particles. J Infect Dis 188, 327-338.

Pinto, L.A., Viscidi, R., Harro, C.D., Kemp, T.J., Garcia-Pineres, A.J., Trivett, M., Demuth, F., Lowy, D.R., Schiller, J.T., Berzofsky, J.A., *et al.* (2006). Cellular immune responses to HPV-18, -31, and -53 in healthy volunteers immunized with recombinant HPV-16 L1 virus-like particles. Virology 353, 451-462.

Pollock, K.G., Kavanagh, K., Potts, A., Love, J., Cuschieri, K., Cubie, H., Robertson, C., Cruickshank, M., Palmer, T.J., Nicoll, S., *et al.* (2014). Reduction of low- and high-grade cervical abnormalities associated with high uptake of the HPV bivalent vaccine in Scotland. Br J Cancer. http://dx.doi.org/10.1000/bjc2014479 [pii] 10.1038/bjc.2014.479 [doi].

Poynten, I.M., Machalek, D., Templeton, D., Jin, F., Hillman, R., Zablotzska, I., Prestage, G., Holt, M., and Grulich, A. (2016). Comparison of age-specific patterns of sexual behaviour and anal HPV prevalence in homosexual men with patterns in women. Sex Transm Infect 92, 228-231. http://dx.doi.org/10.1000/10.1136/sextrans-2015-052032.

Radley, D., Saah, A., and Stanley, M. (2015). Persistent infection with human papillomavirus 16 or 18 is strongly linked with high-grade cervical disease. Human vaccines and immunotherapeutics, 0. http://dx.doi.org/10.1000/10.1080/21645515.2015.1088616.

Rendi-Wagner, P., Shouval, D., Genton, B., Lurie, Y., Rumke, H., Boland, G., Cerny, A., Heim, M., Bach, D., Schroeder, M., *et al.* (2006). Comparative immunogenicity of a PreS/S hepatitis B vaccine in non- and low responders to conventional vaccine. Vaccine 24, 2781-2789. http://dx.doi.org/10.1000/10.1016/j.vaccine.2006.01.007.

Saade, F., Honda-Okubo, Y., Trec, S., and Petrovsky, N. (2013). A novel hepatitis B vaccine containing Advax, a polysaccharide adjuvant derived from delta inulin, induces robust humoral and cellular immunity with minimal reactogenicity in preclinical testing. Vaccine 31, 1999-2007. http://dx.doi.org/10.1000/10.1016/j.vaccine.2012.12.077.

Safaeian, M., Porras, C., Pan, Y., Kreimer, A., Schiller, J.T., Gonzalez, P., Lowy, D.R., Wacholder, S., Schiffman, M., Rodriguez, A.C., *et al.* (2013). Durable antibody responses following one dose of the bivalent human papillomavirus l1 virus-like particle vaccine in the costa rica vaccine trial. Cancer Prev Res (Phila) 6, 1242-1250. http://dx.doi.org/10.1000/10.1158/1940-6207.capr-13-0203.

Sankaranarayanan, R., Prabhu, P.R., Pawlita, M., Gheit, T., Bhatla, N., Muwonge, R., Nene, B.M., Esmy, P.O., Joshi, S., Poli, U.R., *et al.* (2016). Immunogenicity and HPV infection after one, two, and three doses of quadrivalent HPV vaccine in girls in India: a multicentre prospective cohort study. Lancet Oncol 17, 67-77. http://dx.doi.org/10.1000/10.1016/S1470-2045(15)00414-3.

Schellenbacher, C., Kwak, K., Fink, D., Shafti-Keramat, S., Huber, B., Jindra, C., Faust, H., Dillner, J., Roden, R.B., and Kirnbauer, R. (2013). Efficacy of RG1-VLP vaccination against Infections with genital and Cutaneous Human Papillomaviruses. J. Invest. Dermatol. http://dx.doi.org/10.1000/10.1038/jid.2013.253.

Schellenbacher, C., Roden, R., and Kirnbauer, R. (2009). Chimeric L1-L2 Virus-like Particles As Potential Broad-Spectrum Human Papillomavirus Vaccines. J Virol.

Scheller, N.M., Svanstrom, H., Pasternak, B., Arnheim-Dahlstrom, L., Sundstrom, K., Fink, K., and Hviid, A. (2015). Quadrivalent HPV vaccination and risk of multiple sclerosis and other demyelinating diseases of the central nervous system. JAMA: the journal of the American Medical Association 313, 54-61. http://

dx.doi.org/10.1000/2088853 [pii] 10.1001/jama.2014.16946 [doi].

Scherer, E.M., Smith, R.A., Gallego, D.F., Carter, J.J., Wipf, G.C., Hoyos, M., Stern, M., Thurston, T., Trinklein, N.D., Wald, A., *et al.* (2016). A Single Human Papillomavirus Vaccine Dose Improves B Cell Memory in Previously Infected Subjects. EBioMedicine. http://dx.doi.org/10.1000/10.1016/j.ebiom.2016.06.042.

Scherer, E.M., Smith, R.A., Simonich, C.A., Niyonzima, N., Carter, J.J., and Galloway, D.A. (2014). Characteristics of Memory B Cells Elicited by a Highly Efficacious HPV Vaccine in Subjects with No Pre-existing Immunity. PLoS Pathog 10, e1004461. http://dx.doi.org/10.1000/10.1371/journal.ppat.1004461 [doi]
PPATHOGENS-D-14-01551 [pii].

Scherpenisse, M., Schepp, R.M., Mollers, M., Meijer, C.J., Berbers, G.A., and van der Klis, F.R. (2013). Characteristics of HPV-specific antibody responses induced by infection and vaccination: cross-reactivity, neutralizing activity, avidity and IgG subclasses. PLoS One 8, e74797. http://dx.doi.org/10.1000/10.1371/journal.pone.0074797.

Schiller, J.T., Castellsague, X., and Garland, S.M. (2012). A review of clinical trials of human papillomavirus prophylactic vaccines. Vaccine 30 Suppl 5, F123-138. http://dx.doi.org/10.1000/10.1016/j.vaccine.2012.04.108.

Schiller, J.T., and Lowy, D.R. (2015). Raising expectations for subunit vaccine. J Infect Dis 211, 1373-1375. http://dx.doi.org/10.1000/10.1093/infdis/jiu648.

Schwarz, T.F., Huang, L.M., Lin, T.Y., Wittermann, C., Panzer, F., Valencia, A., Suryakiran, P.V., Lin, L., and Descamps, D. (2014). Long-term immunogenicity and safety of the HPV-16/18 AS04-adjuvanted vaccine in 10- to 14-year-old girls: open 6-year follow-up of an initial observer-blinded, randomized trial. Pediatr Infect Dis J 33, 1255-1261. http://dx.doi.org/10.1000/10.1097/inf.0000000000000460.

Serrano, B., de Sanjose, S., Tous, S., Quiros, B., Munoz, N., Bosch, X., and Alemany, L. (2015). Human papillomavirus genotype attribution for HPVs 6, 11, 16, 18, 31, 33, 45, 52 and 58 in female anogenital lesions. Eur J Cancer 51, 1732-1741. http://dx.doi.org/10.1000/10.1016/j.ejca.2015.06.001.

Siegrist, C.A., Lewis, E.M., Eskola, J., Evans, S.J., and Black, S.B. (2007). Human papilloma virus immunization in adolescent and young adults: a cohort study to illustrate what events might be mistaken for adverse reactions. Pediatr Infect Dis J 26, 979-984.

Skinner, S.R., Szarewski, A., Romanowski, B., Garland, S.M., Lazcano-Ponce, E., Salmeron, J., Del Rosario-Raymundo, M.R., Verheijen, R.H., Quek, S.C., da Silva, D.P., *et al.* (2014). Efficacy, safety, and immunogenicity of the human papillomavirus 16/18 AS04-adjuvanted vaccine in women older than 25 years: 4-year interim follow-up of the phase 3, double-blind, randomised controlled VIVIANE study. Lancet 384, 2213-2227. http://dx.doi.org/10.1000/10.1016/S0140-6736(14)60920-X.

Stanley, M. (2012). Perspective: Vaccinate boys too. Nature 488, S10. http://dx.doi.org/10.1000/10.1038/488S10a.

Stanley, M., Lowy, D.R., and Frazer, I. (2006). Chapter 12: Prophylactic HPV vaccines: underlying mechanisms. Vaccine 24 Suppl 3, S3/106-113.

Stanley, M., Pinto, L.A., and Trimble, C. (2012). Human papillomavirus vaccines--immune responses. Vaccine 30 Suppl 5, F83-87. http://dx.doi.org/10.1000/S0264-410X(12)00724-4 [pii]
10.1016/j.vaccine.2012.04.106 [doi].

Stevens, C.E., Taylor, P.E., Tong, M.J., Toy, P.T., Vyas, G.N., Nair, P.V., Weissman, J.Y., and Krugman, S. (1987). Yeast-recombinant hepatitis B vaccine. Efficacy with hepatitis B immune globulin in prevention of perinatal hepatitis B virus transmission. JAMA: the journal of the American Medical Association 257, 2612-2616.

Valenzuela, P., Medina, A., Rutter, W.J., Ammerer, G., and Hall, B.D. (1982). Synthesis and assembly of hepatitis B virus surface antigen particles in yeast. Nature 298, 347-350.

Vichnin, M., Bonanni, P., Klein, N.P., Garland, S.M., Block, S.L., Kjaer, S.K., Sings, H.L., Perez, G., Haupt, R.M., Saah, A.J., *et al.* (2015). An Overview of Quadrivalent Human Papillomavirus Vaccine Safety: 2006 to 2015. Pediatr Infect Dis J 34, 983-991. http://dx.doi.org/10.1000/10.1097/inf.0000000000000793.

Villa, L.L., Ault, K.A., Giuliano, A.R., Costa, R.L., Petta, C.A., Andrade, R.P., Brown, D.R., Ferenczy, A., Harper, D.M., Koutsky, L.A., *et al.* (2006). Immunologic responses following administration of a vaccine targeting human papillomavirus Types 6, 11, 16, and 18. Vaccine 24, 5571-5583.

Walayat, S., Ahmed, Z., Martin, D., Puli, S., Cashman, M., and Dhillon, S. (2015). Recent advances in vaccination of non-responders to standard dose hepatitis B virus vaccine. World J. Hep. 7, 2503-2509. http://dx.doi.org/10.1000/10.4254/wjh.v7.i24.2503.

Wichajarn, K., Kosalaraksa, P., and Wiangnon, S. (2008). Incidence of hepatocellular carcinoma in children in Khon Kaen before and after national hepatitis B vaccine program. Asian Pac J Cancer Prev 9, 507-509.

Williams, O.M., Hart, K.W., Wang, E.C., and Gelder, C.M. (2002). Analysis of CD4(+) T-cell responses to human papillomavirus (HPV) type 11 L1 in healthy adults reveals a high degree of responsiveness and cross-reactivity with other HPV types. J Virol 76, 7418-7429.

Winer, R.L., Lee, S.K., Hughes, J.P., Adam, D.E., Kiviat, N.B., and Koutsky, L.A. (2003). Genital human papillomavirus infection: incidence and risk factors in a cohort of female university students. Am J Epidemiol 157, 218-226.

6

Epstein-Barr Virus and the Pathogenesis of Lymphoma

Paul G. Murray[1]* and Claire Shannon-Lowe[2]*

[1]Institute of Cancer and Genomic Sciences, [2]Institute of Immunology and Immunotherapy, College of Medical and Dental Science, University of Birmingham, Birmingham, UK

**Corresponding authors: p.g.murray@bham.ac.uk, c.shannonlowe@bham.ac.uk*
The authors declare no conflicts of interest

Abstract

Burkitt lymphoma (BL), first recognized by Denis Burkitt in 1958, is a high grade B cell malignancy particularly prevalent in young boys in tropical Africa and New Guinea. The high incidence of BL in areas of holo-endemic malaria prompted the search for a tumour-causing infectious agent transmitted by mosquitoes. This search, led by Anthony Epstein and co-workers, resulted in the discovery in cell lines derived from BL biopsies, of a gamma herpesvirus, later referred to as the Epstein-Barr virus (EBV). Subsequently, EBV was shown to be present in the tumour cells of patients with other forms of B cell lymphoma, such as Hodgkin lymphoma and post-transplant lymphoma, as well as in natural killer/T cell (NK/T) cell lymphomas and in several epithelial cancers such as nasopharyngeal carcinoma and gastric carcinoma. Understanding how EBV contributes to the development of these different forms of cancer has provided fundamental insights into the underlying mechanisms responsible for driving oncogenic processes as well as highlighting opportunities for prophylactic and therapeutic intervention. This chapter will summarise current knowledge of the role of EBV in lymphomagenesis, highlighting the importance of co-factors, including disorders of immunity, which can disrupt the delicate virus-host balance that otherwise ensures asymptomatic virus persistence in normal people.

Discovery of EBV in Burkitt lymphoma biopsies

In March 1961, Denis Burkitt gave a talk entitled 'The commonest children's cancer in tropical Africa: a hitherto unrecognised syndrome'. In the audience was Antony Epstein who later noted –" I was struck from the outset by the highly unusual nature of the new tumour and by the exciting and unprecedented findings on epidemiology which seemed to show that its distribution depended on climatic factors indicating that some biological agent might be involved in the aetiology" (Epstein, 1999). While listening Epstein formulated the hypothesis that "a biological agent must play a part in causation and ... I immediately postulated a climate-dependant arthropod vector spreading a cancer-causing virus. It turned out later that it was a cofactor which was arthropod borne, but my idea focused correctly on the need to search for a viral cause". In discussions after the lecture Burkitt agreed to send fresh biopsies from some of his patients for analysis in Epstein's laboratory.

Following several years without success, in December 1963, a BL biopsy that had been delayed en route, finally reached Epstein's laboratory. In Epstein's own words "As usual the tissue was floating in transit fluid, but unusually this was cloudy. As it was getting late and the cloudiness was likely to be due to bacterial contamination, the feeling was that we could leave the laboratory for the weekend. But instead of discarding the specimen and going home I put a drop of the

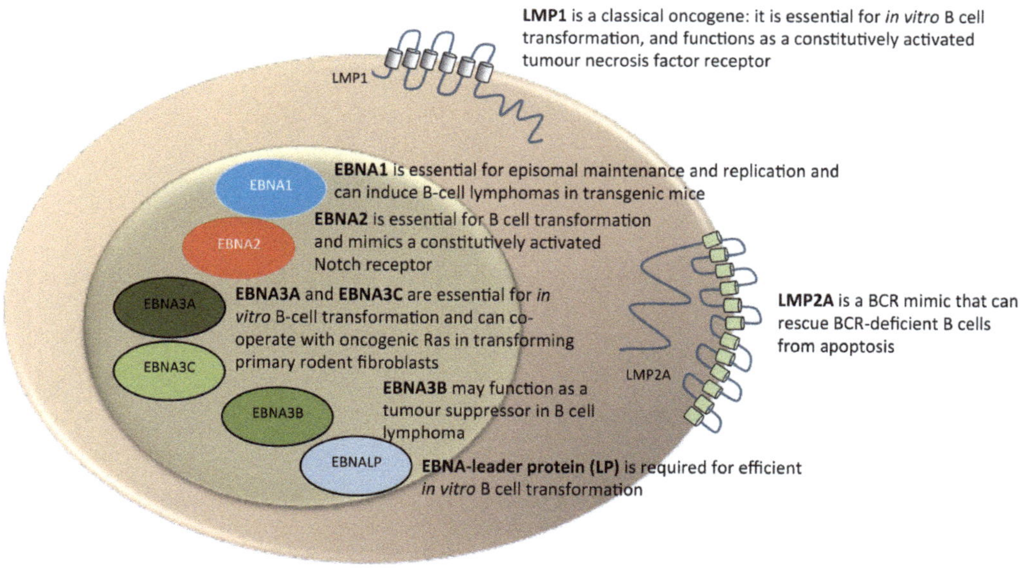

Figure 6.1. The functions of the major EBV latent proteins in B cell transformation/lymphomagenesis. Depicted with their known effects in B cells are the major EBV latent proteins including the Epstein-Barr nuclear antigens (EBNAs) and the latent membrane proteins, LMP1 and LMP2A.

cloudy fluid on a slide and examined it with the light microscope as a wet preparation. Rather than seeing the expected contaminating bacteria I was astonished to find that the cloudiness was due to large numbers of viable-looking free-floating tumour cells. " (Epstein *et al.*, 1964). These cells grew when re-suspended in fresh culture medium, and so the first cell line from a human lymphoma was established (known as the EB1 cell line after Epstein and Yvonne Barr). In February 1964, Epstein prepared cells from the EB1 line for electron microscopy and was 'exhilarated to see unequivocal virus particles in a cultured lymphoma cell in the very first grid square I searched. I was extremely agitated in case the specimen might burn up in the electron beam- I switched off, I walked round the block in the snow without a coat, and when somewhat calmer I returned to record what I had seen. I recognized at once that I was looking at a typical member of the herpesvirus group' (Epstein, 2015).

Establishing the potent transforming ability of EBV in B cells

Following the discovery of EBV in BL, a critical question was whether the virus was transforming for B cells or merely a silent passenger. Volker Diehl co-cultivated cord blood lymphocytes with either an irradiated EBV-positive BL cell line, Jijoye, which produces virus particles, or Raji cells, which do not. He observed growth of the lymphocytes, but only after co-culture with Jijoye cells (Henle *et al.*, 1967). Around the same time, John Pope showed that cell lines could be established from the peripheral blood lymphocytes of EBV-infected individuals (Pope, 1967). These experiments confirmed that EBV can potently transform B cells *in vitro*.

In 1984, the complete sequence of the B95.8 strain of EBV was published, heralding the advent of a post-genomic era in EBV research and one which led directly to the identification and functional characterization of EBV's most important genes (Baer *et al.*, 1984). The EBV genes were broadly separated into the 'latent' genes, which were expressed in a phase of the virus life cycle in which there is no virus replication (latency), and the 'lytic' genes which are expressed during the stepwise progression from latency to virus replication and eventual virion assembly and release. We now know that it is the coordinated expression of the latent genes that is primarily responsible for the *in vitro* immortalization of B cells described above. In vitro immortalized B lymphocytes, which are commonly referred to as lymphoblastoid cell lines (LCL), express all latent genes (referred to as latency III or the 'growth programme') including six nuclear antigens (EBNAs 1, 2, 3A, 3B, 3C and EBNA-LP), two latent membrane proteins (LMP1 and LMP2), the non-coding Epstein-Barr-encoded RNA (EBER1 and EBER2), and a number of viral miRNA (Kerr *et al.*, 1992; Pfeffer, 2004). The functions of the major latent genes are summarised in Figure 6.1.

Asymptomatic infection of B cells is widespread among human populations

The detection of antibodies to EBV proteins in BL patient sera had provided some of the initial evidence to support a pathogenic role for EBV in BL. However, the same antibodies were also present, albeit at lower levels, in most normal healthy people (Henle *et al.*, 1969; Abate *et al.*, 2015). EBV was also found to be the causative agent of infectious mononucleosis (IM) (Henle *et al.*, 1968). Because transforming EBV is detectable at high levels in saliva from IM patients and at lower levels in healthy EBV-seropositive individuals, oral transfer was suspected to be the most likely mode of transmission (Niederman *et al.*, 1976). Kissing is thought to be the major route of EBV transmission among adolescents and young adults. Transmission has also been suggested to occur through intercourse or oral sex (Crawford *et al.*, 2002; Crawford *et al.*, 2006; Higgins *et al.*, 2007). However, the observation that the virus detectable in both male and female genital secretions (Sixbey *et al.*, 1987; Israele *et al.*, 1991; Woodman *et al.*, 2005) is cell-associated rather than free infectious virus, argues against this (Thomas *et al.*, 2006); deep kissing also carries the same risk of primary EBV infection irrespective of whether coitus has taken place (Thomas *et al.*, 2006; Balfour *et al.*, 2012; Balfour *et al.*, 2013).

Although David Thorley-Lawson has elegantly shown that memory B cells are the site of EBV persistence in asymptomatic carriers (Babcock *et al.*, 1998), how the virus gets into the memory B cell pool remains the subject of debate. In one model, memory B cells are directly infected with EBV (Kurth *et al.*, 2000; Kurth *et al.*, 2003). In the other model, proposed by Thorley-Lawson, EBV initially infects naïve B cells which then proliferate as a result of expression of the latency III pattern typical of LCL. The EBV-infected B cells enter germinal centres (GC) - B cell factories in which B cells undergo affinity maturation and class switch recombination. In the GC, the EBV-infected B cells express another form of latency, known as the 'default programme' or latency II; here the two latent membrane proteins, LMP1 and LMP2, are expressed together with EBNA1 driven by the viral Qp promoter (Babcock *et al.*, 2000). LMP1 and LMP2 provide surrogate CD40 and B cell receptor (BCR) signals, respectively, which probably mediate the survival of EBV-infected GC B cells and their subsequent exit from the GC as memory cells (Gires *et al.*, 1997; Caldwell *et al.*, 1998). EBV-infected memory B cells shut down virus gene expression (latency 0) thereby avoiding recognition by the immune system (Babcock *et al.*, 2000). One of the major functions of EBNA1 is to allow replication and segregation of viral episomes during cell division and so when memory B cells are required to proliferate, EBNA1 expression is switched on; this additional form of latency is known as latency I. The activation of EBV-infected memory B cells can lead to plasma cell differentiation which switches on the lytic cycle of EBV with subsequent production of new virions (Laichalk and Thorley-Lawson, 2004).

The origin of EBV-associated B cell lymphomas

Given that the B cell is the site of virus persistence in the asymptomatically infected host then we may think of the development of EBV-positive B cell lymphomas as rare accidents of EBV's ability to colonise the B cell system (Rickinson, 2014). Because malignancy is a relatively rare outcome of infection, co-factors must be involved in the oncogenic process. As we shall see, some of the co-factors are already known, especially in BL, where malaria and HIV, either alone or in combination, act as chronic immune stimulants.

Most EBV-associated B cell lymphomas display evidence of somatic hypermutation and so appear to have arisen from cells that have been through a GC reaction, but which have presumably not fully differentiated to the memory B cell stages (Timms *et al.*, 2003; Bräuninger *et al.*, 2006). The point at which the progenitor B cells are aberrantly arrested in the B cell differentiation pathway varies between the different EBV-associated B cell malignancies and is reflected in the differences in cellular and virus gene expression that characterise each type (Figure 6.2). We now consider in turn each of the major lymphoma subtypes associated with EBV and briefly outline how the virus is thought to contribute to each.

Defining the contribution of EBV to the pathogenesis of Burkitt lymphoma

Two years after the discovery of EBV, antibodies to EBV antigens were detected in the serum of African BL patients (Old *et al.*, 1966). This was followed by reports of higher titres of EBV antibodies in the serum of BL patients compared with age-matched controls from the same area. In 1978, Guy de Thé and colleagues published the results of their serological observations in pre-BL samples taken from over 40,000 Ugandan children (de-Thé *et al.*, 1978). Most of the children who subsequently developed BL had raised EBV antibody titres prior to diagnosis, leading the authors to conclude that 'BL develops in children who have had a long and heavy exposure to EBV' (de-Thé *et al.*, 1978). In 1970, Harald zur Hausen used DNA hybridisation to identify the EBV genome in BL

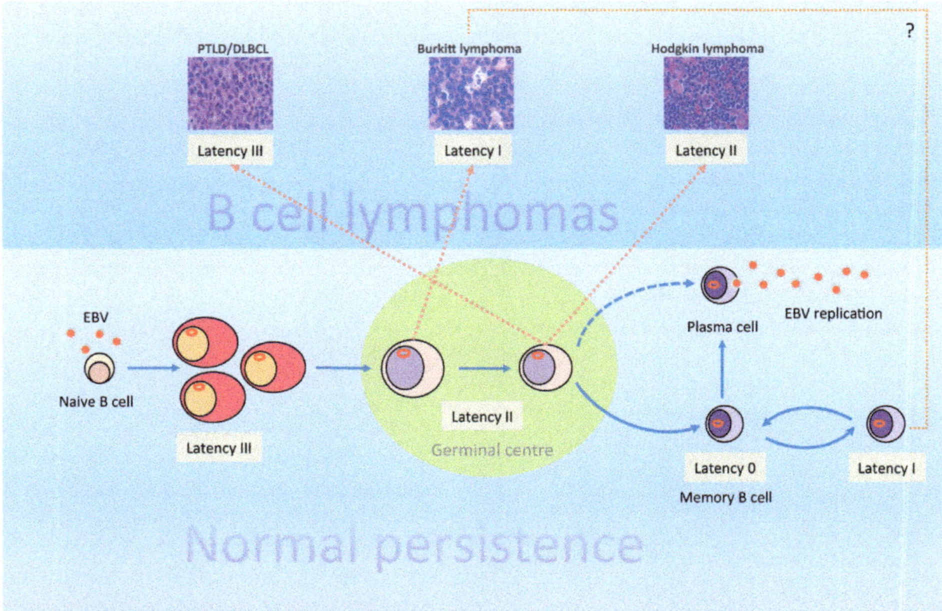

Figure 6.2. Origins of EBV-positive B cell lymphomas. Shown is the germinal centre model of EBV persistence depicting the transition of EBV-infected B cells through a germinal centre reaction and their subsequent differentiation to memory or plasma cells. EBV-infected memory B cells are the site of long-term latency and persistence, whereas virus replication occurs in plasma cells. The potential origin of some of the major EBV-associated B cell lymphomas is also shown. However, there is considerable uncertainty about the precise stage(s) from which each tumour derives.

(zur Hausen *et al*., 1970), and in 1974, Beverley Reedman working in George and Eva Klein's lab', used anti-complement immunofluorescence (ACIF) to show expression of an Epstein-Barr virus nuclear antigen, known as 'EBNA' for short; we now know this protein is EBNA1 (Reedman *et al*., 1974; Dillner *et al*., 1984; Dillner *et al*., 1985).

BL occurs not only in its African or 'endemic' form, but also in a 'sporadic' form at a much lower incidence throughout the world. BL also occurs with an increased incidence in people with HIV, usually occurring before CD4+ T cells counts rapidly decline. Almost all endemic BL is EBV-positive, whereas EBV-positive rates are lower in HIV-associated BL (~40%) and in sporadic BL (15-20%). However, irrespective of EBV status, all forms of BL harbour one of three reciprocal chromosome translocations involving the c-MYC gene on chromosome 8 and either the immunoglobulin heavy chain gene on chromosome 14, or the kappa or lambda light chain genes on chromosomes 2 and 22, respectively (Manolov and Manolova, 1972; Zech *et al*., 1976). These translocations bring c-myc under the transcriptional control of one of the immunoglobulin *loci* resulting in a constitutively elevated c-myc expression which maintains the high proliferative rate that characterises BL (Dean *et al*., 1983; Armelin *et al*., 1984; Einat *et al*., 1985). However, c-myc over-expression also induces apoptosis, and so the BL progenitor must acquire a second independent event that can overcome the apoptosis-inducing effects of c-myc (Evan *et al*., 1992; Milner *et al*., 1993).

The loss of virus episomes from EBV-positive BL cell lines was shown to result in diminished cell survival and provided the first evidence that EBV might provide an anti-apoptotic function to counteract c-myc induced apoptosis (Shimizu *et al*., 1994). Most EBV-positive primary BL and early passage BL cell lines express a latency I pattern of virus gene expression in which there is expression of EBNA1, EBERs and the viral miRNA. Although EBNA1 was shown to mediate some of the anti-apoptotic effects of EBV in BL cells, this virus protein cannot reconstitute complete protection from cell death and other studies suggest that the EBERs and the viral miRNAs have anti-apoptotic functions in BL (Komano *et al*., 1999; Kennedy *et al*., 2003; Vereide *et al*., 2013). In keeping with the requirement for the inhibition of apoptotic pathways, TP53 mutations are present in approximately 30%–40% of BL biopsies (Lindström and Wiman, 2002). Recent global genomic analysis have also identified recurrent mutations in the ID3, TCF3, and CCND3 genes, all of which are involved in the same pathway leading to the activation of proliferation through cyclin D3 and survival through the induction of phosphoinositide 3-kinase (PI3K) signalling (Love *et al.*, 2012; Richter *et al.*, 2012; Schmitz *et al.*, 2012). Mutations in MYC itself, and of ID3, TCF3 and TP53 have been shown to be more common in sporadic BL than in endemic BL, whereas mutations in other genes, including *ARID1A*, *CCNF* and *RHOA*, are less frequently seen in sporadic cases (Abate *et al.*, 2015).

A minority of primary BL and BL cell lines express all the EBNAs with the exception of EBNA2 which is deleted leading to the loss of LMP1 expression (Kelly *et al.*, 2002; Kelly *et al.*, 2005). These unusual cases are known as 'Wp-restricted' BL because expression of the EBNAs is initiated from the Wp promoter as opposed to the Qp promoter that drives EBNA1 expression in conventional BL. It has been shown that Wp-restricted BL cell lines are more resistant to apoptosis than Qp-restricted BL lines, an effect that has been attributed to several possible mechanisms which include the down-regulation of the pro-apoptotic molecule, Bim by EBNA3A and EBNA3C, and the expression of an EBV-encoded homologue of bcl-2, known as BHRF1 (Anderton *et al.*, 2007; Kelly *et al.*, 2009).

It is now well-established that malaria and HIV are co-factors in BL development. Although both act in mechanistically different ways, the outcomes are similar - a sustained increase in EBV loads caused by the polyclonal activation of B-cells which increases the size of the EBV-infected B cell pool. In turn this probably makes the generation of EBV-infected B cells harbouring MYC translocations more likely (Rowe *et al.*, 2014). Although malaria also impairs T-cell immunity to EBV this impairment is transient, whereas the increased EBV loads in circulating memory B cells associated with malarial infection are long-lived (Moss *et al.*, 1983; Njie *et al.*, 2009; Whittle *et al.*, 1984). *P. falciparum* may also act more directly to induce BL since it has been shown to increase expression of the enzyme, activation-induced cytidine deaminase (AID) which can induce BL-associated mutations, including the characteristic translocations (Torgbor *et al.*, 2014).

Post-transplant lymphoproliferative disorders: loss of EBV immunity takes centre stage

Although there had been some reports of an increased risk of lymphoma development in patients suffering from deficiencies of humoral or cell-mediated immunity, it was not until the late 1960s that the importance of immune surveillance in preventing lymphoma development was fully appreciated when several reports of lymphoma in iatrogenically immunosuppressed renal allograft recipients were published (Fialkow, 1967; Doak *et al.*, 1968; Penn *et al.*, 1969).

Raised antibody titres to the viral capsid antigen (VCA) of EBV were initially detected in transplant patients treated with cyclosporine, raising the possibility that EBV was involved (Nagington and Gray, 1980). Confirmation of EBV's involvement came when Dorothy Crawford identfied the 'EBNA' protein in post-transplant lymphoma tissues (Crawford *et al.*, 1980), which was quickly followed by further reports documenting EBV in lymphomas arising in other solid organ transplant recipients, and in bone marrow transplant patients (Crawford *et al.*, 1980; Hanto *et al.*, 1981; Saemundsen Ak *et al.*, 1981; Schubach *et al.*, 1982; Hanto *et al.*, 1983). Surprisingly, most lymphoproliferative diseases arising in bone marrow transplant recipients were not of recipient origin, but derived from the donor B cells (Schubach *et al.*, 1982). Collectively, these tumours were termed 'post-transplant lymphoproliferative disorders' (PTLD) (Starzl *et al.*, 1984). The incidence of PTLD was found to be high in the first weeks to months after transplantation when patients are intensely immunosuppressed. Thereafter, the incidence decreases, but rises again 4–5 years from transplantation (Leblond *et al.*, 2001; Ghobrial *et al.*, 2005a;). These data suggest that early-onset and late-onset tumours are aetiologically distinct. Indeed, early onset PTLD are typically oligoclonal and in many cases may be considered to be the *in vivo* equivalent of LCL - outgrowths of EBV-infected B cells expressing all latent genes (latency III). Late-onset PTLD is less likely to be associated with EBV, and overall is more likely than early-onset PTLD to be extranodal (Leblond *et al.*, 1998; Ghobrial *et al.*, 2005b). Monoclonal PTLD often occur as a type of B cell lymphoma known as diffuse large B cell lymphoma (DLBCL). A similar spectrum of disease is also observed in people with HIV (before the advent of highly active retroviral therapy) and in patients with some primary immunodeficiences (Palendira and Rickinson, 2015). Starzl and colleagues were the first to show that reduction or discontinuance of immunosuppression can lead to regression of PTLD (Starzl *et al.*, 1984). An extension of this approach was pioneered by Cliona Rooney and others using the adoptive transfer of EBV-specific T cells (Rooney *et al.*, 1995; Hislop *et al.*, 2010). These and other immunotherapeutic anti-viral strategies are discussed elsewhere in this volume.

Early-onset PTLD are often thought of as EBV-driven malignancies. However, not all immunosuppressed patients will develop PTLD and so the pathogenesis of these tumours must be dependent upon other co-factors. One striking example of this is the high incidence of early-onset PTLD in the transplanted organ suggesting that chronic antigen stimulation in the graft itself might be involved in the pathogenesis of these lesions. Indeed, T cells are required for the development of PTLD-like tumours in severe combined immunodeficient (SCID) mice, suggesting an important role for T cell help in the growth of PTLD (Johannessen *et al.*, 2000).

EBV-positive DLBCL

As well as occurring during HIV infection and after post-transplant immunosuppressive therapy, EBV-positive DLBCL can also arise in patients without a prior history of immune suppression or lymphoma (Oyama *et al.*, 2003; Oyama *et al.*, 2007; Cohen *et al.*, 2012; Cohen *et al.*, 2014). In older patients senescence of the EBV-specific immune response associated with the normal ageing process is suspected; consistent with this these tumours often display a latency III pattern of virus gene expression (Ok *et al.*, 2013; Ok *et al.*, 2015). EBV-positive DLBCL has also been described in younger patients and in children who are not obviously immunodeficient. The precise nature of the defects in T cell immunity that allow the development of EBV-positive DLBCL have not been properly defined. In one study, compared to age-matched controls, patients with EBV-positive DLBCL showed a narrowed EBV-specific TCR-Vb repertoire, with reduced EBV-specific effector memory CD4+ and CD +8 T cell numbers (Cárdenas *et al.*, 2015). However, more work is required to determine if EBV-specific immune responses are different between patients with EBV-positive and EBV-negative DLBCL, and if these defects remain after successful therapy.

Primary effusion lymphoma

Primary effusion lymphoma (PEL) is a B-cell lymphoma localised predominantly in body cavities (pleural, peritoneal, pericardial) as a neoplastic effusion rather than a solid tumour mass and without recognisable nodal involvement. PEL is characterised by infection of the neoplastic cells with human herpesvirus 8 (HHV8/KSHV) and often also with EBV (Horenstein *et al.*, 1997). The tumour cells express markers of lymphocyte activation (CD30, CD38, CD71, HLA-DR) and plasma cell differentiation (CD138), exhibit clonal immunoglobulin gene rearrangements and show evidence of somatic hypermutation (Jenner *et al.*, 2003; Schulz and Cesarman, 2015), consistent with a plasmablast origin. Although PEL occurs mostly in HIV-infected individuals, cases are also observed following iatrogenic immune suppression for organ transplantation, suggesting this neoplasm is strongly associated with loss of immune control (Boshoff and Weiss, 2001).

HHV8 is thought to be the primary causal agent in PEL because HHV8-encoded genes that can induce cellular proliferation, inhibit apoptosis and contribute to immune evasion are expressed by tumour cells (this is discussed elsewhere in this volume). However, the role of EBV in the pathogenesis of PEL is less well understood. EBV gene expression is largely restricted to the latency I programme, with expression of Qp-driven *EBNA1* and highly variable levels of the *EBER*s. Occasional low levels of *LMP1, LMP2A* and *BZLF1* expression are observed although this is highly variable between and within patients (Szekely *et al.*, 1998; Trivedi *et al.*, 2004). The Wp and Cp promoters are not used, and for this reason expression of *EBNA2* and the *EBNA3*s is not observed (Horenstein *et al.*, 1997). Further study of EBV's contribution to the pathogenesis of PELs is required (Mack and Sugden, 2008).

DLBCL associated with chronic inflammation

DLBCL associated with chronic inflammation are EBV-positive tumours that occur in the context of long-standing chronic inflammation and are derived from late GC or post-GC B cells. They usually present as a tumour mass involving body cavities. Pyrothorax-associated lymphoma (PAL) is the prototypic form. PAL is associated with a history of chronic pyrothorax or chronic pleuritis due to the initiation of a therapeutic artificial pneumothorax for pleuropulmonary tuberculosis, which was used in the past as a surgical treatment for tuberculosis (Loong *et al.*, 2010). The interval between the onset of chronic inflammation and malignant lymphoma is usually over 10 years, (median 37 years) with a median age of 65 to 70 years. Males are more susceptible to this lymphoma, with a male to female ratio of 4:1. Other EBV-positive DLBCL associated with chronic inflammation with similar features to PAL include those associated with metallic implants in bones and joints (Sanchez-Gonzalez *et al.*, 2013), and chronic osteomyelitis or chronic venous ulcers (Copie-Bergman *et al.*, 1997; Cheuk *et al.*, 2005;).

The EBV gene expression profile of PAL is usually latency III. It has been suggested that the EBV-transformed B cells at the site of chronic inflammation are able to escape host immune surveillance through several different mechanisms: Cellular IL-10, an immunosuppressive cytokine (Kanno *et al.*, 1997a; Kanno *et al.*, 1997b), and IL-6 and IL-6 receptor are produced by PAL cell lines and can promote tumour cell growth *in vitro* (Kanno *et al.*, 1996). The down-regulation of HLA class I has also been observed (Kanno *et al.*, 1999) and various mutations of CTL epitopes in EBNA3B, an immunodominant antigen for CTL responses, may also play a role in the immune escape of PAL (Kanno *et al.*, 2000). A microarray analysis of PAL identified the interferon-inducible 27 (IFI27) as one of the most differentially expressed genes in PAL compared to regular DLBCL (Nishiu *et al.*, 2004). Expression of IFI27 is in keeping with the role of chronic inflammation in this condition since it is known to be induced in B lymphocytes following stimulation with interferon-alpha (IFN-α), although the role of IFI27 in lymphomagenesis has yet to be resolved.

Hodgkin lymphoma

Although the incidence of the EBV-positive forms of Hodgkin lymphoma (HL) is also increased in immunosuppressed populations, most cases arise in apparently immunocompetent individuals. HL is distinguished histopathologically from the non-Hodgkin lymphomas, including DLBCL, by virtue of a particularly prominent tumour microenvironment that is especially rich in CD4+ T cells and other inflammatory cells. The tumour microenvironment is so extensive in many cases of HL that the malignant Hodgkin/Reed-Sternberg (HRS) cells are outnumbered and can represent as little as 1% or less of the total tumour mass.

In 1974, two reports were published documenting a significantly increased risk of HL in individuals with a prior history of IM (Connelly and Christine, 1974; Rosdahl *et al.*, 1974). Once again the ACIF assay was employed, this time by Sibrand Poppema, who detected the 'EBNA' protein in the HRS cells of a single patient (Poppema *et al.*, 1985). EBV DNA was then detected in whole HL biopsies and in HRS cells (Weiss *et al.*, 1987; Weiss *et al.*, 1989). Later, RT-PCR and immunohistochemistry showed that EBV-positive HRS cells express a latency II pattern of virus gene expression, characterised by high level expression of *LMP1* and *LMP2* (Wu *et al.*, 1990; Pallesen *et al.*, 1991; Murray *et al.*, 1992; Deacon *et al.*, 1993).

Ralf Kuppers showed that although HRS cells lack markers present on most B cells, they harbour clonal immunoglobulin rearrangements, are somatically hypermutated, and in some cases have mutations that prevent immunoglobulin expression (Kuppers *et al.*, 1994). Thus, HRS cells are thought to be derived from GC B cells lacking a functional BCR and which must have escaped the apoptosis that would otherwise be their normal fate. There is compelling evidence to suggest that EBV can provide the necessary anti-apoptotic functions. Thus, the crippling mutations in immunoglobulin genes have so far almost always only been found in EBV positive cases (Bräuninger *et al.*, 2006). Furthermore, EBV has been shown to immortalise BCR-negative GC B cells *in vitro*; an effect that is dependent upon BCR-like signalling provided by the viral *LMP2A* gene (Caldwell *et al.*, 1998; Bechtel *et al.*, 2005; Chaganti *et al.*, 2005; Mancao *et al.*, 2005). *LMP1* is also probably a major contributor to HRS cell survival because it has been shown to constitutively stimulate several pro-survival pathways, including NF-κB, STAT and PI3K/AKT and can activate an HRS-like transcriptional programme in normal GC B cells (Bargou *et al.*, 1997; Dutton *et al.*, 2005; Holtick *et al.*, 2005; Vockerodt *et al.*, 2008).

X-linked lymphoproliferative disease

Although IM is usually self-limiting, primary EBV infection may be associated with more severe symptoms, typically in immunodeficient patients, and most notably in a condition referred to as 'Duncan's syndrome', later to be called X-linked lymphoproliferative disease-1 (XLP1). Boys with XLP1 develop a severe, often fatal, form of IM usually within weeks following primary EBV infection, and those who survive have a increased risk of lymphoma (Purtilo *et al.*, 1975; Seemayer *et al.*, 1995). David Purtilo established the XLP registry at the University of Nebraska in 1980. In 1995 the registry had recorded data on 80 families including 272 boys; 75% of those with follow-up had died, most before 10 years of age (Seemayer *et al.*, 1995). In 1998, the underlying defect in XLP1 patients was identified as loss of function mutations in the *SH2D1A* gene, resulting in defective forms of its gene product, the signalling lymphocyte activation molecule (SLAM)-associated protein (or SAP for short) (Coffey *et al.*, 1998; Nichols *et al.*, 1998; Sayos *et al.*, 1998). As well as SLAM, SAP binds other proteins including 2B4, NTB-A, CD84, Ly108, and Ly986. XLP1 patients have impaired T-cell and NK cell engagement of EBV-infected B cells rather than a specific inability to cope with EBV itself (Hislop *et al.*, 2010). SAP-defective T and NK cells are also resistant to apoptosis resulting in the massive proliferation of T and NK cells, excessive cytokine production, and hemophagocytic lymphohistiocytosis (HLH) (see below).

Spectrum of EBV-positive NK/T cell disease

The primary cellular targets of EBV during the natural life cycle of the virus are B lymphocytes and

oropharyngeal epithelial cells. However, EBV is seemingly able to infect T cells and NK cells. Because mature NK or T cells do not express the EBV receptor, the mechanism of infection is currently unknown.
The EBV-associated NK/T lymphoproliferative diseases (EBV+ NK/T-LPD) can be subdivided into distinct but overlapping clinical entities, which include: chronic active EBV (CAEBV), haemophagocytic lymphohistiocytosis (HLH), and hydroa vacciniforme (HV). EBV is also associated with a number of highly aggressive malignancies of NK/T cells, including extranodal NK/T cell lymphoma (ENKTL) and aggressive NK leukaemia (ANKL). These malignancies are extremely difficult to treat and as a result have a poor prognosis. Importantly, a proportion of these malignancies develop from the EBV+ NK/T-LPD.

Chronic active EBV

Chronic active EBV (CAEBV) is usually described as a chronic disease of childhood and young adulthood. Patients exhibit persistent IM-like symptoms including fever, hepatosplenomegaly, persistent hepatitis and extensive lymphadenopathy. This is coupled with elevated EBV-DNA loads in peripheral blood mononuclear cells, histological evidence of organ infiltration with EBV-infected cells and high levels of pro-inflammatory cytokines in the blood. CAEBV is considered a chronic disease, yet many patients develop severe, often fatal, complications including multi-organ failure, digestive tract ulceration/ perforation, malignant lymphomas and HLH.

Although CAEBV occurs worldwide, the disease exhibits a distinct geographical distribution with increased frequency observed in East Asia and in native Americans from Central and South America and Mexico. The majority of cases occur in children and young adults with a mean onset of 11.3 years. There is no sex predilection and the frequency of T cell vs. NK cell infection is roughly equal (Kimura *et al.*, 2003). However, the number of cases of CAEBV in adults is increasing. It is unclear if this is a true increase in incidence or merely increased awareness by clinicians. The virus is predominantly found in the T cells of adult cases. EBV infection of T cells is often associated with a worse outcome than infection of NK cells and this does appear to be borne out in the adult onset cases where the disease is more aggressive.

CAEBV was first described in 1978 and yet the pathogenesis of CAEBV is still poorly understood (Virelizier *et al.*, 1978). A pathogenic role for EBV is suggested by the detection of monoclonal episomes in infected cells (Ohga *et al.*, 1999; Oyoshi *et al.*, 2003), which exhibit a latency II viral gene expression profile, including expression of LMP2-TR (LMP2-terminal repeat) (Fox *et al.*, 2010). However, our understanding of the contribution of EBV to CAEBV is limited, partly due to the rarity of the condition and partly because of the inherent difficulty in infecting T cells or NK cells *in vitro*.

Individuals with CAEBV are unable to produce an effective immune response to control EBV. Patients frequently have elevated levels of pro- and anti-inflammatory cytokines including interleukin (IL)-1β, IFN-γ, IL-10, IL-13, IL-15, tumour necrosis factor (TNF)-α and transforming growth factor (TGF)-β (Ohga *et al.*, 2004; Kimura *et al.*, 2005). One study of 11 CAEBV patients reported impaired NK cell activity, lymphokine activated killer activity and EBV-specific CTL activity (Wakiguchi *et al.*, 1988), and another reported very low to undetectable EBV-specific CD8+ T cells (Sugaya *et al.*, 2004). It has been suggested that adoptive T cell therapy may be a valid approach for CAEBV. Although the subset of viral genes expressed in CAEBV elicit only subdominant T cell responses, LMP2-TR specific CD8+ T cells were able to efficiently recognise and kill CAEBV cell lines (Fox *et al.*, 2010) and autologous *ex vivo* expanded cytotoxic CD8+ T cells from CAEBV patients stimulated with LCLs were able to resolve mild/ moderate CAEBV (Savoldo *et al.*, 2002). The association of primary EBV infection with CAEBV and the strong racial predisposition suggests a genetic defect in the infected T- or NK cells, however no consistent mutations or chromosomal aberrations have been identified (Joncas *et al.*, 1989; Fujieda *et al.*, 1993; Sugaya *et al.*, 2004).

CAEBV has also been well documented in the US. These cases differ from those observed in East Asia, because they are predominantly associated with EBV infection of B cells, but not T- or NK cells. The clinical course of the disease is milder, the mean age of clinical onset is 19 years. Unlike cases reported in East Asia, the US cases show a progressive loss of B cells, hypogammaglobulinaemia and reduced NK cell numbers and patients usually succumb to progressive lymphoproliferative disease or infection (Cohen *et al.*, 2009).

In general, the prognosis for CAEBV is poor and the only cure is haematopoietic stem cell transplant. Even following transplant, at least 25% of patients will develop a B-, T- or NK cell malignancy.

Table 6.1. Mutations associated with primary haemophagocytic lymphohistiocytosis.

Primary HLH	Cytogenetic localisation	Mutated gene	Protein	Function
Familial HLH (FHL)				
FHL1	9q21.3-22	Unknown		Unknown
FHL2	10q21-22	*PRF1*	Perforin	Pore formation
FHL3	17q25	*UNC13D*	Munc13-4	Granule priming
FHL4	6q24	*STX11*	Syntaxin-11	Granule fusion
FHL5	19q13	*STXBP2*	Munc 18-2	Granule fusion
Chediak-Higashi syndrome	1q42-43	*LYST*	LYST	Granule trafficking
Griscelli syndrome type 2	15q21	*Rab27A*	Rab27A	Granule docking
XLP type 1	Xq25	*SH2D1A*	SAP	Signalling in T and NK cells
XLP type 2	Xq25	*BIRC4*	XIAP	Signalling pathways involving NF-κB

EBV-related haemophagocytic lymphohistiocytosis (EBV-HLH)

HLH is a life-threatening hyper-inflammatory syndrome that can be classified according to underlying etiology into primary (genetic) or secondary HLH. Primary HLH occurs during infancy and early childhood as the result of mutations affecting the perforin pathway which affect perforin-mediated cytotoxicity of T lymphocytes and NK cells (Table 6.1). Secondary HLH usually occurs in children and adults with no known underlying genetic defects and frequently as a complication of infection, malignancy, autoimmune or rheumatologic disease, immunosuppression or haematopoietic stem cell transplantation.

The diagnosis of HLH is based on fulfilling the diagnostic criteria described in the HLH-2004 protocol (Table 6.2) (Henter *et al.*, 2007). Unfortunately, these diffuse criteria overlap with several other diseases, making the diagnosis of HLH challenging. Delayed diagnosis of EBV-HLH is associated with high mortality, usually within 2 months following the onset of symptoms. Therefore early recognition and prompt treatment is key.

EBV is recognised as the leading infectious cause of secondary HLH (EBV-HLH) (Janka *et al.*, 1998; Rouphael *et al.*, 2007). The highest incidence of EBV-HLH cases is observed in East Asia, including Japan, Taiwan, China and Vietnam where the disease is predominantly observed in children (mean 3.9 years ±2.8 years) (Su *et al.*, 1989; Kawaguchi *et al.*, 1993; Janka *et al.*, 1998). In adults, EBV is associated with approximately 40% of HLH, however, in these cases the HLH usually arises as a complication of EBV-associated lymphomas including Hodgkin lymphoma and extranodal NK/T cell lymphoma. The viral load in peripheral blood mononuclear cells and plasma of EBV-HLH patients is usually very high and monoclonal EBV is found almost exclusively in the NK cells or T cells, the majority of which are clonal.

Secondary HLH is usually treated with the HLH-2004 protocol (etoposide, cyclosporine, dexamethasone) which is administered to eliminate the underlying

Table 6.2. Diagnostic criteria for haemophagocytic lymphohistiocytosis according to the HLH-2004 protocol.

A DIAGNOSIS OF HLH CAN BE MADE IF EITHER CRITERIA 1 OR 2 IS MET:
1. Molecular diagnosis consistent with primary HLH
2. Clinical and laboratory criteria (at least 5/8 criteria should be fulfilled)
Fever
Hepatosplenomegaly
Cytopenia ≥2-3 cell lines in peripheral blood (Haemoglobin <9 g/100mL, platelets <100 x 10^9/L, neutrophils <1.0 x 10^9/L)
Hypertriglyceridemia and/or hypofibrinogenemia (Fasting triglycerides ≥3.0 nmol/L, fibrinogen ≤1.5 g/L)
Haemophagocytosis in bone marrow, spleen, cerebrospinal fluid or lymph nodes. No sign of malignancy
Decreased or absent NK-cell activity
Ferritin ≥500 μg/L
sCD25 (soluble IL2 receptor) ≥2,400 U/mL

trigger of the HLH and to suppress the inflammation and activation of the immune cells. However, the only cure is haematopoietic stem cell transplant.

Like CAEBV, the role of EBV in the pathogenesis of HLH remains unclear. However, cases of both primary and secondary HLH present with the same clinicopathological condition, suggesting that EBV infection could mimic the effects of the genetic mutations observed in primary HLH. Indeed, the clinical manifestations of EBV-HLH and XLP-1 are difficult to distinguish without genetic analysis. As described earlier, the *SH2D1A* gene is mutated in XLP-1 patients (Table 6.1). The absence of SAP induces an exaggerated Th1 response to EBV infection leading to enhanced TNF-α and IFN-γ cytokine secretion, hypercytokinemia and macrophage activation (Veillette, 2010). However, EBV-HLH cases do not have such mutations. Instead, LMP1 inhibits the expression of SAP at the transcriptional level, resulting in enhanced TNF-α and IFN-γ cytokine secretion, akin to that observed in XLP-1 (Chuang *et al.*, 2005).

Hydroa vacciniforme and Hydroa vacciniforme-like lymphoma

Hydroa vacciniforme is a rare chronic paediatric disorder characterised by photodermatosis of unknown etiology, characterised by sensivity to sunlight resulting in edema, vesicles, crusts and large ulcers which cause severe hypopigmented scarring and disfigurement of the skin. The disease usually occurrs in children aged 3-15 years from East Asia, Central and South America and Mexico and remits spontaneously by late adolescence. EBV is nearly always found in the infiltrating T cells or NK cells of the cutaneous lesions and in the blood, with high viral loads in patients with severe disease (Iwatsuki *et al.*, 1999; Cho *et al.*, 2001; Doeden *et al.*, 2008). Classic HV is not however associated with haematologic abnormalities, chromosomal aberrations or mortality, but has been classified within the spectrum of CAEBV.

Hydroa vacciniforme-like lymphoma (HVLL) is defined by the World Health Organisation as an EBV-positive cutaneous T cell lymphoma that occurs mainly in children and adolescents with the same geographical distribution as HV. Here EBV is found predominantly in the malignant CD8 T cells, although infection of NK cells has been reported (Iwatsuki *et al.*, 1999; Cho *et al.*, 2001; Barrionuevo *et al.*, 2002; Doeden *et al.*, 2008). Cases of CD4 T cell infection are rare (Wu *et al.*, 2007; Lee *et al.*, 2012). Patients with HVLL usually have systemic symptoms including lymphadenopathy and hepatosplenomegaly and exhibit a slowly progressive relapsing course. HVLL is considered by some to derive from pre-malignant non-resolving HV over 2 to 10 years, although this point is contentious. The contribution of EBV to HV and the subsequent transformation to HVLL is completely unknown.

Extranodal NK/T-cell lymphoma, nasal type

Extranodal NK/T-cell lymphoma (ENKTL) is a rare but highly aggressive type of non-Hodgkin lymphoma associated with EBV. The lymphoma is marked by extensive necrosis and angioinvasion and usually presents in extranodal sites, predominantly within the upper aerodigestive tract (nasal cavity, nasopharynx, paranasal sinus and palate). Patients can also present with tumours in the skin, soft tissue, respiratory tract, gastrointestinal tract and testis and these are the primary sites of dissemination.

Like the EBV-associated T and NK cell infections described above, ENKTL shows a strong ethnic and geographic predilection. ENKTL occurs with the highest incidence in East Asian, Central and South American, and Mexican populations, accounting for 7% to 10% of non-Hodgkin lymphoma (20% - 30% of peripheral T cell lymphomas). In contrast, ENKTL only accounts for 1% of North American or European cases (Au, 2010). ENKTL usually occurs in adults (median age of presentation 40 to 50 years) with a stronger male predominance. However, it is important to note that the clinicopathological features are the same irrespective of ethnic origin.

ENKTL follows an aggressive clinical course with a poor prognosis. The efficacy of conventional anthracycline-based chemotherapy regimens is poor either with or without radiotherapy, with an overall 5 year survival of 6-25% for patients with advanced disease (Au *et al.*, 2005; Lee *et al.*, 2006; Pagano *et al.*, 2006; Lim *et al.*, 2008; Suzuki *et al.*, 2010; Chauchet *et al.*, 2012). It is thought the poor response to conventional anthracycline-based regimens may be, at least in part, associated with expression of the Multi-Drug Resistance gene (MDR1) encoding P-glycoprotein (Drenou *et al.*, 1997).

Most cases of ENKTL are of NK lineage, as evidenced by positivity for CD56, CD2 and CD3ε, and lack of expression of CD4, CD8 and CD3s and a germline TCR. The remaining cases are of a cytotoxic T cell lineage with clonal TCR rarrangements (Pongpruttipan *et al.*, 2012).

EBV is present in every malignant cell of ENKTL as clonal episomes that display a latency II gene expression pattern (including LMP2-TR).

Chromosome abnormalities are common; deletion of 6q21 is the most frequently observed alteration (Nakashima *et al.*, 2005) and there is loss of expression of tumour suppressor genes including *PRDM1, ATG5, AIM1, FOXO3* and *HACE1* (Jiang *et al.*, 2015). Mutations in TP53, MGA, STAT3, MSN, P21, K-ras, c-kit and β-catenin are also frequently observed (Quintanilla-Martinez *et al.*, 2001). Gene expression profiling has highlighted the activation of several oncogenic pathways in NK/T cell tumours, including NF-κB, MAPK and JAK-STAT (Jiang *et al.*, 2015). Activation of these pathways by LMP1 have been described in considerable detail in nasopharyngeal carcinoma and B cell lymphomas where LMP1 has been shown to drive the canonical and non-canonical NF-κB pathways (Eliopoulos *et al.*, 1997; Blake *et al.*, 2001; Baxendale *et al.*, 2005), the three classic mitogen activated protein kinases (ERK-MAPK, p38 MAPK, JNK/SAPK) (Morris *et al.*, 2009) and the JAK/STAT pathway (Roberts and Cooper, 1998; Dawson *et al.*, 2008; Chen *et al.*, 2010; Morris *et al.*, 2016).

NK/T cell lymphomas also harbour alterations in genes encoding epigenetic modifiers. For example, a recent study of 105 ENKTL patients identified mutations in MLL2, ARID1A, EP300 and ASXL3 (Jiang *et al.*, 2015). This is of interest because data is emerging which links EBV gene expression with the widespread epigenetic modification of the host cell genome in both EBV-associated lymphomas and carcinomas (Kaneda *et al.*, 2012; Leonard *et al.*, 2012; Kreck *et al.*, 2013; Zhang *et al.*, 2014; Niller *et al.*, 2016; Peng *et al.*, 2016; Ramayanti *et al.*, 2016). The most complete study to date was performed in EBV-associated gastric carcinoma where EBV infection is associated with an extensively hypermethylated cellular epigenome (Cancer Genome Atlas Research, 2014). Several studies in both B cell and epithelial cell malignancies have identified the regulation of the cellular DNA methyltransferases DNMT1, DNMT3A and DNMT3B by EBV infection and by LMP1 or LMP2A (Tsai *et al.*, 2006; Hino *et al.*, 2009). Even transient EBV infection of epithelial cells leaves permanent epigenetic marks (Queen *et al.*, 2013; Birdwell *et al.*, 2014). Expression of LMP1 and LMP2A in NK/T cell lymphomas suggests that a hypermethylated epigenome might also be a feature of these tumours.

Conclusions

EBV was first identified as a potent transformer of B cells and to be causally associated with the development of Burkitt lymphoma. Since then EBV has been shown to be associated with a diverse range of different cancer types, including other types of B cell lymphoma and perhaps surprisingly a variety of different lymphoproliferations and malignanices of T and NK cells. Emerging evidence implicates chronic immune stimulation as an important co-factor in the development of many of these lymphomas. Unravelling the complex interplay between these factors and EBV infection and how this contributes to transformation will not only reveal new insights into the pathogenesis of these tumours, but is also likely to lead to the development of novel therapies.

Acknowledgements

The authors are supported by a programme grant from Bloodwise (PGM) and by the Medical Research Council (CS-L).

References

Abate, F., Ambrosio, M.R., Mundo, L., Laginestra, M.A., Fuligni, F., Rossi, M., Zairis, S., Gazaneo, S., De Falco, G., Lazzi, S., *et al.* (2015). Distinct viral and mutational spectrum of endemic Burkitt lymphoma. PLOS Pathogens 11, e1005158.

Anderton, E., Yee, J., Smith, P., Crook, T., White, R.E., and Allday, M.J. (2007). Two Epstein–Barr virus (EBV) oncoproteins cooperate to repress expression of the proapoptotic tumour-suppressor Bim: clues to the pathogenesis of Burkitt's lymphoma. Oncogene 27, 421-433.

Armelin, H.A., Armelin, M.C.S., Kelly, K., Stewart, T., Leder, P., Cochran, B.H., and Stiles, C.D. (1984). Functional role for c-myc in mitogenic response to platelet-derived growth factor. Nature 310, 655-660.

Au, W.Y. (2010). Current management of nasal NK/T-cell lymphoma. Oncology (Williston Park) 24, 352-358.

Au, W.Y., Ma, S.Y., Chim, C.S., Choy, C., Loong, F., Lie, A.K., Lam, C.C., Leung, A.Y., Tse, E., Yau, C.C., *et al.* (2005). Clinicopathologic features and treatment outcome of mature T-cell and natural killer-cell lymphomas diagnosed according to the World Health Organization classification scheme: a single center experience of 10 years. Ann Oncol 16, 206-214.

Babcock, G.J., Decker, L.L., Volk, M., and Thorley-Lawson, D.A. (1998). EBV persistence in memory B cells *in vivo*. Immunity 9, 395-404.

Babcock, G.J., Hochberg, D., and Thorley-Lawson, D.A. (2000). The expression pattern of Epstein-Barr virus latent genes *in vivo* is dependent upon the differentiation stage of the infected B cell. Immunity 13, 497-506.

Baer, R., Bankier, A.T., Biggin, M.D., Deininger, P.L., Farrell, P.J., Gibson, T.J., Hatfull, G., Hudson, G.S., Satchwell, S.C., Séguin, C., *et al.* (1984). DNA sequence and expression of the B95-8 Epstein—Barr virus genome. Nature 310, 207-211.

Balfour, H.H., Odumade, O.A., Schmeling, D.O., Mullan, B.D., Ed, J.A., Knight, J.A., Vezina, H.E., Thomas, W., and Hogquist, K.A. (2012). Behavioral, Virologic, and Immunologic factors associated with acquisition and

severity of primary Epstein-Barr virus infection in university students. J. Infect. Dis. 207, 80-88.
Balfour, H.H., Sifakis, F., Sliman, J.A., Knight, J.A., Schmeling, D.O., and Thomas, W. (2013). Age-specific prevalence of Epstein-Barr virus infection among individuals aged 6-19 years in the United States and factors affecting its acquisition. J. Infect. Dis. 208, 1286-1293.
Bargou, R.C., Emmerich, F., Krappmann, D., Bommert, K., Mapara, M.Y., Arnold, W., Royer, H.D., Grinstein, E., Greiner, A., Scheidereit, C., *et al.* (1997). Constitutive nuclear factor-kappaB-relA activation is required for proliferation and survival of Hodgkin's disease tumor cells. J. Clin. Invest. 100, 2961-2969.
Barrionuevo, C., Anderson, V.M., Zevallos-Giampietri, E., Zaharia, M., Misad, O., Bravo, F., Caceres, H., Taxa, L., Martinez, M.T., Wachtel, A., *et al.* (2002). Hydroa-like cutaneous T-cell lymphoma: a clinicopathologic and molecular genetic study of 16 pediatric cases from Peru. Appl Immunohistochem Mol Morphol 10, 7-14.
Baxendale, A.J., Dawson, C.W., Stewart, S.E., Mudaliar, V., Reynolds, G., Gordon, J., Murray, P.G., Young, L.S., and Eliopoulos, A.G. (2005). Constitutive activation of the CD40 pathway promotes cell transformation and neoplastic growth. Oncogene 24, 7913-7923.
Bechtel, D., Kurth J, J., Unkel, C., and Küppers, R. (2005). Transformation of BCR-deficient germinal-center B cells by EBV supports a major role of the virus in the pathogenesis of Hodgkin and posttransplantation lymphomas. Blood 106, 4345-4350.
Birdwell, C.E., Queen, K.J., Kilgore, P.C., Rollyson, P., Trutschl, M., Cvek, U., and Scott, R.S. (2014). Genome-wide DNA methylation as an epigenetic consequence of Epstein-Barr virus infection of immortalized keratinocytes. J Virol 88, 11442-11458.
Blake, S.M., Eliopoulos, A.G., Dawson, C.W., and Young, L.S. (2001). The transmembrane domains of the EBV-encoded latent membrane protein 1 (LMP1) variant CAO regulate enhanced signalling activity. Virology 282, 278-287.
Boshoff, C., and Weiss, R.A. (2001). Epidemiology and pathogenesis of Kaposi's sarcoma-associated herpesvirus. Philos Trans R Soc Lond B Biol Sci 356, 517-534.
Bräuninger, A., Schmitz, R., Bechtel, D., Renné, C., Hansmann, M.-L., and Küppers, R. (2006). Molecular biology of Hodgkin's and Reed/Sternberg cells in Hodgkin's lymphoma. Int. J. Cancer 118, 1853-1861.
Caldwell, R.G., Wilson, J.B., Anderson, S.J., and Longnecker, R. (1998). Epstein-Barr virus LMP2A drives B cell development and survival in the absence of normal B cell receptor signals. Immunity 9, 405-411.
Cancer Genome Atlas Research, N. (2014). Comprehensive molecular characterization of gastric adenocarcinoma. Nature 513, 202-209.
Cárdenas, D., Vélez, G., Orfao, A., Herrera, M.V., Solano, J., Olaya, M., Uribe, A.M., Saavedra, C., Duarte, M., Rodríguez, M., *et al.* (2015). Epstein-Barr virus-specific CD8 + T lymphocytes from diffuse large B cell lymphoma patients are functionally impaired. Clinical and Experimental Immunology 182, 173-183.
Chaganti, S., Bell Ai, A.I., Pastor, N.B., Milner, A.E., Drayson, M., Gordon, J., and Rickinson, A.B. (2005). Epstein-Barr virus infection *in vitro* can rescue germinal center B cells with inactivated immunoglobulin genes. Blood 106, 4249-4252.
Chauchet, A., Michallet, A.S., Berger, F., Bedgedjian, I., Deconinck, E., Sebban, C., Antal, D., Orfeuvre, H., Corront, B., Petrella, T., *et al.* (2012). Complete remission after first-line radio-chemotherapy as predictor of survival in extranodal NK/T cell lymphoma. J Hematol Oncol 5, 27.
Chen, C.C., Chen, L.C., Liang, Y., Tsang, N.M., and Chang, Y.S. (2010). Epstein-Barr virus latent membrane protein 1 induces the chemotherapeutic target, thymidine phosphorylase, via NF-kappaB and p38 MAPK pathways. Cell Signal 22, 1132-1142.
Cheuk, W., Chan, A.C., Chan, J.K., Lau, G.T., Chan, V.N., and Yiu, H.H. (2005). Metallic implant-associated lymphoma: a distinct subgroup of large B-cell lymphoma related to pyothorax-associated lymphoma? Am J Surg Pathol 29, 832-836.
Cho, K.H., Kim, C.W., Heo, D.S., Lee, D.S., Choi, W.W., Rim, J.H., and Han, W.S. (2001). Epstein-Barr virus-associated peripheral T-cell lymphoma in adults with hydroa vacciniforme-like lesions. Clin Exp Dermatol 26, 242-247.
Chuang, H.C., Lay, J.D., Hsieh, W.C., Wang, H.C., Chang, Y., Chuang, S.E., and Su, I.J. (2005). Epstein-Barr virus LMP1 inhibits the expression of SAP gene and upregulates Th1 cytokines in the pathogenesis of hemophagocytic syndrome. Blood 106, 3090-3096.
Coffey, A.J., Brooksbank, R.A., Brandau, O., Oohashi, T., Howell, G.R., Bye, J.M., Cahn, A.P., Durham, J., Heath, P., Wray, P., *et al.* (1998). Host response to EBV infection in X-linked lymphoproliferative disease results from mutations in an SH2-domain encoding gene. Nature Genetics 20, 129-135.
Cohen, J.I., Kimura, H., Nakamura, S., Ko, Y.H., and Jaffe, E.S. (2009). Epstein-Barr virus-associated lymphoproliferative disease in non-immunocompromised hosts: a status report and summary of an international meeting, 8-9 September 2008. Ann Oncol 20, 1472-1482.
Cohen, M., De Matteo, E., Narbaitz, M., Carreño, F.A., Preciado, M.V., and Chabay, P.A. (2012). Epstein-Barr virus presence in pediatric diffuse large B-cell lymphoma reveals a particular association and latency patterns: Analysis of viral role in tumor microenvironment. Int. J. Cancer 132, 1572-1580.
Cohen, M., Narbaitz, M., Metrebian, F., De Matteo, E., Preciado, M.V., and Chabay, P.A. (2014). Epstein-Barr virus-positive diffuse large B-cell lymphoma association is not only restricted to elderly patients. Int. J. Cancer 135, 2816-2824.
Connelly, R.R., and Christine, B.W. (1974). A cohort study of cancer following infectious mononucleosis. Cancer Research 1172-8 34, 1172-1178.
Copie-Bergman, C., Niedobitek, G., Mangham, D.C., Selves, J., Baloch, K., Diss, T.C., Knowles, D.N., Delsol, G., and Isaacson, P.G. (1997). Epstein-Barr virus in B-cell lymphomas associated with chronic suppurative inflammation. J Pathol 183, 287-292.

Crawford, D.H., Macsween, K.F., Higgins, C.D., Thomas, R., McAulay, K., Williams, H., Harrison, N., Reid, S., Conacher, M., Douglas, J., *et al.* (2006). A cohort study among university students: Identification of risk factors for Epstein–Barr virus Seroconversion and infectious Mononucleosis. Clin. Infect Dis. 43, 276-282.

Crawford, D.H., Swerdlow, A.J., Higgins, C., McAulay, K., Harrison, N., Williams, H., Britton, K., and Macsween, K.F. (2002). Sexual history and Epstein-Barr virus infection. J Infect Dis 186, 731-736.

Crawford, D.H., Thomas, J.A., Janossy, G., Sweny, P., Fernando, O.N., Moorhead, J.F., and Thompson, J.H. (1980). Epstein-Barr virus nuclear antigen positive lymphoma after Cyclosporin A treatment in patients with renal allograft. The Lancet 315, 1355-1356.

Dawson, C.W., Laverick, L., Morris, M.A., Tramoutanis, G., and Young, L.S. (2008). Epstein-Barr virus-encoded LMP1 regulates epithelial cell motility and invasion via the ERK-MAPK pathway. J Virol 82, 3654-3664.

de-Thé, G., Geser, A., Day, N.E., Tukei, P.M., Williams, E.H., Beri, D.P., Smith, P.G., Dean, A.G., Bornkamm, G.W., Feorino, P., *et al.* (1978). Epidemiological evidence for causal relationship between Epstein-Barr virus and Burkitt's lymphoma from Ugandan prospective study. Nature 274, 756-761.

Deacon, E.M., Pallesen, G., Niedobitek, G., Crocker, J., Brooks, L., Rickinson, A.B., and Young Ls, L.S. (1993). Epstein-Barr virus and Hodgkin's disease: transcriptional analysis of virus latency in the malignant cells. J. Exp. Med. 177, 339-349.

Dean, M., Kent, R.B., and Sonenshein, G.E. (1983). Transcriptional activation of immunoglobulin α heavy-chain genes by translocation of the c-myc oncogene. Nature 305, 443-446.

Dillner, J., Kallin, B., Ehlin-Henriksson, B., Timar, L., and Klein, G. (1985). Characterization of a second Epstein-Barr virus-determined nuclear antigen associated with the BamHI WYH region of EBV DNA. Int. J. Cancer 35, 359-366.

Dillner, J., Sternas, L., Kallin, B., Alexander, H., Ehlin-Henriksson, B., Jornvall, H., Klein, G., and Lerner, R. (1984). Antibodies against a synthetic peptide identify the Epstein-Barr virus-determined nuclear antigen. PNAS 81, 4652-4656.

Doak, P.B., Montgomerie, J.Z., North, J.D.K., and Smith, F. (1968). Reticulum Cell Sarcoma after Renal Homotransplantation and Azathioprine and Prednisone Therapy. BMJ 4, 746-748.

Doeden, K., Molina-Kirsch, H., Perez, E., Warnke, R., and Sundram, U. (2008). Hydroa-like lymphoma with CD56 expression. J Cutan Pathol 35, 488-494.

Drenou, B., Lamy, T., Amiot, L., Fardel, O., Caulet-Maugendre, S., Sasportes, M., Diebold, J., Le Prise, P.Y., and Fauchet, R. (1997). CD3- CD56+ non-Hodgkin's lymphomas with an aggressive behavior related to multidrug resistance. Blood 89, 2966-2974.

Dutton, A., Reynolds, G.M., Dawson, C.W., Young, L.S., and Murray, P.G. (2005). Constitutive activation of phosphatidyl-inositide 3 kinase contributes to the survival of Hodgkin's lymphoma cells through a mechanism involving Akt kinase and mTOR. J. Pathol. 205, 498-506.

Einat, M., Resnitzky, D., and Kimchi, A. (1985). Close link between reduction of c-myc expression by interferon and G0/G1 arrest. Nature 313, 597-600.

Eliopoulos, A.G., Stack, M., Dawson, C.W., Kaye, K.M., Hodgkin, L., Sihota, S., Rowe, M., and Young, L.S. (1997). Epstein-Barr virus-encoded LMP1 and CD40 mediate IL-6 production in epithelial cells via an NF-kappaB pathway involving TNF receptor-associated factors. Oncogene 14, 2899-2916.

Epstein, M.A. (2015). Why and How Epstein-Barr Virus Was Discovered 50 Years Ago. Curr Top Microbiol Immunol 390, 3-15.

Epstein, M.A., Achong, B.G., and Barr, Y.M. (1964). Virus particles in cultured lymphoblasts from Burkitt's lymphoma. The Lancet 283, 702-703.

Evan, G.I., Wyllie, A.H., Gilbert, C.S., Littlewood, T.D., Land, H., Brooks, M., Waters, C.M., Penn, L.Z., and Hancock, D.C. (1992). Induction of apoptosis in fibroblasts by c-myc protein. Cell 69, 119-128.

Fialkow, P.J. (1967). "Immunologic" oncogenesis. Blood 30, 388-394.

Fox, C.P., Haigh, T.A., Taylor, G.S., Long, H.M., Lee, S.P., Shannon-Lowe, C., O'Connor, S., Bollard, C.M., Iqbal, J., Chan, W.C., *et al.* (2010). A novel latent membrane 2 transcript expressed in Epstein-Barr virus-positive NK- and T-cell lymphoproliferative disease encodes a target for cellular immunotherapy. Blood 116, 3695-3704.

Fujieda, M., Wakiguchi, H., Hisakawa, H., Kubota, H., and Kurashige, T. (1993). Defective activity of Epstein-Barr virus (EBV) specific cytotoxic T lymphocytes in children with chronic active EBV infection and in their parents. Acta Paediatr Jpn 35, 394-399.

Ghobrial, I.M., Habermann, T.M., Macon, W.R., Ristow, K.M., Larson, T.S., Walker, R.C., Ansell, S.M., Gores, G.J., Stegall, M.D., and McGregor, C.G. (2005a). Differences between early and late Posttransplant Lymphoproliferative disorders in solid organ transplant patients: Are they Two different diseases? Transplantation 79, 244-247.

Ghobrial, I.M., Habermann, T.M., Maurer, M.J., Geyer, S.M., Ristow, K.M., Larson, T.S., Walker, R.C., Ansell, S.M., Macon, W.R., Gores, G.G., *et al.* (2005b). Prognostic analysis for survival in adult solid organ transplant recipients with post-transplantation Lymphoproliferative disorders. J. Clin. Oncol. 23, 7574-7582.

Gires, O., Zimber-Strobl, U., Gonnella, R., Ueffing, M., Marschall, G., Zeidler, R., Pich, D., and Hammerschmidt, W. (1997). Latent membrane protein 1 of Epstein-Barr virus mimics a constitutively active receptor molecule. The EMBO Journal 16, 6131-6140.

Hanto, D.W., Gajl-Peczalska, K.J., Frizzera, G., Arthur, D.C., balfour, H.H., McClain, K.E.N., Simmons, R.L., and Najarian, J.S. (1983). Epstein-Barr Virus (EBV) Induced Polyclonal and Monoclonal B-Cell Lymphoproliferative Diseases Occurring after Renal Transplantation. Annals of Surgery 198, 356-369.

Hanto, D.W., Sakamoto, K., Purtilo, D.T., Simmons, R.L., and Najarian, J.S. (1981). The Epstein-Barr virus in the pathogenesis of posttransplant lymphoproliferative

disorders. Clinical, pathologic, and virologic correlation. Surgery 90, 204-213.
Henle, G., Henle, W., Clifford, P., Diehl, V., Kafuko, G.W., Kirya, B.G., Klein, G., Morrow, R.H., Munube, G.M., Pike, P., *et al.* (1969). Antibodies to Epstein-Barr Virus in Burkitt's Lymphoma and Control Groups. J Natl Cancer Inst 43, 1147-1157.
Henle, G., Henle, W., and Diehl, V. (1968). Relation of Burkitt's tumor-associated herpes-type virus to infectious mononucleosis. PNAS 59, 94-101.
Henle, W., Diehl, V., Kohn, G., zur Hausen, H., and Henle, G. (1967). Herpes-Type virus and chromosome marker in normal Leukocytes after growth with irradiated Burkitt cells. Science 157, 1064-1065.
Henter, J.I., Horne, A., Arico, M., Egeler, R.M., Filipovich, A.H., Imashuku, S., Ladisch, S., McClain, K., Webb, D., Winiarski, J., *et al.* (2007). HLH-2004: Diagnostic and therapeutic guidelines for hemophagocytic lymphohistiocytosis. Pediatr Blood Cancer 48, 124-131.
Higgins, Craig D., Swerdlow, Anthony J., Macsween, Karen F., Harrison, N., Williams, H., McAulay, K., Thomas, R., Reid, S., Conacher, M., Britton, K., *et al.* (2007). A study of risk factors for acquisition of Epstein–Barr virus and its subtypes. J. Infect. Dis. 195, 474-482.
Hino, R., Uozaki, H., Murakami, N., Ushiku, T., Shinozaki, A., Ishikawa, S., Morikawa, T., Nakaya, T., Sakatani, T., Takada, K., *et al.* (2009). Activation of DNA methyltransferase 1 by EBV latent membrane protein 2A leads to promoter hypermethylation of PTEN gene in gastric carcinoma. Cancer Res 69, 2766-2774.
Hislop, A.D., Palendira, U., Leese, A.M., Arkwright, P.D., Rohrlich, P.S., Tangye, S.G., Gaspar, H.B., Lankester, A.C., Moretta, A., and Rickinson, A.B. (2010). Impaired Epstein-Barr virus-specific CD8+ t-cell function in x-linked lymphoproliferative disease is restricted to SLAM family-positive b-cell targets. Blood 116, 3249-3257.
Holtick, U., Vockerodt, M., Pinkert, D., Schoof, N., Stürzenhofecker, B., Kussebi, N., Lauber, K., Wesselborg, S., Löffler, D., Horn, F., *et al.* (2005). STAT3 is essential for Hodgkin lymphoma cell proliferation and is a target of tyrphostin AG17 which confers sensitization for apoptosis. Leukemia 19, 936-944.
Horenstein, M.G., Nador, R.G., Chadburn, A., Hyjek, E.M., Inghirami, G., Knowles, D.M., and Cesarman, E. (1997). Epstein-Barr virus latent gene expression in primary effusion lymphomas containing Kaposi's sarcoma-associated herpesvirus/human herpesvirus-8. Blood 90, 1186-1191.
Israele, V., Shirley, P., and Sixbey, J.W. (1991). Excretion of the Epstein-Barr virus from the genital tract of men. J. Infect. Dis. 163, 1341-1343.
Iwatsuki, K., Xu, Z., Takata, M., Iguchi, M., Ohtsuka, M., Akiba, H., Mitsuhashi, Y., Takenoshita, H., Sugiuchi, R., Tagami, H., *et al.* (1999). The association of latent Epstein-Barr virus infection with hydroa vacciniforme. Br J Dermatol 140, 715-721.
Janka, G., Imashuku, S., Elinder, G., Schneider, M., and Henter, J.I. (1998). Infection- and malignancy-associated hemophagocytic syndromes. Secondary hemophagocytic lymphohistiocytosis. Hematol Oncol Clin North Am 12, 435-444.
Jenner, R.G., Maillard, K., Cattini, N., Weiss, R.A., Boshoff, C., Wooster, R., and Kellam, P. (2003). Kaposi's sarcoma-associated herpesvirus-infected primary effusion lymphoma has a plasma cell gene expression profile. Proc Natl Acad Sci USA 100, 10399-10404.
Jiang, L., Gu, Z.H., Yan, Z.X., Zhao, X., Xie, Y.Y., Zhang, Z.G., Pan, C.M., Hu, Y., Cai, C.P., Dong, Y., *et al.* (2015). Exome sequencing identifies somatic mutations of DDX3X in natural killer/T-cell lymphoma. Nat Genet 47, 1061-1066.
Johannessen, I., Asghar, M., and Crawford, D.H. (2000). Essential role for T cells in human b-cell lymphoproliferative disease development in severe combined immunodeficient mice. British Journal of Haematology 109, 600-610.
Joncas, J., Monczak, Y., Ghibu, F., Alfieri, C., Bonin, A., Ahronheim, G., and Rivard, G. (1989). Brief report: killer cell defect and persistent immunological abnormalities in two patients with chronic active Epstein-Barr virus infection. J Med Virol 28, 110-117.
Kaneda, A., Matsusaka, K., Aburatani, H., and Fukayama, M. (2012). Epstein-Barr virus infection as an epigenetic driver of tumorigenesis. Cancer Res 72, 3445-3450.
Kanno, H., Naka, N., Yasunaga, Y., and Aozasa, K. (1997a). Role of an immunosuppressive cytokine, interleukin-10, in the development of pyothorax-associated lymphoma. Leukemia 11 Suppl 3, 525-526.
Kanno, H., Naka, N., Yasunaga, Y., Iuchi, K., Yamauchi, S., Hashimoto, M., and Aozasa, K. (1997b). Production of the immunosuppressive cytokine interleukin-10 by Epstein-Barr-virus-expressing pyothorax-associated lymphoma: possible role in the development of overt lymphoma in immunocompetent hosts. Am J Pathol 150, 349-357.
Kanno, H., Nakatsuka, S., Iuchi, K., and Aozasa, K. (2000). Sequences of cytotoxic T-lymphocyte epitopes in the Epstein-Barr virus (EBV) nuclear antigen-3B gene in a Japanese population with or without EBV-positive lymphoid malignancies. Int J Cancer 88, 626-632.
Kanno, H., Ohsawa, M., Hashimoto, M., Iuchi, K., Nakajima, Y., and Aozasa, K. (1999). HLA-A alleles of patients with pyothorax-associated lymphoma: anti-Epstein-Barr virus (EBV) host immune responses during the development of EBV latent antigen-positive lymphomas. Int J Cancer 82, 630-634.
Kanno, H., Yasunaga, Y., Iuchi, K., Yamauchi, S., Tatekawa, T., Sugiyama, H., and Aozasa, K. (1996). Interleukin-6-mediated growth enhancement of cell lines derived from pyothorax-associated lymphoma. Lab Invest 75, 167-173.
Kawaguchi, H., Miyashita, T., Herbst, H., Niedobitek, G., Asada, M., Tsuchida, M., Hanada, R., Kinoshita, A., Sakurai, M., Kobayashi, N., *et al.* (1993). Epstein-Barr virus-infected T lymphocytes in Epstein-Barr virus-

associated hemophagocytic syndrome. J Clin Invest 92, 1444-1450.

Kelly, G., Bell, A., and Rickinson, A. (2002). Epstein–Barr virus–associated Burkitt lymphomagenesis selects for downregulation of the nuclear antigen EBNA2. Nature Med. 8, 1098-1104.

Kelly, G.L., Long, H.M., Stylianou, J., Thomas, W.A., Leese, A., Bell, A.I., Bornkamm, G.W., Mautner, J., Rickinson, A.B., and Rowe, M. (2009). An Epstein-Barr Virus Anti-Apoptotic Protein Constitutively Expressed in Transformed Cells and Implicated in Burkitt Lymphomagenesis: The Wp/BHRF1 Link. PLoS Pathogens 5, e1000341.

Kelly, G.L., Milner, A.E., Tierney, R.J., Croom-Carter, D.S.G., Altmann, M., Hammerschmidt, W., Bell, A.I., and Rickinson, A.B. (2005). Epstein-Barr Virus Nuclear Antigen 2 (EBNA2) Gene Deletion Is Consistently Linked with EBNA3A, -3B, and -3C Expression in Burkitt's Lymphoma Cells and with Increased Resistance to Apoptosis. J. Virol. 79, 10709-10717.

Kennedy, G., Komano, J., and Sugden, B. (2003). Epstein-Barr virus provides a survival factor to Burkitt's lymphomas. PNAS 100, 14269-14274.

Kerr, B.M., Lear, A.L., Rowe, M., Croom-Carter, D., Young, L.S., Rookes, S.M., Gallimore, P.H., and Rickinson, A.B. (1992). Three transcriptionally distinct forms of epstein-barr virus latency in somatic cell hybrids: Cell phenotype dependence of virus promoter usage. Virology 187, 189-201.

Kimura, H., Hoshino, Y., Hara, S., Sugaya, N., Kawada, J., Shibata, Y., Kojima, S., Nagasaka, T., Kuzushima, K., and Morishima, T. (2005). Differences between T cell-type and natural killer cell-type chronic active Epstein-Barr virus infection. J Infect Dis 191, 531-539.

Kimura, H., Morishima, T., Kanegane, H., Ohga, S., Hoshino, Y., Maeda, A., Imai, S., Okano, M., Morio, T., Yokota, S., *et al.* (2003). Prognostic factors for chronic active Epstein-Barr virus infection. J Infect Dis 187, 527-533.

Komano, J., Maruo, S., Kurozumi, K., Oda, T., and Takada, K. (1999). Oncogenic role of Epstein-Barr virus-encoded RNAs in Burkitt's lymphoma cell line Akata. J Virol 73, 9827-9831.

Kreck, B., Richter, J., Ammerpohl, O., Barann, M., Esser, D., Petersen, B.S., Vater, I., Murga Penas, E.M., Bormann Chung, C.A., Seisenberger, S., *et al.* (2013). Base-pair resolution DNA methylome of the EBV-positive Endemic Burkitt lymphoma cell line DAUDI determined by SOLiD bisulfite-sequencing. Leukemia 27, 1751-1753.

Kuppers, R., Rajewsky, K., Zhao, M., Simons, G., Laumann, R., Fischer, R., and Hansmann, M.L. (1994). Hodgkin disease: Hodgkin and Reed-Sternberg cells picked from histological sections show clonal immunoglobulin gene rearrangements and appear to be derived from B cells at various stages of development. PNAS 91, 10962-10966.

Kurth, J., Hansmann, M.L., Rajewsky, K., and Kuppers, R. (2003). Epstein-Barr virus-infected B cells expanding in germinal centers of infectious mononucleosis patients do not participate in the germinal center reaction. PNAS 100, 4730-4735.

Kurth, J., Spieker, T., Wustrow, J., Strickler, J.G., Hansmann, M.-L., Rajewsky, K., and Küppers, R. (2000). EBV-Infected B cells in infectious Mononucleosis. Immunity 13, 485-495.

Laichalk, L.L., and Thorley-Lawson, D.A. (2004). Terminal differentiation into plasma cells initiates the Replicative cycle of Epstein-Barr virus *in vivo*. J. Virol. 79, 1296-1307.

Leblond, V., Davi, F., Charlotte, F., Dorent, R., Bitker, M.O., Sutton, L., Gandjbakhch, I., Binet, J.L., and Raphael, M. (1998). Posttransplant lymphoproliferative disorders not associated with Epstein-Barr virus: a distinct entity? J Clin Oncol 16, 2052–2059.

Leblond, V., Dhedin, N., Mamzer Bruneel, M.F., Choquet, C., Hermine, O., Porcher, R., Nguyen Quoc, S., Davi, F., Charlotte, F., Dorent, R., *et al.* (2001). Identification of prognostic factors in 61 patients with posttransplantation lymphoproliferative disorders. J Clin Oncol 19, 772-778.

Lee, H.Y., Baek, J.O., Lee, J.R., Park, S.H., Jeon, I.S., and Roh, J.Y. (2012). Atypical hydroa vacciniforme-like epstein-barr virus associated T/NK-cell lymphoproliferative disorder. Am J Dermatopathol 34, e119-124.

Lee, J., Suh, C., Park, Y.H., Ko, Y.H., Bang, S.M., Lee, J.H., Lee, D.H., Huh, J., Oh, S.Y., Kwon, H.C., *et al.* (2006). Extranodal natural killer T-cell lymphoma, nasal-type: a prognostic model from a retrospective multicenter study. J Clin Oncol 24, 612-618.

Leonard, S., Gordon, N., Smith, N., Rowe, M., Murray, P.G., and Woodman, C.B. (2012). Arginine Methyltransferases Are Regulated by Epstein-Barr Virus in B Cells and Are Differentially Expressed in Hodgkin's Lymphoma. Pathogens 1, 52-64.

Lim, S.T., Hee, S.W., Quek, R., Lim, L.C., Yap, S.P., Loong, E.L., Sng, I., Tan, L.H., Ang, M.K., Ngeow, J., *et al.* (2008). Comparative analysis of extra-nodal NK/T-cell lymphoma and peripheral T-cell lymphoma: significant differences in clinical characteristics and prognosis. Eur J Haematol 80, 55-60.

Lindström, M.S., and Wiman, K.G. (2002). Role of genetic and epigenetic changes in Burkitt lymphoma. Seminars in Cancer Biology 12, 381-387.

Loong, F., Chan, A.C., Ho, B.C., Chau, Y.P., Lee, H.Y., Cheuk, W., Yuen, W.K., Ng, W.S., Cheung, H.L., and Chan, J.K. (2010). Diffuse large B-cell lymphoma associated with chronic inflammation as an incidental finding and new clinical scenarios. Mod Pathol 23, 493-501.

Love, C., Sun, Z., Jima, D., Li, G., Zhang, J., Miles, R., Richards, K.L., Dunphy, C.H., Choi, W.W.L., Srivastava, G., *et al.* (2012). The genetic landscape of mutations in Burkitt lymphoma. Nature Genetics 44, 1321-1325.

Mack, A.A., and Sugden, B. (2008). EBV is necessary for proliferation of dually infected primary effusion lymphoma cells. Cancer Res 68, 6963-6968.

Mancao, C., Altmann, M., Jungnickel, B., and Hammerschmidt, W. (2005). Rescue of "crippled" germinal center B cells from apoptosis by Epstein-Barr virus. Blood 106, 4339-4343.

Manolov, Y., and Manolova, G. (1972). Marker band in one chromosome 14 from Burkitt lymphomas. Nature 237:, 33-34.

Milner, A.E., Grand, R.J., Waters, C.M., and Gregory, C.D. (1993). Apoptosis in Burkitt lymphoma cells is driven by c-myc. Oncogene 8, 3385-3391.

Morris, M.A., Dawson, C.W., Laverick, L., Davis, A.M., Dudman, J.P., Raveenthiraraj, S., Ahmad, Z., Yap, L.F., and Young, L.S. (2016). The Epstein-Barr virus encoded LMP1 oncoprotein modulates cell adhesion via regulation of activin A/TGFbeta and beta1 integrin signalling. Sci Rep 6, 19533.

Morris, M.A., Dawson, C.W., and Young, L.S. (2009). Role of the Epstein-Barr virus-encoded latent membrane protein-1, LMP1, in the pathogenesis of nasopharyngeal carcinoma. Future Oncol 5, 811-825.

Moss, D.J., Burrows, S.R., Castelino, D.J., Kane, R.G., Pope, J.H., Rickinson, A.B., Alpers, M.P., and Heywood, P.P. (1983). A comparison of epstein-barr virus-specific t-cell immunity in malaria-endemic and -nonendemic regions of Papua New Guinea. Int. J. Cancer 31, 727-732.

Murray, P.G., Young, L.S., Rowe, M., and Crocker, J. (1992). Immunohistochemical demonstration of the Epstein-Barr virus-encoded latent membrane protein in paraffin sections of Hodgkin's disease. J. Pathol. 166, 1-5.

Nagington, J., and Gray, J. (1980). Cyclosporin A, immunosuppression, Epstein-Barr antibody, and lymphoma. The Lancet 315, 536-537.

Nakashima, Y., Tagawa, H., Suzuki, R., Karnan, S., Karube, K., Ohshima, K., Muta, K., Nawata, H., Morishima, Y., Nakamura, S., *et al.* (2005). Genome-wide array-based comparative genomic hybridization of natural killer cell lymphoma/leukemia: different genomic alteration patterns of aggressive NK-cell leukemia and extranodal Nk/T-cell lymphoma, nasal type. Genes Chromosomes Cancer 44, 247-255.

Nichols, K.E., Harkin, D.P., Levitz, S., Krainer, M., Kolquist, K.A., Genovese, C., Bernard, A., Ferguson, M., Zuo, L., Snyder, E., *et al.* (1998). Inactivating mutations in an SH2 domain-encoding gene in x-linked lymphoproliferative syndrome. PNAS 95, 13765-13770.

Niederman, J.C., Miller, G., Pearson, H.A., Pagano, J.S., and Dowaliby, J.M. (1976). Infectious mononucleosis. Epstein-Barr-virus shedding in saliva and the oropharynx. N Engl J Med 2941, 355-359.

Niller, H.H., Banati, F., Salamon, D., and Minarovits, J. (2016). Epigenetic Alterations in Epstein-Barr Virus-Associated Diseases. Adv Exp Med Biol 879, 39-69.

Nishiu, M., Tomita, Y., Nakatsuka, S., Takakuwa, T., Iizuka, N., Hoshida, Y., Ikeda, J., Iuchi, K., Yanagawa, R., Nakamura, Y., *et al.* (2004). Distinct pattern of gene expression in pyothorax-associated lymphoma (PAL), a lymphoma developing in long-standing inflammation. Cancer Sci 95, 828-834.

Njie, R., Bell, A.I., Jia, H., Croom-Carter, D., Chaganti, S., Hislop, A.D., Whittle, H., and Rickinson, A.B. (2009). The effects of acute malaria on Epstein-Barr virus (EBV) load and EBV-Specific T cell immunity in Gambian children. J. Infect. Dis. 199, 31-38.

Ohga, S., Kimura, N., Takada, H., Nagano, M., Ohshima, K., Nomura, A., Muraoka, K., Take, H., Yamamori, S., and Hara, T. (1999). Restricted diversification of T-cells in chronic active Epstein-Barr virus infection: potential inclination to T-lymphoproliferative disease. Am J Hematol 61, 26-33.

Ohga, S., Nomura, A., Takada, H., Tanaka, T., Furuno, K., Takahata, Y., Kinukawa, N., Fukushima, N., Imai, S., and Hara, T. (2004). Dominant expression of interleukin-10 and transforming growth factor-beta genes in activated T-cells of chronic active Epstein-Barr virus infection. J Med Virol 74, 449-458.

Ok, C.Y., Papathomas, T.G., Medeiros, L.J., and Young, K.H. (2013). EBV-positive diffuse large b-cell lymphoma of the elderly. Blood 122, 328-340.

Ok, C.Y., Ye, Q., Li, L., Manyam, G.C., Deng, L., Goswami, R.R., Wang, X., Montes-Moreno, S., Visco, C., Tzankov, A., *et al.* (2015). Age cutoff for Epstein-Barr virus-positive diffuse large b-cell lymphoma–is it necessary. Oncotarget 6, 13933-13945.

Old, L.J., Boyse, E.A., Oettgen, H.F., Harven, E.D., Geering, G., Williamson, B., and Clifford, P. (1966). PRECIPITATING ANTIBODY IN HUMAN SERUM TO AN ANTIGEN PRESENT IN CULTURED BURKITT'S LYMPHOMA CELLS. PNAS 56, 1699-1704.

Oyama, T., Ichimura, K., Suzuki, R., Suzumiya, J., Ohshima, K., Yatabe, Y., Yokoi, T., Kojima, M., Kamiya, Y., Taji, H., *et al.* (2003). Senile EBV+ B-Cell Lymphoproliferative Disorders. The American Journal of Surgical Pathology 27, 16-26.

Oyama, T., Yamamoto, K., Asano, N., Oshiro, A., Suzuki, R., Kagami, Y., Morishima, Y., Takeuchi, K., Izumo, T., Mori, S., *et al.* (2007). Age-Related EBV-Associated B-Cell Lymphoproliferative Disorders Constitute a Distinct Clinicopathologic Group: A Study of 96 Patients. Clinical Cancer Research 13, 5124-5132.

Oyoshi, M.K., Nagata, H., Kimura, N., Zhang, Y., Demachi, A., Hara, T., Kanegane, H., Matsuo, Y., Yamaguchi, T., Morio, T., *et al.* (2003). Preferential expansion of Vgamma9-JgammaP/Vdelta2-Jdelta3 gammadelta T cells in nasal T-cell lymphoma and chronic active Epstein-Barr virus infection. Am J Pathol 162, 1629-1638.

Pagano, L., Gallamini, A., Trape, G., Fianchi, L., Mattei, D., Todeschini, G., Spadea, A., Cinieri, S., Iannitto, E., Martelli, M., *et al.* (2006). NK/T-cell lymphomas 'nasal type': an Italian multicentric retrospective survey. Ann Oncol 17, 794-800.

Palendira, U., and Rickinson, A.B. (2015). Primary immunodeficiencies and the control of Epstein-Barr virus infection. Annals of the New York Academy of Sciences 1356, 22-44.

Pallesen, G., Hamilton-Dutoit, S.J., Rowe, M., and Young, L.S. (1991). Expression of Epstein-Barr virus latent gene products in tumour cells of Hodgkin's disease. The Lancet 337, 320-322.

Peng, H., Chen, Y., Gong, P., Cai, L., Lyu, X., Jiang, Q., Wang, J., Lu, J., Yao, K., Liu, K., *et al.* (2016). Higher methylation intensity induced by EBV LMP1 via NF-kappaB/ DNMT3b signaling contributes to silencing of PTEN gene. Oncotarget.

Penn, I., Hammond W, W., Brettschneider, L., and Starzl, T.E. (1969). Malignant lymphomas in transplantation patients. Transplant Proc 1, 106-112.

Pfeffer, S. (2004). Identification of virus-encoded MicroRNAs. Science 304, 734-736.

Pongpruttipan, T., Sukpanichnant, S., Assanasen, T., Wannakrairot, P., Boonsakan, P., Kanoksil, W., Kayasut, K., Mitarnun, W., Khuhapinant, A., Bunworasate, U., *et al.* (2012). Extranodal NK/T-cell lymphoma, nasal type, includes cases of natural killer cell and alphabeta, gammadelta, and alphabeta/gammadelta T-cell origin: a comprehensive clinicopathologic and phenotypic study. Am J Surg Pathol 36, 481-499.

Pope, J.H. (1967). Establishment of cell lines from peripheral Leucocytes in infectious Mononucleosis. Nature 216, 810-811.

Poppema, S., van Imhoff, G., Torensma, R., and Smit, J. (1985). Lymphadenopathy morphologically consistent with Hodgkin's disease associated with Epstein-Barr virus infection. American J. Clin. Pathol. 84, 385-390.

Purtilo, D.T., Yang, J.P.S., Cassel, C.K., Harper, R., Stephenson, S.R., Landing, B.H., and Vawter, G.F. (1975). X-LINKED RECESSIVE PROGRESSIVE COMBINED VARIABLE IMMUNODEFICIENCY (DUNCAN'S DISEASE). The Lancet 305, 935-941.

Queen, K.J., Shi, M., Zhang, F., Cvek, U., and Scott, R.S. (2013). Epstein-Barr virus-induced epigenetic alterations following transient infection. Int J Cancer 132, 2076-2086.

Quintanilla-Martinez, L., Kremer, M., Keller, G., Nathrath, M., Gamboa-Dominguez, A., Meneses, A., Luna-Contreras, L., Cabras, A., Hoefler, H., Mohar, A., *et al.* (2001). p53 Mutations in nasal natural killer/T-cell lymphoma from Mexico: association with large cell morphology and advanced disease. Am J Pathol 159, 2095-2105.

Ramayanti, O., Juwana, H., Verkuijlen, S.A., Adham, M., Pegtel, M.D., Greijer, A.E., and Middeldorp, J.M. (2016). Epstein-Barr virus mRNA profiles and viral DNA methylation status in nasopharyngeal brushings from nasopharyngeal carcinoma patients reflect tumor origin. Int J Cancer.

Reedman, B.M., Klein, G., Pope, J.H., Walters, M.K., Hilgers, J., Singh, S., and Johansson, B. (1974). Epstein-Barr virus-associated complement-fixing and nuclear antigens in Burkitt lymphoma biopsies. Int. J. Cancer 13, 755-763.

Richter, J., Schlesner, M., Hoffmann, S., Kreuz, M., Leich, E., Burkhardt, B., Rosolowski, M., Ammerpohl, O., Wagener, R., Bernhart, S.H., *et al.* (2012). Recurrent mutation of the ID3 gene in Burkitt lymphoma identified by integrated genome, exome and transcriptome sequencing. Nature Genetics 44, 1316-1320.

Rickinson, A.B. (2014). Co-infections, inflammation and oncogenesis: Future directions for EBV research. Seminars in Cancer Biology 26, 99-115.

Roberts, M.L., and Cooper, N.R. (1998). Activation of a ras-MAPK-dependent pathway by Epstein-Barr virus latent membrane protein 1 is essential for cellular transformation. Virology 240, 93-99.

Rooney, C.M., Ng, C.C., Loftin, S., Smith, C.A., Li, C., Krance, R.A., Brenner, M.K., and Heslop, H.E. (1995). Use of gene-modified virus-specific T lymphocytes to control Epstein-Barr-virus-related lymphoproliferation. The Lancet 345, 9-13.

Rosdahl, N., Larsen, S.O., and Clemmesen, J. (1974). Hodgkin's Disease in Patients with Previous Infectious Mononucleosis: 30 Years' Experience. BMJ 2, 253-256.

Rouphael, N.G., Talati, N.J., Vaughan, C., Cunningham, K., Moreira, R., and Gould, C. (2007). Infections associated with haemophagocytic syndrome. Lancet Infect Dis 7, 814-822.

Rowe, M., Fitzsimmons, L., and Bell, A.I. (2014). Epstein-Barr virus and Burkitt lymphoma. Chinese J. Cancer.

Saemundsen Ak, A.K., Purtilo Dt, D.T., Sakamoto K, K., Sullivan Jl, J.L., Synnerholm Ac, A.C., Hanto D, D., Simmons R, R., Anvret M, M., Collins R, R., and Klein G, G. (1981). Documentation of Epstein-Barr virus infection in immunodeficient patients with life-threatening lymphoproliferative diseases by Epstein-Barr virus complementary RNA/DNA and viral DNA/DNA hybridization. Cancer Res 41, 4237-4242.

Sanchez-Gonzalez, B., Garcia, M., Montserrat, F., Sanchez, M., Angona, A., Solano, A., and Salar, A. (2013). Diffuse large B-cell lymphoma associated with chronic inflammation in metallic implant. J Clin Oncol 31, e148-151.

Savoldo, B., Huls, M.H., Liu, Z., Okamura, T., Volk, H.D., Reinke, P., Sabat, R., Babel, N., Jones, J.F., Webster-Cyriaque, J., *et al.* (2002). Autologous Epstein-Barr virus (EBV)-specific cytotoxic T cells for the treatment of persistent active EBV infection. Blood 100, 4059-4066.

Sayos, J., Wu, C., Morra, M., Wang, N., Zhang, X., Allen, D., van Schaik, S., Notarangelo, L., Geha, R., Roncarolo, M.G., *et al.* (1998). The X-linked lymphoproliferative-disease gene product SAP regulates signals induced through the co-receptor SLAM. Nature 395, 462-469.

Schmitz, R., Young, R.M., Ceribelli, M., Jhavar, S., Xiao, W., Zhang, M., Wright, G., Shaffer, A.L., Hodson, D.J., Buras, E., *et al.* (2012). Burkitt lymphoma pathogenesis and therapeutic targets from structural and functional genomics. Nature 490, 116-120.

Schubach, W.H., Hackman, R., Neiman, P.E., Miller, G., and Thomas, E.D. (1982). A monoclonal immunoblastic sarcoma in donor cells bearing Epstein-Barr virus genomes following allogeneic marrow grafting for acute lymphoblastic leukemia. Blood 60, 180-187.

Schulz, T.F., and Cesarman, E. (2015). Kaposi Sarcoma-associated Herpesvirus: mechanisms of oncogenesis. Curr Opin Virol 14, 116-128.

Seemayer, T.A., Gross, T.G., Egeler, R.M., Pirruccello, S.J., Davis, J.R., Kelly, C.M., Okano, M., Lanyi, A., and Sumegi, J. (1995). X-linked Lymphoproliferative disease: Twenty-Five years after the discovery. Pediatric Research 38, 471-478.

Shimizu, N., Tanabe-Tochikura, A., Kuroiwa, Y., and Takada, K. (1994). Isolation of Epstein-Barr virus (EBV)-negative cell clones from the EBV-positive Burkitt's lymphoma (BL) line Akata: malignant phenotypes of BL cells are dependent on EBV. J Virol 1994 Sep;68(9):6069-73 68, 6069-6073.

Sixbey, J.W., Lemon, S.M., and Pagano, J.S. (1987). A second site for Epstein-Barr virus shedding: The uterine cervix. International Journal of Gynecology and Obstetrics 25, 425-426.

Starzl, T.E., Porter, K.A., Iwatsuki, S., Rosenthal, J.T., Shaw, B.W., Atchison, R.W., Nalesnik, M.A., Ho, M., Griffith, B.P., Hakala, T.R., *et al.* (1984). Reversibility of lymphomas and lymphoproliferative lesions developing under cyclosporin-steroid therapy. The Lancet 323, 583-587.

Su, I.J., Hsieh, H.J., and Lee, C.Y. (1989). Histiocytic medullary reticulosis: a lethal form of primary EBV infection in young children in Taiwan. Lancet 1, 389.

Sugaya, N., Kimura, H., Hara, S., Hoshino, Y., Kojima, S., Morishima, T., Tsurumi, T., and Kuzushima, K. (2004). Quantitative analysis of Epstein-Barr virus (EBV)-specific CD8+ T cells in patients with chronic active EBV infection. J Infect Dis 190, 985-988.

Suzuki, R., Suzumiya, J., Yamaguchi, M., Nakamura, S., Kameoka, J., Kojima, H., Abe, M., Kinoshita, T., Yoshino, T., Iwatsuki, K., *et al.* (2010). Prognostic factors for mature natural killer (NK) cell neoplasms: aggressive NK cell leukemia and extranodal NK cell lymphoma, nasal type. Ann Oncol 21, 1032-1040.

Szekely, L., Chen, F., Teramoto, N., Ehlin-Henriksson, B., Pokrovskaja, K., Szeles, A., Manneborg-Sandlund, A., Lowbeer, M., Lennette, E.T., and Klein, G. (1998). Restricted expression of Epstein-Barr virus (EBV)-encoded, growth transformation-associated antigens in an EBV- and human herpesvirus type 8-carrying body cavity lymphoma line. J Gen Virol 79 (Pt 6), 1445-1452.

Thomas, R., Macsween, K.F., McAulay, K., Clutterbuck, D., Anderson, R., Reid, S., Higgins, C.D., Swerdlow, A.J., Harrison, N., Williams, H., *et al.* (2006). Evidence of shared Epstein-Barr viral isolates between sexual partners, and low level EBV in genital secretions. Journal of Medical Virology 78, 1204-1209.

Timms, J.M., Bell, A., Flavell, J.R., Murray, P.G., Rickinson, A.B., Traverse-Glehen, A., Berger, F., and Delecluse, H.-J. (2003). Target cells of Epstein-Barr-virus (EBV)-positive post-transplant lymphoproliferative disease: Similarities to EBV-positive Hodgkin's lymphoma. The Lancet 361, 217-223.

Torgbor, C., Awuah, P., Deitsch, K., Kalantari, P., Duca, K.A., and Thorley-Lawson, D.A. (2014). A Multifactorial role for P. *Falciparum* malaria in endemic Burkitt's lymphoma pathogenesis. PLoS Pathogens 10, e1004170.

Trivedi, P., Takazawa, K., Zompetta, C., Cuomo, L., Anastasiadou, E., Carbone, A., Uccini, S., Belardelli, F., Takada, K., Frati, L., *et al.* (2004). Infection of HHV-8+ primary effusion lymphoma cells with a recombinant Epstein-Barr virus leads to restricted EBV latency, altered phenotype, and increased tumorigenicity without affecting TCL1 expression. Blood 103, 313-316.

Tsai, C.L., Li, H.P., Lu, Y.J., Hsueh, C., Liang, Y., Chen, C.L., Tsao, S.W., Tse, K.P., Yu, J.S., and Chang, Y.S. (2006). Activation of DNA methyltransferase 1 by EBV LMP1 Involves c-Jun NH(2)-terminal kinase signaling. Cancer Res 66, 11668-11676.

Veillette, A. (2010). SLAM-family receptors: immune regulators with or without SAP-family adaptors. Cold Spring Harb Perspect Biol 2, a002469.

Vereide, D.T., Seto, E., Chiu, Y.F., Hayes, M., Tagawa, T., Grundhoff, A., Hammerschmidt, W., and Sugden, B. (2013). Epstein–Barr virus maintains lymphomas via its miRNAs. Oncogene 33, 1258-1264.

Virelizier, J.L., Lenoir, G., and Griscelli, C. (1978). Persistent Epstein-Barr virus infection in a child with hypergammaglobulinaemia and immunoblastic proliferation associated with a selective defect in immune interferon secretion. Lancet 2, 231-234.

Vockerodt, M., Morgan, S.L., Kuo, M., Wei, W., Chukwuma, M.B., Arrand, J.R., Kube, D., Gordon, J., Young, L.S., Woodman, C.B., *et al.* (2008). The Epstein-Barr virus oncoprotein, latent membrane protein-1, reprograms germinal centre B cells towards a Hodgkin's Reed-Sternberg-like phenotype. J. Pathol. 216, 83-92.

Wakiguchi, H., Fujieda, M., Matsumoto, K., Ohara, Y., Wakiguchi, A., and Kurashige, T. (1988). Defective killer cell activity in patients with chronic active Epstein-Barr virus infection. Acta Med Okayama 42, 137-142.

Weiss, L.M., Movahed, L.A., Warnke, R.A., and Sklar, J. (1989). Detection of Epstein–Barr Viral Genomes in Reed–Sternberg Cells of Hodgkin's Disease. New England Journal of Medicine 320, 502-506.

Weiss, L.M., Strickler, J.G., Warnke, R.A., Purtilo, D.T., and Sklar, J. (1987). Epstein-Barr viral DNA in tissues of Hodgkin's disease. American Journal of Pathology 129, 86-91.

Whittle, H.C., Brown, J., Marsh, K., Greenwood, B.M., Seidelin, P., Tighe, H., and Wedderburn, L. (1984). T-cell control of Epstein–Barr virus-infected B cells is lost during *P. falciparum* malaria. Nature 312, 449-450.

Woodman, C.B.J., Collins, S.I., Vavrusova, N., Rao, A., Middeldorp, J.M., Kolar, Z., Kumari, A., Nelson, P., Young, L.S., and Murray, P.G. (2005). Role of sexual behavior in the acquisition of asymptomatic Epstein-Barr virus infection. The Pediatric Infectious Disease Journal 24, 498-502.

Wu, T.-C., Mann, R.B., Charache, P., Hayward, S.D., Staal, S., Lambe, B.C., and Ambinder, R.F. (1990). Detection of EBV gene expression in Reed-Sternberg cells of Hodgkin's disease. Int. J. Cancer 46, 801-804.

Wu, Y.H., Chen, H.C., Hsiao, P.F., Tu, M.I., Lin, Y.C., and Wang, T.Y. (2007). Hydroa vacciniforme-like Epstein-Barr virus-associated monoclonal T-lymphoproliferative disorder in a child. Int J Dermatol 46, 1081-1086.

Zech, L., Haglund, U., and Nilsson, K. (1976). Characteristic chromosomal abnormalities in biopsies and lymphoid-cell lines from patients with Burkitt and non-Burkitt lymphomas. Int J Cancer 17, 47-56.

Zhang, T., Ma, J., Nie, K., Yan, J., Liu, Y., Bacchi, C.E., Queiroga, E.M., Gualco, G., Sample, J.T., Orazi, A., *et al.* (2014). Hypermethylation of the tumor suppressor gene PRDM1/Blimp-1 supports a pathogenetic role in

EBV-positive Burkitt lymphoma. Blood Cancer J 4, e261.
zur Hausen, H., Schulte-Holthausen H, H., Klein, G., Henle, W., Henle, G., Clifford, P., and Santesson, L. (1970). EBV DNA in biopsies of Burkitt tumours and anaplastic carcinomas of the nasopharynx. Nature 228, 1056-1058.

7

Epstein-Barr Virus and Epithelial Carcinogenesis

Lee Fah Yap[1]* and Kwok-Wai Lo[2]

[1]Department of Oral and Craniofacial Sciences and Oral Cancer Research and Coordinating Centre, Faculty of Dentistry, University of Malaya, Kuala Lumpur, Malaysia
[2]Department of Anatomical and Cellular Pathology, State Key Laboratory in Oncology in South China and Li Ka Shing Institute of Health Science, The Chinese University of Hong Kong, Hong Kong

**Corresponding author: yapleefah@um.edu.my*
No conflicts of interest were declared

Abstract

While the contribution of Epstein-Barr virus (EBV) to the development of B cell tumours is well-studied, its precise role in the pathogenesis of epithelial cancers, such as nasopharyngeal carcinoma (NPC) and EBV-associated gastric cancer (EBVaGC), remains largely undefined. Most of our current understanding of the role of EBV in epithelial carcinogenesis comes from the studies on NPC because almost 98% of NPCs (compared to approximately 10% of gastric cancer) are EBV-positive. Indeed, many unique characteristics of NPC could be attributable to the virus and hence NPC is classified as a distinct type of head and neck cancer. Establishment of latent EBV infection in epithelial cells is thought to be an initiating event in preneoplastic cells, as well as the driving force for disease progression together with additional somatic changes as the tumour evolves. This chapter summarizes the major milestones in EBV research relating to epithelial cancers and describes some recent findings which help further our understanding of the role of EBV in epithelial carcinogenesis.

Introduction

Epstein-Barr virus (EBV) was the first human tumour virus to be discovered (Epstein *et al.*, 1964) and is now classified as a Group I carcinogen. It is estimated that EBV accounts for more than 200,000 cases of cancer every year and that 1.8% of all cancer deaths are due to EBV-attributable malignancies (de Martel *et al.*, 2012; Khan and Hashim, 2014). EBV is the most common persistent virus infection in humans and more than 90% of the world's adult population is asymptomatic life-long carriers of EBV. EBV resides in memory B cells of healthy individuals and disruption of this unique interaction can result in B cell malignancies (Vockerodt *et al.*, 2015) (discussed elsewhere in this volume). The role of EBV in human cancers has been studied extensively since its discovery more than 50 years ago. However, the contribution of EBV to epithelial carcinogenesis is enigmatic and, specifically, the role of EBV in the pathogenesis of non-keratinizing nasopharyngeal carcinoma (NPC) and EBV-associated gastric cancer (EBVaGC), remains to be fully elucidated. In this chapter, we provide an overview of our current understanding of the role of EBV in epithelial carcinogenesis and highlight recent findings that provide new insights into its role in the pathogenesis of these diseases.

Historical background

In 1966, two years after the discovery of EBV in Burkitt lymphoma (BL), the link between EBV and epithelial tumours was revealed unexpectedly from an observation that BL antigen-specific antibodies were also detectable in the 'control' sera of patients with 'post-nasal space carcinoma' [later known as NPC] (Old *et al.*, 1966). The report was followed by studies showing that only patients with NPC but not other head and neck cancers, had elevated EBV antibody levels (de Schryver *et al.*, 1969; Henle *et al.*, 1970). However, the association between EBV and NPC did not gain full credibility until 1973 when EBV genomes were detected by *in situ* hybridization in NPC tumour cells, but not in the infiltrating lymphoid cells (Wolf *et al.*, 1973). In the following year, anti-complement immunofluorescence (ACIF) assays were used to show for the first time expression of EBV nuclear antigen in sera of NPC patients (Reedman *et al.*, 1974). Subsequently, a series of serological studies were initiated, including mass serological screening programmes in China involving more than 160,000 individuals. An important aspect of these analyses was that IgA titers to the viral capsid antigen (VCA) and early antigen (EA) were specific to NPC and could serve as prognostic, early diagnostic and predictive indicators (Henle and Henle, 1976; Henle *et al.*, 1977; Zeng *et al.*, 1982; Zeng *et al.*, 1983; Zeng *et al.*, 1985). Therefore, by 1985 the link between EBV and NPC was confirmed, and in addition to B lymphocytes, EBV could also infect epithelial cells.

Compared to NPC, the connection of EBV with GC was revealed much later. In 1990, EBV DNA was detected in a case of undifferentiated lymphoepithelioma-like gastric carcinoma by polymerase chain reaction (Burke *et al.*, 1990). Soon after, Shibata and colleagues used *in situ* hybridization techniques to show that EBV genomes were consistently present in the cancer cells of lymphoepithelioma-like gastric carcinomas (Shibata *et al.*, 1991). It was subsequently confirmed in 1992 that EBV genomes were also detected in the tumour cells of typical gastric adenocarcinomas (Shibata and Weiss, 1992). While almost 98% of NPC cases are associated with EBV, only approximately 10% of all GCs are EBV positive.

Incidence and histology

NPC is described as a distinct type of head and neck cancer, partly ascribable to its unique relationship with EBV infection. It is the most common cancer originating in the nasopharynx, particularly the fossa of Resenmüller. NPC is a geographically skewed disease, with exceptionally high incidence rates in Southern China and South East Asia but rarely seen in Western countries (70,108 of 86,691 cases in Asia in 2012). According to the IARC Globocan 2012 (http://globocan.iarc.fr), it was estimated that Malaysia had the highest aged-standardised rate (ARS) of NPC in the world. NPC is also closely related to ethnicity. In multi-cultural countries like Malaysia and Singapore, NPC is commonly found in Chinese, Malays and natives of Sabah and Sarawak, but very rare in Indians (Zainal Ariffin and Nor Saleha, 2011). EBVaGC displays a predominant location at the proximal stomach and the remnant stomach after partial gastrectomy. Unlike NPC, EBVaGC is not an endemic disease and with similar incidence rates among Asian and Western populations (Murphy *et al.*, 2009; Camargo *et al.*, 2014). Based on the worldwide prevalence of gastric cancer (952,000 cases in 2012), almost 70,000 – 80,000 people per year are estimated to develop EBVaGC. Further details of EBV-associated cancer rates discussed elsewhere in this volume.

The World Health Organisation (WHO) classifies NPC into two major histological subtypes: keratinizing squamous cell carcinoma and non-keratinising carcinoma (differentiated or undifferentiated) (Chan *et al.*, 2005). Non-keratinising NPC accounts for more than 90% of NPC cases in endemic regions and is invariably associated with EBV. It is also referred as lympho-epithelioma of the nasopharynx because of its prominent lymphocytic infiltration (Figure 7.1). In contrast, keratinizing NPC is usually seen in non-endemic regions and its association with EBV is variable. Nonetheless, keratinizing NPC in endemic regions is always EBV positive. Histologically, EBVaGC is divided into two types - conventional adenocarcinoma type and lymphoepithelioma-like carcinoma (LELC) type. Among EBVaGC, EBV-associated gastric adenocarcinoma is the most common subtype although it was only found in 16% of all conventional gastric adenocarcinomas (Shibata and Weiss, 1992). EBV-associated gastric adenocarcinoma shows well to moderately differentiated adenocarcinoma with variable amounts of infiltrating lymphocytes (Lauren, 1965) (Figure 7.1). On the other hand, for the rare lymphoepithelioma-like gastric carcinoma subtype, 89% of tumours are associated with EBV infection (Shibata *et al.*, 1991). Lymphoepithelioma-like carcinomas (LELCs) are rare tumor types with morphologic features similar to NPC (undifferentiated carcinoma with intense lymphoid infiltration) that occur outside the nasopharynx. EBV is also consistently demonstrated in LELCs of the salivary gland, thymus, lung and intrahepatic biliary epithelium (Iezzoni *et al.*, 1995; Chan *et al.*, 2014), implying that an immune microenvironment facilitates

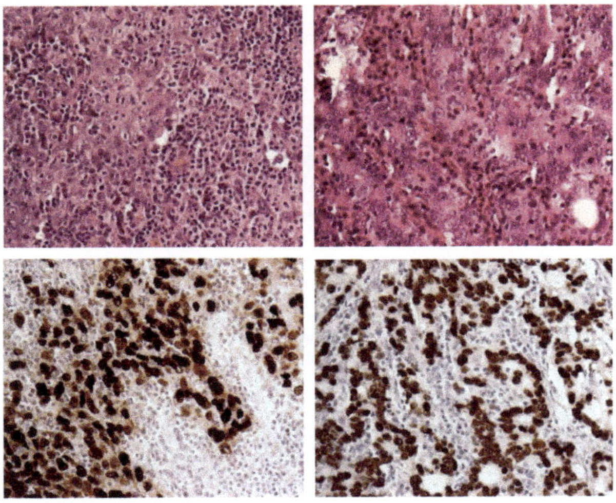

Figure 7.1. Histopathology of NPC and EBVaGC. Left panels: Haematoxylin and eosin (H&E) staining of NPC shows a prominent tumoral lymphocytic infiltrate. The presence of EBV genomes in cancer cells of NPC is detected by EBER *in situ* hybridization, the gold standard method to detect EBV infection in pathology samples. Right panels: Moderate lymphoid infiltration and EBER expression are shown in EBV-positive gastric adenocarcinomas.

the oncogenic properties of EBV in undifferentiated carcinomas.

EBV infection in epithelial cells

Dual tropism of EBV infection

A breakthrough in our understanding of primary EBV infection in healthy individuals came from the discovery that EBV is the causative agent of infectious mononucleosis (IM) (Henle *et al.*, 1968). The presence of EBV in saliva and oropharynx of IM patients suggested that EBV is orally transmitted (Niederman *et al.*, 1976). It is proposed that EBV initially infects naïve B cells in the tonsil and drives their proliferation to become activated B blasts. Some EBV-infected cells escape primary T cell response and undergo a germinal centre reaction. Through a series of viral latency programmes, the EBV-infected B cells eventually reside in a stable reservoir of resting memory B cells to achieve a life-long infection (Babcock *et al.*, 1998; Thorley-Lawson and Gross, 2004). The differentiation of EBV-infected memory B cells into plasma cells switches on the EBV lytic cycle to release new virions for transmission to other susceptible hosts (Laichalk and Thorley-Lawson, 2005). The epithelium of the oropharynx and salivary glands appear to be the major sites for viral replication and indeed, EBV replication was found within the epithelial cells in oral 'hairy' leukoplakia of individuals with human immunodeficiency virus (Greenspan *et al.*, 1985). A recent study has also demonstrated that EBV can establish a productive infection in the suprabasal layers of stratified epithelium in an organotypic oral keratinocyte culture system, supporting a role for epithelial cells in the replication and spread of EBV (Temple *et al.*, 2014). Thus, the dual tropism of EBV infection appears to be vital for the virus to establish a persistent infection in humans.

EBV entry mechanisms

EBV infects B cells and epithelial cells using distinct mechanisms. Entry of EBV into B cells is initiated by the attachment of the viral envelope glycoprotein, gp350/220, to the complement receptor 2 (CR2/CD21) on the surface of B cells (Nemerow *et al.*, 1985). This allows the virus to approach closer to the cell membrane where another viral glycoprotein, gp42, can interact with the human leukocyte antigen (HLA) class II molecule on the B cell surface. gp42 also binds directly to another viral glycoprotein, gH, a component of the viral "core fusion machinery" (consists of a homotrimer gB and a heterodimer gHgL). These interactions activate the core fusion machinery and trigger the fusion of the virion envelope with the cellular plasma membrane (Hutt-Fletcher, 2007; Chesnokova and Hutt-Fletcher, 2014).

EBV infection of human epithelial cells *in vitro* is much less efficient, as epithelial cells express neither CR2 nor HLA class II molecules. Fusion of EBV with an epithelial cell is initiated differently. Instead of using gp42 as an intermediary, gHgL directly interacts with the host integrin complexes, $\alpha v\beta 5$, $\alpha v\beta 6$ and $\alpha v\beta 8$, to trigger the entry into epithelial cells (Chesnokova *et al.*, 2009; Chesnokova and Hutt-Fletcher, 2011). It is of interest that the presence of gp42 impedes the interaction between gHgL and integrins (Wang *et al.*, 1998). Therefore, the EBV virion carries both the trimeric complex gp42/gHgL and also the dimeric complexes of gHgL that lack gp42. Indeed, viral particles originating in epithelial cells are rich in gp42 that facilitate an efficient infection of B cells whereas B cell-derived virus lacks gp42 and therefore better infects epithelial cells (Borza and Hutt-Fletcher, 2002; Borza *et al.*, 2004). On polarised epithelial cells, which resemble more closely the *in vivo* environment, an interaction between the viral protein BMRF2 and the host β1 integrins has been implicated in the entry of virus through basolateral membranes (Tugizov *et al.*, 2003; Xiao *et al.*, 2008). Another mechanism of EBV entry into polarised epithelial cells via the basolateral surface was described as a process of 'transfer infection' whereby EBV-loaded CD11b+ B cells form conjugates with epithelial cells by binding to the CD44v3 and lymphocyte-endothelial-epithelial cell adhesion molecule (LEEP-CAM) (Shannon-Lowe and Rowe, 2011). Recently, two cellular molecules, neuropilin 1 and nonmuscle myosin heavy chain IIA, have been identified as EBV entry factors into epithelial cells by interacting with gHgL (Wang *et al.*, 2015; Xiong *et al.*, 2015). Furthermore, a novel "in-cell infection" mechanism for EBV infection of nasopharyngeal epithelial cells was described, which occurs through the invasion of EBV-positive B cells into epithelial cells by forming cell-in-cell structures that subsequently results in the release of virions into epithelial cells (Ni *et al.*, 2015).

EBV latent infection

EBV is a potent transforming virus in cell culture, robustly converting B cells into immortalised lymphoblastoid cell lines (LCLs) in which all latent genes are expressed (latency III), including six Epstein-Barr nuclear antigens (EBNAs 1, 2, 3A, 3B, 3C and LP), three latent membrane proteins (LMP1, LMP2A and LMP2B), non-coding Epstein-Barr virus-encoded RNAs (EBER1 and EBER2), BamHI-A rightward transcripts (BARTs), microRNAs (miRNAs) of BARTs and BamHI fragment H rightward open reading frame 1 (BHRF1). During different stages of B cell differentiation *in vivo*, EBV expresses either the latency III programme, or a more restricted pattern of latency known as latency II (only EBNA1, LMPs, EBERs, BARTs, miR-BARTs are expressed) or latency I (only EBNA1, EBERs, BARTs and miR-BARTs are expressed) (Rowe *et al.*, 1992). NPC expresses a latency II programme and EBVaGC displays a latency I/II pattern; both tumour types express an additional latent protein called BamHI-A fragment rightward reading frame 1 (BARF1) (Young *et al.*, 1988; Imai *et al.*, 1994; Seto *et al.*, 2005).

LCLs provide useful *in vitro* tools to study the infection of B cells with EBV, however the inability of the virus to transform epithelial cells and the challenges in generating EBV-positive epithelial cell lines have hampered progress in understanding the mechanisms of EBV latent infection in epithelial cells. A number of EBV-negative NPC and GC cell lines can be infected with EBV and express type II latency genes, but these models are not ideal to elucidate the mechanisms by which the virus drives malignant transformation. Although latent gene expression was observed in EBV infection of primary keratinocytes, the infected cells failed to proliferate or the EBV genome was lost during prolonged culture (Pegtel *et al.*, 2004; Feederle *et al.*, 2007; Shannon-Lowe *et al.*, 2009; Temple *et al.*, 2014). Using primary or immortalized nasopharyngeal epithelial cells as more relevant cell models, EBV infection of these cells results in growth inhibition or cellular senescence, as indicated by the increased levels of p16 and p21 (Tsang *et al.*, 2012). It is of great significance that these growth-inhibitory effects of EBV infection can be overridden in cells displaying over-expression of cyclin D1 and/or p16 inactivation, two common features of NPC tumours (Tsang *et al.*, 2012). Given that EBV infection is closely associated with a number of undifferentiated carcinomas, it is possible that the undifferentiated properties of epithelial cells may be a prerequisite for persistence EBV latent infection. Of note, over-expression of cyclin D1 or a CDK4 mutant (resistant to p16) suppressed serum-induced differentiation (Tsang *et al.*, 2012), implying that the differentiation status of epithelial cells is important in the establishment of persistence EBV latent infection.

EBV-associated epithelial cancers

Nasopharyngeal carcinoma

Genetic susceptibility

Aetiological studies of NPC have revealed the contribution of genetic susceptibility, EBV infection and environmental factors to the development of NPC (Chang and Adami, 2006). The high incidence rates of NPC in populations of Southeast Asia and Southern Chinese, but low in those of Western countries strongly suggested the involvement of genetic predisposition in the tumourigenesis. Notably, familial aggregation of NPC and occurrence of multiple cases of the disease in the first degree relatives have been well documented in the Chinese populations in endemic regions (Yu and Yuan, 2002; Lo *et al.*, 2004b; Chang and Adami, 2006), further consolidating the importance of genetic susceptibility in high-risk populations.

The link between variants of major histocompatibility complex (MHC) class I genes and the risk of NPC has been demonstrated in a number of comprehensive studies for the past three decades (Goldsmith *et al.*,

2002). The MHC class I genes, including human leucocyte antigen *(HLA)-A* and *HLA-B*, encode proteins required for the identification and presentation of foreign antigens (for example EBV peptides) to the cytotoxic T cells, triggering host immune responses against targeted cells. High-resolution genotyping studies of Chinese population show that different alleles of the *HLA* genes are associated with different susceptibilities to NPC. Significantly, an increased risk of NPC is found in individuals possessing the HLA-A*02:07 and B*38:02 alleles whereas those with the HLA-A*11:01, HLA-A*31:01, B*13:01 and B*55:02 alleles have a decreased risk of the disease (Tang *et al.*, 2012; Tian *et al.*, 2015). Further, multiple large scale genome-wide association studies (GWAS) have consistently affirmed the strong association between NPC and genes within the MHC region on chromosome 6p21 (Tse *et al.*, 2009; Bei *et al.*, 2010; Tang *et al.*, 2012; Zhao *et al.*, 2012). Independent strong association signals were particularly demonstrated in the *HLA-A, HLA-B* and *HLA-C* loci, supporting the major role of *HLA* genes in the inheritance of NPC. Nevertheless, the association between NPC risk and *HLA* genes may also be interpreted as representing linkage disequilibrium (LD) with other NPC susceptible loci in the MHC region. In this regards, several genes adjacent to the *HLA* genes, including *GABBR1, HCG9* and *MICA*, also show strong susceptibilities to NPC. In addition, these GWAS studies have also discovered several new susceptibility loci, including *TNFRS19* on 13p12, *MECOM* on 3q26 and *CDKN2A-CDKN2B* gene cluster on 9p21. Recently, a meta-analysis of four NPC GWAS studies has not only further confirmed the association of these 3 loci with NPC, but also identified a novel association between genetic variation in the *CLPTM1L/TET* region on 5p15.33 and NPC (Bei *et al.*, 2016). Further, genome wide studies of copy number variations (CNV) have consistently found a specific association of CNV on 6p21.3, which includes the major histocompatibility complex Class I chain-related (MIC) A (*MICA*) and HLA Complex P5 (*HCP5*) genes located adjacent to the *HLA-B/C* genes, with NPC susceptibility (Tse *et al.*, 2011). The study on Malaysian Chinese populations also discovered a candidate NPC susceptibility locus on 11q14.3 (Low *et al.*, 2016). With the advent of whole exome sequencing (WES) technologies, a number of novel susceptibility genes were identified. Sequencing of germline DNAs from three closely related family members with NPC revealed three unique variants (T316S, S772L and Y816*) in the *MLL3* gene (Sasaki *et al.*, 2015). Given the high homology between *MLL3* and a pseudogene, *psiTPTE22* on chromosome 22, the functional role of these mutations needs to be carefully verified. In a cohort of 161 NPC patients and 895 controls from Southern China, Dai and colleagues identified macrophage-stimulating 1 receptor (*MST1R*) on chromosome 3p21.3 as a novel susceptibility gene of NPC patients with early-age onset (age of ≤20y) (Dai *et al.*, 2016). Taken together, identification of NPC susceptibility loci and elucidation of their biological function would greatly enhance our understanding of the contribution of genetic factors to NPC tumourigenesis.

In contrast to NPC, little is known about the genetic susceptibility of EBVaGC. Although several GWAS of gastric cancer have been published (Mocellin *et al.*, 2015), none of these studies focused on the potential genetic factors that are associated with the EBV-positive subgroup, probably due to the lack of obvious relationship between the incidence rates of EBVaGC and ethnicity, or family history. Nonetheless, investigation on the genetic susceptibility of EBVaGC would be informative and of interest to the field of EBV biology.

Genomic landscape

Although EBV latent infection is believed to be an early and crucial event in the pathogenesis of NPC, EBV latent gene products may provide only a subset of the oncogenic hits and additional somatic genetic alterations are required to complete the process of malignant transformation. In our earlier studies, we have revealed several crucial genetic changes, including chromosomal 3p deletion, *CDKN2A/p16* inactivation and *CCND1/cyclin D1* amplification/over-expression in pre-cancerous lesions within the nasopharynx without the evidence of EBV infection (Chan *et al.*, 2000; Chan *et al.*, 2002; Tsang *et al.*, 2012). Whilst the candidate gene(s) on chromosome 3p and its role in NPC tumourigenesis are yet to be fully explored, the necessity of *CDKN2A/p16* inactivation or cyclin D1 overexpression in persistent EBV infection has been clearly demonstrated (Tsang *et al.*, 2012). It is now believed that epithelial cells displaying unique genetic changes are prone to stable EBV latent infection and subsequent additional somatic alterations may co-operate with EBV to drive the development of NPC.

Multiple genetic and epigenetic changes have been identified in the tumourigenesis of NPC. In the past fifteen years, several high throughput genome-wide studies, including allelotyping and microarray-based comparative genome hybridization (aCGH) analyses, have been conducted to decipher the NPC genome. Focal losses on chromosomes 3p, 9p, 11q and 14q which contain tumour suppressor genes such as *CDKN2A/p16, RASSF1A* and *TRAF3*, have been

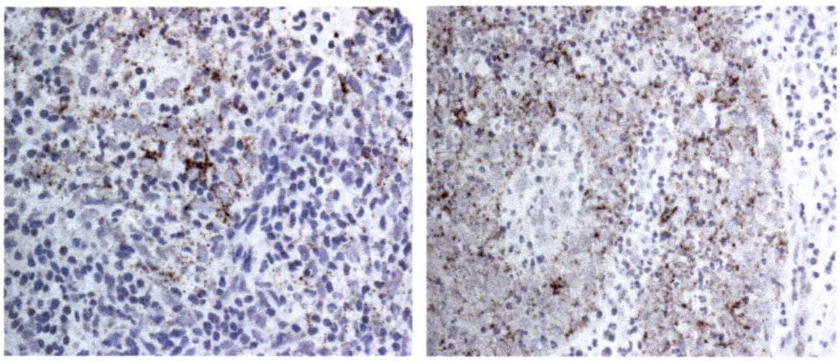

Figure 7.2. High expression of LMP1 in two NPC cases. Variable levels of LMP1 protein are detected in primary NPCs with approximately 30-70% of cases are LMP1-positive.

detected in > 80% of primary NPC tumours (Hui *et al.*, 1999; Lo *et al.*, 2000; Hui *et al.*, 2005; Lo *et al.*, 2012; Chung *et al.*, 2013a). In particular, silencing of *CDKN2A/p16* on 9p21 by homozygous deletion and/or promoter hypermethylation is consistently shown in almost all NPC cases (Lo *et al.*, 1995; Lo *et al.*, 1996). Collectively, these findings indicate that *p16* inactivation perhaps is the most crucial genetic change in the development of this disease. In addition, recurrent amplifications of several oncogenes including *CCND1*, *LTBR* and *PIK3CA* are identified in approximately 8-20% of primary NPC tumours, suggesting the involvement of activated NF-κB and PI3K-AKT signalling pathways in the tumorigenic process (Hui *et al.*, 2005; Or *et al.*, 2006; Or *et al.*, 2010). It is of interest that we recently identified a novel fusion gene *UBR5-ZNF423* that possesses a transforming ability in 8.3% (12/144) of primary NPC cases, mainly in advanced stage of tumours (Chung *et al.*, 2013b). A well-known transforming fusion gene *FGFR3-TACC3* was also recurrently detected in 4/159 cases (2.5%) (Yuan *et al.*, 2014). Given their transforming abilities, these oncogenic fusion genes are believed to act as "drivers" in promoting disease development in a subset of NPC and therefore may be potential therapeutic targets for this group of patients.

Lin and colleagues used a combination of WES, target deep sequencing and single nucleotide polymorphism (SNP) array analyses to decipher a mutational landscape of NPC in the first genome wide study on somatic mutations (Lin *et al.*, 2014a). WES of 56 primary NPC samples and five NPC cell lines revealed distinct mutational signatures, a low mutation rate and wide mutation diversity in the NPC genome. Further integrated analyses have identified nine significantly mutated genes including *TP53, BAP1, MLL2, TSHZ3, PIK3CA, ERBB3, ERBB2, KRAS* and *NRAS*. Except for *TP53* with a mutation frequency of 8.5%, other significantly mutated genes were only detected in <5% of the tumours analysed. Nonetheless, informatics analyses have pinpointed the widespread genetic alterations targeting the chromatin modification pathways (e.g. *ARID1A, MLL2, BAP1, MLL3*) and ERBB-PI3K signalling pathway (e.g. *PIK3CA, PTEN, AKT2, ERBB2, ERBB3*). Furthermore, recurrent mutations were also reported in NOTCH (e.g. *NOTCH1, NOTCH3*) and FAT (e.g. *FAT1, FAT2, FAT3*) gene families. Very recently, Zheng and colleagues further reported the recurrent loss-of-function mutations of multiple NF-κB pathway regulators (*NFKBIA, CYLD, TNFAIP3*) in NPC using WES and target deep sequencing analyses (Zheng *et al.*, 2016). In contrast to these two studies, our recent WES analyses using a larger cohort of micro-dissected NPC tumours revealed a relative higher mutational rate and a distinct genome landscape (Li *et al.*, 2017). The overall non-synonymous mutation rate of EBV-positive NPC was similar to that of EBV-positive gastric adenocarcinoma (Cancer Genome Atlas Research, 2014; Bruce *et al.*, 2015). Interestingly, an increase in mutational burden in primary tumours was significantly correlated with poorer patient outcome. As shown in other WES studies, *TP53* mutations were confirmed in approximately 10% of NPC samples, mainly in advanced and recurrent cases, suggesting its role in the tumour progression. Recurrent mutations of multiple components of the chromatin modification pathways and ERBB-PI3K signalling pathway are also commonly found in our samples. Significantly, we identified frequent genomic aberrations in multiple negative regulators of the NF-κB pathway, including *CYLD, TRAF3, NFKBIA* and *NLRC5*, in more than 40% of the primary and recurrent NPC cases. These data are in line with our previous studies showing recurrent *TRAF3* and *NFKBIA* alterations in EBV-positive NPC cell lines and xenografts (Chung *et al.*, 2013a). It is well-known that the oncogenic EBV-encoded LMP1 activates the NF-κB pathway via the binding of TRAFs (TRAF2, TRAF3, TRAF6) (Dawson *et al.*, 2012; Raab-Traub, 2015). Although the expression of LMP1 protein is variable in primary NPC with high levels only being detected in approximately 30% of NPC cases (Figure 7.2), we show that its expression is mutually exclusive with the presence of the mutated NF-κB regulators. Therefore, these findings imply that the occurrence of mutations in the negative regulators of NK-κB pathway supplants the need for LMP1-mediated NK-κB activation during the progression of NPC. Taken together, these

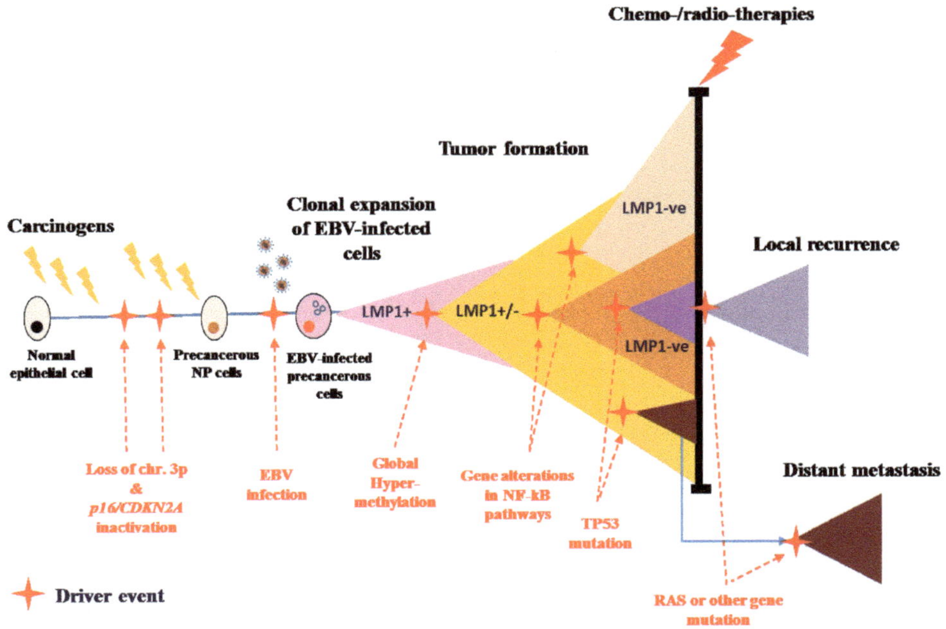

Figure 7.3. Schematic representation of the progressive genomic changes in NPC. Long-term exposure to carcinogens (such as volatile nitrosamines in the preserved foods, or cigarette smoking) may induce DNA damage and field cancerization in nasopharyngeal epithelium. Epithelial cells carrying genetic alterations may acquire growth advantages and expand into pre-cancerous lesions. It is hypothesized that the pre-cancerous lesions with loss of chromosome 3p and inactivation of *p16/CDKN2A* are susceptible to persistent EBV infection, resulting in clonal expansion of the EBV-infected cells and the initiation of the transformation process. The expression of EBV latent gene products (EBNA1, LMP1, LMP2 and various noncoding RNAs) may then play driver roles in transforming the infected cells via de-regulating multiple signalling pathways, interfering with cellular mechanisms and inducing global hypermethylation. During tumour progression, the expression of viral genes may be modulated in a subset of cells such that additional somatic genetic changes may supplant the need for viral gene expression. For example, high expression of LMP1 exhibits cytotoxic effects in epithelial cells (Tsao *et al.*, 2002), therefore tumour clones may use an alternative strategy for progression by down-regulating LMP1 expression and acquiring mutations in NF-κB regulators. During tumour evolution, expansion of malignant sub-clones with additional somatic genetic changes (for example, TP53 and RAS mutations) drives the development of metastatic lesions or local recurrent disease after treatment.

observations suggest that the NF-κB pathway is constitutively activated in NPC and its activation is mediated by either LMP1 or somatic genetic changes, to confer an inflammatory signalling that is crucial in NPC tumourigenesis. Notably, we also identified a high frequency of genetic alterations targeting the antigen presentation mechanisms in NPC. Somatic mutations and rearrangement of multiple MHC class I genes (*HLA-A, HLA-B, HLA-C, B2M*) and its regulator (*NLRC5*) were found in 30% of the NPC cases. Significantly, patients harbour these MHC class I aberrations showed an inferior outcome. It is likely that these acquired somatic changes allow the NPC cells to escape from the host immuosurveillance and result in a rapid tumour progression.

In addition to genetic alterations, DNA methylation is wide-spread in the NPC genome. A number of comparative methylome studies have revealed a global hypermethylation of cellular genes in NPC (Li *et al.*, 2015). It is suggested that the CpG island methylator phenotype (CIMP) of NPC is attributed to the up-regulation of DNA methyltransferases (DNMTs) by EBV-encoded LMP1 (Tsai *et al.*, 2002; Tsai *et al.*, 2006). Epigenetic silencing of genes involving in cell growth/cell cycle regulation (*RASSF1A, CHFR, TSLC1, DLEC1*), invasion/migration (*CDH1, CDH13, PCDH10*), DNA damage response (*OPCML, UCLH1*), apoptosis (*DAPK*), methylation regulation (*TET1*) and a variety of signalling pathways (*RASAL, DLC1, DAB2, WIF1*), was also demonstrated (Li *et al.*, 2011; Lo *et al.*, 2012). It is now recognised that inducing global epigenetic alterations may be an important strategy for EBV to drive the clonal expansion of the infected epithelial cells by developing a wide variety of new cellular properties for rapidly responding the challenges of intrinsic stress within the tumour microenvironment. A proposed model of the progressive genomic changes in NPC is illustrated in Figure 7.3.

Table 7.1. Comparison of characteristics of NPC and EBVaGC.

	NPC	EBVaGC
Histological feature	Squamous cell carcinoma; Intensive lymphoid infiltration	Adenocarcinoma; Moderate lymphoid infiltration
EBV latent gene expression	EBERs, EBNA1, LMP1 (30-70%), LMP2, BARTs, miR-BARTs	EBERs, EBNA1, LMP2A, BARTs, miBARTs
Epigenome	Extreme CIMP	Extreme CIMP
***CDKN2A* inactivation**	~100% promoter hypermethylation	~40-50% homozygous deletion; ~ 40%-50% promoter hypermethylation
***TP53* inactivation**	~9-10%	Rare
Mutations in NF-κB pathway	>40%	Rare
***PIK3CA* mutations**	~4-5%	>70%
***ARID1A* inactivating mutations**	Rare	55%
***BCOR* mutation**	Rare	23%
Amplification of *PDL1, PDL2, JAK2*	Rare	15%
MHC Class I gene mutations	30%	? not common

EBV-associated gastric cancer

Since EBVaGC only constitutes a small set (~10%) of all types of gastric cancer, relatively few studies have focused on the identification of somatic genetic changes in this specific tumour subtype. Nevertheless, epigenetic changes in EBVaGC are well-documented and EBV-encoded LMP2A is thought to induce the wide-spread hypermethylation by up-regulating cellular DNMT1 expression through the activation of STAT3 signalling (Hino *et al.*, 2009; Li *et al.*, 2014). Importantly, a TCGA genome study has comprehensively characterised the genome landscape of EBVaGC and showed that all cases uniquely displayed a CIMP and promoter hypermethylation of *pCDKN2A/p16* (Cancer Genome Atlas Research, 2014). These findings support that *p16* silencing and CIMP are common features that are critical for EBV transformation of epithelial cells. Notably, a salient trait of EBVaGC is the high frequency of *PIK3CA* mutations, being detected in 80% of cases (Cancer Genome Atlas Research, 2014). Frequent *ARID1A* (55%) and *BCOR* (23%) mutations were also found in this tumour type. Of note, no *TP53* mutations were detected, supporting the hypothesis that EBV infection interferes with the function of p53 during the development of EBVaGC. Unlike NPC, amplification of chromosome 9p24.1 which consists of *JAK2, CD274 (PD-L1)* and *PDCD1LG2 (PD-L2)* genes is commonly found in EBVaGC (Cancer Genome Atlas Research, 2014). These could result in the over-expression of the co-inhibitory receptors PD-L1 and PD-L2 and subsequently inhibition of EBV-infected cells killing by cytotoxic T cells. In summary, although several features are shared by these major types of EBV-associated epithelial cancers, it appears that NPC and EBVaGC are driven by distinct signalling pathways, namely NF-κB and PI3K-AKT pathways, respectively (Table 7.1).

Function of EBV latent genes

Latent proteins

Epstein-Barr virus nuclear antigen 1

Following the discovery of a "complement-fixing" antigen that seemed to be a serological marker for EBV infection in BL (Pope *et al.*, 1969), EBNA1 became the focus of much attention in the search for clues linking EBV infection to tumour growth. In 1974, Reedman and colleagues used ACIF to show for the first time expression of an EBV protein (we now know this protein is EBNA1) in every BL tumour cell (Reedman *et al.*, 1974). EBNA1 was subsequently shown to be indispensable for the persistence of EBV genomes in all infected cells by governing the maintenance and replication of the EBV episomes in daughter cells during mitosis (Yates *et al.*, 1985). It also functions as a viral transcriptional transactivator to regulate its own expression and that of other EBV latency genes (*EBNAs* and *LMP1*) required for cell immortalization (Westhoff Smith and Sugden, 2013). EBNA1 is the only EBV-encoded protein that is consistently expressed in all types of latency, pointing to a crucial role in contributing to the malignant phenotypes (Frappier, 2012). It has been shown that EBNA1 enhances epithelial cell survival following DNA damage by destabilizing p53 through the disruption of promyelocytic leukaemia (PML) nuclear bodies (Sivachandran *et al.*, 2008). Further, EBNA1 induces genetic instability in infected cells through a variety of mechanisms, including inducing DNA damage and oxidative stress (Gruhne *et al.*, 2009a; Gruhne *et al.*, 2009b; Cao *et al.*, 2012). EBNA1 may also contribute to oncogenesis through its ability to modulate various signalling pathways, including TGF-β1, AP-1, NF-κB and IL6 signalling (Wood *et al.*, 2007; O'Neil *et al.*, 2008; Valentine *et al.*, 2010).

Notably, EBNA1 plays a central role in the viral life cycle. Two distinct roles for EBNA1 have been identified: suppressing spontaneous lytic reactivation and facilitating lytic infection once the lytic switch is initiated (Sivachandran *et al.*, 2012). A recent report further showed that EBNA1 maintained EBV latency by inducing host let-7 miRNAs which, in turn, reduces the levels of Dicer, a protein that promotes EBV reactivation (Mansouri *et al.*, 2014). Since EBNA1 is expressed in every EBV-positive tumour, targeting EBNA1 is believed to be an efficient therapeutic approach for clinical intervention. A number of EBNA1 specific inhibitors have been developed and showed promising antitumor effects in the pilot pre-clinical studies using various EBV-positive epithelial cancer models (Thompson *et al.*, 2010; Gianti *et al.*, 2016).

Latent membrane proteins 1 and 2

A landmark study in 1985 demonstrated that LMP1 functions as a classical oncogene in rodent fibroblast transformation assays (Wang *et al.*, 1985), and its essential role in EBV-induced B cell transformation *in vitro* was subsequently confirmed a few years later (Kaye *et al.*, 1993). Functionally, LMP1 mimics the tumour necrosis receptor (TNFR) family members, TNFR1 and CD40, but signals in a constitutive, ligand independent manner (Mosialos *et al.*, 1995; Uchida *et al.*, 1999). LMP1 is a 66kDa integral membrane protein comprising three domains: a short amino acid cytoplasmic N-terminus, six hydrophobic alpha-helical transmembrane loops and a long cytoplasmic C-terminal tail. Most of the signalling capacity of the protein is attributed to the C-terminal tail, which contains three functional domains termed C-terminal activating region 1-3 (CTAR 1-3). These domains serve as docking sites for several signalling adaptor proteins, in particular, TNFR-associated factors (TRAFs), TNFR-associated death domain (TRADD), BS69 and Janus kinase (JAK)-3 proteins. These interactions activate multiple signalling pathways, including NF-κB, JNK-p38, PI3K-Akt, ERK-MAPK and JAK-STAT that collectively contribute to a range of growth-promoting phenotypic effects (Brennan, 2001; Dawson *et al.*, 2012).

Whilst expression of LMP1 in human epithelial cells is not normally associated with growth transformation, LMP1 does exert an array of effects that favour tumourigenesis. To enhance cell survival, LMP1 impairs apoptosis by regulating the expression of a number of anti-apoptotic and pro-apoptotic proteins (e.g. Bcl-2, TNFAIP3/A20, survivin, Bmi-1, Bad, Foxo3a) (Henderson *et al.*, 1991; Laherty *et al.*, 1992; Dutton *et al.*, 2007; Dawson *et al.*, 2012). However, high levels of LMP1 can induce apoptosis in both B and epithelial cells, suggesting the expression of LMP1 in the cells has to be tightly controlled (Pratt *et al.*, 2012; Tsao *et al.*, 2015). LMP1 has been shown to stimulate epithelial cell growth by suppressing cell cycle inhibitors (e.g. p16, p21) and inducing growth factor receptors (e.g. EGFR) or inhibitor of differentiation-1 (Id-1) (Miller *et al.*, 1995; Lo *et al.*, 2010). A key aspect of LMP1's transforming ability is to modulate epithelial cell morphology and promote tumour metastasis. LMP1 can induce epithelial-mesenchyme-transition (EMT) and regulate the expression of many molecules that are involved in cell motility and invasion (e.g. MMPs, Cdc42) (Horikawa *et al.*, 2007; Dawson *et al.*, 2012; Liu *et al.*, 2012). Notably, cells expressing LMP1 acquired stem cell/ progenitor cell-like properties, possibly through the activation of the Hedgehog signalling pathway (Kondo *et al.*, 2011; Port *et al.*, 2013). Recent reports also revealed the involvement of LMP1 in energy metabolism of epithelial cells by inhibiting the tumour suppressive LKB1-AMPK pathway or promoting the aerobic glycolysis pathway (Lo *et al.*, 2013; Xiao *et al.*, 2014; Lo *et al.*, 2015). In addition, the ability of LMP1 in promoting angiogenesis is mediated through the induction of VEGF, HIF-1α and IL-8 (Murono *et al.*, 2001; Yoshizaki *et al.*, 2001; Wakisaka *et al.*, 2004). Importantly, LMP1 can also induce the production of pro-inflammatory cytokines (e.g. IL-1, IL-6) (Eliopoulos *et al.*, 1997; Huang *et al.*, 2010), implying a possible role in modulating the tumour microenvironment. In addition, LMP1 has been identified as a modulator of miRNA expression in both B and epithelial cells, which contributes to its involvement in cell invasion and hypoxic regulation (Li *et al.*, 2010; Du *et al.*, 2011).

The *LMP2* gene encodes two proteins, LMP2A and LMP2B, that share 12 transmembrane domains and a short cytoplasmic C-terminus, but it appears that many functional effects of LMP2 proteins are ascribed to the immunoreceptor tyrosine activation (ITAM) motif within the N-terminal cytoplasmic domain of LMP2A. Similar to LMP1, LMP2A modulates multiple signalling pathways, in particular PI3K-Akt signalling, in both B and epithelial cells (Fukuda and Longnecker, 2007). During epithelial carcinogenesis, LMP2A induces anchorage-independent growth, enhances cell adhesion and motility, inhibits cell differentiation, counteracts growth inhibitory and apoptotic effects of TGF-β1 (Scholle *et al.*, 2000; Fukuda and Longnecker, 2004; Allen *et al.*, 2005; Morrison and Raab-Traub, 2005; Fotheringham *et al.*, 2012; Fukuda and Kawaguchi, 2014; Zhou *et al.*, 2014). A recent report revealed a novel function of LMP2A in which it cooperates with LMP1 to modulate DNA damage signalling to provide survival advantage for NPC cells (Wasil *et al.*, 2015). In addition, LMP2A induces EMT, which is associated with the acquisition of stem cell-like properties and the induction of metastatic tumour antigen 1 (MTA1) through mTOR pathway (Kong *et al.*, 2010; Lin *et al.*, 2014b). Several studies have revealed the ability of LMP2A in activating other oncogenic signalling pathways (e.g. Notch, ERK and LPA) in carcinoma cells, highlighting its importance in the pathogenesis of EBV-associated cancers (Anderson and Longnecker, 2008; Lan *et al.*, 2012; Iwakiri *et al.*, 2013; Yap *et al.*, 2015). In addition, a role for LMP2A in suppressing host immunity has been reported. Both LMP2A and LMP2B confer resistance to the antiviral effects of interferon α/β in epithelial cells (Shah *et al.*, 2009), an effect that can be linked to the modulation of NF-κB and IL-6/STAT3 signalling by LMP2A (Stewart *et al.*, 2004). LMP2A has also recently been shown to down-regulate MHC class I and class II expression and to impair the recognition of EBV-infected cells by cytotoxic T cells (Deb Pal and Banerjee, 2015; Lin *et al.*, 2015; Rancan *et al.*, 2015).

BamHI-A fragment rightward reading frame 1 (BARF1)

BARF1 was originally thought to be a lytic gene because it is only expressed upon the induction of lytic cycle in BL cell lines (Zhang *et al.*, 1988). However, subsequent studies confirmed that *BARF1* is a latent gene in NPC and EBVaGC that encodes a secreted protein (Seto *et al.*, 2005). *BARF1* is regarded as an oncogene through its abilities to induce malignant transformation of rodent fibroblasts (Wei and Ooka, 1989; Sheng *et al.*, 2001) and immortalize primary monkey kidney epithelial cells (Wei *et al.*, 1997), as well as enhance the tumorigenicity of EBV-negative BL and NPC cell lines (Wei *et al.*, 1994; Sheng *et al.*, 2003; Seto *et al.*, 2008). BARF1 is a decoy receptor for macrophage colony-stimulating factor (M-CSF/CSF-1) that can block the CSF-1 signaling and hamper the functions of macrophages, thereby demonstrating an immune-evasive role for BARF1 in acute and persistent EBV infection (Hoebe *et al.*, 2012; Ohashi *et al.*, 2012; Shim *et al.*, 2012). In EBVaGC cells, BARF1 exerts an anti-apoptotic role by increasing Bcl-2 to Bax ratio and facilitates cell proliferation by upregulating the expression of NF-κB RelA as well as NF-κB-dependent cyclin D1 proteins (Wang *et al.*, 2006; Chang *et al.*, 2013). Very recently, BARF1 is shown to suppress SMAD4 expression and inactivate TGF-β signalling through the induction of miR-146a expression in EBVaGC (Kim *et al.*, 2016).

Non-coding RNAs

BamHI-A rightward transcripts (BART) miRNAs

EBV was the first human virus found to express miRNAs (Pfeffer *et al.*, 2004). EBV-encoded miRNAs emanate from two regions of the EBV genome: BamHI-A rightward transcripts (BART) and BamHI fragment H rightward open reading frame 1 (BHRF1). To date, 48 mature EBV-encoded miRNAs (25 precursor miRNAs) have been conclusively verified, with 44 of them being BART miRNAs (Kozomara and Griffiths-Jones, 2014). While BART transcripts can be detected in a wide range of EBV-associated cancers with particularly high levels observed in epithelial malignancies such as NPC (Gilligan *et al.*, 1991), the expression of functional BART-encoded protein(s) *in vivo* has yet to be confirmed. Therefore the finding that the BARTs can encode miRNAs has been a major advance. Although the abundance of individual miRNAs can vary greatly, BART miRNAs (miR-BARTs) are generally detectable in all forms of EBV latency, and particularly abundant in latently infected epithelial cells (Cai *et al.*, 2006; Marquitz and Raab-Traub, 2012). In contrast, the expression of *BHRF1* miRNAs appears to be more restricted in B cells undergoing a latency III programme (Cai *et al.*, 2006).

Several viral and host targets of the miR-BARTs have been identified. Like other herpesviruses, evidence has emerged to show that EBV-encoded miRNAs maintain latency by targeting EBV lytic genes. The miR-BARTs limit the expression of viral DNA polymerase BALF5 (Barth *et al.*, 2008) and EBV immediate early genes, *BZLF1 and BRLF1* (Iizasa *et al.*, 2010; Jung *et al.*, 2014). In latently infected memory B cells, miR-BART18 favours latency by suppressing lytic viral replication through the down-regulation of a critical signalling molecule MAP3K2 (Qiu and Thorley-Lawson, 2014). Further, a number of other BART miRNAs can suppress the expression of LMP1 and LMP2A in NPC cells (Lo *et al.*, 2007; Lung *et al.*, 2009), suggesting that these miRNAs might play key roles in the tumorigenic process by regulating tissue homeostasis and immune evasion.

Compared to the viral transcript targets, more progress has been achieved in identifying cellular targets of these EBV-encoded miRNAs. Promisingly, one of the major functions of BART miRNAs is to prevent the apoptosis of malignant epithelial cells. Multiple miR-BARTs have been shown to impair apoptosis by targeting pro-apoptotic proteins, including PUMA/BBC3 (Choy *et al.*, 2008), BIM/BCL2L11 (Marquitz *et al.*, 2011), TOMM22 (Dolken *et al.*, 2010), CASP3 (Vereide *et al.*, 2014) and BID (Shinozaki-Ushiku *et al.*, 2015). The anti-apoptotic activity of miR-BARTs was further validated in a recent study identifying a substantial number of novel cellular transcript targets for these miRNAs (Kang *et al.*, 2015). Perhaps acting in concert to target LMP2A, cellular targets that impair host immune responses, namely MICB which is involved in eliminating infected cells (Nachmani *et al.*, 2009) and IPO7 which is implicated in the production of IL-6 (Dolken *et al.*, 2010), have been identified. The miR-BARTs also appear crucial in the regulation of Wnt signalling, targeting both Wnt inhibitory (WIF1 and APC) and activating (CAPRIN2) molecules (Riley *et al.*, 2012; Wong *et al.*, 2012). Significantly, recent studies have demonstrated that miR-BARTs can promote NPC tumourigenesis by targeting PTEN-dependent pathways (Cai *et al.*, 2015a; Cai *et al.*, 2015b). Other cellular targets of miR-BARTs include tumour suppressor genes DICE1, E-cadherin and NDRG1, all of which are involved in tumour growth and metastasis (Lei *et al.*, 2013b; Hsu *et al.*, 2014; Kanda *et al.*, 2015). Notably, miR-BARTs have recently been shown to drive tumour growth *in vivo* (Cai *et al.*, 2015a; Qiu *et al.*, 2015), further consolidating the contribution of these miRNAs to EBV oncogenesis. It is now clear that EBV-encoded miRNAs, miR-BARTs in particular, are crucial in the pathogenesis of EBV-associated epithelial cancers, however, consensus among different studies could be difficult to achieve in identifying specific target genes ascribed to individual miRNAs (Klinke *et al.*, 2014).

In addition to the miRNAs produced from the introns, multiple alternatively spliced transcripts in the BARTs region have been demonstrated to function as long non-coding RNAs (lncRNAs) in the AGS gastric cancer cell line. These abundantly expressed viral lncRNAs were shown to promote a transformation phenotype by regulating cell growth and host transcriptional profile, data that warrant further investigation (Marquitz *et al.*, 2015).

EBV-encoded RNAs

EBERs are the most abundant viral transcripts in cells with latent EBV infection and serve as sensitive targets for detecting EBV infection in cells and tissues. EBERs exist as EBER1 and EBER2 which are non-polyadenylated, noncoding RNAs that form stem-loop structures by intermolecular base-pairing giving rise to double-stranded RNA (dsRNA)-like molecules. Although EBERs are not essential for EBV-induced transformation of primary B lymphocytes, growth stimulatory properties of EBERs have been demonstrated (Takada, 2012; Kang and Kieff, 2015). EBERs exert a range of biological effects through forming complexes with various cellular proteins, such as RNA-activated protein kinase (PKR), ribosomal protein L22, lupus erythematosus-associated antigen (La) and retinoic acid-inducible gene 1 (RIG-1). EBERs induce interleukin 10 (IL-10) and type I interferon (IFN-α and IFN-β) through RIG-1 signalling, contributing to the activation of innate immunity in BL (Samanta *et al.*, 2006; Samanta *et al.*, 2008). Further, the induction of type I interferon has also been shown to be mediated by the activation of toll-like receptor 3 (TLR3) through the interaction of EBERs with La (Iwakiri *et al.*, 2009). On the other hand, in epithelial cells, EBERs counteract the antiviral and antiproliferative effects of IFNs by blocking PKR pathway through direct interaction, conferring resistance to IFN-induced and Fas-mediated apoptosis (Nanbo *et al.*, 2005).

EBV sequence variations

The first complete EBV sequence of the B95-8 strain was published in 1984 (Baer *et al.*, 1984), representing the largest DNA sequence that had been determined at that time. In 2003, a "wild type" EBV sequence (EBVwt) was constructed using B95-8 as a backbone while a 12-kb deleted segment (encoding some of the miR-BARTs and one of the origins of lytic replication) was provided by the Raji sequences (de Jesus *et al.*, 2003). The current reference EBV sequence was

Table 7.2. Complete genomes of EBV strains from biopsies of NPC and EBVaGC published to date.

EBV strain	Origin	Disease	Year	Reference
GD1	China	NPC	2005	(Zeng et al., 2005)
GD2	China	NPC	2011	(Liu et al., 2011)
HKNPC1	Hong Kong	NPC	2012	(Kwok et al., 2012)
M81	Hong Kong	NPC	2013	(Tsai et al., 2013)
HKNPC2 - HKNPC9	Hong Kong	NPC	2014	(Kwok et al., 2014)
D3201.2	China	NPC	2015	(Palser et al., 2015)
EBVaGC1 - EBVaGC9	China	EBVaGC	2016	(Liu et al., 2016)

released in 2010, termed RefSeq HHV4 (EBV) sequence (GenBank accession number NC_007605). EBV strains are classified into type 1 and type 2 (also known as type A and type B) primarily based on the differences in the sequences of EBNA2 and EBNA3 genes (Dambaugh *et al.*, 1984; Sample *et al.*, 1990). Although there is little evidence for a relationship between disease and EBV types, the main phenotypic difference is that type 1 strains are more prevalent worldwide and have greater transforming potential (Zimber *et al.*, 1986; Rickinson *et al.*, 1987).

The incidence of EBV-associated cancers varies remarkably around the world and it is hypothesized that EBV strain variations account for this puzzling phenomenon. With the recent advances in next-generation sequencing (NGS) technology, almost 100 EBV isolates from all over the world were sequenced (Zeng *et al.*, 2005; Dolan *et al.*, 2006; Liu *et al.*, 2011; Kwok *et al.*, 2012; Lei *et al.*, 2013a; Lin *et al.*, 2013; Tsai *et al.*, 2013; Kwok *et al.*, 2014; Santpere *et al.*, 2014; Palser *et al.*, 2015; Liu *et al.*, 2016; Wang *et al.*, 2016), including 13 and 9 strains from biopsies of NPC and EBVaGC, respectively (Table 7.2), and all of which belong to EBV type 1. It is of interest that an isolate from NPC cells, M81 strain, showed a strong tropism to epithelial cells (Tsai *et al.*, 2013). This EBV strain is spontaneously replicated in B cells and displays high propensity to infect epithelial cells. It has now become clear that while there is a high level of overall similarity among the virus strains, variations exist in some viral genes that might result in functional differences. In general, latent genes harbour the highest variation and, *LMP*1 is of particular interest. Seven sequence variants of *LMP1* have been previously identified, however, little evidence for variant-specific disease associations was observed (Edwards *et al.*, 1999; Mainou and Raab-Traub, 2006). Nonetheless, a LMP1 variant containing a 10-amino acid (30 bp) deletion at a region upstream of CTAR2 (residues 346 to 355) was commonly found in NPC from the endemic regions and it has been linked to an augmented transforming or oncogenic ability compared to the prototype B95.8-LMP1 variant (Hu *et al.*, 1993; Li *et al.*, 1996; Lo *et al.*, 2004a). Many variations, ranging from single base mutations to extensive insertions and deletions, have been identified in *LMP1* from the NGS analyses (Liu *et al.*, 2011; Kwok *et al.*, 2012; Lin *et al.*, 2013; Kwok *et al.*, 2014; Palser *et al.*, 2015). Notably, variations identified on T-cell epitope sequences in latent genes such as *EBNA1*, *EBNA3* and *LMP2A* have been implicated in immune surveillance (Kwok *et al.*, 2012; Kwok *et al.*, 2014; Palser *et al.*, 2015; Liu *et al.*, 2016). However, the differences in the biological properties of these variations and/or their disease association are yet to be determined.

Concluding remarks

EBV has provided a paradigm for our understanding of cancer pathogenesis and treatment since it was discovered over 50 years ago (Young *et al.*, 2016). The ability of EBV to transform B cells *in vitro* has contributed to a deep understanding of EBV-induced lymphomagenesis, however, the precise role of EBV in the pathogenesis of epithelial cancers, mainly NPC and EBVaGC, remains unclear. Nonetheless, accumulating evidence from many years of study into the role of EBV in NPC has started to reveal the mechanisms of EBV oncogenesis in epithelial cells. Latent EBV infection is consistently detected in primary NPC and pre-invasive nasopharyngeal mucosal lesions, but not in normal nasopharyngeal epithelium (Pathmanathan *et al.*, 1995). It is proposed that carcinogenic insults from the environment (such as nitrosamines in salted fish) induce genetic changes in epithelial cells that render these cells to become susceptible to persistent EBV infection. EBV latent genes may subsequently provide, or act in concert with additional genetic changes to drive the disease development. Indeed, NPC tumour cells have been shown to carry clonal EBV genomes (Pathmanathan *et al.*, 1995), indicating that EBV infection must have occurred before expansion of the malignant cell clone. As the tumour evolves, the expression of viral genes may be modulated in a subset of cells such that additional somatic genetic changes may supplant the need for viral gene expression (for example, *LMP1*). Continued efforts in identifying genetic factors that regulate stable EBV infection and elucidating the relationship between viral and host gene expression programmes will facilitate progress in understanding the role of EBV in epithelial carcinogenesis. The prominent lymphocytic infiltrate in EBV-associated epithelial cancers highlights the importance of the tumour microenvironment in the pathogenesis of these diseases. The frequent somatic alterations targeting the antigen presentation mechanisms and immune check points in EBV-associated epithelial cancers have provided new evidence for the crucial role of immune evasion in the progression of tumours with heavy cytotoxic T cells infiltration. Defining the role of inflammatory stroma will not only unravel the complex interplay between the virus and the host cells, but also improve the efficacy of EBV-based therapies for patient benefit. With the advent of NGS technologies, it has now become possible to address the question as to the role of EBV strain variation in carcinogenesis. Functional characterization of the pathogenic strains will be critical to develop appropriate *in vitro* systems and provide insights into the mechanisms of EBV-induced epithelial malignancies. Our growing understanding of the role of EBV in epithelial cancers will continue to shed further light on EBV oncogenesis in general.

Acknowledgement

LFY is supported by Fundamental Research Grant Scheme (FP013-2016) from the Ministry of Higher Education Malaysia and the Newton-Ungku Omar Fund MR/P013201/1 (IF016-2017) from the Academy of Sciences Malaysia and Medical Research Council. KWL is funded by Research Grant Council, Hong Kong Theme-based Research Scheme (T12-401/13-R), General Research Fund (#471413, 471211), Focused Innovations Scheme and Faculty Strategic Research (4620513) of Faculty of Medicine, and VC's One-off Discretionary Fund (VCF2014017, VCF2014015), The Chinese University of Hong Kong.

References

Allen, M.D., Young, L.S., and Dawson, C.W. (2005). The Epstein-Barr virus-encoded LMP2A and LMP2B proteins promote epithelial cell spreading and motility. J. Virol. 79, 1789-1802. http://dx.doi.org/10.1128/JVI.79.3.1789-1802.2005

Anderson, L.J., and Longnecker, R. (2008). An auto-regulatory loop for EBV LMP2A involves activation of Notch. Virology 371, 257-266. http://dx.doi.org/10.1016/j.virol.2007.10.009

Babcock, G.J., Decker, L.L., Volk, M., and Thorley-Lawson, D.A. (1998). EBV persistence in memory B cells *in vivo*. Immunity 9, 395-404.

Baer, R., Bankier, A.T., Biggin, M.D., Deininger, P.L., Farrell, P.J., Gibson, T.J., Hatfull, G., Hudson, G.S., Satchwell, S.C., Seguin, C., *et al.* (1984). DNA sequence and expression of the B95-8 Epstein-Barr virus genome. Nature 310, 207-211.

Barth, S., Pfuhl, T., Mamiani, A., Ehses, C., Roemer, K., Kremmer, E., Jaker, C., Hock, J., Meister, G., and Grasser, F.A. (2008). Epstein-Barr virus-encoded microRNA miR-BART2 down-regulates the viral DNA polymerase BALF5. Nucleic Acids Res. 36, 666-675. http://dx.doi.org/10.1093/nar/gkm1080

Bei, J.X., Li, Y., Jia, W.H., Feng, B.J., Zhou, G., Chen, L.Z., Feng, Q.S., Low, H.Q., Zhang, H., He, F., *et al.* (2010). A genome-wide association study of nasopharyngeal carcinoma identifies three new susceptibility loci. Nat. Genet. 42, 599-603. http://dx.doi.org/10.1038/ng.601

Bei, J.X., Su, W.H., Ng, C.C., Yu, K., Chin, Y.M., Lou, P.J., Hsu, W.L., McKay, J.D., Chen, C.J., Chang, Y.S., *et al.* (2016). A GWAS Meta-analysis and Replication Study Identifies a Novel Locus within CLPTM1L/TERT Associated with Nasopharyngeal Carcinoma in Individuals of Chinese Ancestry. Cancer Epidemiol. Biomarkers Prev. 25, 188-192. http://dx.doi.org/10.1158/1055-9965.EPI-15-0144

Borza, C.M., and Hutt-Fletcher, L.M. (2002). Alternate replication in B cells and epithelial cells switches tropism of Epstein-Barr virus. Nat. Med. 8, 594-599. http://dx.doi.org/10.1038/nm0602-594

Borza, C.M., Morgan, A.J., Turk, S.M., and Hutt-Fletcher, L.M. (2004). Use of gHgL for attachment of Epstein-Barr virus to epithelial cells compromises infection. J. Virol. 78, 5007-5014.

Brennan, P. (2001). Signalling events regulating lymphoid growth and survival. Semin. Cancer Biol. 11, 415-421. http://dx.doi.org/10.1006/scbi.2001.0408

Bruce, J.P., Yip, K., Bratman, S.V., Ito, E., and Liu, F.F. (2015). Nasopharyngeal Cancer: Molecular Landscape. J. Clin. Oncol. 33, 3346-3355. http://dx.doi.org/10.1200/JCO.2015.60.7846

Burke, A.P., Yen, T.S., Shekitka, K.M., and Sobin, L.H. (1990). Lymphoepithelial carcinoma of the stomach with Epstein-Barr virus demonstrated by polymerase chain reaction. Mod. Pathol. 3, 377-380.

Cai, L., Li, J., Zhang, X., Lu, Y., Wang, J., Lyu, X., Chen, Y., Liu, J., Cai, H., Wang, Y., *et al.* (2015a). Gold nano-particles (AuNPs) carrying anti-EBV-miR-BART7-3p inhibit growth of EBV-positive nasopharyngeal carcinoma. Oncotarget 6, 7838-7850.

Cai, L., Ye, Y., Jiang, Q., Chen, Y., Lyu, X., Li, J., Wang, S., Liu, T., Cai, H., Yao, K., *et al.* (2015b). Epstein-Barr virus-encoded microRNA BART1 induces tumour metastasis by regulating PTEN-dependent pathways in nasopharyngeal carcinoma. Nature communications 6, 7353. http://dx.doi.org/10.1038/ncomms8353

Cai, X., Schafer, A., Lu, S., Bilello, J.P., Desrosiers, R.C., Edwards, R., Raab-Traub, N., and Cullen, B.R. (2006). Epstein-Barr virus microRNAs are evolutionarily conserved and differentially expressed. PLoS Pathog. 2, e23. http://dx.doi.org/10.1371/journal.ppat.0020023

Camargo, M.C., Kim, W.H., Chiaravalli, A.M., Kim, K.M., Corvalan, A.H., Matsuo, K., Yu, J., Sung, J.J., Herrera-Goepfert, R., Meneses-Gonzalez, F., *et al.* (2014). Improved survival of gastric cancer with tumour Epstein-Barr virus positivity: an international pooled analysis. Gut 63, 236-243. http://dx.doi.org/10.1136/gutjnl-2013-304531

Cancer Genome Atlas Research, N. (2014). Comprehensive molecular characterization of gastric adenocarcinoma. Nature 513, 202-209. http://dx.doi.org/10.1038/nature13480

Cao, J.Y., Mansouri, S., and Frappier, L. (2012). Changes in the nasopharyngeal carcinoma nuclear proteome induced by the EBNA1 protein of Epstein-Barr virus reveal potential roles for EBNA1 in metastasis and oxidative stress responses. J. Virol. 86, 382-394. http://dx.doi.org/10.1128/JVI.05648-11

Chan, A.S., To, K.F., Lo, K.W., Ding, M., Li, X., Johnson, P., and Huang, D.P. (2002). Frequent chromosome 9p losses in histologically normal nasopharyngeal epithelia from southern Chinese. Int. J. Cancer 102, 300-303. http://dx.doi.org/10.1002/ijc.10689

Chan, A.S., To, K.F., Lo, K.W., Mak, K.F., Pak, W., Chiu, B., Tse, G.M., Ding, M., Li, X., Lee, J.C., *et al.* (2000). High frequency of chromosome 3p deletion in histologically normal nasopharyngeal epithelia from southern Chinese. Cancer Res. 60, 5365-5370.

Chan, A.W., Tong, J.H., Sung, M.Y., Lai, P.B., and To, K.F. (2014). Epstein-Barr virus-associated lymphoepithelioma-like cholangiocarcinoma: a rare variant of intrahepatic cholangiocarcinoma with favourable outcome. Histopathology 65, 674-683. http://dx.doi.org/10.1111/his.12455

Chan, J.K.C., Bray, F., McCarron, P., Foo, W., A.W.M., L., Yip, T., Kuo, T.T., Pilch, B.Z., Wenig, B.M., Huang, D., *et al.* (2005). Nasopharyngeal carcinoma. In World Health Organisation Classification of Tumours: Pathology and Genetics. Head and Neck Tumours. Barnes, L., Eveson, J.W., Reichart, P., Sidransky, D. (eds) IARC Press: Lyon, 85-97.

Chang, E.T., and Adami, H.O. (2006). The enigmatic epidemiology of nasopharyngeal carcinoma. Cancer Epidemiol. Biomarkers Prev. 15, 1765-1777. http://dx.doi.org/10.1158/1055-9965.EPI-06-0353

Chang, M.S., Kim, D.H., Roh, J.K., Middeldorp, J.M., Kim, Y.S., Kim, S., Han, S., Kim, C.W., Lee, B.L., Kim, W.H., *et al.* (2013). Epstein-Barr virus-encoded BARF1 promotes proliferation of gastric carcinoma cells through regulation of NF-kappaB. J. Virol. 87, 10515-10523. http://dx.doi.org/10.1128/JVI.00955-13

Chesnokova, L.S., and Hutt-Fletcher, L.M. (2011). Fusion of Epstein-Barr virus with epithelial cells can be triggered by alphavbeta5 in addition to alphavbeta6 and alphavbeta8, and integrin binding triggers a conformational change in glycoproteins gHgL. J. Virol. 85, 13214-13223. http://dx.doi.org/10.1128/JVI.05580-11

Chesnokova, L.S., and Hutt-Fletcher, L.M. (2014). Epstein-Barr virus infection mechanisms. Chin. J. Cancer 33, 545-548. http://dx.doi.org/10.5732/cjc.014.10168

Chesnokova, L.S., Nishimura, S.L., and Hutt-Fletcher, L.M. (2009). Fusion of epithelial cells by Epstein-Barr virus proteins is triggered by binding of viral glycoproteins gHgL to integrins alphavbeta6 or alphavbeta8. Proc Natl Acad Sci USA 106, 20464-20469. http://dx.doi.org/10.1073/pnas.0907508106

Choy, E.Y., Siu, K.L., Kok, K.H., Lung, R.W., Tsang, C.M., To, K.F., Kwong, D.L., Tsao, S.W., and Jin, D.Y. (2008). An Epstein-Barr virus-encoded microRNA targets PUMA to promote host cell survival. J. Exp. Med. 205, 2551-2560. http://dx.doi.org/10.1084/jem.20072581

Chung, G.T., Lou, W.P., Chow, C., To, K.F., Choy, K.W., Leung, A.W., Tong, C.Y., Yuen, J.W., Ko, C.W., Yip, T.T., *et al.* (2013a). Constitutive activation of distinct NF-kappaB signals in EBV-associated nasopharyngeal carcinoma. J. Pathol. 231, 311-322. http://dx.doi.org/10.1002/path.4239

Chung, G.T., Lung, R.W., Hui, A.B., Yip, K.Y., Woo, J.K., Chow, C., Tong, C.Y., Lee, S.D., Yuen, J.W., Lun, S.W., *et al.* (2013b). Identification of a recurrent transforming UBR5-ZNF423 fusion gene in EBV-associated nasopharyngeal carcinoma. J. Pathol. 231, 158-167. http://dx.doi.org/10.1002/path.4240

Dai, W., Zheng, H., Cheung, A.K., Tang, C.S., Ko, J.M., Wong, B.W., Leong, M.M., Sham, P.C., Cheung, F., Kwong, D.L., *et al.* (2016). Whole-exome sequencing identifies MST1R as a genetic susceptibility gene in nasopharyngeal carcinoma. Proc. Natl. Acad. Sci. U. S. A. 113, 3317-3322. http://dx.doi.org/10.1073/pnas.1523436113

Dambaugh, T., Hennessy, K., Chamnankit, L., and Kieff, E. (1984). U2 region of Epstein-Barr virus DNA may encode Epstein-Barr nuclear antigen 2. Proc. Natl. Acad. Sci. U. S. A. 81, 7632-7636.

Dawson, C.W., Port, R.J., and Young, L.S. (2012). The role of the EBV-encoded latent membrane proteins LMP1 and LMP2 in the pathogenesis of nasopharyngeal carcinoma (NPC). Semin. Cancer Biol. 22, 144-153. http://dx.doi.org/10.1016/j.semcancer.2012.01.004

de Jesus, O., Smith, P.R., Spender, L.C., Elgueta Karstegl, C., Niller, H.H., Huang, D., and Farrell, P.J. (2003). Updated Epstein-Barr virus (EBV) DNA sequence and analysis of a promoter for the BART (CST, BARF0) RNAs of EBV. J. Gen. Virol. 84, 1443-1450.

de Martel, C., Ferlay, J., Franceschi, S., Vignat, J., Bray, F., Forman, D., and Plummer, M. (2012). Global burden of cancers attributable to infections in 2008: a review and synthetic analysis. Lancet Oncol. 13, 607-615. http://dx.doi.org/10.1016/S1470-2045(12)70137-7

de Schryver, A., Friberg, S., Jr., Klein, G., Henle, W., Henle, G., De-The, G., Clifford, P., and Ho, H.C. (1969). Epstein-Barr virus-associated antibody patterns in carcinoma of the post-nasal space. Clin. Exp. Immunol. 5, 443-459.

Deb Pal, A., and Banerjee, S. (2015). Epstein-Barr virus latent membrane protein 2A mediated activation of Sonic Hedgehog pathway induces HLA class Ia downregulation in gastric cancer cells. Virology 484, 22-32. http://dx.doi.org/10.1016/j.virol.2015.05.007

Dolan, A., Addison, C., Gatherer, D., Davison, A.J., and McGeoch, D.J. (2006). The genome of Epstein-Barr virus type 2 strain AG876. Virology 350, 164-170. http://dx.doi.org/10.1016/j.virol.2006.01.015

Dolken, L., Malterer, G., Erhard, F., Kothe, S., Friedel, C.C., Suffert, G., Marcinowski, L., Motsch, N., Barth, S., Beitzinger, M., *et al.* (2010). Systematic analysis of viral and cellular microRNA targets in cells latently infected with human gamma-herpesviruses by RISC immunoprecipitation assay. Cell host and microbe 7, 324-334. http://dx.doi.org/10.1016/j.chom.2010.03.008

Du, Z.M., Hu, L.F., Wang, H.Y., Yan, L.X., Zeng, Y.X., Shao, J.Y., and Ernberg, I. (2011). Upregulation of MiR-155 in nasopharyngeal carcinoma is partly driven by LMP1 and LMP2A and downregulates a negative prognostic marker JMJD1A. PLoS One 6, e19137. http://dx.doi.org/10.1371/journal.pone.0019137

Dutton, A., Woodman, C.B., Chukwuma, M.B., Last, J.I., Wei, W., Vockerodt, M., Baumforth, K.R., Flavell, J.R., Rowe, M., Taylor, A.M., *et al.* (2007). Bmi-1 is induced by the Epstein-Barr virus oncogene LMP1 and regulates the expression of viral target genes in Hodgkin lymphoma cells. Blood 109, 2597-2603. http://dx.doi.org/10.1182/blood-2006-05-020545

Edwards, R.H., Seillier-Moiseiwitsch, F., and Raab-Traub, N. (1999). Signature amino acid changes in latent membrane protein 1 distinguish Epstein-Barr virus strains. Virology 261, 79-95. http://dx.doi.org/10.1006/viro.1999.9855

Eliopoulos, A.G., Stack, M., Dawson, C.W., Kaye, K.M., Hodgkin, L., Sihota, S., Rowe, M., and Young, L.S. (1997). Epstein-Barr virus-encoded LMP1 and CD40 mediate IL-6 production in epithelial cells via an NF-kappaB pathway involving TNF receptor-associated factors. Oncogene 14, 2899-2916. http://dx.doi.org/10.1038/sj.onc.1201258

Epstein, M.A., Achong, B.G., and Barr, Y.M. (1964). Virus Particles in Cultured Lymphoblasts from Burkitt's Lymphoma. Lancet 1, 702-703.

Feederle, R., Neuhierl, B., Bannert, H., Geletneky, K., Shannon-Lowe, C., and Delecluse, H.J. (2007). Epstein-Barr virus B95.8 produced in 293 cells shows marked tropism for differentiated primary epithelial cells and reveals interindividual variation in susceptibility to viral infection. Int. J. Cancer 121, 588-594. http://dx.doi.org/10.1002/ijc.22727

Fotheringham, J.A., Coalson, N.E., and Raab-Traub, N. (2012). Epstein-Barr virus latent membrane protein-2A induces ITAM/Syk- and Akt-dependent epithelial migration through alphav-integrin membrane translocation. J. Virol. 86, 10308-10320. http://dx.doi.org/10.1128/JVI.00853-12

Frappier, L. (2012). Role of EBNA1 in NPC tumourigenesis. Semin. Cancer Biol. 22, 154-161. http://dx.doi.org/10.1016/j.semcancer.2011.12.002

Fukuda, M., and Kawaguchi, Y. (2014). Role of the immunoreceptor tyrosine-based activation motif of latent membrane protein 2A (LMP2A) in Epstein-Barr virus LMP2A-induced cell transformation. J. Virol. 88, 5189-5194. http://dx.doi.org/10.1128/JVI.03714-13

Fukuda, M., and Longnecker, R. (2004). Latent membrane protein 2A inhibits transforming growth factor-beta 1-induced apoptosis through the phosphatidylinositol 3-kinase/Akt pathway. J. Virol. 78, 1697-1705.

Fukuda, M., and Longnecker, R. (2007). Epstein-Barr virus latent membrane protein 2A mediates transformation through constitutive activation of the Ras/PI3-K/Akt Pathway. J. Virol. 81, 9299-9306. http://dx.doi.org/10.1128/JVI.00537-07

Gianti, E., Messick, T.E., Lieberman, P.M., and Zauhar, R.J. (2016). Computational analysis of EBNA1 "druggability" suggests novel insights for Epstein-Barr virus inhibitor design. J. Comput. Aided Mol. Des. 30, 285-303. http://dx.doi.org/10.1007/s10822-016-9899-y

Gilligan, K.J., Rajadurai, P., Lin, J.C., Busson, P., Abdel-Hamid, M., Prasad, U., Tursz, T., and Raab-Traub, N. (1991). Expression of the Epstein-Barr virus BamHI A fragment in nasopharyngeal carcinoma: evidence for a viral protein expressed *in vivo*. J. Virol. 65, 6252-6259.

Goldsmith, D.B., West, T.M., and Morton, R. (2002). HLA associations with nasopharyngeal carcinoma in Southern Chinese: a meta-analysis. Clin. Otolaryngol. Allied Sci. 27, 61-67.

Greenspan, J.S., Greenspan, D., Lennette, E.T., Abrams, D.I., Conant, M.A., Petersen, V., and Freese, U.K. (1985). Replication of Epstein-Barr virus within the epithelial cells of oral "hairy" leukoplakia, an AIDS-associated lesion. N. Engl. J. Med. 313, 1564-1571. http://dx.doi.org/10.1056/NEJM198512193132502

Gruhne, B., Sompallae, R., Marescotti, D., Kamranvar, S.A., Gastaldello, S., and Masucci, M.G. (2009a). The Epstein-Barr virus nuclear antigen-1 promotes genomic instability via induction of reactive oxygen species. Proc. Natl. Acad. Sci. U. S. A. 106, 2313-2318. http://dx.doi.org/10.1073/pnas.0810619106

Gruhne, B., Sompallae, R., and Masucci, M.G. (2009b). Three Epstein-Barr virus latency proteins independently promote genomic instability by inducing DNA damage, inhibiting DNA repair and inactivating cell cycle checkpoints. Oncogene 28, 3997-4008. http://dx.doi.org/10.1038/onc.2009.258

Henderson, S., Rowe, M., Gregory, C., Croom-Carter, D., Wang, F., Longnecker, R., Kieff, E., and Rickinson, A. (1991). Induction of bcl-2 expression by Epstein-Barr virus latent membrane protein 1 protects infected B cells from programmed cell death. Cell 65, 1107-1115.

Henle, G., and Henle, W. (1976). Epstein-Barr virus-specific IgA serum antibodies as an outstanding feature of nasopharyngeal carcinoma. Int. J. Cancer 17, 1-7.

Henle, G., Henle, W., and Diehl, V. (1968). Relation of Burkitt's tumor-associated herpes-ytpe virus to infectious mononucleosis. Proc. Natl. Acad. Sci. U. S. A. 59, 94-101.

Henle, W., Henle, G., Ho, H.C., Burtin, P., Cachin, Y., Clifford, P., de Schryver, A., de-The, G., Diehl, V., and Klein, G. (1970). Antibodies to Epstein-Barr virus in nasopharyngeal carcinoma, other head and neck neoplasms, and control groups. J. Natl. Cancer Inst. 44, 225-231.

Henle, W., Ho, J.H., Henle, G., Chau, J.C., and Kwan, H.C. (1977). Nasopharyngeal carcinoma: significance of changes in Epstein-Barr virus-related antibody patterns following therapy. Int. J. Cancer 20, 663-672.

Hino, R., Uozaki, H., Murakami, N., Ushiku, T., Shinozaki, A., Ishikawa, S., Morikawa, T., Nakaya, T., Sakatani, T., Takada, K., *et al.* (2009). Activation of DNA methyltransferase 1 by EBV latent membrane protein 2A leads to promoter hypermethylation of PTEN gene in gastric carcinoma. Cancer Res. 69, 2766-2774. http://dx.doi.org/10.1158/0008-5472.CAN-08-3070

Hoebe, E.K., Le Large, T.Y., Tarbouriech, N., Oosterhoff, D., De Gruijl, T.D., Middeldorp, J.M., and Greijer, A.E. (2012). Epstein-Barr virus-encoded BARF1 protein is a decoy receptor for macrophage colony stimulating factor and interferes with macrophage differentiation and activation. Viral Immunol. 25, 461-470. http://dx.doi.org/10.1089/vim.2012.0034

Horikawa, T., Yang, J., Kondo, S., Yoshizaki, T., Joab, I., Furukawa, M., and Pagano, J.S. (2007). Twist and epithelial-mesenchymal transition are induced by the EBV oncoprotein latent membrane protein 1 and are associated with metastatic nasopharyngeal carcinoma. Cancer Res. 67, 1970-1978. http://dx.doi.org/10.1158/0008-5472.CAN-06-3933

Hsu, C.Y., Yi, Y.H., Chang, K.P., Chang, Y.S., Chen, S.J., and Chen, H.C. (2014). The Epstein-Barr virus-encoded microRNA MiR-BART9 promotes tumor metastasis by targeting E-cadherin in nasopharyngeal carcinoma. PLoS Pathog. 10, e1003974. http://dx.doi.org/10.1371/journal.ppat.1003974

Hu, L.F., Chen, F., Zheng, X., Ernberg, I., Cao, S.L., Christensson, B., Klein, G., and Winberg, G. (1993). Clonability and tumorigenicity of human epithelial cells expressing the EBV encoded membrane protein LMP1. Oncogene 8, 1575-1583.

Huang, Y.T., Liu, M.Y., Tsai, C.H., and Yeh, T.H. (2010). Upregulation of interleukin-1 by Epstein-Barr virus latent membrane protein 1 and its possible role in nasopharyngeal carcinoma cell growth. Head Neck 32, 869-876. http://dx.doi.org/10.1002/hed.21270

Hui, A.B., Lo, K.W., Leung, S.F., Teo, P., Fung, M.K., To, K.F., Wong, N., Choi, P.H., Lee, J.C., and Huang, D.P. (1999). Detection of recurrent chromosomal gains and losses in primary nasopharyngeal carcinoma by comparative genomic hybridisation. Int. J. Cancer 82, 498-503.

Hui, A.B., Or, Y.Y., Takano, H., Tsang, R.K., To, K.F., Guan, X.Y., Sham, J.S., Hung, K.W., Lam, C.N., van Hasselt, C.A., *et al.* (2005). Array-based comparative genomic hybridization analysis identified cyclin D1 as a target oncogene at 11q13.3 in nasopharyngeal carcinoma. Cancer Res. 65, 8125-8133. http://dx.doi.org/10.1158/0008-5472.CAN-05-0648

Hutt-Fletcher, L.M. (2007). Epstein-Barr virus entry. J. Virol. 81, 7825-7832. http://dx.doi.org/10.1128/JVI.00445-07

Iezzoni, J.C., Gaffey, M.J., and Weiss, L.M. (1995). The role of Epstein-Barr virus in lymphoepithelioma-like carcinomas. Am. J. Clin. Pathol. 103, 308-315.

Iizasa, H., Wulff, B.E., Alla, N.R., Maragkakis, M., Megraw, M., Hatzigeorgiou, A., Iwakiri, D., Takada, K., Wiedmer, A., Showe, L., *et al.* (2010). Editing of Epstein-Barr virus-encoded BART6 microRNAs controls their dicer targeting and consequently affects viral latency. J. Biol. Chem. 285, 33358-33370. http://dx.doi.org/10.1074/jbc.M110.138362

Imai, S., Koizumi, S., Sugiura, M., Tokunaga, M., Uemura, Y., Yamamoto, N., Tanaka, S., Sato, E., and Osato, T. (1994). Gastric carcinoma: monoclonal epithelial malignant cells expressing Epstein-Barr virus latent infection protein. Proc. Natl. Acad. Sci. U. S. A. 91, 9131-9135.

Iwakiri, D., Minamitani, T., and Samanta, M. (2013). Epstein-Barr virus latent membrane protein 2A contributes to anoikis resistance through ERK activation. J. Virol. 87, 8227-8234. http://dx.doi.org/10.1128/JVI.01089-13

Iwakiri, D., Zhou, L., Samanta, M., Matsumoto, M., Ebihara, T., Seya, T., Imai, S., Fujieda, M., Kawa, K., and Takada, K. (2009). Epstein-Barr virus (EBV)-encoded small RNA is released from EBV-infected cells and activates signaling from Toll-like receptor 3. J. Exp. Med. 206, 2091-2099. http://dx.doi.org/10.1084/jem.20081761

Jung, Y.J., Choi, H., Kim, H., and Lee, S.K. (2014). MicroRNA miR-BART20-5p stabilizes Epstein-Barr virus latency by directly targeting BZLF1 and BRLF1. J. Virol. 88, 9027-9037. http://dx.doi.org/10.1128/JVI.00721-14

Kanda, T., Miyata, M., Kano, M., Kondo, S., Yoshizaki, T., and Iizasa, H. (2015). Clustered microRNAs of the Epstein-Barr virus cooperatively downregulate an epithelial cell-specific metastasis suppressor. J. Virol. 89, 2684-2697. http://dx.doi.org/10.1128/JVI.03189-14

Kang, D., Skalsky, R.L., and Cullen, B.R. (2015). EBV BART MicroRNAs Target Multiple Pro-apoptotic Cellular Genes to Promote Epithelial Cell Survival. PLoS Pathog. 11, e1004979. http://dx.doi.org/10.1371/journal.ppat.1004979

Kang, M.S., and Kieff, E. (2015). Epstein-Barr virus latent genes. Exp. Mol. Med. 47, e131. http://dx.doi.org/10.1038/emm.2014.84

Kaye, K.M., Izumi, K.M., and Kieff, E. (1993). Epstein-Barr virus latent membrane protein 1 is essential for B-lymphocyte growth transformation. Proc. Natl. Acad. Sci. U. S. A. 90, 9150-9154.

Khan, G., and Hashim, M.J. (2014). Global burden of deaths from Epstein-Barr virus attributable malignancies 1990-2010. Infect. Agent. Cancer 9, 38. http://dx.doi.org/10.1186/1750-9378-9-38

Kim, D.H., Chang, M.S., Yoon, C.J., Middeldorp, J.M., Martinez, O.M., Byeon, S.J., Rha, S.Y., Kim, S.H., Kim, Y.S., and Woo, J.H. (2016). Epstein-Barr virus

BARF1-induced NFkappaB/miR-146a/SMAD4 alterations in stomach cancer cells. Oncotarget. http://dx.doi.org/10.18632/oncotarget.10511

Klinke, O., Feederle, R., and Delecluse, H.J. (2014). Genetics of Epstein-Barr virus microRNAs. Semin. Cancer Biol. 26, 52-59. http://dx.doi.org/10.1016/j.semcancer.2014.02.002

Kondo, S., Wakisaka, N., Muramatsu, M., Zen, Y., Endo, K., Murono, S., Sugimoto, H., Yamaoka, S., Pagano, J.S., and Yoshizaki, T. (2011). Epstein-Barr virus latent membrane protein 1 induces cancer stem/progenitor-like cells in nasopharyngeal epithelial cell lines. J. Virol. 85, 11255-11264. http://dx.doi.org/10.1128/JVI.00188-11

Kong, Q.L., Hu, L.J., Cao, J.Y., Huang, Y.J., Xu, L.H., Liang, Y., Xiong, D., Guan, S., Guo, B.H., Mai, H.Q., *et al.* (2010). Epstein-Barr virus-encoded LMP2A induces an epithelial-mesenchymal transition and increases the number of side population stem-like cancer cells in nasopharyngeal carcinoma. PLoS Pathog. 6, e1000940. http://dx.doi.org/10.1371/journal.ppat.1000940

Kozomara, A., and Griffiths-Jones, S. (2014). miRBase: annotating high confidence microRNAs using deep sequencing data. Nucleic Acids Res. 42, D68-73. http://dx.doi.org/10.1093/nar/gkt1181

Kwok, H., Tong, A.H., Lin, C.H., Lok, S., Farrell, P.J., Kwong, D.L., and Chiang, A.K. (2012). Genomic sequencing and comparative analysis of Epstein-Barr virus genome isolated from primary nasopharyngeal carcinoma biopsy. PLoS One 7, e36939. http://dx.doi.org/10.1371/journal.pone.0036939

Kwok, H., Wu, C.W., Palser, A.L., Kellam, P., Sham, P.C., Kwong, D.L., and Chiang, A.K. (2014). Genomic diversity of Epstein-Barr virus genomes isolated from primary nasopharyngeal carcinoma biopsy samples. J. Virol. 88, 10662-10672. http://dx.doi.org/10.1128/JVI.01665-14

Laherty, C.D., Hu, H.M., Opipari, A.W., Wang, F., and Dixit, V.M. (1992). The Epstein-Barr virus LMP1 gene product induces A20 zinc finger protein expression by activating nuclear factor kappa B. J. Biol. Chem. 267, 24157-24160.

Laichalk, L.L., and Thorley-Lawson, D.A. (2005). Terminal differentiation into plasma cells initiates the replicative cycle of Epstein-Barr virus *in vivo*. J. Virol. 79, 1296-1307. http://dx.doi.org/10.1128/JVI.79.2.1296-1307.2005

Lan, Y.Y., Hsiao, J.R., Chang, K.C., Chang, J.S., Chen, C.W., Lai, H.C., Wu, S.Y., Yeh, T.H., Chang, F.H., Lin, W.H., *et al.* (2012). Epstein-Barr virus latent membrane protein 2A promotes invasion of nasopharyngeal carcinoma cells through ERK/Fra-1-mediated induction of matrix metalloproteinase 9. J. Virol. 86, 6656-6667. http://dx.doi.org/10.1128/JVI.00174-12

Lauren, P. (1965). The Two Histological Main Types of Gastric Carcinoma: Diffuse and So-Called Intestinal-Type Carcinoma. An Attempt at a Histo-Clinical Classification. Acta Pathol. Microbiol. Scand. 64, 31-49.

Lei, H., Li, T., Hung, G.C., Li, B., Tsai, S., and Lo, S.C. (2013a). Identification and characterization of EBV genomes in spontaneously immortalized human peripheral blood B lymphocytes by NGS technology. BMC Genomics 14, 804. http://dx.doi.org/10.1186/1471-2164-14-804

Lei, T., Yuen, K.S., Xu, R., Tsao, S.W., Chen, H., Li, M., Kok, K.H., and Jin, D.Y. (2013b). Targeting of DICE1 tumor suppressor by Epstein-Barr virus-encoded miR-BART3* microRNA in nasopharyngeal carcinoma. Int. J. Cancer 133, 79-87. http://dx.doi.org/10.1002/ijc.28007

Li, G., Wu, Z., Peng, Y., Liu, X., Lu, J., Wang, L., Pan, Q., He, M.L., and Li, X.P. (2010). MicroRNA-10b induced by Epstein-Barr virus-encoded latent membrane protein-1 promotes the metastasis of human nasopharyngeal carcinoma cells. Cancer Lett. 299, 29-36. http://dx.doi.org/10.1016/j.canlet.2010.07.021

Li, L., Zhang, Y., Fan, Y., Sun, K., Su, X., Du, Z., Tsao, S.W., Loh, T.K., Sun, H., Chan, A.T., *et al.* (2015). Characterization of the nasopharyngeal carcinoma methylome identifies aberrant disruption of key signaling pathways and methylated tumor suppressor genes. Epigenomics 7, 155-173. http://dx.doi.org/10.2217/epi.14.79

Li, L., Zhang, Y., Guo, B.B., Chan, F.K., and Tao, Q. (2014). Oncogenic induction of cellular high CpG methylation by Epstein-Barr virus in malignant epithelial cells. Chin. J. Cancer 33, 604-608. http://dx.doi.org/10.5732/cjc.014.10191

Li, L.L., Shu, X.S., Wang, Z.H., Cao, Y., and Tao, Q. (2011). Epigenetic disruption of cell signaling in nasopharyngeal carcinoma. Chin. J. Cancer 30, 231-239.

Li, S.N., Chang, Y.S., and Liu, S.T. (1996). Effect of a 10-amino acid deletion on the oncogenic activity of latent membrane protein 1 of Epstein-Barr virus. Oncogene 12, 2129-2135.

Li, Y.Y., Chung, G.T., Lui, V.W., To, K.F., Ma, B.B., Chow, C., Woo, J.K., Yip, K.Y., Seo, J., Hui, E.P., *et al.* (2017). Exome and genome sequencing of nasopharynx cancer identifies NF-kappaB pathway activating mutations. Nature communications 8, 14121. http://dx.doi.org/10.1038/ncomms14121

Lin, D.C., Meng, X., Hazawa, M., Nagata, Y., Varela, A.M., Xu, L., Sato, Y., Liu, L.Z., Ding, L.W., Sharma, A., *et al.* (2014a). The genomic landscape of nasopharyngeal carcinoma. Nat. Genet. 46, 866-871. http://dx.doi.org/10.1038/ng.3006

Lin, J.H., Lin, J.Y., Chou, Y.C., Chen, M.R., Yeh, T.H., Lin, C.W., Lin, S.J., and Tsai, C.H. (2015). Epstein-Barr virus LMP2A suppresses MHC class II expression by regulating the B-cell transcription factors E47 and PU.1. Blood 125, 2228-2238. http://dx.doi.org/10.1182/blood-2014-08-594689

Lin, Z., Wan, X., Jiang, R., Deng, L., Gao, Y., Tang, J., Yang, Y., Zhao, W., Yan, X., Yao, K., *et al.* (2014b). Epstein-Barr virus-encoded latent membrane protein 2A promotes the epithelial-mesenchymal transition in nasopharyngeal carcinoma via metastatic tumor antigen 1 and mechanistic target of rapamycin signaling induction. J. Virol. 88, 11872-11885. http://dx.doi.org/10.1128/JVI.01867-14

Lin, Z., Wang, X., Strong, M.J., Concha, M., Baddoo, M., Xu, G., Baribault, C., Fewell, C., Hulme, W., Hedges, D., *et al.* (2013). Whole-genome sequencing of the Akata and Mutu Epstein-Barr virus strains. J. Virol. 87, 1172-1182. http://dx.doi.org/10.1128/JVI.02517-12

Liu, H.P., Chen, C.C., Wu, C.C., Huang, Y.C., Liu, S.C., Liang, Y., Chang, K.P., and Chang, Y.S. (2012). Epstein-Barr virus-encoded LMP1 interacts with FGD4 to activate Cdc42 and thereby promote migration of nasopharyngeal carcinoma cells. PLoS Pathog. 8, e1002690. http://dx.doi.org/10.1371/journal.ppat.1002690

Liu, P., Fang, X., Feng, Z., Guo, Y.M., Peng, R.J., Liu, T., Huang, Z., Feng, Y., Sun, X., Xiong, Z., *et al.* (2011). Direct sequencing and characterization of a clinical isolate of Epstein-Barr virus from nasopharyngeal carcinoma tissue by using next-generation sequencing technology. J. Virol. 85, 11291-11299. http://dx.doi.org/10.1128/JVI.00823-11

Liu, Y., Yang, W., Pan, Y., Ji, J., Lu, Z., and Ke, Y. (2016). Genome-wide analysis of Epstein-Barr virus (EBV) isolated from EBV-associated gastric carcinoma (EBVaGC). Oncotarget 7, 4903-4914. http://dx.doi.org/10.18632/oncotarget.6751

Lo, A.K., Dawson, C.W., Lo, K.W., Yu, Y., and Young, L.S. (2010). Upregulation of Id1 by Epstein-Barr virus-encoded LMP1 confers resistance to TGFbeta-mediated growth inhibition. Mol. Cancer 9, 155. http://dx.doi.org/10.1186/1476-4598-9-155

Lo, A.K., Dawson, C.W., Young, L.S., Ko, C.W., Hau, P.M., and Lo, K.W. (2015). Activation of the FGFR1 signalling pathway by the Epstein-Barr Virus-encoded LMP1 promotes aerobic glycolysis and transformation of human nasopharyngeal epithelial cells. J. Pathol. http://dx.doi.org/10.1002/path.4575

Lo, A.K., Huang, D.P., Lo, K.W., Chui, Y.L., Li, H.M., Pang, J.C., and Tsao, S.W. (2004a). Phenotypic alterations induced by the Hong Kong-prevalent Epstein-Barr virus-encoded LMP1 variant (2117-LMP1) in nasopharyngeal epithelial cells. Int. J. Cancer 109, 919-925. http://dx.doi.org/10.1002/ijc.20051

Lo, A.K., Lo, K.W., Ko, C.W., Young, L.S., and Dawson, C.W. (2013). Inhibition of the LKB1-AMPK pathway by the Epstein-Barr virus-encoded LMP1 promotes proliferation and transformation of human nasopharyngeal epithelial cells. J. Pathol. 230, 336-346. http://dx.doi.org/10.1002/path.4201

Lo, A.K., To, K.F., Lo, K.W., Lung, R.W., Hui, J.W., Liao, G., and Hayward, S.D. (2007). Modulation of LMP1 protein expression by EBV-encoded microRNAs. Proc. Natl. Acad. Sci. U. S. A. 104, 16164-16169. http://dx.doi.org/10.1073/pnas.0702896104

Lo, K.W., Cheung, S.T., Leung, S.F., van Hasselt, A., Tsang, Y.S., Mak, K.F., Chung, Y.F., Woo, J.K., Lee, J.C., and Huang, D.P. (1996). Hypermethylation of the p16 gene in nasopharyngeal carcinoma. Cancer Res. 56, 2721-2725.

Lo, K.W., Chung, G.T., and To, K.F. (2012). Deciphering the molecular genetic basis of NPC through molecular, cytogenetic, and epigenetic approaches. Semin. Cancer Biol. 22, 79-86. http://dx.doi.org/10.1016/j.semcancer.2011.12.011

Lo, K.W., Huang, D.P., and Lau, K.M. (1995). p16 gene alterations in nasopharyngeal carcinoma. Cancer Res. 55, 2039-2043.

Lo, K.W., Teo, P.M., Hui, A.B., To, K.F., Tsang, Y.S., Chan, S.Y., Mak, K.F., Lee, J.C., and Huang, D.P. (2000). High resolution allelotype of microdissected primary nasopharyngeal carcinoma. Cancer Res. 60, 3348-3353.

Lo, K.W., To, K.F., and Huang, D.P. (2004b). Focus on nasopharyngeal carcinoma. Cancer Cell 5, 423-428.

Low, J.S., Chin, Y.M., Mushiroda, T., Kubo, M., Govindasamy, G.K., Pua, K.C., Yap, Y.Y., Yap, L.F., Subramaniam, S.K., Ong, C.A., *et al.* (2016). A Genome Wide Study of Copy Number Variation Associated with Nasopharyngeal Carcinoma in Malaysian Chinese Identifies CNVs at 11q14.3 and 6p21.3 as Candidate Loci. PLoS One 11, e0145774. http://dx.doi.org/10.1371/journal.pone.0145774

Lung, R.W., Tong, J.H., Sung, Y.M., Leung, P.S., Ng, D.C., Chau, S.L., Chan, A.W., Ng, E.K., Lo, K.W., and To, K.F. (2009). Modulation of LMP2A expression by a newly identified Epstein-Barr virus-encoded microRNA miR-BART22. Neoplasia 11, 1174-1184.

Mainou, B.A., and Raab-Traub, N. (2006). LMP1 strain variants: biological and molecular properties. J. Virol. 80, 6458-6468. http://dx.doi.org/10.1128/JVI.00135-06

Mansouri, S., Pan, Q., Blencowe, B.J., Claycomb, J.M., and Frappier, L. (2014). Epstein-Barr virus EBNA1 protein regulates viral latency through effects on let-7 microRNA and dicer. J. Virol. 88, 11166-11177. http://dx.doi.org/10.1128/JVI.01785-14

Marquitz, A.R., Mathur, A., Edwards, R.H., and Raab-Traub, N. (2015). Host Gene Expression Is Regulated by Two Types of Noncoding RNAs Transcribed from the Epstein-Barr Virus BamHI A Rightward Transcript Region. J. Virol. 89, 11256-11268. http://dx.doi.org/10.1128/JVI.01492-15

Marquitz, A.R., Mathur, A., Nam, C.S., and Raab-Traub, N. (2011). The Epstein-Barr Virus BART microRNAs target the pro-apoptotic protein Bim. Virology 412, 392-400. http://dx.doi.org/10.1016/j.virol.2011.01.028

Marquitz, A.R., and Raab-Traub, N. (2012). The role of miRNAs and EBV BARTs in NPC. Semin. Cancer Biol. 22, 166-172. http://dx.doi.org/10.1016/j.semcancer.2011.12.001

Miller, W.E., Earp, H.S., and Raab-Traub, N. (1995). The Epstein-Barr virus latent membrane protein 1 induces expression of the epidermal growth factor receptor. J. Virol. 69, 4390-4398.

Mocellin, S., Verdi, D., Pooley, K.A., and Nitti, D. (2015). Genetic variation and gastric cancer risk: a field synopsis and meta-analysis. Gut 64, 1209-1219. http://dx.doi.org/10.1136/gutjnl-2015-309168

Morrison, J.A., and Raab-Traub, N. (2005). Roles of the ITAM and PY motifs of Epstein-Barr virus latent membrane protein 2A in the inhibition of epithelial cell differentiation and activation of {beta}-catenin signaling. J. Virol. 79, 2375-2382. http://dx.doi.org/10.1128/JVI.79.4.2375-2382.2005

Mosialos, G., Birkenbach, M., Yalamanchili, R., VanArsdale, T., Ware, C., and Kieff, E. (1995). The Epstein-Barr virus transforming protein LMP1 engages signaling proteins for the tumor necrosis factor receptor family. Cell 80, 389-399.

Murono, S., Inoue, H., Tanabe, T., Joab, I., Yoshizaki, T., Furukawa, M., and Pagano, J.S. (2001). Induction of cyclooxygenase-2 by Epstein-Barr virus latent membrane protein 1 is involved in vascular endothelial growth factor production in nasopharyngeal carcinoma cells. Proc. Natl. Acad. Sci. U. S. A. 98, 6905-6910. http://dx.doi.org/10.1073/pnas.121016998

Murphy, G., Pfeiffer, R., Camargo, M.C., and Rabkin, C.S. (2009). Meta-analysis shows that prevalence of Epstein-Barr virus-positive gastric cancer differs based on sex and anatomic location. Gastroenterology 137, 824-833. http://dx.doi.org/10.1053/j.gastro.2009.05.001

Nachmani, D., Stern-Ginossar, N., Sarid, R., and Mandelboim, O. (2009). Diverse herpesvirus microRNAs target the stress-induced immune ligand MICB to escape recognition by natural killer cells. Cell host and microbe 5, 376-385. http://dx.doi.org/10.1016/j.chom.2009.03.003

Nanbo, A., Yoshiyama, H., and Takada, K. (2005). Epstein-Barr virus-encoded poly(A)- RNA confers resistance to apoptosis mediated through Fas by blocking the PKR pathway in human epithelial intestine 407 cells. J. Virol. 79, 12280-12285. http://dx.doi.org/10.1128/JVI.79.19.12280-12285.2005

Nemerow, G.R., Wolfert, R., McNaughton, M.E., and Cooper, N.R. (1985). Identification and characterization of the Epstein-Barr virus receptor on human B lymphocytes and its relationship to the C3d complement receptor (CR2). J. Virol. 55, 347-351.

Ni, C., Chen, Y., Zeng, M., Pei, R., Du, Y., Tang, L., Wang, M., Hu, Y., Zhu, H., He, M., *et al.* (2015). In-cell infection: a novel pathway for Epstein-Barr virus infection mediated by cell-in-cell structures. Cell Res. 25, 785-800. http://dx.doi.org/10.1038/cr.2015.50

Niederman, J.C., Miller, G., Pearson, H.A., Pagano, J.S., and Dowaliby, J.M. (1976). Infectious mononucleosis. Epstein-Barr-virus shedding in saliva and the oropharynx. N. Engl. J. Med. 294, 1355-1359. http://dx.doi.org/10.1056/NEJM197606172942501

O'Neil, J.D., Owen, T.J., Wood, V.H., Date, K.L., Valentine, R., Chukwuma, M.B., Arrand, J.R., Dawson, C.W., and Young, L.S. (2008). Epstein-Barr virus-encoded EBNA1 modulates the AP-1 transcription factor pathway in nasopharyngeal carcinoma cells and enhances angiogenesis *in vitro*. J. Gen. Virol. 89, 2833-2842. http://dx.doi.org/10.1099/vir.0.2008/003392-0

Ohashi, M., Fogg, M.H., Orlova, N., Quink, C., and Wang, F. (2012). An Epstein-Barr virus encoded inhibitor of Colony Stimulating Factor-1 signaling is an important determinant for acute and persistent EBV infection. PLoS Pathog. 8, e1003095. http://dx.doi.org/10.1371/journal.ppat.1003095

Old, L.J., Boyse, E.A., Oettgen, H.F., Harven, E.D., Geering, G., Williamson, B., and Clifford, P. (1966). Precipitating antibody in human serum to an antigen present in cultured burkitt's lymphoma cells. Proc. Natl. Acad. Sci. U. S. A. 56, 1699-1704.

Or, Y.Y., Chung, G.T., To, K.F., Chow, C., Choy, K.W., Tong, C.Y., Leung, A.W., Hui, A.B., Tsao, S.W., Ng, H.K., *et al.* (2010). Identification of a novel 12p13.3 amplicon in nasopharyngeal carcinoma. J. Pathol. 220, 97-107. http://dx.doi.org/10.1002/path.2609

Or, Y.Y., Hui, A.B., To, K.F., Lam, C.N., and Lo, K.W. (2006). PIK3CA mutations in nasopharyngeal carcinoma. Int. J. Cancer 118, 1065-1067. http://dx.doi.org/10.1002/ijc.21444

Palser, A.L., Grayson, N.E., White, R.E., Corton, C., Correia, S., Ba Abdullah, M.M., Watson, S.J., Cotten, M., Arrand, J.R., Murray, P.G., *et al.* (2015). Genome diversity of Epstein-Barr virus from multiple tumor types and normal infection. J. Virol. 89, 5222-5237. http://dx.doi.org/10.1128/JVI.03614-14

Pathmanathan, R., Prasad, U., Sadler, R., Flynn, K., and Raab-Traub, N. (1995). Clonal proliferations of cells infected with Epstein-Barr virus in preinvasive lesions related to nasopharyngeal carcinoma. N. Engl. J. Med. 333, 693-698. http://dx.doi.org/10.1056/NEJM199509143331103

Pegtel, D.M., Middeldorp, J., and Thorley-Lawson, D.A. (2004). Epstein-Barr virus infection in *ex vivo* tonsil epithelial cell cultures of asymptomatic carriers. J. Virol. 78, 12613-12624. http://dx.doi.org/10.1128/JVI.78.22.12613-12624.2004

Pfeffer, S., Zavolan, M., Grasser, F.A., Chien, M., Russo, J.J., Ju, J., John, B., Enright, A.J., Marks, D., Sander, C., *et al.* (2004). Identification of virus-encoded microRNAs. Science 304, 734-736. http://dx.doi.org/10.1126/science.1096781

Pope, J.H., Horne, M.K., and Wetters, E.J. (1969). Significance of a complement-fixing antigen associated with herpes-like virus and detected in the Raji cell line. Nature 222, 186-187.

Port, R.J., Pinheiro-Maia, S., Hu, C., Arrand, J.R., Wei, W., Young, L.S., and Dawson, C.W. (2013). Epstein-Barr virus induction of the Hedgehog signalling pathway imposes a stem cell phenotype on human epithelial cells. J. Pathol. 231, 367-377. http://dx.doi.org/10.1002/path.4245

Pratt, Z.L., Zhang, J., and Sugden, B. (2012). The latent membrane protein 1 (LMP1) oncogene of Epstein-Barr virus can simultaneously induce and inhibit apoptosis in B cells. J. Virol. 86, 4380-4393. http://dx.doi.org/10.1128/JVI.06966-11

Qiu, J., Smith, P., Leahy, L., and Thorley-Lawson, D.A. (2015). The Epstein-Barr virus encoded BART miRNAs potentiate tumor growth *in vivo*. PLoS Pathog. 11, e1004561. http://dx.doi.org/10.1371/journal.ppat.1004561

Qiu, J., and Thorley-Lawson, D.A. (2014). EBV microRNA BART 18-5p targets MAP3K2 to facilitate persistence *in vivo* by inhibiting viral replication in B cells. Proc. Natl. Acad. Sci. U. S. A. 111, 11157-11162. http://dx.doi.org/10.1073/pnas.1406136111

Raab-Traub, N. (2015). Nasopharyngeal Carcinoma: An Evolving Role for the Epstein-Barr Virus. Curr. Top. Microbiol. Immunol. 390, 339-363. http://dx.doi.org/10.1007/978-3-319-22822-8_14

Rancan, C., Schirrmann, L., Huls, C., Zeidler, R., and Moosmann, A. (2015). Latent Membrane Protein LMP2A Impairs Recognition of EBV-Infected Cells by

CD8+ T Cells. PLoS Pathog. 11, e1004906. http://dx.doi.org/10.1371/journal.ppat.1004906

Reedman, B.M., Klein, G., Pope, J.H., Walters, M.K., Hilgers, J., Singh, S., and Johansson, B. (1974). Epstein-Barr virus-associated complement-fixing and nuclear antigens in Burkitt lymphoma biopsies. Int. J. Cancer 13, 755-763.

Rickinson, A.B., Young, L.S., and Rowe, M. (1987). Influence of the Epstein-Barr virus nuclear antigen EBNA 2 on the growth phenotype of virus-transformed B cells. J. Virol. 61, 1310-1317.

Riley, K.J., Rabinowitz, G.S., Yario, T.A., Luna, J.M., Darnell, R.B., and Steitz, J.A. (2012). EBV and human microRNAs co-target oncogenic and apoptotic viral and human genes during latency. EMBO J. 31, 2207-2221. http://dx.doi.org/10.1038/emboj.2012.63

Rowe, M., Lear, A.L., Croom-Carter, D., Davies, A.H., and Rickinson, A.B. (1992). Three pathways of Epstein-Barr virus gene activation from EBNA1-positive latency in B lymphocytes. J. Virol. 66, 122-131.

Samanta, M., Iwakiri, D., Kanda, T., Imaizumi, T., and Takada, K. (2006). EB virus-encoded RNAs are recognized by RIG-I and activate signaling to induce type I IFN. EMBO J. 25, 4207-4214. http://dx.doi.org/10.1038/sj.emboj.7601314

Samanta, M., Iwakiri, D., and Takada, K. (2008). Epstein-Barr virus-encoded small RNA induces IL-10 through RIG-I-mediated IRF-3 signaling. Oncogene 27, 4150-4160. http://dx.doi.org/10.1038/onc.2008.75

Sample, J., Young, L., Martin, B., Chatman, T., Kieff, E., Rickinson, A., and Kieff, E. (1990). Epstein-Barr virus types 1 and 2 differ in their EBNA-3A, EBNA-3B, and EBNA-3C genes. J. Virol. 64, 4084-4092.

Santpere, G., Darre, F., Blanco, S., Alcami, A., Villoslada, P., Mar Alba, M., and Navarro, A. (2014). Genome-wide analysis of wild-type Epstein-Barr virus genomes derived from healthy individuals of the 1,000 Genomes Project. Genome Biol. Evol. 6, 846-860. http://dx.doi.org/10.1093/gbe/evu054

Sasaki, M.M., Skol, A.D., Bao, R., Rhodes, L.V., Chambers, R., Vokes, E.E., Cohen, E.E., and Onel, K. (2015). Integrated genomic analysis suggests MLL3 is a novel candidate susceptibility gene for familial nasopharyngeal carcinoma. Cancer Epidemiol. Biomarkers Prev. 24, 1222-1228. http://dx.doi.org/10.1158/1055-9965.EPI-15-0275

Scholle, F., Bendt, K.M., and Raab-Traub, N. (2000). Epstein-Barr virus LMP2A transforms epithelial cells, inhibits cell differentiation, and activates Akt. J. Virol. 74, 10681-10689.

Seto, E., Ooka, T., Middeldorp, J., and Takada, K. (2008). Reconstitution of nasopharyngeal carcinoma-type EBV infection induces tumorigenicity. Cancer Res. 68, 1030-1036. http://dx.doi.org/10.1158/0008-5472.CAN-07-5252

Seto, E., Yang, L., Middeldorp, J., Sheen, T.S., Chen, J.Y., Fukayama, M., Eizuru, Y., Ooka, T., and Takada, K. (2005). Epstein-Barr virus (EBV)-encoded BARF1 gene is expressed in nasopharyngeal carcinoma and EBV-associated gastric carcinoma tissues in the absence of lytic gene expression. J. Med. Virol. 76, 82-88. http://dx.doi.org/10.1002/jmv.20327

Shah, K.M., Stewart, S.E., Wei, W., Woodman, C.B., O'Neil, J.D., Dawson, C.W., and Young, L.S. (2009). The EBV-encoded latent membrane proteins, LMP2A and LMP2B, limit the actions of interferon by targeting interferon receptors for degradation. Oncogene 28, 3903-3914. http://dx.doi.org/10.1038/onc.2009.249

Shannon-Lowe, C., Adland, E., Bell, A.I., Delecluse, H.J., Rickinson, A.B., and Rowe, M. (2009). Features distinguishing Epstein-Barr virus infections of epithelial cells and B cells: viral genome expression, genome maintenance, and genome amplification. J. Virol. 83, 7749-7760. http://dx.doi.org/10.1128/JVI.00108-09

Shannon-Lowe, C., and Rowe, M. (2011). Epstein-Barr virus infection of polarized epithelial cells via the basolateral surface by memory B cell-mediated transfer infection. PLoS Pathog. 7, e1001338. http://dx.doi.org/10.1371/journal.ppat.1001338

Sheng, W., Decaussin, G., Ligout, A., Takada, K., and Ooka, T. (2003). Malignant transformation of Epstein-Barr virus-negative Akata cells by introduction of the BARF1 gene carried by Epstein-Barr virus. J. Virol. 77, 3859-3865.

Sheng, W., Decaussin, G., Sumner, S., and Ooka, T. (2001). N-terminal domain of BARF1 gene encoded by Epstein-Barr virus is essential for malignant transformation of rodent fibroblasts and activation of BCL-2. Oncogene 20, 1176-1185. http://dx.doi.org/10.1038/sj.onc.1204217

Shibata, D., Tokunaga, M., Uemura, Y., Sato, E., Tanaka, S., and Weiss, L.M. (1991). Association of Epstein-Barr virus with undifferentiated gastric carcinomas with intense lymphoid infiltration. Lymphoepithelioma-like carcinoma. Am. J. Pathol. 139, 469-474.

Shibata, D., and Weiss, L.M. (1992). Epstein-Barr virus-associated gastric adenocarcinoma. Am. J. Pathol. 140, 769-774.

Shim, A.H., Chang, R.A., Chen, X., Longnecker, R., and He, X. (2012). Multipronged attenuation of macrophage-colony stimulating factor signaling by Epstein-Barr virus BARF1. Proc. Natl. Acad. Sci. U. S. A. 109, 12962-12967. http://dx.doi.org/10.1073/pnas.1205309109

Shinozaki-Ushiku, A., Kunita, A., Isogai, M., Hibiya, T., Ushiku, T., Takada, K., and Fukayama, M. (2015). Profiling of Virus-Encoded MicroRNAs in Epstein-Barr Virus-Associated Gastric Carcinoma and Their Roles in Gastric Carcinogenesis. J. Virol. 89, 5581-5591. http://dx.doi.org/10.1128/JVI.03639-14

Sivachandran, N., Sarkari, F., and Frappier, L. (2008). Epstein-Barr nuclear antigen 1 contributes to nasopharyngeal carcinoma through disruption of PML nuclear bodies. PLoS Pathog. 4, e1000170. http://dx.doi.org/10.1371/journal.ppat.1000170

Sivachandran, N., Wang, X., and Frappier, L. (2012). Functions of the Epstein-Barr virus EBNA1 protein in viral reactivation and lytic infection. J. Virol. 86, 6146-6158. http://dx.doi.org/10.1128/JVI.00013-12

Stewart, S., Dawson, C.W., Takada, K., Curnow, J., Moody, C.A., Sixbey, J.W., and Young, L.S. (2004).

Epstein-Barr virus-encoded LMP2A regulates viral and cellular gene expression by modulation of the NF-kappaB transcription factor pathway. Proc. Natl. Acad. Sci. U. S. A. 101, 15730-15735. http://dx.doi.org/10.1073/pnas.0402135101

Takada, K. (2012). Role of EBER and BARF1 in nasopharyngeal carcinoma (NPC) tumorigenesis. Semin. Cancer Biol. 22, 162-165. http://dx.doi.org/10.1016/j.semcancer.2011.12.007

Tang, M., Lautenberger, J.A., Gao, X., Sezgin, E., Hendrickson, S.L., Troyer, J.L., David, V.A., Guan, L., McIntosh, C.E., Guo, X., *et al.* (2012). The principal genetic determinants for nasopharyngeal carcinoma in China involve the HLA class I antigen recognition groove. PLoS Genet 8, e1003103. http://dx.doi.org/10.1371/journal.pgen.1003103

Temple, R.M., Zhu, J., Budgeon, L., Christensen, N.D., Meyers, C., and Sample, C.E. (2014). Efficient replication of Epstein-Barr virus in stratified epithelium *in vitro*. Proc. Natl. Acad. Sci. U. S. A. 111, 16544-16549. http://dx.doi.org/10.1073/pnas.1400818111

Thompson, S., Messick, T., Schultz, D.C., Reichman, M., and Lieberman, P.M. (2010). Development of a high-throughput screen for inhibitors of Epstein-Barr virus EBNA1. J Biomol Screen 15, 1107-1115. http://dx.doi.org/10.1177/1087057110379154

Thorley-Lawson, D.A., and Gross, A. (2004). Persistence of the Epstein-Barr virus and the origins of associated lymphomas. N. Engl. J. Med. 350, 1328-1337. http://dx.doi.org/10.1056/NEJMra032015

Tian, W., Zhu, F.M., Wang, W.Y., Cai, J.H., Zhang, W., Li, L.X., Liu, K.L., Jin, H.K., and Wang, F. (2015). Sequence-based typing of HLA-A gene in 930 patients with nasopharyngeal carcinoma in Hunan province, southern China. Tissue Antigens 86, 15-20. http://dx.doi.org/10.1111/tan.12576

Tsai, C.L., Li, H.P., Lu, Y.J., Hsueh, C., Liang, Y., Chen, C.L., Tsao, S.W., Tse, K.P., Yu, J.S., and Chang, Y.S. (2006). Activation of DNA methyltransferase 1 by EBV LMP1 Involves c-Jun NH(2)-terminal kinase signaling. Cancer Res. 66, 11668-11676. http://dx.doi.org/10.1158/0008-5472.CAN-06-2194

Tsai, C.N., Tsai, C.L., Tse, K.P., Chang, H.Y., and Chang, Y.S. (2002). The Epstein-Barr virus oncogene product, latent membrane protein 1, induces the downregulation of E-cadherin gene expression via activation of DNA methyltransferases. Proc. Natl. Acad. Sci. U. S. A. 99, 10084-10089. http://dx.doi.org/10.1073/pnas.152059399

Tsai, M.H., Raykova, A., Klinke, O., Bernhardt, K., Gartner, K., Leung, C.S., Geletneky, K., Sertel, S., Munz, C., Feederle, R., *et al.* (2013). Spontaneous lytic replication and epitheliotropism define an Epstein-Barr virus strain found in carcinomas. Cell reports 5, 458-470. http://dx.doi.org/10.1016/j.celrep.2013.09.012

Tsang, C.M., Yip, Y.L., Lo, K.W., Deng, W., To, K.F., Hau, P.M., Lau, V.M., Takada, K., Lui, V.W., Lung, M.L., *et al.* (2012). Cyclin D1 overexpression supports stable EBV infection in nasopharyngeal epithelial cells. Proc. Natl. Acad. Sci. U. S. A. 109, E3473-3482. http://dx.doi.org/10.1073/pnas.1202637109

Tsao, S.W., Tramoutanis, G., Dawson, C.W., Lo, A.K., and Huang, D.P. (2002). The significance of LMP1 expression in nasopharyngeal carcinoma. Semin. Cancer Biol. 12, 473-487.

Tsao, S.W., Tsang, C.M., To, K.F., and Lo, K.W. (2015). The role of Epstein-Barr virus in epithelial malignancies. J. Pathol. 235, 323-333. http://dx.doi.org/10.1002/path.4448

Tse, K.P., Su, W.H., Chang, K.P., Tsang, N.M., Yu, C.J., Tang, P., See, L.C., Hsueh, C., Yang, M.L., Hao, S.P., *et al.* (2009). Genome-wide association study reveals multiple nasopharyngeal carcinoma-associated *loci* within the *HLA* region at chromosome 6p21.3. Am. J. Hum. Genet. 85, 194-203. http://dx.doi.org/10.1016/j.ajhg.2009.07.007

Tse, K.P., Su, W.H., Yang, M.L., Cheng, H.Y., Tsang, N.M., Chang, K.P., Hao, S.P., Yao Shugart, Y., and Chang, Y.S. (2011). A gender-specific association of CNV at 6p21.3 with NPC susceptibility. Hum. Mol. Genet. 20, 2889-2896. http://dx.doi.org/10.1093/hmg/ddr191

Tugizov, S.M., Berline, J.W., and Palefsky, J.M. (2003). Epstein-Barr virus infection of polarized tongue and nasopharyngeal epithelial cells. Nature Med. 9, 307-314. http://dx.doi.org/10.1038/nm830

Uchida, J., Yasui, T., Takaoka-Shichijo, Y., Muraoka, M., Kulwichit, W., Raab-Traub, N., and Kikutani, H. (1999). Mimicry of CD40 signals by Epstein-Barr virus LMP1 in B lymphocyte responses. Science 286, 300-303.

Valentine, R., Dawson, C.W., Hu, C., Shah, K.M., Owen, T.J., Date, K.L., Maia, S.P., Shao, J., Arrand, J.R., Young, L.S., *et al.* (2010). Epstein-Barr virus-encoded EBNA1 inhibits the canonical NF-kappaB pathway in carcinoma cells by inhibiting IKK phosphorylation. Mol. Cancer 9, 1. http://dx.doi.org/10.1186/1476-4598-9-1

Vereide, D.T., Seto, E., Chiu, Y.F., Hayes, M., Tagawa, T., Grundhoff, A., Hammerschmidt, W., and Sugden, B. (2014). Epstein-Barr virus maintains lymphomas via its miRNAs. Oncogene 33, 1258-1264. http://dx.doi.org/10.1038/onc.2013.71

Vockerodt, M., Yap, L.F., Shannon-Lowe, C., Curley, H., Wei, W., Vrzalikova, K., and Murray, P.G. (2015). The Epstein-Barr virus and the pathogenesis of lymphoma. J. Pathol. 235, 312-322. http://dx.doi.org/10.1002/path.4459

Wakisaka, N., Kondo, S., Yoshizaki, T., Murono, S., Furukawa, M., and Pagano, J.S. (2004). Epstein-Barr virus latent membrane protein 1 induces synthesis of hypoxia-inducible factor 1 alpha. Mol. Cell. Biol. 24, 5223-5234. http://dx.doi.org/10.1128/MCB.24.12.5223-5234.2004

Wang, D., Liebowitz, D., and Kieff, E. (1985). An EBV membrane protein expressed in immortalized lymphocytes transforms established rodent cells. Cell 43, 831-840.

Wang, H.B., Zhang, H., Zhang, J.P., Li, Y., Zhao, B., Feng, G.K., Du, Y., Xiong, D., Zhong, Q., Liu, W.L., *et al.* (2015). Neuropilin 1 is an entry factor that promotes EBV infection of nasopharyngeal epithelial cells.

Nature communications 6, 6240. http://dx.doi.org/10.1038/ncomms7240

Wang, Q., Tsao, S.W., Ooka, T., Nicholls, J.M., Cheung, H.W., Fu, S., Wong, Y.C., and Wang, X. (2006). Anti-apoptotic role of BARF1 in gastric cancer cells. Cancer Lett. 238, 90-103. http://dx.doi.org/10.1016/j.canlet.2005.06.023

Wang, S., Xiong, H., Yan, S., Wu, N., and Lu, Z. (2016). Identification and Characterization of Epstein-Barr Virus Genomes in Lung Carcinoma Biopsy Samples by Next-Generation Sequencing Technology. Sci. Rep. 6, 26156. http://dx.doi.org/10.1038/srep26156

Wang, X., Kenyon, W.J., Li, Q., Mullberg, J., and Hutt-Fletcher, L.M. (1998). Epstein-Barr virus uses different complexes of glycoproteins gH and gL to infect B lymphocytes and epithelial cells. J. Virol. 72, 5552-5558.

Wasil, L.R., Wei, L., Chang, C., Lan, L., and Shair, K.H. (2015). Regulation of DNA Damage Signaling and Cell Death Responses by Epstein-Barr Virus Latent Membrane Protein 1 (LMP1) and LMP2A in Nasopharyngeal Carcinoma Cells. J. Virol. 89, 7612-7624. http://dx.doi.org/10.1128/JVI.00958-15

Wei, M.X., de Turenne-Tessier, M., Decaussin, G., Benet, G., and Ooka, T. (1997). Establishment of a monkey kidney epithelial cell line with the BARF1 open reading frame from Epstein-Barr virus. Oncogene 14, 3073-3081. http://dx.doi.org/10.1038/sj.onc.1201128

Wei, M.X., Moulin, J.C., Decaussin, G., Berger, F., and Ooka, T. (1994). Expression and tumorigenicity of the Epstein-Barr virus BARF1 gene in human Louckes B-lymphocyte cell line. Cancer Res. 54, 1843-1848. http

Wei, M.X., and Ooka, T. (1989). A transforming function of the BARF1 gene encoded by Epstein-Barr virus. EMBO J. 8, 2897-2903.

Westhoff Smith, D., and Sugden, B. (2013). Potential cellular functions of Epstein-Barr Nuclear Antigen 1 (EBNA1) of Epstein-Barr Virus. Viruses 5, 226-240. http://dx.doi.org/10.3390/v5010226

Wolf, H., zur Hausen, H., and Becker, V. (1973). EB viral genomes in epithelial nasopharyngeal carcinoma cells. Nat. New Biol. 244, 245-247.

Wong, A.M., Kong, K.L., Tsang, J.W., Kwong, D.L., and Guan, X.Y. (2012). Profiling of Epstein-Barr virus-encoded microRNAs in nasopharyngeal carcinoma reveals potential biomarkers and oncomirs. Cancer 118, 698-710. http://dx.doi.org/10.1002/cncr.26309

Wood, V.H., O'Neil, J.D., Wei, W., Stewart, S.E., Dawson, C.W., and Young, L.S. (2007). Epstein-Barr virus-encoded EBNA1 regulates cellular gene transcription and modulates the STAT1 and TGFbeta signaling pathways. Oncogene 26, 4135-4147. http://dx.doi.org/10.1038/sj.onc.1210496

Xiao, J., Palefsky, J.M., Herrera, R., Berline, J., and Tugizov, S.M. (2008). The Epstein-Barr virus BMRF-2 protein facilitates virus attachment to oral epithelial cells. Virology 370, 430-442. http://dx.doi.org/10.1016/j.virol.2007.09.012

Xiao, L., Hu, Z.Y., Dong, X., Tan, Z., Li, W., Tang, M., Chen, L., Yang, L., Tao, Y., Jiang, Y., *et al.* (2014). Targeting Epstein-Barr virus oncoprotein LMP1-mediated glycolysis sensitizes nasopharyngeal carcinoma to radiation therapy. Oncogene 33, 4568-4578. http://dx.doi.org/10.1038/onc.2014.32

Xiong, D., Du, Y., Wang, H.B., Zhao, B., Zhang, H., Li, Y., Hu, L.J., Cao, J.Y., Zhong, Q., Liu, W.L., *et al.* (2015). Nonmuscle myosin heavy chain IIA mediates Epstein-Barr virus infection of nasopharyngeal epithelial cells. Proc Natl Acad Sci USA 112, 11036-11041. http://dx.doi.org/10.1073/pnas.1513359112

Yap, L.F., Velapasamy, S., Lee, H.M., Thavaraj, S., Rajadurai, P., Wei, W., Vrzalikova, K., Ibrahim, M.H., Khoo, A.S., Tsao, S.W., *et al.* (2015). Down-regulation of LPA receptor 5 contributes to aberrant LPA signalling in EBV-associated nasopharyngeal carcinoma. J. Pathol. 235, 456-465. http://dx.doi.org/10.1002/path.4460

Yates, J.L., Warren, N., and Sugden, B. (1985). Stable replication of plasmids derived from Epstein-Barr virus in various mammalian cells. Nature 313, 812-815.

Yoshizaki, T., Horikawa, T., Qing-Chun, R., Wakisaka, N., Takeshita, H., Sheen, T.S., Lee, S.Y., Sato, H., and Furukawa, M. (2001). Induction of interleukin-8 by Epstein-Barr virus latent membrane protein-1 and its correlation to angiogenesis in nasopharyngeal carcinoma. Clin. Cancer Res. 7, 1946-1951.

Young, L.S., Dawson, C.W., Clark, D., Rupani, H., Busson, P., Tursz, T., Johnson, A., and Rickinson, A.B. (1988). Epstein-Barr virus gene expression in nasopharyngeal carcinoma. J. Gen. Virol. 69 (Pt 5), 1051-1065. http://dx.doi.org/10.1099/0022-1317-69-5-1051

Young, L.S., Yap, L.F., and Murray, P.G. (2016). Epstein-Barr virus: more than 50 years old and still providing surprises. Nat. Rev. Cancer 16, 789-802. http://dx.doi.org/10.1038/nrc.2016.92

Yu, M.C., and Yuan, J.M. (2002). Epidemiology of nasopharyngeal carcinoma. Semin. Cancer Biol. 12, 421-429.

Yuan, L., Liu, Z.H., Lin, Z.R., Xu, L.H., Zhong, Q., and Zeng, M.S. (2014). Recurrent FGFR3-TACC3 fusion gene in nasopharyngeal carcinoma. Cancer Biol. Ther. 15, 1613-1621. http://dx.doi.org/10.4161/15384047.2014.961874

Zainal Ariffin, O., and Nor Saleha, I.T. (2011). National Cancer Registry Report 2007. Ministry of Health, Malaysia.

Zeng, M.S., Li, D.J., Liu, Q.L., Song, L.B., Li, M.Z., Zhang, R.H., Yu, X.J., Wang, H.M., Ernberg, I., and Zeng, Y.X. (2005). Genomic sequence analysis of Epstein-Barr virus strain GD1 from a nasopharyngeal carcinoma patient. J. Virol. 79, 15323-15330. http://dx.doi.org/10.1128/JVI.79.24.15323-15330.2005

Zeng, Y., Zhang, L.G., Li, H.Y., Jan, M.G., Zhang, Q., Wu, Y.C., Wang, Y.S., and Su, G.R. (1982). Serological mass survey for early detection of nasopharyngeal carcinoma in Wuzhou City, China. Int. J. Cancer 29, 139-141.

Zeng, Y., Zhang, L.G., Wu, Y.C., Huang, Y.S., Huang, N.Q., Li, J.Y., Wang, Y.B., Jiang, M.K., Fang, Z., and Meng, N.N. (1985). Prospective studies on nasopharyngeal carcinoma in Epstein-Barr virus IgA/

VCA antibody-positive persons in Wuzhou City, China. Int. J. Cancer 36, 545-547.

Zeng, Y., Zhong, J.M., Li, L.Y., Wang, P.Z., Tang, H., Ma, Y.R., Zhu, J.S., Pan, W.J., Liu, Y.X., Wei, Z.N., *et al.* (1983). Follow-up studies on Epstein-Barr virus IgA/VCA antibody-positive persons in Zangwu County, China. Intervirology 20, 190-194.

Zhang, C.X., Decaussin, G., Daillie, J., and Ooka, T. (1988). Altered expression of two Epstein-Barr virus early genes localized in BamHI-A in nonproducer Raji cells. J. Virol. 62, 1862-1869.

Zhao, M., Cai, H., Li, X., Zheng, H., Yang, X., Fang, W., Zhang, L., Wei, G., Li, M., Yao, K., *et al.* (2012). Further evidence for the existence of major susceptibility of nasopharyngeal carcinoma in the region near HLA-A locus in Southern Chinese. J. Transl. Med. 10, 57. http://dx.doi.org/10.1186/1479-5876-10-57

Zheng, H., Dai, W., Cheung, A.K., Ko, J.M., Kan, R., Wong, B.W., Leong, M.M., Deng, M., Kwok, T.C., Chan, J.Y., *et al.* (2016). Whole-exome sequencing identifies multiple loss-of-function mutations of NF-kappaB pathway regulators in nasopharyngeal carcinoma. Proc. Natl. Acad. Sci. U. S. A. http://dx.doi.org/10.1073/pnas.1607606113

Zhou, X., Matskova, L., Rathje, L.Z., Xiao, X., Gish, G., Werner, M., Ignatyev, I., Yu, N., Zhao, W., Tian, F., *et al.* (2014). SYK interaction with ITGbeta4 suppressed by Epstein-Barr virus LMP2A modulates migration and invasion of nasopharyngeal carcinoma cells. Oncogene. http://dx.doi.org/10.1038/onc.2014.380

Zimber, U., Adldinger, H.K., Lenoir, G.M., Vuillaume, M., Knebel-Doeberitz, M.V., Laux, G., Desgranges, C., Wittmann, P., Freese, U.K., Schneider, U., *et al.* (1986). Geographical prevalence of two types of Epstein-Barr virus. Virology 154, 56-66.

8

The Development of Prophylactic and Therapeutic Vaccines for Epstein-Barr Virus and its Associated Malignancies

Alexander C. Dowell and Graham S. Taylor*

Institute of Immunology and Immunotherapy, University of Birmingham, Edgbaston, Birmingham B15 2TT, UK

**Corresponding author: g.s.taylor@bham.ac.uk*

Abstract

Epstein-Barr virus (EBV), a member of the herpesvirus family, infects 90-95% of the world's population. Primary infection is usually silent but in some cases is associated with acute pathology that usually self-resolves. Following infection, EBV is carried for life as latent infection in memory B-cells with viral reactivation in the oropharynx generating infectious progeny for spread to new hosts. Although life-long infection with EBV is usually asymptomatic, the virus is associated with a range of lymphoid and epithelial cancers that total over 200,000 cases each year worldwide. Evidence that EBV infection is linked to autoimmune disorders, particularly multiple sclerosis, is also accumulating. Targeting the virus, by prophylactic or therapeutic vaccination, represents a rational therapeutic strategy. This chapter focuses on the development of such vaccines, drawing parallels with other herpesviruses and other oncogenic viruses. The current status of EBV vaccine candidates and the signals of efficacy generated by laboratory studies and clinical trials are described.

Introduction

Derived from the Greek word herpein ("to creep") in reference to their ability to establish lifelong latent infection in animals as well as humans, the Herpesviridae are a large group of double stranded DNA viruses. In total, eight herpesviruses infect humans: herpes simplex viruses (HSV) 1 and 2, varicella-zoster virus (VZV), human cytomegalovirus (CMV), human herpesvirus 6, human herpesvirus 7, Kaposi's sarcoma-associated herpesvirus and Epstein-Barr virus (EBV) (Arvin *et al.*, 2007). Here we will discuss vaccination against these viruses in reference to development of a prophylactic vaccine and the prospects for the development of a therapeutic vaccine to treat the thousands of cases of EBV-associated cancer that occur each year worldwide (de Martel *et al.*, 2012).

EBV has co-evolved with our species for millennia and is a highly successful pathogen infecting 90-95% of people worldwide (de-The *et al.*, 1975). The virus is acquired orally and undergoes lytic cycle replication within epithelial and B-cells in the oropharynx, with some 80 viral genes expressed in a temporally regulated manner to produce progeny virus to help establish the infection within the host and to allow horizontal transmission to new hosts (McKenzie and El-Guindy, 2015). EBV also drives the proliferation of infected B-cells through the expression of eight latent cycle proteins. In some of these cells the virus downregulates its growth transforming program to establish life-long persistence in a small proportion of the host's resting memory B-cells in the absence of detectable viral gene expression (Hochberg *et al.*, 2004). These latently-infected B-cells occasionally divide in response to homeostatic cues whereupon expression of the EBV genome maintenance protein

EBNA1 occurs, presumably to partition the viral genome into both daughter cells (Hochberg *et al.*, 2004). Over time, some of the latently infected B-cells reactivate in the oropharynx (Laichalk and Thorley-Lawson, 2005), entering lytic cycle to produce infectious virions for transmission to new hosts (Hadinoto *et al.*, 2009).

Infection with EBV typically occurs during early childhood and in this circumstance is usually asymptomatic. In developed nations, however, the time of first encounter and infection has shifted to adolescence or early adulthood. In such cases there is significant risk of developing infectious mononucleosis (IM – also known as glandular fever) with symptoms of pharyngitis, fever, lymphoadenopathy and malaise. Mononucleosis-like illnesses can have multiple causes but EBV is by far the most common, accounting for over 90% of cases (Hurt and Tammaro, 2007). A recent prospective study of college students aged 18-22 in the United States reports that in this population 75% of EBV infections cause IM (Balfour *et al.*, 2013). Although IM is usually self-limiting, some 10% of people have fatigue lasting longer than 6 months and 1% have severe complications that may comprise neurological, bone marrow or liver disease (Jenson, 2000). Although there is a clear association between the age of EBV infection and risk of developing IM, the reasons why IM is more common in later life and why some people develop it upon EBV infection, and others of similar age do not, are still unknown.

The immune response to EBV infection

Infection with EBV provokes a diverse range of immune responses. Studies in IM patients have allowed the early innate and adaptive immune responses to EBV to be defined, albeit in a pathological setting. During IM the frequency and absolute number of NK cells in the blood increase (Tomkinson *et al.*, 1987) although reports differ as to whether this increase is correlated or inversely correlated with IM symptomology (Balfour *et al.*, 2013; Williams *et al.*, 2005). Recent work in a humanised mouse model of EBV infection supports a role for NK cells in limiting EBV infection (Chijioke *et al.*, 2016). Far greater expansions of CD8+ T-cells occur in IM, with studies using MHC-I tetramers showing that responses to individual EBV T-cell epitopes may comprise up to 50% of the expanded T-cell population (Hislop *et al.*, 2007). It is worth noting that although most of our understanding of the immune response in early infection comes from IM, a small number of studies have been able to study asymptomatic EBV infection. Expansion of EBV specific CD8 T cells also occurs in such cases but without perturbation of the total CD8 T cell number, suggesting that the drastic expansion of CD8 T cells seen in IM is characteristic of the pathological state rather than necessary for response and control of the virus *per se* (Jayasooriya *et al.*, 2015; Silins *et al.*, 2001).

The EBV-specific CD8+ T-cell response shows a marked hierarchy, focussing predominantly on responses to antigens expressed early in lytic infection with smaller responses to proteins expressed later in the cycle (Pudney *et al.*, 2005; Steven *et al.*, 1997). This pattern may be due to increasing expression of immune evasion proteins during EBV lytic cycle in turn reducing the ability of cells to efficiently present antigen (Quinn *et al.*, 2014). CD8+ T-cell responses to latent cycle proteins can also be detected during IM although these are numerically much smaller compared to the lytic cycle responses (Hislop *et al.*, 2007). More recent work has explored the CD4+ T-cell response and although there is no absolute increase in CD4+ T-cell numbers during IM, EBV-specific CD4+ T-cell responses may reach up to 1% of the CD4+ T-cell pool (Long *et al.*, 2013). Compared to the CD8+ T-cell response the specificity of the CD4+ T-cell response is different. Latent antigen responses are stronger than lytic antigen responses and the latter are directed against a broader range of proteins expressed at all stages of lytic cycle (Taylor *et al.*, 2015).

Patients with EBV also develop antibody responses to a range of EBV proteins and these can be used to confirm EBV as the etiological agent of IM symptoms and the stage of the infection (De Paschale and Clerici, 2012). Thus, the initial antibody response is mounted to viral capsid antigen (VCA) with rapid production of IgM antibodies, which decline over several weeks, and IgG antibodies which persist for life. An IgG antibody response to the EBNA1 protein also develops but only three to six months after infection (Odumade *et al.*, 2011). Testing for these three antibodies (VCA IgM, VCA IgG, EBNA1 IgG) is sufficient to define the stage of infection in the large majority of cases (De Paschale and Clerici, 2012).

Following EBV infection antibodies to other EBV proteins also develop, some of which have neutralising activity. Most of the neutralising activity present in serum is directed against gp350, the most abundant of the EBV surface glycoproteins (Johannsen *et al.*, 2004). The gp350 protein binds to CD21 on the surface of B-cells enabling attachment and internalisation of the virus. Loss of gp350 greatly reduces EBV infection of B-cells although infection of epithelial cells is unaffected (Janz *et al.*, 2000). The IgG antibody response to gp350 develops as early as

Table 8.1. Estimated new cases of EBV-associated cancers worldwide per year.

Cancer	Number of cases	Number of cases attributable to EBV
Burkitt lymphoma		
- Developed countries	400	100
- Less-developed countries	7800	6600
Hodgkin lymphoma	62,400	28,600
Diffuse Large B cell Lymphoma	147,691*	14,769**
Gastric carcinoma	933,900	84,050
Nasopharyngeal carcinoma	80,000	78,100
Total		212,219

Table adapted from Cohen et al. (2011). *Estimated incidence from IARC Globocan 2012 dataset assuming 40% of NHL are DLBCL. **Estimated assuming 10% of DLBCL are EBV-positive.

three weeks post infection, peaking between three and twelve months although there is no association between neutralizing antibody titres and symptoms of IM (Bu *et al.*, 2016). Antibodies to other EBV glycoproteins, including gp42, gH, gL and BMRF2, also exhibit neutralising activity (Li *et al.*, 1995; Tugizov *et al.*, 2003).

The worldwide burden of EBV-associated cancer

Although EBV is carried by most people with no long-term consequences, the vital role of the EBV-specific cellular immune response in maintaining this state is clearly demonstrated by post-transplant lymphoproliferative disease, where iatrogenic immunosuppression permits the development of EBV driven lymphomas (Dierickx *et al.*, 2015). These lymphomas are highly responsive to adoptively transferred EBV-specific T-cell preparations which, in effect, restore immune control over the virus (Heslop *et al.*, 2010). A large body of evidence has shown EBV is linked to several other malignancies leading to its classification in 1997 as a group I carcinogen by the International Agency for Research on Cancer (Bouvard *et al.*, 2009). Given its potent ability to transform B-cells *in vitro*, it is perhaps no surprise that EBV is linked to several lymphomas of B-cell origin. The virus was first discovered in a biopsy sample from an African child with Burkitt lymphoma (Epstein *et al.*, 1964) and almost all cases of the endemic form of Burkitt lymphoma are EBV positive. Subsequent research identified the virus in one third of classical Hodgkin Lymphoma cases and, more recently, in 10% of diffuse large B-cell lymphoma cases (Oyama *et al.*, 2003; Oyama *et al.*, 2007).

EBV is also associated with several non-B-cell malignancies. The greatest burden of morbidity and mortality arise from two epithelial malignancies: nasopharyngeal carcinoma (NPC) and gastric cancer. Almost all cases of NPC, some 80,000 cases each year, are EBV positive (Bouvard *et al.*, 2009). Although only 10% of gastric cancers are EBV positive, the higher prevalence of this disease means that an estimated 80,000 cases each year are also EBV positive. Finally, in rare circumstances EBV can infect other cell lineages as demonstrated by extranodal NK/T-cell lymphoma, a rare but highly aggressive tumour with an NK or T-cell phenotype, with all such cases EBV-positive (Fox *et al.*, 2011). In total, taking into account both B-cell and non-B-cell malignancies an estimated 200,000 cases of EBV-associated cancer occur each year worldwide (see Table 8.1). The roles of EBV in these various malignancies are discussed elsewhere in the volume.

Several observations support the concept that EBV is a driver of oncogenesis. For example, the virus has potent growth transforming activity *in vitro*, a property harnessed by many laboratories world-wide to produce immortalised lymphoblastoid cell lines, replicated *in vivo* in post-transplant lymphoma patients. In nasopharyngeal carcinoma the tumour cells carry monoclonal viral genomes which is consistent with EBV infection occurring before the establishment of the malignant cell clone (Raab-Traub and Flynn, 1986). Compared to their virus-negative counterparts, EBV-associated gastric cancer and DLBCL have distinctive patterns of driver mutations (Cancer Genome Atlas Research, 2014; Gebauer *et al.*, 2015). The corollary of EBV playing an essential role in the formation of these cancers is that preventing viral infection would prevent cancer development. Proof of principle for this concept is provided by two other human oncoviruses, hepatitis B virus and human papillomavirus. The introduction of the hepatitis B vaccine has substantially reduced the incidence of liver cirrhosis and hepatocellular carcinoma, even in regions where perinatal infection occurs (Gust, 1996; Orlando *et al.*, 2015). The more recently introduced human papillomavirus vaccine has resulted in decreased rates of cervical abnormalities and a concomitant reduction of cervical cancer rates is anticipated (Lowy and Schiller, 2012). These two prophylactic vaccines are discussed elsewhere in this volume.

Development of an EBV vaccine - lessons from the human herpesviruses

Preventing EBV-associated cancers by prophylactic vaccination is, of course, entirely dependent on the development of a suitable vaccine. In this regard it is useful to consider the development of vaccines to prevent infection with other human herpesviruses. At present only one licensed herpesvirus vaccine exists, a live attenuated vaccine against varicella zoster virus (VZV), developed in 1974 by serial passage *in vitro* of the Oka strain of VZV (Takahashi *et al.*, 1974). Demonstration of efficacy in several small trials on leukemic children, in whom VZV infection can be fatal, eventually led to licencing of the vaccine for universal immunization of children in the United States in 1995 (Arvin and Gershon, 1996). Because attenuated vaccines replicate in their host, the immunising dose was thought not to matter and a single vaccination was used. Experience from breakthrough outbreaks showed the vaccine to be only 76% effective as a single dose, resulting in refinement of the vaccination schedule to provide a second dose resulting in near complete protection. Real world data now shows a reduction of ~80% in the incidence of VZV in the vaccinated population. Incidence in unvaccinated people has also decreased presumably via herd immunity (Gershon and Katz, 2008). Importantly there has also been an associated drop in hospitalisation (88%) and mortality (66%). In older adults, waning of VZV-specific cellular immunity may allow reactivation of latent VZV to occur causing herpes zoster (shingles) which can significantly reduce quality of life (Burke *et al.*, 1982; Dworkin and Schmader, 2003). Administering the attenuated VZV vaccine to older adults, using a dose 10-fold greater than that given to children, has been shown to boost VZV cellular immunity and in a Phase III trial reduced the incidence of herpes zoster by half (Oxman *et al.*, 2005). The high dose Oka strain vaccine is licensed in the United States and Europe for the prevention of herpes zoster in older adults (Ansaldi *et al.*, 2016). It is important to note that the Oka VZV vaccine does not produce sterilising immunity (Gershon, 2013). Nevertheless, the vaccine's ability to reduce symptomatic infection in children and reawaken immunity in the elderly has made it a valuable part of the prophylactic vaccine armamentarium for many years.

An alternative VZV vaccine, HZ/su, has recently been tested in clinical trials. This subunit vaccine consists of the most abundant VZV surface glycoprotein in combination with GlaxoSmithKline's AS01B adjuvant which can stimulate both antibody and T-cell responses (Cunningham, 2016). In placebo controlled trials HZ/su showed much higher levels of protection against herpes zoster in adults older than 50 years of age compared to the historic level reported for Oka (Lal *et al.*, 2015). In contrast to Oka, which confers lower levels of herpes zoster protection with advancing age, the level of protection achieved by HZ/su was the same for all ages (in the 50-70 year range) that were recruited to the clinical trials (Lal *et al.*, 2015). Although HZ/Su confers higher levels of protection, at least over the three year follow up period of the trials, it also caused frequent injections site reactions and systemic reactogenicity (Lal *et al.*, 2015). Several key questions remain to be answered regarding the HZ/su vaccine. Is protection maintained beyond three years? Do breakthrough cases develop and, if so, are these reduced in severity? How does protection by HZ/su compare to the attenuated vaccine? The latter question is currently being investigated by an on-going Phase I trial (NCT 02114333) that is directly comparing the two vaccines. Regardless of the outcome of this study, a key advantage of the subunit vaccine will be its safety in severely immunocompromised patients who are at particular risk of primary VZV infection or herpes zoster reactivation and in whom the attenuated Oka vaccine is contraindicated.

Although a licensed HSV vaccine is currently unavailable, significant progress has been made towards this goal. HSV infection is widespread in the population and, as is the case with all herpesviruses, persists for the lifetime of the host (Arvin *et al.*, 2007). HSV-1 is commonly thought to cause orolabial herpes although it is increasingly responsible for genital herpes. HSV-2 is sexually transmitted, the predominant cause of genital ulcers, and can cause serious infections of neonates that become exposed to the virus during birth. HSV-2 infection also increases the risk of human immunodeficiency virus (HIV) transmission between sexual partners, with some estimates reporting up to 8-fold increase in risk (Buve, 2010). The seroprevalence of HSV-2 ranges from 16% in the United States to over 80% in sub-Saharan Africa. In some populations prevalence is almost 100% suggesting susceptibility to infection is universal. Current HSV treatment options are limited; antivirals are commonly used in the developed world, however, HSV seroprevalance rates in the United States have changed little since their introduction, despite widespread access to antivirals in this country. As such antivirals are useful only on a case-by-case basis and are not a global solution due to availability and compliance (Johnston *et al.*, 2016).

An effective HSV-2 vaccine would therefore make a significant impact and a variety of different strategies have been pursued. The candidates include live attenuated, whole inactivated, and subunit vaccines, which are reviewed in detail elsewhere (Johnston *et al.*, 2016). Subunit approaches have tended to focus on glycoprotein D-2 (gD2), the major target of the HSV-specific neutralising antibody response. Phase III trials in HSV discordant couples have shown that gD2 in combination with the AS04 adjuvant shows some protection against infection in seronegative women, but not in men or HSV-1 seropositive women (Stanberry *et al.*, 2002). A subsequent double blind randomised efficacy trial performed in 8323 HSV-1 and HSV-2 seronegative women subsequently showed the vaccine reduced HSV-1 infection and genital ulcers but failed to prevent infection with HSV-2 (Belshe *et al.*, 2012). Interestingly, protection against HSV-1 infection correlated with anti-gD2 antibody titres but not with cellular immune responses (Belshe *et al.*, 2012). A more recent vaccine candidate, GEN-003, combines gD2 with the HSV ICP4 protein, a target of CD8+ T-cell responses in HSV-2 exposed or infected individuals (Skoberne *et al.*, 2013). In a phase II placebo controlled trial in seropositive individuals, GEN-003 reduced HSV shedding by 58% and lesions by up to 69% (Fife *et al.*, 2016). To put this result into context it has been estimated in mathematical models that reducing HSV shedding by 75% could result in a 30% decrease in incidence of new infections over 10 years (Alsallaq *et al.*, 2010).

Development of an EBV prophylactic vaccine – experience from clinical trials.

The development pathway of a prophylactic EBV vaccine has, as has been the case for VZV and HSV, largely focussed on virion glycoproteins. The identification of the EBV gp350 protein as a neutralising antigen led to vaccine studies in cottontop tamarins (*Saguinus oedipus oedipus*), old world monkeys that develop lymphomas following the administration of high doses of EBV (Cleary *et al.*, 1985). Vaccination with purified gp350, recombinant gp350-expressing adenovirus or recombinant gp350-expressing vaccinia virus protected animals against EBV challenge (Epstein *et al.*, 1985; Morgan *et al.*, 1988a; Morgan *et al.*, 1988b; Ragot *et al.*, 1993). It is interesting to note that the gp350-expressing vaccinia virus protected animals from infection in the absence of a detectable gp350-specific antibody response suggesting the cell-mediated immunity may by itself confer protection (Morgan *et al.*, 1988b). The tamarin model, however, uses high doses of EBV administered parenterally which may render the virus more susceptible to cellular immunity. Although this is not the natural route of EBV infection in humans, it does resemble infection of EBV naïve solid organ transplant recipients by virus present in the transplanted graft. Such patients are at significant risk of developing post-transplant lymphoproliferative disease and establishing protection in this setting may depend on inducing appropriate cellular immunity, rather than humoural immunity.

The first human trial of a prophylactic EBV vaccine was a small Phase I trial in EBV seronegative children in China using the Tien Tan strain of vaccinia virus that had been genetically engineered to express gp350 (Gu *et al.*, 1995). Although a similar virus had not induced antibodies in the tamarin model, all children in the trial made a gp350-specific antibody response. Sixteen months after vaccination, all ten of the unvaccinated children enrolled on the trial had become infected with EBV, whereas only three of the nine vaccinated children had been infected (Gu *et al.*, 1995). No further trials of this vaccine have taken place despite this preliminary evidence of protection.

Three other gp350 vaccine trials in healthy volunteers have been reported. The immunogen in all three cases was gp350 protein overexpressed and purified from Chinese Hamster Ovary cells. Two initial Phase I and Phase I/II trials were performed in 148 young adults, regardless of EBV sero-status, to compare the immunogenicity of purified gp350 protein either alone or in combination with different adjuvants (Moutschen *et al.*, 2007). Although all patients had an antibody response to vaccination, the use of alum or AS04 adjuvant significantly increased the titre of anti-gp350 antibodies. The AS04 adjuvant also yielded higher levels of neutralising antibodies and this formulation was selected for further development (Moutschen *et al.*, 2007). Vaccine-induced T-cell responses, measured using regression assays, showed increases in T-cell activity in up to 41% of patients, with AS04 again being the most immunogenic adjuvant (Moutschen *et al.*, 2007). These results led to a subsequent Phase II randomised double blind placebo-controlled trial of the gp350/AS04 formulation in 181 EBV seronegative healthy young adults (Sokal *et al.*, 2007). Following three doses of vaccine, at 0, 1 and 6 months, anti-gp350 antibodies were detected in 98% of vaccinees with titres five-fold higher than those measured in people in the placebo group who became infected with EBV. The efficacy of the vaccine, in terms of reducing IM, was 78% although there was no apparent decrease in the rate of EBV infection in the vaccine group compared to the placebo. It should be noted that the two cases of IM that were detected in the vaccine group both occurred prior to the third dose of vaccine being given and it is therefore possible that full vaccine-mediated protection had not yet been established. Optimising the vaccination schedule to speed up the development of a protective immune response would reduce the time period in which people are susceptible to EBV infection after starting their course of vaccinations, thereby enhancing the efficacy of the vaccine.

The gp350-based vaccine strategy has also been tested in a cohort of children awaiting kidney transplantation (Rees *et al.*, 2009). The aim here was to reduce the risk of post-transplant lymphoma, which is particularly high for EBV-naïve recipients, typically children, who become infected in the peritransplant period (Sampaio *et al.*, 2012). Infection can occur via the oral route but, as described earlier, the donor organ itself can also act as the source of infection if it is from an EBV-positive donor. Regardless of the route of infection, the lack of pre-existing host immunity and use of iatrogenic immunosuppression generally allows a high-level of infection of the recipient's B-cell pool to occur. The Phase I trial vaccinated 13 EBV-seronegative children with up to four cycles of purified gp350 at a lower dose than that used in the healthy volunteer trials (Moutschen *et al.*, 2007; Sokal *et al.*, 2007). All children made an anti-gp350 antibody response after vaccination but after the third vaccination only four had developed a neutralising antibody response. One patient still had low levels of antibodies after three cycles and was therefore eligible to receive a fourth vaccination. This stimulated a rapid and substantial increases in their anti-gp350 antibody levels and a neutralising antibody response could also then be detected for the first time (Rees *et al.*, 2009). This result again suggests there may be scope to improve the immunogenicity of gp350 protein-based vaccines by optimising the vaccination regimen. Although the vaccine clearly stimulated antibody responses in this cohort of patients, there was no effect on levels of virus in the blood following transplantation and subsequent infection with EBV (Rees *et al.*, 2009). This may be due to several reasons. The vaccine-stimulated humoral response declined over time and may therefore not have been sufficient to afford protection by the time transplantation occurred. There may also be a greater need for T-cell responses to control infection following transplantation which allows the virus to bypass the mucosa; induction of gp350-specific T-cell responses was not assessed in the trial.

An alternative EBV vaccine strategy could be to induce an EBV latent antigen-specific T-cell response to prevent symptomatic IM upon EBV infection. This approach has been investigated in a randomised Phase I placebo controlled trial. The immunogen was a single chemically synthesised HLA-B8 restricted T-cell epitope peptide from the latent cycle protein EBNA3A in the water-in-oil adjuvant Montanide ISA720 (Elliott *et al.*, 2008). Using *ex vivo* ELIspot assays, T-cells specific for the vaccine peptide could be detected in eight of the nine EBV-seronegative young adults who were vaccinated. The EBV seroconversion rate was the same for the vaccine and placebo groups, although none of the four vaccinees who seroconverted developed IM compared with one of two individuals in the placebo group (Elliott *et al.*, 2008).

Development of an EBV prophylactic vaccine - Towards a second generation candidate.

The clinical trials described above provide clear proof of principle that it is possible to modulate the course of EBV infection and, as demonstrated by the decreased incidence of IM in the vaccinees, reduce the impact of primary EBV infection on the host. A history of IM increases the risk of developing EBV-associated Hodgkin lymphoma fourfold over the following decade (Hjalgrim *et al.*, 2003) and elevates the lifetime risk of developing multiple sclerosis by two to three-fold (Pakpoor *et al.*, 2013). If IM is the cause of these increased risks, rather than a surrogate marker of increased susceptibility to these conditions, then preventing IM could prevent these long-term sequaleae. A vaccine that did not confer complete protection, but nevertheless prevented symptomatic IM, could be beneficial in this respect. A study in rhesus macaques suggests that such a vaccine might also decrease the risk of developing other EBV-associated malignancies by decreasing the level of virus carriage in the B-cell system following infection with EBV (Sashihara *et al.*, 2011). A partially protective EBV vaccine could therefore be beneficial.

The development of a truly sterilising EBV vaccine would, however, unequivocally interrupt the pathological chain of events that lead to the development of an EBV-associated cancer. How could such a vaccine be achieved? One way could be to enhance the immunogenicity of soluble gp350 protein-based vaccines. Although the AS04 adjuvant used in the Phase II trial was more immunogenic than alum, other adjuvant systems such as AS01B, AS01E and AS03 show even greater potency. A Phase II trial using hepatitis B surface antigen as a model antigen showed these more potent adjuvants elicited 5-fold higher numbers of CD4+ T-cells and 10- to 100-fold higher levels of antibody compared to levels seen using AS04 (Leroux-Roels *et al.*, 2016). It would not be unreasonable to expect similar increases in immunogenicity to extend to a gp350 subunit vaccine.

An alternative, and potentially complementary approach, is to increase immunogenicity of the gp350 protein itself. It has long been recognised that presenting multiple copies of an antigen in an organised particulate form can substantially enhance immunogenicity and a large number of polymeric presentation systems have been developed to exploit this effect (Plummer and Manchester, 2011). A number of groups have used this strategy to enhance the immunogenicity of EBV antigens. The gp350 protein is a 907 amino acid long type I transmembrane protein and fusion of the ectodomain (amino acids 1-864) to a Newcastle disease virus protein produced virus like particles that retained the cell binding properties of gp350 and were capable of stimulating EBV neutralising antibodies in mice (Ogembo *et al.*, 2015). A smaller region of gp350 (amino acids 1-470) fused to a leucine zipper sequence was able to self-assemble into tetramers that, compared to the monmeric form of gp350, stimulated up to 20-fold higher titres of gp350-specific antibodies and neutralising antibodies in mice (Cui *et al.*, 2013). Furthermore, DNA encoding the gp350 tetramer was immunogenic in mice, eliciting eight-fold more antibodies than DNA encoding the monomer suggesting that an EBV vaccine based on a polymeric presentation system could be delivered via different formulations. Two fusion proteins, each consisting of the gp350 ectodomain (amino acids 1-425) joined to ferritin or encapsulin, were able to spontaneously self-assemble into nanoparticles of 20-50 nm diameter (Kanekiyo *et al.*, 2015). Both constructs were significantly more immunogenic in mice or macaques than soluble gp350. In both species the ferritin construct elicited slightly higher titres of antibody suggesting that the packing of the gp350 moiety on the nanoparticle surface may influence immunogenicity; understanding this effect may provide a way to further increase immunogenicity.

Finally, the polymeric presentation system strategy has now been extended to the gH, gL, and gB proteins that are also located on the outer surface of EBV, albeit at lower levels than gp350. Immunisation of rabbits with trimeric gH/gL and trimeric gB stimulated the production of higher titres of neutralising antibodies than the tetrameric gp350 protein that was described earlier (Cui *et al.*, 2016; Cui *et al.*, 2013). This is an intriguing observation, although whether it is a property of the polymeric presentation system *per se* or whether these alternative antigens are inherently more protective than gp350 remains to be seen. Finally, although an attenuated EBV vaccine is unlikely to ever be licensed due to the virus's oncogenicity, it may be possible to produce EBV virus like particles (VLPs) that are devoid of viral DNA. Such VLPs have been produced and are immunogenic in mice, stimulating broad humoral and cellular immune responses to structural proteins but also non-structural latent cycle proteins such as EBNA1 and LMP2 (Ruiss *et al.*, 2011). Taken together, the work outlined above strongly suggests that there is significant potential to boost the immunogenicity of candidate EBV prophylactic vaccines beyond the levels achieved by the soluble gp350 formulation used in the Phase II trial. Although no plans have yet been announced for a clinical trial of a second-generation prophylactic vaccine there is increasing interest once again in the possibility of eradicating EBV-associated cancers from the world (Cohen *et al.*, 2011).

Therapeutic vaccination to treat EBV-associated cancers

Even if a highly protective EBV prophylactic vaccine was deployed tomorrow, billions of people worldwide are already infected with EBV and therefore at risk of developing an EBV-associated cancer. There is therefore an urgent need to develop improved treatments for these diseases. The consistent expression of EBV-proteins in these tumour cells represents a therapeutic opportunity. The immune system is capable of exquisite sensitivity and selectivity and multiple clinical trials of immunotherapies targeting EBV proteins have been reported. Due to their tumour restricted expression, therapies targeting these proteins have the potential to be highly selective with minimal risk of off-target effects. Most work to date has focussed on adoptive T-cell therapy for the treatment of post-transplant lymphoma and here there is overwhelming evidence of both efficacy and safety (Heslop *et al.*, 2010). Several clinical trials of adoptive-T-cell transfer for other EBV-associated lymphomas and NPC have been reported with evidence of clinical responses in some patients, despite the more limited range of EBV proteins expressed in the tumour cells (Bollard *et al.*, 2014; Chia *et al.*, 2014; Comoli *et al.*, 2005; Louis *et al.*, 2010; Louis *et al.*, 2009; Secondino *et al.*, 2012; Smith *et al.*, 2012). Therefore, there is clear rationale to develop and implement T-cell based immunotherapies to treat EBV-associated cancers.

While adoptive T-cell therapy is a powerful technology for boosting tumour-specific T-cells in cancer patients, customised cell therapy approaches require an advanced healthcare infrastructure and highly trained staff. This may restrict their use to high resource countries and even here availability may be limited by the increasing cost pressures faced by healthcare providers (Smith and Hillner, 2011). Alternative ways to boost the numbers of therapeutically relevant T-cells in patients are therefore needed. In contrast to custom cell products, vaccines have a high fixed cost and a low marginal cost of production. This means that vaccines benefit from economies of scale; as more doses of a vaccine are made the cost per patient decreases. An affordable and therefore widely implemented therapeutic EBV cancer vaccine, capable of boosting appropriate components of the immune system, could make a global impact on the EBV-associated cancers in the short to medium term.

In many cancers the tumour antigen targets are self-encoded proteins and cognate T-cell responses are low avidity; high-avidity effectors are deleted by central tolerance during immune development. This is not the case for T-cells specific for the foreign EBV antigens present in the tumour cells of EBV-associated cancers (Haigh *et al.*, 2008; Khanna *et al.*, 1998; Khanna *et al.*, 1995; Kobayashi *et al.*, 2008; Lee *et al.*, 2004; Lee *et al.*, 2000; Lee *et al.*, 1993; Lee *et al.*, 1997; Nikiforow *et al.*, 2003; Paludan *et al.*, 2002; Straathof *et al.*, 2005). Therapeutic vaccination in this context therefore has the potential to elicit a more effective tumour-specific T-cell response. This is not to say that these diseases will be easy to treat. In contrast to post-transplant lymphoma, the other EBV-associated cancers usually develop in hosts that are not obviously immunosuppressed. The tumour cells and/or the tumour microenvironment are therefore likely to possess one or more immune escape mechanisms which may need to be countered in order for vaccine-elicited immune responses to yield clinical benefit.

NPC provides a useful setting in order to consider these issues as the tumours show evidence of a range of evasion mechanisms that can vary from patient to patient. In terms of general immune status, EBV-specific T-cell responses can be detected in NPC patients although some studies suggest that the frequency of these cells is reduced (Fogg *et al.*, 2013; Fogg *et al.*, 2009; Lee *et al.*, 2000; Li *et al.*, 2007; Lin *et al.*, 2008). The frequency of regulatory T-cells appears to be elevated in the blood of newly-diagnosed patients and those who have completed standard of care first-line chemoradiotherapy (Lau *et al.*, 2007; Taylor *et al.*, 2014). Focusing on the tumour itself, immunohistochemistry suggests that HLA-class I is downregulated or lost in up to one third of cases and HLA-G and galectin-9, which can inhibit immune cell function, are expressed in some cases (Cai *et al.*, 2012; Klibi *et al.*, 2009; Yao *et al.*, 2000). Indoleamine 2,3-dioxygenase, which can deplete tryptophan levels in tumours thereby suppressing T-cell function, is present in 75% of biopsies although it is currently unclear whether its presence indicates the presence of infiltrating myeloid derived suppressor cells or whether the enzyme is being produced by the tumour cells themselves (Ben-Haj-Ayed *et al.*, 2016). NPC tumours often contain a rich infiltrate of lymphoid cells, including T-cells, NK cells, B-cells and regulatory T-cells, which appear to be enriched compared with adjacent non-cancerous tissue (Gourzones *et al.*, 2012; Lau *et al.*, 2007; Yip *et al.*, 2009). Finally, the negative immune regulator programmed death ligand 1 (PD-L1) is expressed in up to 90% of NPC tumours, although it is currently unclear whether this expression stems from inherent immune resistance in which expression is driven by perturbed cell signalling pathways, or is a result of adaptive immune resistance where expression is a response to cytokines released by tumour-infiltrating immune effectors (Chen and Mellman, 2013; Fang *et al.*, 2014; Hsu *et al.*, 2015). Several clinical responses in a trial of pembrolizumab, an inhibitor of PD-1 (the receptor for PD-L1) , strongly suggest that, for some patients at least, effector cell function is blocked by PD-L1 (Hsu *et al.*, 2015). Using vaccination to boost the number of effectors, combined with PD-1/L1 inhibitors to allow these cells to function in the tumours, could be a way to increase the number of clinical responses.

At the time of writing, therapeutic vaccination to treat EBV-associated cancer has been tested only for NPC. The first vaccination clinical trial that was performed used autologous dendritic cells (DCs), generated *in vitro* from monocytes, which were loaded with LMP2-specific MHC-I restricted epitope peptides (Lin *et al.*, 2002). Before vaccination the frequency of LMP2-

specific effectors as measured by *ex vivo* ELIspot assays was low or undetectable for most patients. Of the 16 patients vaccinated, increased frequencies of LMP2-specific T-cell were observed in nine patients. Of the seven who lacked a detectable immune response, four were leukopenic when recruited to the study. Encouragingly, among the nine patients who had a vaccine-stimulated immune response two patients had partial clinical responses (Lin *et al.*, 2002). A similar trial administered LMP2 peptide-loaded DCs to 16 patients (Li *et al.*, 2013). Although clinical responses were not observed in this study, several patients experienced decreases in their serum EBV DNA levels, a surrogate marker of tumour burden. Interestingly, these decreases correlated with delayed type hypersensitivity reactions to the immunising peptides. Most immune studies test for immunogenicity in the blood yet only 2% of lymphocytes are in this compartment (Blum and Pabst, 2007); blood-based immunoassays may therefore underestimate the true immunogenicity of immune-based therapies.

An alternative source of antigen, a recombinant adenovirus encoding truncated LMP1 and full-length LMP2 has also been used to prepare DCs for clinical trial. This method has the advantage of allowing the DCs to process and present these antigens which means they can stimulate the widest possible range of CD8+ and, potentially, CD4+ T-cells in each patient, rather than a limited repertoire of the most common defined epitope peptides. Following vaccination of sixteen patients, one patient achieved a partial clinical response and two patients stable disease. Again none of the patients had a detectable increase of T-cell responses in the blood, as measured by *ex vivo* ELISpot assays, but nine patients exhibited delayed type hypersensitivity reactions to intradermally injected transduced DCs, again suggesting that measuring responses in the blood is likely to underestimate immunogenicity (Chia *et al.*, 2012).

The experience from the dendritic cell vaccine trials performed in NPC patients clearly show it is possible to boost EBV-specific T-cell responses in patients who have an EBV-associated cancer. DC based vaccines, like adoptive T-cell therapies, are custom cell products requiring specialised facilities and staff and therefore suffer from the same limitations of complexity and cost. An 'off-the-shelf' vaccine able to similarly boost immune responses, but which could also be manufactured to large scales to maximise economies of scale, would offer a significant advantage. With this concept in mind we produced such a vaccine based on the modified vaccinia Ankara vector (MVA). MVA was successfully administered to more than 120,000 people at the end of the smallpox eradication campaign, therefore there is strong data showing MVA is safe in the human population (Cottingham and Carroll, 2013). MVA has a wide tropism and readily infects human cells. It does not replicate in these cells but nevertheless efficiently expresses its transgene payload.

We designed the MVA-EBNA1/LMP2 vaccine to boost immunity to EBV proteins that are present in all EBV-associated malignancies. When administered to patients via intradermal injection, the most immunogenic route for MVA-based vaccines, it expresses a fusion protein comprising the carboxy-terminal half of the EBNA1 protein and full-length LMP2, the latter modified to remove its signalling function (Taylor *et al.*, 2004). This design removes the large internal glycine/alanine repeat domain from EBNA1 that would otherwise interfere with expression and antigen processing yet contains almost all of the known EBNA1 and all LMP2 MHC-I and MHC-II epitope peptides (Taylor *et al.*, 2015). Another feature of the construct is that the LMP2 portion of the fusion protein directs some of the EBNA1 protein into the HLA-II processing pathway thereby enhancing its visibility to the CD4+ T-cell system and potentially increasing the vaccine's immunogenicity (Taylor *et al.*, 2004). There is increasing evidence that CD4+ T-cells play important roles in tumour immunotherapy. First, although the patients are EBV carriers and therefore already possess EBV-specific T-cell responses it is possible that vaccination may induce new immune responses and in this situation CD4+ T-cell help is important for establishing effective T-cell memory. Second, during a recall response, which may be the more relevant situation in patients with EBV-specific malignancy who have some degree of pre-existing EBV-specific immunity, CD4+ T-cells help promote CD8+ T-cell survival. Third, in addition to providing essential helper functions CD4+ T-cells can act as direct effectors (Perez-Diez *et al.*, 2007; Quezada *et al.*, 2010), and in this respect it is interesting to note that the majority of post-transplant lymphoma patients who were infused with EBNA1-specific CD4 T-cells experienced clinical and virological responses (Icheva *et al.*, 2013).

Two dose escalation trials of MVA-EBNA1/LMP2, conducted in NPC patients in the United Kingdom and Hong Kong, have demonstrated that the vaccine is safe and well tolerated (Hui *et al.*, 2013; Taylor *et al.*, 2014). Side effects are predominantly low-grade injection site reactions and 'flu like' symptoms. In terms of its immunogenicity, pooled analysis of the two trials showed that of 27 patients for whom *ex vivo* ELIspot data was available, 18 and 12 had increases in

EBNA1- and LMP2-specific T-cells respectively (Hui *et al.*, 2013; Taylor *et al.*, 2014). These T-cell responses were observed in Chinese and Western patients and consisted of both CD8+ and CD4+ T-cells restricted by a range of common HLA-class I and class II alleles. The strength of the immune response in one patient was sufficient to allow analysis by intracellular flow cytometry (Taylor *et al.*, 2014) which although less sensitive than *ex vivo* ELIspot assays has the advantage of allowing the functional properties of the antigen-specific cells to be studied in depth. CD8+ T-cells are generally thought to act as cytotoxic effectors, releasing perforin and granzymes that lyse target cells, while CD4+ T-cells are thought to secrete cytokines that direct the immune response. As might therefore be expected, the LMP2-specific CD8+ T-cells and EBNA1-specific CD4+ T-cells boosted by vaccination possessed distinct functional properties. The former degranulated upon antigen stimulation *in vitro*, a surrogate marker of perforin-mediated cytotoxicity, and secreted interferon-gamma and tumour necrosis factor-alpha. By contrast, EBNA1-specific CD4+ T-cells did not degranulate but produced interleukin-2 in addition to the interferon-gamma and tumour necrosis factor-alpha produced by the CD8+ T-cells. Note that the lack of degranulation exhibited by the CD4+ T-cells does not necessarily mean they lacked cytotoxic activity since EBNA1-specific CD4+ T-cells can potentially utilise the FAS/Fas-Ligand pathway (Nikiforow *et al.*, 2003) which was not measured by the flow cytometric assay. These data also showed that, in addition to increasing the size of the immune response, vaccination also increased the number of effector functions possessed by the EBNA1- and LMP2-specific T-cells. Such increases in T-cell polyfunctionality are thought to indicate a better quality of immune response (Seder *et al.*, 2008). Importantly, there was no signs that the LMP2 and EBNA1-specific T-cells were being driven to terminally differentiate by the three cycles of vaccination the patient received. Taken together, these data support our strategy of harnessing both CD8+ and CD4+ T-cell subsets and have led to two further trials of the vaccine. In the UK, a Phase IB trial is examining vaccine immunogenicity in finer detail whereas in Hong Kong, a Phase II trial is measuring clinical responses in patients with persistent, recurrent or metastatic NPC.

Conclusions

EBV is an important human pathogen associated with over 200,000 cases of cancer each year. Experience from hepatitis B virus shows it is possible to reduce the incidence of virus-associated cancers through the development and implementation of a safe and effective prophylactic vaccine. The human papillomavirus vaccine has been shown to reduce the incidence of genital warts and is anticipated to also reduce the incidence of HPV-associated cancers over the longer term. Accordingly, there is again a growing desire to develop a prophylactic vaccine for EBV. Initial forays toward this goal have been encouraging, delivering a measure of protection against IM, although the goal of preventing infection was not achieved. Considerable progress has now been made towards second-generation prophylactic vaccine candidates that possess improved immunogenicity and, potentially, comprise a wider range of viral antigens. The consistent presence of EBV antigens in the cells of EBV-associated cancers provides a therapeutic opportunity that is currently being harnessed by a range of immunotherapies including therapeutic vaccines. There is considerable scope for further improvement in clinical outcomes by combining therapeutic vaccination with other agents that overcome immunosuppression within the tumour. The next few years are likely to see significant advances in prophylactic and therapeutic vaccination against EBV and its cancers.

References

Alsallaq, R.A., Schiffer, J.T., Longini, I.M., Jr., Wald, A., Corey, L., and Abu-Raddad, L.J. (2010). Population level impact of an imperfect prophylactic vaccine for herpes simplex virus-2. Sex. Trans. Dis. 37, 290-297.

Ansaldi, F., Trucchi, C., Alicino, C., Paganino, C., Orsi, A., and Icardi, G. (2016). Real-World Effectiveness and Safety of a Live-Attenuated Herpes Zoster Vaccine: A Comprehensive Review. Advances in therapy 33, 1094-1104.

Arvin, A., Campadelli-Fiume, G., Mocarski, E., P.S., M., Roizman, B., Whitley, R., and Yamanishi, K. (2007). Human Herpesviruses: Biology, Therapy, and Immunoprophylaxis (Cambridge University Press).

Arvin, A.M., and Gershon, A.A. (1996). Live attenuated varicella vaccine. Annual review of microbiology 50, 59-100.

Balfour, H.H., Odumade, O.A., Schmeling, D.O., Mullan, B.D., Ed, J.A., Knight, J.A., Vezina, H.E., Thomas, W., and Hogquist, K.A. (2013). Behavioral, Virologic, and Immunologic Factors Associated With Acquisition and Severity of Primary Epstein–Barr Virus Infection in University Students. J. Infect. Dis. 207, 80-88.

Belshe, R.B., Leone, P.A., Bernstein, D.I., Wald, A., Levin, M.J., Stapleton, J.T., Gorfinkel, I., Morrow, R.L., Ewell, M.G., Stokes-Riner, A., *et al.* (2012). Efficacy results of a trial of a herpes simplex vaccine. New Engl. J. Med. 366, 34-43.

Ben-Haj-Ayed, A., Moussa, A., Ghedira, R., Gabbouj, S., Miled, S., Bouzid, N., Tebra-Mrad, S., Bouaouina, N., Chouchane, L., Zakhama, A., *et al.* (2016). Prognostic value of indoleamine 2,3-dioxygenase activity and expression in nasopharyngeal carcinoma. Immunology letters 169, 23-32.

Blum, K.S., and Pabst, R. (2007). Lymphocyte numbers and subsets in the human blood. Do they mirror the situation in all organs? Immunology letters 108, 45-51.

Bollard, C.M., Gottschalk, S., Torrano, V., Diouf, O., Ku, S., Hazrat, Y., Carrum, G., Ramos, C., Fayad, L., Shpall, E.J., *et al.* (2014). Sustained complete responses in patients with lymphoma receiving autologous cytotoxic T lymphocytes targeting Epstein-Barr virus latent membrane proteins. J. Clin. Oncol.: Offical J. Am Soc. Clin. Oncol. 32, 798-808.

Bouvard, V., Baan, R., Straif, K., Grosse, Y., Secretan, B., El Ghissassi, F., Benbrahim-Tallaa, L., Guha, N., Freeman, C., Galichet, L., *et al.* (2009). A review of human carcinogens--Part B: biological agents. Lancet Oncol. 10, 321-322.

Bu, W., Hayes, G.M., Liu, H., Gemmell, L., Schmeling, D.O., Radecki, P., Aguilar, F., Burbelo, P.D., Woo, J., Balfour, H.H., Jr., *et al.* (2016). Kinetics of Epstein-Barr Virus (EBV) Neutralizing and Virus-Specific Antibodies after Primary Infection with EBV. Clin Vaccine Immunol 23, 363-369.

Burke, B.L., Steele, R.W., Beard, O.W., Wood, J.S., Cain, T.D., and Marmer, D.J. (1982). Immune responses to varicella-zoster in the aged. Archives of internal medicine 142, 291-293.

Buve, A. (2010). Can we reduce the spread of HIV infection by suppressing herpes simplex virus type 2 infection? F1000 medicine reports 2.

Cai, M.B., Han, H.Q., Bei, J.X., Liu, C.C., Lei, J.J., Cui, Q., Feng, Q.S., Wang, H.Y., Zhang, J.X., Liang, Y., *et al.* (2012). Expression of human leukocyte antigen G is associated with prognosis in nasopharyngeal carcinoma. International journal of biological sciences 8, 891-900.

Cancer Genome Atlas Research, N. (2014). Comprehensive molecular characterization of gastric adenocarcinoma. Nature 513, 202-209.

Chen, D.S., and Mellman, I. (2013). Oncology meets immunology: the cancer-immunity cycle. Immunity 39, 1-10.

Chia, W.K., Teo, M., Wang, W.W., Lee, B., Ang, S.F., Tai, W.M., Chee, C.L., Ng, J., Kan, R., Lim, W.T., *et al.* (2014). Adoptive T-cell transfer and chemotherapy in the first-line treatment of metastatic and/or locally recurrent nasopharyngeal carcinoma. Mol. Ther.: J. Am. Soc. Gene Ther. 22, 132-139.

Chia, W.K., Wang, W.W., Teo, M., Tai, W.M., Lim, W.T., Tan, E.H., Leong, S.S., Sun, L., Chen, J.J., Gottschalk, S., *et al.* (2012). A phase II study evaluating the safety and efficacy of an adenovirus-DeltaLMP1-LMP2 transduced dendritic cell vaccine in patients with advanced metastatic nasopharyngeal carcinoma. Annal Oncol. 23, 997-1005.

Chijioke, O., Landtwing, V., and Munz, C. (2016). NK Cell Influence on the Outcome of Primary Epstein-Barr Virus Infection. Frontiers in immunology 7, 323.

Cleary, M.L., Epstein, M.A., Finerty, S., Dorfman, R.F., Bornkamm, G.W., Kirkwood, J.K., Morgan, A.J., and Sklar, J. (1985). Individual tumors of multifocal EB virus-induced malignant lymphomas in tamarins arise from different B-cell clones. Science (New York, NY) 228, 722-724.

Cohen, J.I., Fauci, A.S., Varmus, H., and Nabel, G.J. (2011). Epstein-Barr virus: an important vaccine target for cancer prevention. Science translational medicine 3, 107fs107.

Comoli, P., Pedrazzoli, P., Maccario, R., Basso, S., Carminati, O., Labirio, M., Schiavo, R., Secondino, S., Frasson, C., Perotti, C., *et al.* (2005). Cell therapy of stage IV nasopharyngeal carcinoma with autologous Epstein-Barr virus-targeted cytotoxic T lymphocytes. J. Clin. Oncol.: Offical J. Am Soc. Clin. Oncol. 23, 8942-8949.

Cottingham, M.G., and Carroll, M.W. (2013). Recombinant MVA vaccines: dispelling the myths. Vaccine 31, 4247-4251.

Cui, X., Cao, Z., Chen, Q., Arjunaraja, S., Snow, A.L., and Snapper, C.M. (2016). Rabbits immunized with Epstein-Barr virus gH/gL or gB recombinant proteins elicit higher serum virus neutralizing activity than gp350. Vaccine 34, 4050-4055.

Cui, X., Cao, Z., Sen, G., Chattopadhyay, G., Fuller, D.H., Fuller, J.T., Snapper, D.M., Snow, A., Mond, J.J., and Snapper, C.M. (2013). A novel tetrameric gp350(1-470) as a potential Epstein-Barr virus vaccine. Vaccine 31, 3039-3045.

Cunningham, A.L. (2016). The herpes zoster subunit vaccine. Expert Opin. Biol. Ther. 16, 265-271.

de-The, G., Day, N.E., Geser, A., Lavoue, M.F., Ho, J.H., Simons, M.J., Sohier, R., Tukei, P., Vonka, V., and Zavadova, H. (1975). Sero-epidemiology of the Epstein-Barr virus: preliminary analysis of an international study - a review. IARC scientific publications, 3-16.

de Martel, C., Ferlay, J., Franceschi, S., Vignat, J., Bray, F., Forman, D., and Plummer, M. (2012). Global burden of cancers attributable to infections in 2008: a review and synthetic analysis. Lancet Oncol. 13, 607-615.

De Paschale, M., and Clerici, P. (2012). Serological diagnosis of Epstein-Barr virus infection: Problems and solutions. World J. Virol. 1, 31-43.

Dierickx, D., Tousseyn, T., and Gheysens, O. (2015). How I treat posttransplant lymphoproliferative disorders. Blood 126, 2274-2283.

Dworkin, R.H., and Schmader, K.E. (2003). Treatment and prevention of postherpetic neuralgia. Clin. Infect Dis. 36, 877-882.

Elliott, S.L., Suhrbier, A., Miles, J.J., Lawrence, G., Pye, S.J., Le, T.T., Rosenstengel, A., Nguyen, T., Allworth, A., Burrows, S.R., *et al.* (2008). Phase I Trial of a CD8(+) T-Cell Peptide Epitope-Based Vaccine for Infectious Mononucleosis. J. Virol. 82, 1448-1457.

Epstein, M.A., Achong, B.G., and Barr, Y.M. (1964). Virus particles in cultured lymphoblasts from Burkitt's lymphoma. Lancet (London, England) 1, 702-703.

Epstein, M.A., Morgan, A.J., Finerty, S., Randle, B.J., and Kirkwood, J.K. (1985). Protection of cottontop tamarins against Epstein-Barr virus-induced malignant lymphoma by a prototype subunit vaccine. Nature 318, 287-289.

Fang, W., Zhang, J., Hong, S., Zhan, J., Chen, N., Qin, T., Tang, Y., Zhang, Y., Kang, S., Zhou, T., *et al.* (2014).

EBV-driven LMP1 and IFN-gamma up-regulate PD-L1 in nasopharyngeal carcinoma: Implications for oncotargeted therapy. Oncotarget 5, 12189-12202.

Fife, K., Van Wagoner, N., Leone, P., Bernstein, D., Warren, T., Panther, L., Novak, R., Beigi, R., Kriesel, J., Tyring, S., *et al.* (2016). GEN-003, a Therapeutic Vaccine for Genital Herpes, Significantly Reduces Viral Shedding and Lesions for at Least 6 Months. In ASM Microbe 2016 (Boston, MA).

Fogg, M., Murphy, J.R., Lorch, J., Posner, M., and Wang, F. (2013). Therapeutic targeting of regulatory T cells enhances tumor-specific CD8+ T cell responses in Epstein-Barr virus associated nasopharyngeal carcinoma. Virology 441, 107-113.

Fogg, M.H., Wirth, L.J., Posner, M., and Wang, F. (2009). Decreased EBNA-1-specific CD8+ T cells in patients with Epstein-Barr virus-associated nasopharyngeal carcinoma. Proc Natl Acad Sci USA 106, 3318-3323.

Fox, C.P., Shannon-Lowe, C., and Rowe, M. (2011). Deciphering the role of Epstein-Barr virus in the pathogenesis of T and NK cell lymphoproliferations. Herpesviridae 2, 8.

Gebauer, N., Gebauer, J., Hardel, T.T., Bernard, V., Biersack, H., Lehnert, H., Rades, D., Feller, A.C., and Thorns, C. (2015). Prevalence of targetable oncogenic mutations and genomic alterations in Epstein-Barr virus-associated diffuse large B-cell lymphoma of the elderly. Leukemia and lymphoma 56, 1100-1106.

Gershon, A.A. (2013). Varicella zoster vaccines and their implications for development of HSV vaccines. Virology 435, 29-36.

Gershon, A.A., and Katz, S.L. (2008). Perspective on live varicella vaccine. J Infect Dis 197 Suppl 2, S242-245.

Gourzones, C., Barjon, C., and Busson, P. (2012). Host-tumor interactions in nasopharyngeal carcinomas. Seminars in cancer biology 22, 127-136.

Gu, S.Y., Huang, T.M., Ruan, L., Miao, Y.H., Lu, H., Chu, C.M., Motz, M., and Wolf, H. (1995). First EBV vaccine trial in humans using recombinant vaccinia virus expressing the major membrane antigen. DEv. Biol. Standard. 84, 171-177.

Gust, I.D. (1996). Immunisation against hepatitis B in Taiwan. Gut 38 Suppl 2, S67-68.

Hadinoto, V., Shapiro, M., Sun, C.C., and Thorley-Lawson, D.A. (2009). The dynamics of EBV shedding implicate a central role for epithelial cells in amplifying viral output. PLoS Pathog 5, e1000496.

Haigh, T.A., Lin, X., Jia, H., Hui, E.P., Chan, A.T., Rickinson, A.B., and Taylor, G.S. (2008). EBV latent membrane proteins (LMPs) 1 and 2 as immunotherapeutic targets: LMP-specific CD4+ cytotoxic T cell recognition of EBV-transformed B cell lines. J. Immunol. (Baltimore, Md: 1950) 180, 1643-1654.

Heslop, H.E., Slobod, K.S., Pule, M.A., Hale, G.A., Rousseau, A., Smith, C.A., Bollard, C.M., Liu, H., Wu, M.F., Rochester, R.J., *et al.* (2010). Long-term outcome of EBV-specific T-cell infusions to prevent or treat EBV-related lymphoproliferative disease in transplant recipients. Blood 115, 925-935.

Hislop, A.D., Taylor, G.S., Sauce, D., and Rickinson, A.B. (2007). Cellular Responses to Viral Infection in Humans: Lessons from Epstein-Barr Virus. Ann. REv. Immunol. 25, 587-617.

Hjalgrim, H., Askling, J., Rostgaard, K., Hamilton-Dutoit, S., Frisch, M., Zhang, J.S., Madsen, M., Rosdahl, N., Konradsen, H.B., Storm, H.H., *et al.* (2003). Characteristics of Hodgkin's lymphoma after infectious mononucleosis. New Engl. J. Med. 349, 1324-1332.

Hochberg, D., Middeldorp, J.M., Catalina, M., Sullivan, J.L., Luzuriaga, K., and Thorley-Lawson, D.A. (2004). Demonstration of the Burkitt's lymphoma Epstein-Barr virus phenotype in dividing latently infected memory cells *in vivo*. Proc Natl Acad Sci USA 101, 239-244.

Hsu, C., Lee, S.H., Ejadi, S., Even, C., Cohen, R., Le Tourneau, C., Mehnert, J., Algazi, A., Van Brummelen, E., Yuan, S.S., *et al.* (2015). Antitumor activity and safety of pembrolizumab in patients with PD-L1-positive nasopharyngeal carcinoma: Interim results from a phase 1b study. Eur. J. Cancer 51, S558.

Hui, E.P., Taylor, G.S., Jia, H., Ma, B.B., Chan, S.L., Ho, R., Wong, W.L., Wilson, S., Johnson, B.F., Edwards, C., *et al.* (2013). Phase I trial of recombinant modified vaccinia ankara encoding Epstein-Barr viral tumor antigens in nasopharyngeal carcinoma patients. Cancer research 73, 1676-1688.

Hurt, C., and Tammaro, D. (2007). Diagnostic Evaluation of Mononucleosis-Like Illnesses. Am. J. Med. 120, 911.e911-911.e918.

Icheva, V., Kayser, S., Wolff, D., Tuve, S., Kyzirakos, C., Bethge, W., Greil, J., Albert, M.H., Schwinger, W., Nathrath, M., *et al.* (2013). Adoptive transfer of epstein-barr virus (EBV) nuclear antigen 1-specific t cells as treatment for EBV reactivation and lymphoproliferative disorders after allogeneic stem-cell transplantation. J. Clin. Oncol.: Offical J. Am Soc. Clin. Oncol. 31, 39-48.

Janz, A., Oezel, M., Kurzeder, C., Mautner, J., Pich, D., Kost, M., Hammerschmidt, W., and Delecluse, H.J. (2000). Infectious Epstein-Barr virus lacking major glycoprotein BLLF1 (gp350/220) demonstrates the existence of additional viral ligands. J Virol 74, 10142-10152.

Jayasooriya, S., de Silva, T.I., Njie-jobe, J., Sanyang, C., Leese, A.M., Bell, A.I., McAulay, K.A., Yanchun, P., Long, H.M., Dong, T., *et al.* (2015). Early Virological and Immunological Events in Asymptomatic Epstein-Barr Virus Infection in African Children. PLoS Pathogens 11, e1004746.

Jenson, H.B. (2000). Acute complications of Epstein-Barr virus infectious mononucleosis. Current opinion in pediatrics 12, 263-268.

Johannsen, E., Luftig, M., Chase, M.R., Weicksel, S., Cahir-McFarland, E., Illanes, D., Sarracino, D., and Kieff, E. (2004). Proteins of purified Epstein-Barr virus. Proc Natl Acad Sci USA 101, 16286-16291.

Johnston, C., Gottlieb, S.L., and Wald, A. (2016). Status of vaccine research and development of vaccines for herpes simplex virus. Vaccine 34, 2948-2952.

Kanekiyo, M., Bu, W., Joyce, M.G., Meng, G., Whittle, J.R.R., Baxa, U., Yamamoto, T., Narpala, S., Todd, J.-P., Rao, S.S., *et al.* (2015). Rational Design of an Epstein-Barr Virus Vaccine Targeting the Receptor-Binding Site. Cell 162, 1090-1100.

Khanna, R., Burrows, S.R., Nicholls, J., and Poulsen, L.M. (1998). Identification of cytotoxic T cell epitopes within Epstein-Barr virus (EBV) oncogene latent membrane protein 1 (LMP1): evidence for HLA A2 supertype-restricted immune recognition of EBV-infected cells by LMP1-specific cytotoxic T lymphocytes. Eur. J. Immunol. 28, 451-458.

Khanna, R., Burrows, S.R., Steigerwald-Mullen, P.M., Thomson, S.A., Kurilla, M.G., and Moss, D.J. (1995). Isolation of cytotoxic T lymphocytes from healthy seropositive individuals specific for peptide epitopes from Epstein-Barr virus nuclear antigen 1: implications for viral persistence and tumor surveillance. Virology 214, 633-637.

Klibi, J., Niki, T., Riedel, A., Pioche-Durieu, C., Souquere, S., Rubinstein, E., Le Moulec, S., Guigay, J., Hirashima, M., Guemira, F., *et al.* (2009). Blood diffusion and Th1-suppressive effects of galectin-9-containing exosomes released by Epstein-Barr virus-infected nasopharyngeal carcinoma cells. Blood 113, 1957-1966.

Kobayashi, H., Nagato, T., Takahara, M., Sato, K., Kimura, S., Aoki, N., Azumi, M., Tateno, M., Harabuchi, Y., and Celis, E. (2008). Induction of EBV-latent membrane protein 1-specific MHC class II-restricted T-cell responses against natural killer lymphoma cells. Cancer research 68, 901-908.

Laichalk, L.L., and Thorley-Lawson, D.A. (2005). Terminal differentiation into plasma cells initiates the replicative cycle of Epstein-Barr virus *in vivo*. J Virol 79, 1296-1307.

Lal, H., Cunningham, A.L., Godeaux, O., Chlibek, R., Diez-Domingo, J., Hwang, S.J., Levin, M.J., McElhaney, J.E., Poder, A., Puig-Barbera, J., *et al.* (2015). Efficacy of an adjuvanted herpes zoster subunit vaccine in older adults. New Engl. J. Med. 372, 2087-2096.

Lau, K.M., Cheng, S.H., Lo, K.W., Lee, S.A., Woo, J.K., van Hasselt, C.A., Lee, S.P., Rickinson, A.B., and Ng, M.H. (2007). Increase in circulating Foxp3+CD4+CD25(high) regulatory T cells in nasopharyngeal carcinoma patients. Brit. J. Cancer 96, 617-622.

Lee, S.P., Brooks, J.M., Al-Jarrah, H., Thomas, W.A., Haigh, T.A., Taylor, G.S., Humme, S., Schepers, A., Hammerschmidt, W., Yates, J.L., *et al.* (2004). CD8 T cell recognition of endogenously expressed epstein-barr virus nuclear antigen 1. J. Exp. Med. 199, 1409-1420.

Lee, S.P., Chan, A.T., Cheung, S.T., Thomas, W.A., CroomCarter, D., Dawson, C.W., Tsai, C.H., Leung, S.F., Johnson, P.J., and Huang, D.P. (2000). CTL control of EBV in nasopharyngeal carcinoma (NPC): EBV-specific CTL responses in the blood and tumors of NPC patients and the antigen-processing function of the tumor cells. J. Immunol. (Baltimore, Md: 1950) 165, 573-582.

Lee, S.P., Thomas, W.A., Murray, R.J., Khanim, F., Kaur, S., Young, L.S., Rowe, M., Kurilla, M., and Rickinson, A.B. (1993). HLA A2.1-restricted cytotoxic T cells recognizing a range of Epstein-Barr virus isolates through a defined epitope in latent membrane protein LMP2. J Virol 67, 7428-7435.

Lee, S.P., Tierney, R.J., Thomas, W.A., Brooks, J.M., and Rickinson, A.B. (1997). Conserved CTL epitopes within EBV latent membrane protein 2: a potential target for CTL-based tumor therapy. J. Immunol. (Baltimore, Md: 1950) 158, 3325-3334.

Leroux-Roels, G., Marchant, A., Levy, J., Van Damme, P., Schwarz, T.F., Horsmans, Y., Jilg, W., Kremsner, P.G., Haelterman, E., Clement, F., *et al.* (2016). Impact of adjuvants on CD4(+) T cell and B cell responses to a protein antigen vaccine: Results from a phase II, randomized, multicenter trial. Clinical immunology (Orlando, Fla) 169, 16-27.

Li, F., Song, D., Lu, Y., Zhu, H., Chen, Z., and He, X. (2013). Delayed-type hypersensitivity (DTH) immune response related with EBV-DNA in nasopharyngeal carcinoma treated with autologous dendritic cell vaccination after radiotherapy. Journal of immunotherapy (Hagerstown, Md: 1997) 36, 208-214.

Li, J., Zeng, X.H., Mo, H.Y., Rolen, U., Gao, Y.F., Zhang, X.S., Chen, Q.Y., Zhang, L., Zeng, M.S., Li, M.Z., *et al.* (2007). Functional inactivation of EBV-specific T-lymphocytes in nasopharyngeal carcinoma: implications for tumor immunotherapy. PloS one 2, e1122.

Li, Q., Turk, S.M., and Hutt-Fletcher, L.M. (1995). The Epstein-Barr virus (EBV) BZLF2 gene product associates with the gH and gL homologs of EBV and carries an epitope critical to infection of B cells but not of epithelial cells. J Virol 69, 3987-3994.

Lin, C.L., Lo, W.F., Lee, T.H., Ren, Y., Hwang, S.L., Cheng, Y.F., Chen, C.L., Chang, Y.S., Lee, S.P., Rickinson, A.B., *et al.* (2002). Immunization with Epstein-Barr Virus (EBV) peptide-pulsed dendritic cells induces functional CD8+ T-cell immunity and may lead to tumor regression in patients with EBV-positive nasopharyngeal carcinoma. Cancer research 62, 6952-6958.

Lin, X., Gudgeon, N.H., Hui, E.P., Jia, H., Qun, X., Taylor, G.S., Barnardo, M.C., Lin, C.K., Rickinson, A.B., and Chan, A.T. (2008). CD4 and CD8 T cell responses to tumour-associated Epstein-Barr virus antigens in nasopharyngeal carcinoma patients. Cancer immunology, immunotherapy: CII 57, 963-975.

Long, H.M., Chagoury, O.L., Leese, A.M., Ryan, G.B., James, E., Morton, L.T., Abbott, R.J., Sabbah, S., Kwok, W., and Rickinson, A.B. (2013). MHC II tetramers visualize human CD4+ T cell responses to Epstein-Barr virus infection and demonstrate atypical kinetics of the nuclear antigen EBNA1 response. J. Exp. Med. 210, 933-949.

Louis, C.U., Straathof, K., Bollard, C.M., Ennamuri, S., Gerken, C., Lopez, T.T., Huls, M.H., Sheehan, A., Wu, M.F., Liu, H., *et al.* (2010). Adoptive transfer of EBV-specific T cells results in sustained clinical responses in patients with locoregional nasopharyngeal carcinoma. Journal of immunotherapy (Hagerstown, Md: 1997) 33, 983-990.

Louis, C.U., Straathof, K., Bollard, C.M., Gerken, C., Huls, M.H., Gresik, M.V., Wu, M.F., Weiss, H.L., Gee, A.P., Brenner, M.K., *et al.* (2009). Enhancing the *in vivo* expansion of adoptively transferred EBV-specific

CTL with lymphodepleting CD45 monoclonal antibodies in NPC patients. Blood 113, 2442-2450.

Lowy, D.R., and Schiller, J.T. (2012). Reducing HPV-associated cancer globally. Cancer prevention research (Philadelphia, Pa) 5, 18-23.

McKenzie, J., and El-Guindy, A. (2015). Epstein-Barr Virus Lytic Cycle Reactivation. Current topics in microbiology and immunology 391, 237-261.

Morgan, A.J., Finerty, S., Lovgren, K., Scullion, F.T., and Morein, B. (1988a). Prevention of Epstein-Barr (EB) virus-induced lymphoma in cottontop tamarins by vaccination with the EB virus envelope glycoprotein gp340 incorporated into immune-stimulating complexes. J. Gen. Virol. 69 (Pt 8), 2093-2096.

Morgan, A.J., Mackett, M., Finerty, S., Arrand, J.R., Scullion, F.T., and Epstein, M.A. (1988b). Recombinant vaccinia virus expressing Epstein-Barr virus glycoprotein gp340 protects cottontop tamarins against EB virus-induced malignant lymphomas. Journal of medical virology 25, 189-195.

Moutschen, M., Leonard, P., Sokal, E.M., Smets, F., Haumont, M., Mazzu, P., Bollen, A., Denamur, F., Peeters, P., Dubin, G., *et al.* (2007). Phase I/II studies to evaluate safety and immunogenicity of a recombinant gp350 Epstein-Barr virus vaccine in healthy adults. Vaccine 25, 4697-4705.

Nikiforow, S., Bottomly, K., Miller, G., and Munz, C. (2003). Cytolytic CD4(+)-T-cell clones reactive to EBNA1 inhibit Epstein-Barr virus-induced B-cell proliferation. J Virol 77, 12088-12104.

Odumade, O.A., Hogquist, K.A., and Balfour, H.H., Jr. (2011). Progress and problems in understanding and managing primary Epstein-Barr virus infections. Clin Microbiol Rev 24, 193-209.

Ogembo, J.G., Muraswki, M.R., McGinnes, L.W., Parcharidou, A., Sutiwisesak, R., Tison, T., Avendano, J., Agnani, D., Finberg, R.W., Morrison, T.G., *et al.* (2015). A chimeric EBV gp350/220-based VLP replicates the virion B-cell attachment mechanism and elicits long-lasting neutralizing antibodies in mice. Journal of Translational Medicine 13, 50.

Orlando, R., Foggia, M., Maraolo, A.E., Mascolo, S., Palmiero, G., Tambaro, O., and Tosone, G. (2015). Prevention of hepatitis B virus infection: from the past to the future. Eur. J. Clin. Microbiol. Infect. Dis. 34, 1059-1070.

Oxman, M.N., Levin, M.J., Johnson, G.R., Schmader, K.E., Straus, S.E., Gelb, L.D., Arbeit, R.D., Simberkoff, M.S., Gershon, A.A., Davis, L.E., *et al.* (2005). A vaccine to prevent herpes zoster and postherpetic neuralgia in older adults. New Engl. J. Med. 352, 2271-2284.

Oyama, T., Ichimura, K., Suzuki, R., Suzumiya, J., Ohshima, K., Yatabe, Y., Yokoi, T., Kojima, M., Kamiya, Y., Taji, H., *et al.* (2003). Senile EBV+ B-cell lymphoproliferative disorders: a clinicopathologic study of 22 patients. The American journal of surgical pathology 27, 16-26.

Oyama, T., Yamamoto, K., Asano, N., Oshiro, A., Suzuki, R., Kagami, Y., Morishima, Y., Takeuchi, K., Izumo, T., Mori, S., *et al.* (2007). Age-related EBV-associated B-cell lymphoproliferative disorders constitute a distinct clinicopathologic group: a study of 96 patients. Clinical cancer research: an official journal of the American Association for Cancer Research 13, 5124-5132.

Pakpoor, J., Disanto, G., Gerber, J.E., Dobson, R., Meier, U.C., Giovannoni, G., and Ramagopalan, S.V. (2013). The risk of developing multiple sclerosis in individuals seronegative for Epstein-Barr virus: a meta-analysis. Multiple sclerosis (Houndmills, Basingstoke, England) 19, 162-166.

Paludan, C., Bickham, K., Nikiforow, S., Tsang, M.L., Goodman, K., Hanekom, W.A., Fonteneau, J.F., Stevanovic, S., and Munz, C. (2002). Epstein-Barr nuclear antigen 1-specific CD4(+) Th1 cells kill Burkitt's lymphoma cells. J. Immunol. (Baltimore, Md: 1950) 169, 1593-1603.

Perez-Diez, A., Joncker, N.T., Choi, K., Chan, W.F., Anderson, C.C., Lantz, O., and Matzinger, P. (2007). CD4 cells can be more efficient at tumor rejection than CD8 cells. Blood 109, 5346-5354.

Plummer, E.M., and Manchester, M. (2011). Viral nanoparticles and virus-like particles: platforms for contemporary vaccine design. Wiley Interdiscip Rev Nanomed Nanobiotechnol. 3, 174-196.

Pudney, V.A., Leese, A.M., Rickinson, A.B., and Hislop, A.D. (2005). CD8+ immunodominance among Epstein-Barr virus lytic cycle antigens directly reflects the efficiency of antigen presentation in lytically infected cells. J. Exp. Med. 201, 349-360.

Quezada, S.A., Simpson, T.R., Peggs, K.S., Merghoub, T., Vider, J., Fan, X., Blasberg, R., Yagita, H., Muranski, P., Antony, P.A., *et al.* (2010). Tumor-reactive CD4(+) T cells develop cytotoxic activity and eradicate large established melanoma after transfer into lymphopenic hosts. J. Exp. Med. 207, 637-650.

Quinn, L.L., Zuo, J., Abbott, R.J.M., Shannon-Lowe, C., Tierney, R.J., Hislop, A.D., and Rowe, M. (2014). Cooperation between Epstein-Barr Virus Immune Evasion Proteins Spreads Protection from CD8(+) T Cell Recognition across All Three Phases of the Lytic Cycle. PLoS Pathogens 10, e1004322.

Raab-Traub, N., and Flynn, K. (1986). The structure of the termini of the Epstein-Barr virus as a marker of clonal cellular proliferation. Cell 47, 883-889.

Ragot, T., Finerty, S., Watkins, P.E., Perricaudet, M., and Morgan, A.J. (1993). Replication-defective recombinant adenovirus expressing the Epstein—Barr virus (EBV) envelope glycoprotein gp340/220 induces protective immunity against EBV-induced lymphomas in the cottontop tamarin. Journal of General Virology 74, 501-507.

Rees, L., Tizard, E.J., Morgan, A.J., Cubitt, W.D., Finerty, S., Oyewole-Eletu, T.A., Owen, K., Royed, C., Stevens, S.J., Shroff, R.C., *et al.* (2009). A phase I trial of epstein-barr virus gp350 vaccine for children with chronic kidney disease awaiting transplantation. Transplantation 88, 1025-1029.

Ruiss, R., Jochum, S., Wanner, G., Reisbach, G., Hammerschmidt, W., and Zeidler, R. (2011). A virus-like particle-based Epstein-Barr virus vaccine. J Virol 85, 13105-13113.

Sampaio, M.S., Cho, Y.W., Shah, T., Bunnapradist, S., and Hutchinson, I.V. (2012). Impact of Epstein-Barr virus donor and recipient serostatus on the incidence of post-transplant lymphoproliferative disorder in kidney transplant recipients. Nephrol. Dialysis Transplan. 27, 2971-2979.

Sashihara, J., Hoshino, Y., Bowman, J.J., Krogmann, T., Burbelo, P.D., Coffield, V.M., Kamrud, K., and Cohen, J.I. (2011). Soluble Rhesus Lymphocryptovirus gp350 Protects against Infection and Reduces Viral Loads in Animals that Become Infected with Virus after Challenge. PLoS Pathogens 7, e1002308.

Secondino, S., Zecca, M., Licitra, L., Gurrado, A., Schiavetto, I., Bossi, P., Locati, L., Schiavo, R., Basso, S., Baldanti, F., *et al.* (2012). T-cell therapy for EBV-associated nasopharyngeal carcinoma: preparative lymphodepleting chemotherapy does not improve clinical results. Annal Oncol. 23, 435-441.

Seder, R.A., Darrah, P.A., and Roederer, M. (2008). T-cell quality in memory and protection: implications for vaccine design. Nature reviews Immunology 8, 247-258.

Silins, S.L., Sherritt, M.A., Silleri, J.M., Cross, S.M., Elliott, S.L., Bharadwaj, M., Le, T.T.T., Morrison, L.E., Khanna, R., Moss, D.J., *et al.* (2001). Asymptomatic primary Epstein-Barr virus infection occurs in the absence of blood T-cell repertoire perturbations despite high levels of systemic viral load. Blood 98, 3739.

Skoberne, M., Cardin, R., Lee, A., Kazimirova, A., Zielinski, V., Garvie, D., Lundberg, A., Larson, S., Bravo, F.J., Bernstein, D.I., *et al.* (2013). An adjuvanted herpes simplex virus 2 subunit vaccine elicits a T cell response in mice and is an effective therapeutic vaccine in Guinea pigs. J Virol 87, 3930-3942.

Smith, C., Tsang, J., Beagley, L., Chua, D., Lee, V., Li, V., Moss, D.J., Coman, W., Chan, K.H., Nicholls, J., *et al.* (2012). Effective treatment of metastatic forms of Epstein-Barr virus-associated nasopharyngeal carcinoma with a novel adenovirus-based adoptive immunotherapy. Cancer research 72, 1116-1125.

Smith, T.J., and Hillner, B.E. (2011). Bending the cost curve in cancer care. New Engl. J. Med. 364, 2060-2065.

Sokal, E.M., Hoppenbrouwers, K., Vandermeulen, C., Moutschen, M., Leonard, P., Moreels, A., Haumont, M., Bollen, A., Smets, F., and Denis, M. (2007). Recombinant gp350 vaccine for infectious mononucleosis: a phase 2, randomized, double-blind, placebo-controlled trial to evaluate the safety, immunogenicity, and efficacy of an Epstein-Barr virus vaccine in healthy young adults. J Infect Dis 196, 1749-1753.

Stanberry, L.R., Spruance, S.L., Cunningham, A.L., Bernstein, D.I., Mindel, A., Sacks, S., Tyring, S., Aoki, F.Y., Slaoui, M., Denis, M., *et al.* (2002). Glycoprotein-D-adjuvant vaccine to prevent genital herpes. New Engl. J. Med. 347, 1652-1661.

Steven, N.M., Annels, N.E., Kumar, A., Leese, A.M., Kurilla, M.G., and Rickinson, A.B. (1997). Immediate early and early lytic cycle proteins are frequent targets of the Epstein-Barr virus-induced cytotoxic T cell response. J. Exp. Med. 185, 1605-1617.

Straathof, K.C., Leen, A.M., Buza, E.L., Taylor, G., Huls, M.H., Heslop, H.E., Rooney, C.M., and Bollard, C.M. (2005). Characterization of latent membrane protein 2 specificity in CTL lines from patients with EBV-positive nasopharyngeal carcinoma and lymphoma. J. Immunol. (Baltimore, Md: 1950) 175, 4137-4147.

Takahashi, M., Otsuka, T., Okuno, Y., Asano, Y., and Yazaki, T. (1974). Live vaccine used to prevent the spread of varicella in children in hospital. Lancet (London, England) 2, 1288-1290.

Taylor, G.S., Haigh, T.A., Gudgeon, N.H., Phelps, R.J., Lee, S.P., Steven, N.M., and Rickinson, A.B. (2004). Dual stimulation of Epstein-Barr Virus (EBV)-specific CD4+- and CD8+-T-cell responses by a chimeric antigen construct: potential therapeutic vaccine for EBV-positive nasopharyngeal carcinoma. J Virol 78, 768-778.

Taylor, G.S., Jia, H., Harrington, K., Lee, L.W., Turner, J., Ladell, K., Price, D.A., Tanday, M., Matthews, J., Roberts, C., *et al.* (2014). A recombinant modified vaccinia ankara vaccine encoding Epstein-Barr Virus (EBV) target antigens: a phase I trial in UK patients with EBV-positive cancer. Clinical cancer research: an official journal of the American Association for Cancer Research 20, 5009-5022.

Taylor, G.S., Long, H.M., Brooks, J.M., Rickinson, A.B., and Hislop, A.D. (2015). The immunology of Epstein-Barr virus-induced disease. Annu Rev Immunol 33, 787-821.

Tomkinson, B.E., Wagner, D.K., Nelson, D.L., and Sullivan, J.L. (1987). Activated lymphocytes during acute Epstein-Barr virus infection. J. Immunol. (Baltimore, Md: 1950) 139, 3802-3807.

Tugizov, S.M., Berline, J.W., and Palefsky, J.M. (2003). Epstein-Barr virus infection of polarized tongue and nasopharyngeal epithelial cells. Nature Med. 9, 307-314.

Williams, H., McAulay, K., Macsween, K.F., Gallacher, N.J., Higgins, C.D., Harrison, N., Swerdlow, A.J., and Crawford, D.H. (2005). The immune response to primary EBV infection: a role for natural killer cells. British Journal of Haematology 129, 266-274.

Yao, Y., Minter, H.A., Chen, X., Reynolds, G.M., Bromley, M., and Arrand, J.R. (2000). Heterogeneity of HLA and EBER expression in Epstein-Barr virus-associated nasopharyngeal carcinoma. Int. J. Cancer 88, 949-955.

Yip, W.K., Abdullah, M.A., Yusoff, S.M., and Seow, H.F. (2009). Increase in tumour-infiltrating lymphocytes with regulatory T cell immunophenotypes and reduced zeta-chain expression in nasopharyngeal carcinoma patients. Clinical and experimental immunology 155, 412-422.

9

Molecular Pathogenesis of Kaposi Sarcoma Herpes Virus-Associated Cancers

Silvia Gramolelli[1]*, Giuseppe Balistreri[1], and Päivi M. Ojala[1,2,3]*

[1]Research Programs Unit, Translational Cancer Biology, University of Helsinki, Finland
[2]Foundation for the Finnish Cancer Institute, Helsinki, Finland
[3]Section of Virology, Division of Infectious Diseases, Department of Medicine, Imperial College London, London, UK

**Corresponding authors: silvia.gramolelli@helsinki.fi, p.ojala@imperial.ac.uk*

Abstract

Kaposi sarcoma herpesvirus (KSHV) is the causative agent of three malignancies, namely Kaposi sarcoma (KS), primary effusion lymphoma (PEL) and multicentric Castelman disease (MCD) and it is also the etiological agent of an inflammatory cytokine syndrome, KICS. In order to ensure its own survival and the successful accomplishment of its two modes of life cycle, a latent and a lytic phase, KSHV subverts several cellular processes. This chapter focuses on virus-host interactions that include manipulation of the cell cycle and apoptotic cascade, atypical differentiation of infected cells, aberrant angiogenesis and KSHV-induced metabolic alterations. By deregulating these cellular events, KSHV modulates the infected cell microenviroment and promotes viral tumourigenesis. Lastly, we present an overview of the current therapies for KSHV-associated malignancies and provide examples of how the study of KSHV molecular pathology could pave the way for new therapeutic approaches.

KSHV-associated cancers and clinical manifestations

Kaposi sarcoma herpesvirus (KSHV) is a human virus classified as a group 1 carcinogen (International Agency for Research on Cancer, 2012), and the etiological agent of three human cancers; Kaposi sarcoma (KS), primary effusion lymphoma (PEL) and multicentric Castleman disease (MCD) (Chang *et al.*, 1994; Cesarman *et al.*, 1995; Soulier *et al.*, 1995). KSHV is found virtually in 100% of the cases in KS and PEL, and about 50% of MCD patients are infected with KSHV (Dittmer *et al.*, 1998; Dupin *et al.*, 2000; Dittmer, 2003; Cronin and Warnke, 2009).

Kaposi sarcoma

KS is classified into four different forms: i) classic, ii) endemic African, iii) iatrogenic/immunosuppressive, and iv) AIDS-related KS (AIDS-KS) (Antman and Chang, 2000). Classic and endemic KS have a specific geographical distribution, the former affecting elderly individuals from the Mediterranean countries and the latter those living in sub-Saharan Africa. In particular, in the sub-Saharan belt KS represents the most common tumour among men and one of the most frequent cancers in children. In the general population, KS is a rather rare disease, but the incidence increases significantly in immunosuppressed sub-groups. Iatrogenic KS occurs in patients undergoing immunosuppressive therapies following organ transplantation. AIDS-KS is a common AIDS-associated cancer and represents the most aggressive variant of KS (see below), involving also visceral organs.

Although epidemiologically different, all four KS variants are histologically identical. KS tumours consist of multifocal lesions that display stage-specific appearances. The early lesions are composed of red maculae characterized by irregular dilated vascular spaces (patch lesions) that then progress through a plaque-like stage to the nodular stage, in which accumulation of the spindle-like KS tumour cells leads to the generation of large tumour nodules (Ganem, 2006).

Is KS a tumour or a hyperplasia?

Several strands of evidence suggest that KS is not a tumour *per se* that results from the transformation of infected cells, but is rather an angiogenic and inflammatory hyperplasia of the endothelium. Unlike other cancer cells, KS cells isolated from the primary lesion: i) do not form new tumours in nude mice, ii) do not have increased growth rate, iii) are strictly dependent on extracellular growth stimuli for their survival, iv) are diploid and do not show chromosomal alterations (Ensoli and Sturzl, 1998, 1999; Samaniego *et al.*, 1998 Ensoli *et al.*, 2001). Furthermore, contrarily to other cancers, in KS the angiogenesis occurs prior to the growth of the tumour mass and it is present already at the early phases (patch stage) of the disease. The fact that iatrogenic KS spontaneously resolves after the restoration of the immune system and that the lesions often arise in sites of traumas or unresolved surgical scars, where local inflammation takes place, (Janier *et al.*, 1990; Maral, 2000) also supports the hypothesis that a deregulated and/or persistent inflammatory response and angiogenesis sustain the development of KS (Connick *et al.*, 2004).

The clonality of KS is a matter of debate as well. KS has been reported to be mono- as well as oligo/polyclonal (Rabkin *et al.*, 1997; Gill *et al.*, 1998; Judde *et al.*, 2000; Pyakurel *et al.*, 2006). It is tempting to speculate that on one hand, polyclonal KS does not represent a true malignancy but is rather an inflammatory hyperplasia that is strictly dependent on the inflammatory microenvironment (see below). On the other hand, under the effect of inflammatory and proliferating stimuli, infected endothelial cells within a lesion could acquire genetic alterations that probably drive a clonal expansion, thus leading to the formation of monoclonal lesions. Although transforming events have not been observed in KS lesions, KSHV-induced endothelial cell transformation has been shown to occur *in vitro* in human immortalized (An *et al.*, 2006) and in murine endothelial cells (Mutlu *et al.*, 2007) as well as in rat mesenchymal fibroblasts (Jones *et al.*, 2012), thus confirming the transforming potential of KSHV in certain contexts.

Primary effusion lymphoma

Primary effusion lymphoma is a KSHV-associated non-Hogkin lymphoma mainly diagnosed in HIV-infected individuals. In 70% of cases co-infection with another gammaherpesvirus, Epstein Barr virus (EBV) is also observed (Ansari *et al.*, 1996; Nador *et al.*, 1996). PEL tumours usually present as pleural, peritoneal and pericardial effusions (Gaidano and Carbone, 2001), but, in rare cases, contemporary nodal involvement and solid tumour mass can be found (Chadburn *et al.*, 2004; Carbone *et al.*, 2005).

The cell of origin in PELs is of B-cell lineage. Although positive for differentiated plasma cells markers (CD138, VS38c, MUM1/IRF4) and for activation markers (CD30, CD38, CD71), the tumour cells lack expression of common B cell-associated antigens (Arvanitakis *et al.*, 1996). PEL is consistently composed of a monoclonal B-cell population, as shown by immunoglobulin gene rearrangement (Judde *et al.*, 2000). Interestingly, few sporadic cases of PEL with T-cell antigen expression have also been described (Said *et al.*, 1999; Coupland *et al.*, 2005; Nepka *et al.*, 2012; Goto *et al.*, 2013;). Continuos cell lines established from the malignant lesions of patients with PEL have proven an indispensable tool for the understanding of KSHV biology and its impact on the clinical manifestation of PEL (Carbone *et al.*, 2010).

Castleman disease

Clinically, Castleman disease manifests in several forms that are distinguished based on the number of lymph nodes involved. The KSHV-associated form appears to represent a distinct entity from the other forms of MCD (Oksenhendler *et al.*, 1996). It involves multiple enlarged lymph nodes, and is therefore called "multicentric" Castleman disease (MCD) (Cronin and Warnke, 2009). Approximately half of MCD cases from immunocompetent individuals are associated with KSHV infection, but the incidence approaches 100% in individuals with HIV. These patients display larger, KSHV-positive cells, called plasmablasts, almost invariably expressing IgM lambda (Du *et al.*, 2001). Patients with KSHV-MCD show increased viral loads and elevated peripheral levels of both virus-encoded and cellular pro- and anti-inflammatory cytokines (Polizzotto *et al.*, 2013).

KSHV inflammatory cytokine syndrome

In 2010 a novel inflammatory symptom complex associated with KSHV infection was described (Uldrick *et al.*, 2010). The disease, named KSHV inflammatory cytokine syndrome (KICS), clinically resembles KSHV-MCD, but without the involvement

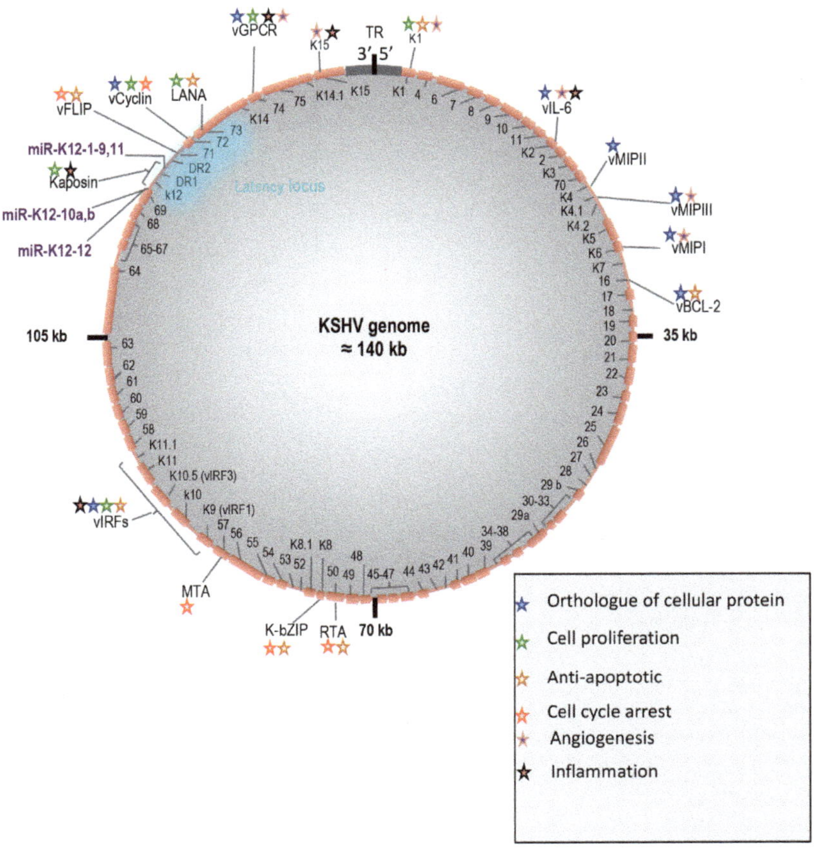

Figure 9.1. KSHV episomal DNA genome. KSHV Functional Map: schematic representation of the KSHV circular double-stranded DNA episome. Solid arrows represent ORFs, the direction indicates the direction of transcription. The latency locus is marked in light blue and miRNAs are indicated in purple. Many viral gene products have cellular orthologs (blue star). Examples of viral proteins that display multiple functions are indicated: stimulation of cell proliferation (green star), anti-apoptotic (orange star), cell cycle arrest (red star), angiogenesis (pink lined blue star), immune evasion (yellow star) and inflammation (black lined red star). TR=terminal repeats (grey boxes).

of lymphadenopathy or pathologic nodal changes. The patients are often co-infected with HIV, and have severe KS. Similar to KSHV-MCD, KICS is also associated with elevated viral loads and viral and cellular cytokine storms (Polizzotto *et al.*, 2012). Moreover, KICS may play a role in some KS patients who appear to be refractory to the current therapies.

Co-infection with HIV enhances KSHV-induced pathogenesis

KSHV associated malignancies affect HIV positive individuals at a high rate. As already mentioned above, a more severe and disseminated form of KS is linked to HIV infection. In addition, KSHV associated PEL and MCD also occur more often in HIV positive individuals. Although the immunosuppression induced by HIV strongly contributes to these associations, several evidences support a synergistic effect between these two viruses in enhancing KSHV pathogenicity. For instance, KS incidence among HIV positive individuals is 20,000 fold higher than in the general population and 300 fold higher than among transplant recipients (who undergo immunosuppressive therapy to prevent transplant rejection) (Beral *et al.*, 1990). Furthermore, HAART therapy, that has helped to turn HIV-1 infection a chronic disease, has significantly reduced KS by inducing both regression of the lesions and a decrease in viral loads.

At a molecular level, the HIV encoded proteins Tat and Nef exacerbate some of the pathologic features induced by KSHV infection such as increased angiogenesis (da Silva and de Oliveira, 2011) and might explain the more severe prognosis of AIDS-associated KS.

Kaposi sarcoma herpesvirus

After the discovery from a biopsy of an AIDS-KS patient in 1994 (Chang *et al.*, 1994) KSHV has been actively investigated. The viral double-stranded DNA genome was sequenced from a PEL cell line, BC-1, in 1996 revealing a 140.5-kb region coding 81 open reading frames (ORFs) flanked by GC-rich terminal repeat sequences (Russo *et al.*, 1996) (Figure 9.1). The recent mapping and annotation of the KSHV genome using ribosome profiling (Arias *et al.*, 2014) has demonstrated that the virus genome posesses 45% higher coding capacity than initially anticipated by bioinformatic analyses alone, including at least 63 new ORFs, miRNA-encoding genes as well as numerous long non-coding RNAs.

The viral capsid structure was solved by cryo-EM and demonstrated a typical herpesvirus icosahedral capsid shell composed of four structural proteins (Wu *et al.*, 2000; Nealon *et al.*, 2001; Trus *et al.*, 2001). KSHV entry is a complex multistep process and involves interaction of viral envelope glycoproteins with several cell surface molecules. The virus shows a relatively broad cell tropism mostly due to its ability to bind to different cellular receptors on the target cells. Heparan sulfate was first shown to contribute to KSHV adsorbtion and entry into the target cell (Akula *et al.*, 2001a; Akula *et al.*, 2001b; Birkmann *et al.*, 2001) followed by indentification of α3β1 integrin as a KSHV receptor in 2002 (Akula *et al.*, 2002). The α3β1 integrin is expressed abundantly in endothelial, epithelial, and B cells, all naturally infected by KSHV. Active research for other KSHV receptors led to identification of DC-SIGN in hematopoietic cells (such as activated B and dendritic cells) (Rappocciolo *et al.*, 2006), the cystine/glutamate exchange transporter xCT (Kaleeba and Berger, 2006), αVβ3 integrin (Garrigues *et al.*, 2008; Veettil *et al.*, 2008) and the ephrin receptor tyrosine kinase A2 in endothelial cells (Chakraborty *et al.*, 2012; Hahn *et al.*, 2012). KSHV binding to the cellular receptors activates host cell signalling pathways that in turn trigger a cascade of events enabling the rapid entry of virus particles. The entry through the plasma membrane occurs either through clathrin-mediated endocytosis (Akula *et al.*, 2003; Kaleeba and Berger, 2006) or by macropinocytosis (Raghu *et al.*, 2009). The viral capsids are then transported through the cytoplasm into the nucleus where viral gene expression takes place.

KSHV replication programs

KSHV, as other herpesviruses, has two distinct replication programs, the latency and the lytic replication phase. After the primary infection, there is a short period when both latent and lytic genes are expressed (known as "lytic burst") (Krishnan *et al.*, 2004), but following a few rounds of replication, the lytic gene expression diminishes and latency becomes the dominant phase. In the following sections the latent and lytic programs will be described together with their contribution to virus-induced tumourigenesis.

Latency

During latency the KSHV genome persists in the host cell without producing infectious progeny and expressing only a few proteins and microRNAs (Figure 9.1). Of those, latency-associated nuclear antigen (LANA), viral FADD-like interleukin-1β-converting enzyme (FLICE/caspase 8)-inhibitory protein (vFLIP), a viral homolog of cellular cyclin D2 (vCyclin), K12 (Kaposin), v-IRF3 (also called LANA2) and multiple microRNAs are consistently detectable in KSHV-infected B cells (Cai *et al.*, 2005; Pfeffer *et al.*, 2005; Samols *et al.*, 2005). Expression of LANA is needed for the establishment of latency (Ballestas *et al.*, 1999; Ballestas and Kaye, 2001). LANA exists as dimers that were recently shown to oligomerize and form higher order structures that facilitate its binding to DNA (Domsic *et al.*, 2013; Hellert *et al.*, 2013; Hellert *et al.*, 2015; Li *et al.*, 2015; Ponnusamy *et al.*, 2015). The C-terminus of LANA interacts with LANA-binding sites in the viral genome, while the N-terminus mediates LANA binding to the host chromosomes (Barbera *et al.*, 2004; Barbera *et al.*, 2006; Ottinger *et al.*, 2006), thereby tethering the viral genomes to host chromosomes during mitosis to enable the efficient segregation of KSHV episomes to progeny cells.

In PEL, latency represents the main modality of KSHV persistence in the host and the expression of latent proteins enables the transformation of the infected B cells. However, in some contexts the canonical classification of an infected cell in a lytic or latent stage, at least in the currently defined terms, is not always applicable to KSHV infection. This concerns particularly the KSHV-infected lymphatic endothelial cells (LECs). In LECs a "leaky" and unique viral expression program is observed. Apart from the recognized latent proteins (see above), in KSHV-infected LECs several lytic proteins are detected, among others ORF45, vGPCR and K-bZIP (Chang and Ganem, 2013) without the induction of a fully productive lytic cycle. Futhermore, viral ORFs previously classified as lytic, K1, K15 and vIL6, have been detected also during latency in endothelial cells and fibroblasts (Chandriani and Ganem, 2010; Gramolelli *et al.*, 2015). These findings pave the way to the possibility that, similarly to EBV, also KSHV displays more than one latency program whose expression might be context and cell type dependent.

Lytic replication phase

The lytic replication cascade is initiated through the induction of the viral replication and transcription activator RTA, encoded by the ORF 50 gene (Lukac *et al.*, 1998; Lukac *et al.*, 1999; Gradoville *et al.*, 2000). Induction of RTA and thus lytic replication can be experimentally initiated by chemicals such as 12-O-tetradecanoylphorbol-13-acetate, valproic acid, and sodium butyrate (Miller *et al.*, 2007). Physiological activators of lytic replication include cellular stress, hypoxia and certain cytokines.

During lytic replication virtually all viral ORFs are expressed in a tightly regulated cascade which is divided into three phases: immediate early (IE), delayed early (DE), and late (Sarid *et al.*, 1998; Jenner *et al.*, 2001; Dittmer, 2003; Arias *et al.*, 2014). The transcription of IE and DE genes does not require viral DNA replication. However, the late genes are expressed after the initiation of viral DNA replication and encode for structural proteins required for assembling new virions. Viral replication inhibitors (e.g. viral polymerase inhibitor ganciclovir, foscarnet or cidofovir) can prevent the production of infectious progeny virions (Kedes and Ganem, 1997).

Contrary to PEL cells, in which latent infection is sufficient for tumourigenesis, lytic infection is required to sustain KS progression (Alagiozoglou *et al.*, 2003; Song *et al.*, 2004; Laney *et al.*, 2007) as well as KSHV-MCD and KICS (Polizzotto *et al.*, 2012). In the case of endothelial cells, newly infected cells as well as cells isolated from KS lesion lose the episomes after several passages in the absence of drug selection (Grundhoff and Ganem, 2004). This suggests that, in order to maintain a stable infection in proliferating endothelial cells, continuos production of virus and reinfection are required. Indeed, within a KS lesion, a subset of infected cells shows positivity for viral lytic markers (Staskus *et al.*, 1997) and high KSHV viral loads correlate with KS progression (Laney *et al.*, 2007). In addition, impairing KSHV lytic replication by specific inhibitors of viral DNA replication (e.g. ganciclovir and foscarnet) is beneficial for KS prognosis (Glesby *et al.*, 1996; Mocroft *et al.*, 1996; Martin and Gutkind, 2008).

Apart from its role in the maintenance of a population of infected cells, lytic reactivation in endothelial cells contributes to KS development by inducing the secretion of a plethora of growth factors, chemo- and cytokines of both cellular and viral origin which contribute in a paracrine fashion to the establishment of a pro-tumourigenic and pro-inflammatory micro-environment (Chandriani and Ganem, 2007).

Several KSHV lytic proteins have been shown to modify the extracellular environment either by being secreted or by triggering the secretion of pro-inflammatory cellular cytokines and growth factors, of which some examples are given below. During the lytic replication phase, KSHV expresses three homologues of cellular cytokines, namely vMIPI, vMIPII, and vMIPIII (Kledal *et al.*, 1997) (Figure 9.1). These genes have been acquired by KSHV through the process of molecular piracy and share 40-50% of sequence identity with cellular MIP1α. Unlike their cellular counterparts, vMIPs bind very efficiently their cognate receptor, in particular vMIPI is recruited by CCR8, vMIPII binds to a broad spectrum of chemokine receptors and has an antagonistic effect, and vMIPIII interacts and activates CCR4 (Stine *et al.*, 2000; Weber *et al.*, 2001). Once secreted in the extracellular environment, all vMIPs exert a proangiogenic effect, which is crucial for KSHV-induced tumourigenesis. Furthermore, vMIPIII induces a chemotactic effect for Th2 type memory T cells. Since the recruitment of Th2, rather than Th1 cells favours tumour progression and viral survival, the contribution of these proteins to KSHV-induced tumourigenesis and immune modulation has gained a prominent role (Stine *et al.*, 2000).

Another viral protein pirated by KSHV is the vIL6, which shares 25% identity with the human IL6 (Moore *et al.*, 1996). In contrast to the cellular IL6 which needs to recruit both IL6Rα and the co-receptor gp130, common to many interleukin receptors for efficient signaling, KSHV-encoded vIL6 can trigger signalling by binding to gp130 only, thus having a more pleiotropic effect (Molden *et al.*, 1997). vIL6 expression supports PEL cell survival and has a protective effect against the antiviral activity of IFNα (Sakakibara and Tosato, 2011). vIL6 is inefficiently secreted and a major fraction is retained in the ER where it can signal intracellularly through gp130 and enhance the expression and the secretion of cellular IL6 thereby contributing to the generation of the inflammatory microenvironment (Mori *et al.*, 2000; Choi and Nicholas, 2008;).

Among the viral proteins that can twist the cellular secretome towards a more inflammatory phenotype, is the viral G-protein coupled receptor vGPCR (Cesarman *et al.*, 1996). This ligand-independent and constitutively active homologue of the cellular IL8R has been shown to be sufficient to induce KS-like tumours in several mouse models (the CD2-vGPCR mouse) (Yang *et al.*, 2000; Montaner *et al.*, 2003; Sodhi *et al.*, 2004; Mutlu *et al.*, 2007; Ma *et al.*, 2010). When ectopically expressed, and presumably leading to higher expression levels than in the *in vivo* situation, vGPCR induces the secretion of numerous cytokines, chemokines and growth factors (IL8, IL-1β, IL-2, IL-4, IL-6, TNFα, CCL3/MIPI, bFGF, VEGF) thereby increasing the survival of the neighboring cells and their angiogenic potential (Montaner *et al.*, 2003; Ma *et al.*, 2010). Being an early lytic protein, vGPCR is expressed at low levels only by a minority of infected cells within a KS lesion, thus making the detection and the study of this protein in the context of the viral infection, particularly challenging. Therefore, the actual role of vGPCR in sarcomagenesis needs further investigations.

Despite the abundant inflammatory infiltrate within a KS lesion, some cell types are absent. This is the case for neutrophils and NK cells (Yamin *et al.*, 2013), whereas B cells and macrophages are particularly abundant. In particular, vMIPIII promotes monocyte chemotaxis (Nakano *et al.*, 2003) and vIL6 is able to inhibit neutrofil recruitment, as shown in an *in vivo* mouse model (Fielding *et al.*, 2005).

KSHV manipulation of the cell cycle and apoptosis

Like other herpesviruses, KSHV can regulate the cell cycle. Latently infected cells are 'forced' to proliferate and bypass cell-cycle checkpoints, and are predisposed to uncontrolled growth. During latency, KSHV, in order to ensure its own survival, must also prevent the infected host cell from undergoing premature apoptosis. On the other hand, upon reactivation, the lytic cascade leads to cell cycle arrest and eventually cell death when the lytic replication cycle and production of new progeny virions is completed (Moore, 2007; Banerjee *et al.*, 2016).

During latency, KSHV promotes cell proliferation and prevents apoptosis by constitutively expressing a subset of viral proteins and microRNAs. Their targets are cellular proteins that play crucial roles in cell cycle regulation, such as cyclin-dependend kinases (CDKs) and their inhibitors (CDKI), the retinoblastoma protein (pRB) and the transcription factor p53, among others.

One of the KSHV proteins that manipulates the cell cycle is vCyclin (ORF72) (Cesarman *et al.*, 1996; Chang *et al.*, 1996). This protein possesses sequence similarity to the cellular D-type cyclins, which target CDK4 and CDK6 to phosphorylate pRb and thereby activate E2F-dependent transcription. KSHV vCyclin is expressed from a bicistronic mRNA along with vFLIP. Although during latency vCyclin primarily partners with CDK6 and phosphorylates pRb, it can also bind CDK4 and CDK2 (Godden-Kent *et al.*, 1997; Platt *et al.*, 2000; Kaldis *et al.*, 2001; Pekkonen *et al.*, 2014). Unlike D-type cyclins, the vCyclin/CDK6 complex phosphorylates a broader range of proteins that actually are more typical targets of other cyclins/CDKs (e.g cyclins A and E and CDK2). These targets include the CDK inhibitors (CDKIs) $p27^{Kip1}$ and $p21^{Cip1}$, histone H1, Cdc25A, as well as Cdc6 and ORC1 (Ellis *et al.*, 1999; Mann *et al.*, 1999; Laman *et al.*, 2001; Jarviluoma *et al.*, 2004; Sarek *et al.*, 2006), reviewed in (Verschuren *et al.*, 2004). Unlike the cellular counterpart that requires CDK6 phosphorylation by a CDK-activating kinase (CAK), the vCyclin/CDK6 complex is fully active in a non-phosphorylated form (Child and Mann, 2001; Kaldis *et al.*, 2001). Importantly, vCyclin lacks the cyclin destruction box that targets cellular cyclins for rapid protein degradation (Klotzbucher *et al.*, 1996), suggesting that the viral protein may be active throughout the different stages of the cell cycle (Van Dross *et al.*, 2005). Bypassing the cell cycle arrest is an important aspect of tumourigenesis and the vCyclin/CDK complexes can stimulate G1/S cell cycle progression even under stress conditions since they are insensitive to CDKIs such as $p21^{Cip1}$, $p27^{Kip1}$ and $p16^{INK4}$ (Jarviluoma *et al.*, 2006; Sarek *et al.*, 2006; Swanton *et al.*, 1997). In addition to direct inhibition of CDKIs, vCyclin has been shown to bypass the G1-arrest triggered by cell-cell contact and induce cellular transformation (Jones *et al.*, 2014).

The inactivation of pRb is a conserved function of herpesviruses (Hume *et al.*, 2008) and KSHV has at least two gene products that are able to perform this task, vCyclin and LANA. Similarly to its cellular counterparts, the vCyclin/CDK6 complex phosphorylates and inactivates pRb, allowing cells to progress into S-phase (Chang *et al.*, 1996; Godden-Kent *et al.*, 1997; Li *et al.*, 1997). LANA, on the other hand, interferes with the function of both p53 and pRb (rewieved in Wei *et al.*, 2016). LANA binds to the pocket region of the non-phosphorylated (active) form of pRb. This interaction inhibits the pRb-mediated repression of the E2F protein family (Radkov *et al.*, 2000). By activating the promoter of cell-cycle progression genes, LANA can overcome serum starvation, overexpression of the cyclin-dependent kinase inhibitor $p16^{INK4}$, as well as the BRD4 and BRD2/RING3-induced G1 cell cycle arrest (Platt *et al.*, 2000; Direkze and Laman, 2004; An *et al.*, 2005; Viejo-Borbolla *et al.*, 2005; Ottinger *et al.*, 2006).

In normal cells, deregulation of the cell cycle results in the activation of the transcription factor p53, which, depending on the severity of the stress, can result in cell cycle arrest or apoptosis. LANA was the first p53 inhibitor discovered in KSHV (Friborg *et al.*, 1999). The C-terminus of the viral protein interacts with p53 and inhibits its function as a transcription activator. LANA forms a ternary complex with p53 and its negative regulator MDM2. Inhibitors of the p53-MDM2 interaction, such as Nutlin-3, disrupt the ternary complex leading to apoptosis of PEL cells (Petre *et al.*, 2007; Platt *et al.*, 2000; Sarek *et al.*, 2007; Sarek and Ojala, 2007; Chen *et al.*, 2010; Ye *et al.*, 2012). In addition to direct inhibition of p53, recent studies showed that LANA inhibits apoptosis by binding to the aurora kinase B, inducing phosphorylation of survivin and cell proliferation (Lu *et al.*, 2014), and the multifunctional angiogenin to utilize its anti-apoptotic functions (Paudel *et al.*, 2012).

The importance of inhibiting apoptosis during latency is further demonstrated by a number of alternative strategies that KSHV utilizes to pursue this scope. While inhibition of p53 by LANA blocks the early stages of the apoptotic program, another viral latent protein, vFLIP, is able to prevent the execution of apoptosis at both early and late stages. Moreover, the viral protein interacts with IκB kinase (IKK) and thereby activates the NF-κB pathway (Liu *et al.*, 2002; Field *et al.*, 2003;), which may have great importance for the survival of the infected tumour cell (Chaudhary *et al.*, 1999). Accordingly, small molecule inhibitors of NF-κB or silencing of vFLIP expression rapidly induce PEL cell apoptosis (Keller *et al.*, 2000; Guasparri *et al.*, 2004). In addition, vFLIP induction of NF-κB impairs autophagosome elongation in latently infected cells, thus suppressing both starvation- and rapamycin-induced cell death of KSHV infected B-lymphocytes (Lee *et al.*, 2009). More recently, vFLIP was also demonstrated to protect the latently infected cells from oncogene-induced senescence elicited by vCyclin (Leidal *et al.*, 2012). Thus, vFLIP serves as both an anti-apoptotic as well as an anti-autophagic viral protein.

In addition to proteins, a set of virally encoded micro RNAs (miRNAs), have been shown to influence different cellular functions including cell growth and apoptosis, (reviewed in Ramalingam *et al.*, 2012; Zhu *et al.*, 2013)

Inhibition of apoptosis during the lytic replication phase

The progression of the lytic replication program coincides with an increase of cellular stress signals including the accumulation of reactive oxygen species (Li *et al.*, 2011; Ye *et al.*, 2011; Ma *et al.*, 2013), DNA damage (Luftig, 2014; Hollingworth *et al.*, 2015; Balistreri *et al.*, 2016), shut-off of host protein synthesis (Glaunsinger and Ganem, 2004; Luftig, 2014) and eventually cell death. To prolong the survival of the host cell, KSHV devotes a large number of viral lytic proteins to ensure the survival of the host cell until the completion of the lytic cycle.

The inhibition of p53 is maintained during the lytic phase by at least four lytic proteins, RTA, K-bZIP, and the interferon regulatory factor homologues vIRF1 and 3. The inhibition of p53 transcriptional activity occurs when the viral proteins bind to some histone acetyltransferases (HATs) thereby preventing them from functioning as p53 transcriptional co-adaptors (Li *et al.*, 2000; Park *et al.*, 2000; Gwack *et al.*, 2001a; Gwack *et al.*, 2001b; Hwang *et al.*, 2001; Lin *et al.*, 2001; Nakamura *et al.*, 2001). Both vIRF1 and -3 inhibit p53-induced apoptosis by interacting directly with the central DNA-binding domain (DBD) of p53 and ATM kinases, which results in reduced levels of p53 phosphorylation on the serine residue S15 (Shin *et al.*, 2006; Baresova *et al.*, 2014;). This causes an increase in p53 proteasomal degradation via the MDM2-mediated ubiquitination. In addition, K-bZIP has been shown to sequester p53 to promyelocytic leukemia (PML) bodies possibly contributing to K-bZIP-mediated inhibition of p53 transcription during lytic cycle (Park *et al.*2000) and thereby delaying the onset of the apoptotic signalling cascade (Katano *et al.*, 2001).

As described above, one viral protein can lead to the activation of large number of cellular proteins that have anti-apoptotic functions. However, when the extensive stresses shift the balance towards cell death, some cellular anti-apoptotic proteins can be converted to pro-apoptotic factors. This is the case for Bcl-2, which upon caspase cleavage turns its function from anti- to pro-apoptotic (Kirsch *et al.*, 1999; Guerrero *et al.*, 2012;). Interestingly, KSHV encodes an homolog of the cellular Bcl-2, vBcl2 (Sarid *et al.*, 1997), which is expressed as an early lytic gene and interferes with apoptosis through the inhibition of BH3 domain-containing proteins that have pro-apoptotic functions (Sun *et al.*, 1999; Flanagan and Letai, 2008;). Importantly, whereas cellular Bcl-2 can be cleaved by cellular caspases and converted to a pro-apoptotic version, KSHV vBcl-2 lacks this cleavage site and thus evades cellular regulation (Cheng *et al.*, 1997). Also, vBcl-2 escapes inactivation by the KSHV vCyclin protein which otherwise inactivates the cellular Bcl-2 by a CDK6-mediated phosphorylation (Ojala *et al.*, 2000). In addition, it has been shown that vBcl-2 can also inhibit autophagy via direct interaction with the host protein Beclin 1 (Pattingre *et al.*, 2005).

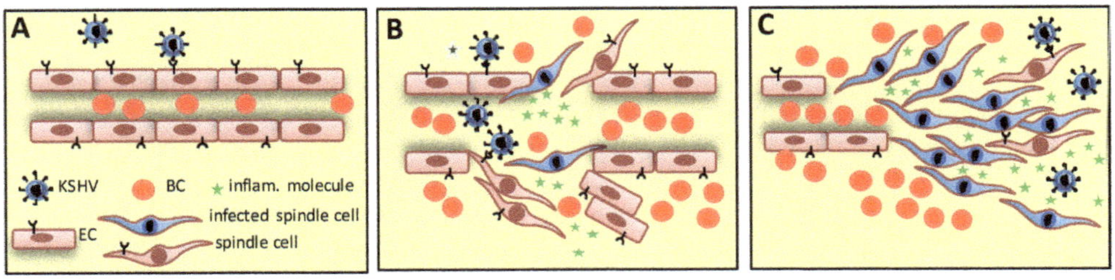

Figure 9.2. Schematic diagram of a KS lesion progression. **A)** KSHV virions infect endothelial cells. **B)** Early KS lesion: few KSHV infected endothelial cells (blue cells) produce proinflammatory molecules that affect the migratory and angiogenic potential of the surrounding uninfected cells. The vessel becomes leaky and red blood cells extravasate and accumulate in the extracellular space **C)** Late KS lesion. The majority of the cells are spindling and infected by KSHV. These cells proliferate and represent the main element of the lesion.

Lytic phase and cell cycle arrest

When expressed outside the context of infection, many viral proteins have been shown to induce cell cycle arrest. RTA and K-bZIP induce a G1-arrest (Wu *et al.*, 2002; Izumiya *et al.*, 2003; Wang *et al.*, 2003a; Wang *et al.*, 2003b; Wu *et al.*, 2003; Kumar and Wood, 2013), while S- and G2-phase arrests have been described for cells expressing vCyclin (Koopal *et al.*, 2007) and MTA (Jackson *et al.*, 2014), respectively. Caution should, however, be taken in inferring that this is their function during infection since the outcome of a full infection cycle is the result of cumulative, often compensatory, effects of many viral products. For instance, as we now know, the cell cycle arrest obtained upon overexpression of vCyclin is the consequence of an excessive stimulation of cell proliferation and DNA damage - the main function of this protein- rather than direct inhibition of cell cycle progression (Koopal *et al.*, 2007; Leidal *et al.*, 2012).

Analysis of the cell cycle progression during the lytic phase has shown that cells undergo cell cycle arrest prior to lysis. This is in striking contrast to the proliferative properties of latently infected cells. While early studies indicated that chemically reactivated cells accumulated in G1 (Izumiya *et al.*, 2003), more recent works performed in cell lines that more efficiently sustain KSHV reactivation have shown that the reactivated cells arrest at a stage of the cell cycle more similar to late S- or early G2 phase (Hollingworth *et al.*, 2015; Balistreri *et al.*, 2016). Similar results have been obtained for EBV (Gastaldello *et al.*, 2010). Using markers of reactivation and single-cell analysis, it was shown that reactivated cells are able to bypass a pre-imposed G1/S arrest and to enter S-phase (Balistreri *et al.*, 2016). Converging evidences suggest that DNA damage response (DDR) could play a role in this process. When DNA damage occurs, cells arrest in early G2 phase and try to repair the damage before entering mitosis (Branzei and Foiani, 2008). Interestingly, depleting or inhibiting cellular factors that are involved in the G2/M arrest (i.e. p53 and $p21^{Cip1}$) impaired KSHV viral lytic gene expression (Balistreri *et al.*, 2016), suggesting that the cellular environment of cells arrested in S/G2-phase favors the virus lytic cycle. The reasons are not yet clear. Future work will hopefully clarify how and why γ-herpesviruses induce G2-arrest during the lytic phase.

Atypical differentiation, abnormal angiogenesis and altered metabolism of KSHV-infected endothelial cells

Atypical differentiation

The predominant cell type in advanced KS lesions is the spindle cell (SC) (Figure 9.2). The origin of those cells is still controversial. Despite SCs expressing markers of both lymphatic and blood endothelial cell types they also display mesenchymal markers (α-SMA, vimentin and PDGFRα), as well as those of dendritic cells and macrophages (Weich *et al.*, 1991; Kaaya *et al.*, 1995; Sturzl *et al.*, 1995). A gene expression study comparing the expression pattern of different cell types to that of the so-called KS signature (i.e. a set of 1482 genes differentially expressed in KS tissues as compared to normal skin), indicated that the KS SCs are of endothelial origin, with a profile closer to those of lymphatic rather than blood endothelial cells (LEC and BEC, respectively) (Wang *et al.*, 2004a).

The expression of lineage specific markers like LYVE1, podoplanin and VEGFR3 in SC supports their LEC origin (Carroll *et al.*, 2004; Hong *et al.*, 2004; Wang *et al.*, 2004a). In vitro, LECs were found to be more susceptible to KSHV infection, to sustain the viral replication and to maintain the episome more efficiently than BECs (Wang *et al.*, 2004a; Chang and Ganem, 2013). Furthermore, infection of LECs by KSHV induces morphological changes strongly reminiscent of the SCs observed in KS tumours. However, the expression pattern of KS SCs does not faithfully represent those of uninfected LECs. Many BEC markers were expressed also in KS SCs (Wang *et al.*, 2004a; Cheng *et al.*, 2011) and *in vitro* it has been extensively shown that KSHV can induce cellular reprogramming of LECs towards a BEC phenotype and, vice versa, KSHV infected BECs display the expression of several LEC specific markers (Carroll *et al.*, 2004; Hong *et al.*, 2004; Wang *et al.*, 2004a; Cheng *et al.*, 2011).

The investigation of the mechanisms hijacked by the virus to induce such a reprogramming in both LECs and BECs pointed to Prox1 as a major player (Hong *et al.*, 2004; Yoo *et al.*, 2010; Aguilar *et al.*, 2012; Cancian *et al.*, 2013). Prox1 is a transcription factor expressed in the central nervous system, in the liver and in colon cancer stem cells (Elsir *et al.*, 2012; Wiener *et al.*, 2014). Considered also as the master regulator of the lymphatic cell fate, PROX1 is highly expressed in LECs and is required for the maintenance of the lymphatic cellular identity (Johnson *et al.*, 2008; Bernaudin *et al.*, 2013). Ectopic expression of Prox1 in BECs triggers their differentiation into LECs (Hong *et al.*, 2002; Petrova *et al.*, 2002; Mishima *et al.*, 2007). KSHV, by modulating PROX1 expression in both LECs and BECs, can modify their cell lineage identity. In particular, upon KSHV infection of LECs, PROX1 expression is significantly downregulated due to vFLIP and vGPCR-mediated activation of the Notch signalling pathway (Hansen *et al.*, 2010; Aguilar *et al.*, 2012;). This results in the downregulation of markers of the lymphatic lineage identity. When BECs are infected with KSHV, the increased expression of IL3Rα induces Prox1 upregulation with the subsequent induction of LEC-specific markers (Yoo *et al.*, 2012).

By culturing KSHV-infected LECs in a three-dimensional (3D) culture model as spheroids, it was recently shown that KSHV-infected LECs lose *PROX1* expression, dedifferentiate, start expressing mesenchymal genes and acquire invasive properties (Cheng *et al.*, 2011). Further studies using this 3D model elucidated the molecular mechanism of KSHV-induced LEC reprogramming, which involves vFLIP and vGPCR-mediated Notch signalling pathway activation and a subsequent increase of the membrane-bound metalloprotease MMP14 expression which has been shown to be required for the increased invasiveness of infected endothelial cells (Cheng *et al.*, 2011).

Abnormal angiogenesis

Aberrant angiogenesis is another feature of KS and it is macroscopically visible already at the first stages of the disease. The red-purplish color of the lesions is in fact attributable to the presence of an aberrant and leaky vasculature throughout the lesion. This allows the extravasation of red blood cells and their accumulation in the extracellular space (Orenstein, 2008) (Figure 9.2). Several KSHV-encoded proteins, expressed in both lytic and latent phase, contribute to the angiogenic features of KS (Figure 9.1).

At the left end of KSHV genome, right after the terminal repeats (Figure 9.1), K1 ORF encodes a single pass transmembrane protein, which has been shown to encompass protumourigenic and angiogenic properties. K1, an immunoglobulin-like protein expressed mainly during the KSHV lytic cycle, induces the secretion of VEGF and MMP9 in HUVECs and has a general angiogenic and pro-survival effect (Wang *et al.*, 2004b; Wang *et al.*, 2006). When ectopically expressed in epithelial cells, K1 increases the transcription and the secretion of all active forms of VEGF. Thereby, VEGF by stimulating its cognate receptor VEGFR2 can trigger angiogenesis through the activation of PI3K and the downstream Akt signalling (Tomlinson and Damania, 2004; Wang *et al.*, 2004b). Although it does not possess transforming potential on its own, K1 can exacerbate the tumourigenicity of the cervical carcinoma-derived C33A cells in a mouse model. When these cells, stably expressing K1, were injected into nude mice, the vascularization and the size of the tumours were significantly higher as compared to tumours caused by GFP-expressing C33A cells (Wang *et al.*, 2006).

Another proangiogenic, viral membrane protein encoded by an ORF at the other extremity of the viral genome is K15 (Figure 9.1) (Glenn *et al.*, 1999; Poole *et al.*, 1999; Choi *et al.*, 2000). Although able to induce the secretion of proangiogenic molecules (IL8, IL6, IL1, CCL20, CCL2) (Brinkmann *et al.*, 2007), the main angiogenesis-promoting effect of K15 is displayed in an autocrine fashion. K15 seems in fact to mimic a constitutively active VEGFR2 by recruiting and triggering the phosphorylation of PLCγ1 resulting in the NFAT-dependent activation of DSCR1 in a VEGFR2-independent manner (Bala *et al.*, 2012). K15 increases also the invasive properties of endothelial cells and its expression is crucial for virus-induced angiogenesis and invasiveness. Depletion of the K15 ORF from the KSHV genome reduces the *in vitro* angiogenesis and invasiveness of endothelial cells (Bala *et al.*, 2012; Gramolelli *et al.*, 2015).

KSHV infection of endothelial cells triggers a massive secretion of VEGFA, which by activating its cognate receptor in the surrounding cells, induces angiogenesis (Masood *et al.*, 2002; Sivakumar *et al.*, 2008; Wang and Damania, 2008). The mechanism of KSHV-mediated induction of VEGFA has been extensively investigated and HIF1α upregulation is a major factor involved in this phenomenon (Glaunsinger and Ganem, 2004; Carroll *et al.*, 2006; Ma *et al.*, 2015). In particular, KSHV infection increases the levels of this transcription factor, which, during normoxia is rapidly degraded (Carroll *et al.*, 2006). Multiple viral proteins contribute to the increase of HIF1α half-life. LANA induces the degradation of two HIF1α repressors: the VHL protein and p53, and furthermore by direct binding to HIF1α increases its transcriptional activity (Cai *et al.*, 2006). Similarly, also vIRF3 recruits and triggers the accumulation of HIF1α (Shin *et al.*, 2008). In addition, vGPCR, through the activation of the MAPK and p38 signalling axis, induces HIF1α phosphorylation and enhances its transcriptional activity (Sodhi *et al.*, 2000). Remarkably, the degree of HIF1α expression was correlated with tumour progression in KS lesions of HIV-positive patients, thus pointing to this transcription factor as a key player in KS pathogenesis and indicating the potential use of HIF1α as a marker of disease progression, at least in AIDS-associated KS (Long *et al.*, 2009).

During the latent phase, KSHV expresses a set of 12 miRNAs, which have a role in the fine-tuning of host cell expression and they contribute, at least *in vitro*, to the proangiogenic phenotype promoted by KSHV (Gallaher *et al.*, 2013). These viral-encoded miRNAs have also a role in the increase of HIF1α induced by KSHV (see below).

Alteration of cell metabolism

KSHV infection of endothelial cells triggers a change in cell metabolism known as Warburg effect (Delgado *et al.*, 2010; Delgado *et al.*, 2012; Ma *et al.*, 2015). Observed in many types of cancer, the Warburg effect consists in a metabolic switch towards lactic acid production and anaerobic glycolysis, which becomes the main source of energy even in the presence of normal oxygen levels (Warburg, 1956a, b). The general effect is a downregulation of the mitochondrial oxidative phosphorylation and a decrease in oxygen consumption. In the case of KSHV-infected cells, this phenomenon is essential for cell survival, in fact, inhibitors of glycolysis cause apoptosis in latently infected cells (Delgado *et al.*, 2010). Furthermore, KSHV induced alteration of cell metabolism is important for the maintenance of the latent phase (Delgado *et al.*, 2010).

This metabolic switch is, at least in part, due to the effect of the virus-encoded miRNAs. When ectopically expressed in LECs, they alter the host metabolism through two independent mechanisms (Yogev *et al.*, 2014). On one hand, KSHV miRNAs target and reduce the levels of the mitochondrial heat shock protein HSPA9, important for mitochondrial viability, thus affecting the number of mitochondria and contributing to the virus-induced metabolic switch (Yogev *et al.*, 2014). On the other hand, several KSHV miRNAs target the prolyl hydroxylase EGLN2, which is responsible for the degradation of HIF1α. The resulting downregulation of EGLN2 leads to the accumulation of HIF1α, responsible for the increase in anerobic glycolysis. The mechanism of HIF1α involvement in KSHV-mediated Warburg effect in latently infected endothelial cells entails also the increased expression of its metabolic effector pyruvate kinase 2 (PK2), which, in turn, is essential for maintaining anaerobic glycolysis in KSHV infected cells (Ma *et al.*, 2015).

Prospects of clinical translation and future trends

To date, no standard therapeutic guidelines exist for the treatment of KSHV-associated malignancies. In KS patients, the therapy can be different depending on the location and variant of the malignancy as well as immune competence status of the patient. In classic KS, the therapy ranges from no treatment to surgical excision of the lesion, chemo- and radiotherapy (Schwartz, 2004). Iatrogenic KS often regresses with the interruption or reduction of the immunosuppressive therapy (Hengge *et al.*, 2002). Systemic chemotherapy is normally used in endemic KS (Hengge *et al.*, 2002). The therapy of choice in AIDS-KS relies on the HAART combined with pegylated liposomal doxorubicin or other liposomal anthracyclines (Bower *et al.*, 2008). PEL prognosis is very poor (less than 6 months) and is usually treated with chemotherapy (Simonelli *et al.*, 2003a; Simonelli *et al.*, 2003b). MCD therapy is based on tocilizumab, an IL6 receptor antibody and radiotherapy (Song *et al.*, 2010). Ganciclovir, an inhibitor of viral replication, also showed beneficial effects (Casper *et al.*, 2004).

The high costs and the adverse effects of these current therapies together with the high number of cases intractable, has prompted reserchers for the identification of novel therapeutic agents for the cure of KSHV-associated malignancies.

The studies of the molecular mechanisms of KSHV-induced pathogenesis have given the basis for a better understanding of the virus-induced manipulation of the infected cells, and importantly, also helped in the identification of new druggable targets for therapies. One possible therapeutic strategy consists in targeting cellular pathways essential for KSHV survival but dispensable for the host. Such an approch would specifically affect the survival of infected cells without influencing the viability of the uninfected ones. Sirolimus, an inhibitor of mTOR, which is activated by KSHV, has been shown *in vitro* to inhibit the expression of KSHV early lytic genes and in a pilot study was successfully used for treating iatrogenic KS (Krown *et al.*, 2012). Another suitable cellular target might be c-myc. In cancers where this protein is overexpressed, bromodomain (brd) inhibitors have shown promising results. In PEL, c-myc is deregulated at the post-translational level and is essential for cell cycle progression and proliferation of infected cells as well as for lymphoma progression in a mouse model (Sin *et al.*, 2015). Treatment with the brd extraterminal domain (BET) inhibitor JQ1 suppressed the expression of c-myc and induced growth arrest both *in vitro* and in a xenograft PEL model (Tolani *et al.*, 2014). Inhibition of MDM2-mediated p53 degradation by the small molecule Nutlin-3 impaired the formation of the MDM2-p53-LANA ternary complex (see above), induced selective apoptosis and showed antitumour effect in a PEL mouse xenograft model (Sarek *et al.*, 2007; Sarek and Ojala, 2007). Currently, studies focussed on the generation of more stable and effective inhibitors of the MDM2-p53 interaction thus paving the way for more efficent treatments of KSHV associated malignancies.

KSHV relies on different cellular kinases for the successfull initiation and completion of lytic reactivation. As already mentioned, lytic reactivation sustains the pathogenic process induced by KSHV in KS and MCD. Therefore, identifying and targeting cellular kinases essential for this process can reveal new possible therapeutic modalities. Screens seeking for the identification of such kinases have identified Raf/MEK/ERK, Pim kinases, tousled-like kinase and the MAP4K4 kinase as essential players required for successfull lytic replication and as possible novel therapeutic targets (Yu *et al.*, 2007; Varjosalo *et al.*, 2008; Cheng *et al.*, 2009; Dillon *et al.*, 2013; Haas *et al.*, 2013).

Another therapeutic approach that is showing encouraging results involves the use of inhibitors of receptor tyrosine kinases (RTKs), some of which have been linked to KS pathogenesis. Treatment with imatinib, a broad RTK inhibitor targeting e.g. BCR-Abl and c-Kit, induced KS regression in one third of the patients (Koon *et al.*, 2005; Koon *et al.*, 2014) Sorafenib, which targets VEGF and PDGF, although used in a single patient study, showed complete regression of KS (Ardavanis *et al.*, 2008).

References

Biological agents. Volume 100 B. A review of human carcinogens. IARC monographs on the evaluation of carcinogenic risks to humans / World Health Organization, International Agency for Research on Cancer 100, 1-441.

Aguilar, B., Choi, I., Choi, D., Chung, H.K., Lee, S., Yoo, J., Lee, Y.S., Maeng, Y.S., Lee, H.N., Park, E., *et al.* (2012). Lymphatic reprogramming by Kaposi sarcoma herpes virus promotes the oncogenic activity of the virus-encoded G-protein-coupled receptor. Cancer research 72, 5833-5842. 10.1158/0008-5472.CAN-12-1229

Akula, S.M., Naranatt, P.P., Walia, N.S., Wang, F.Z., Fegley, B., and Chandran, B. (2003). Kaposi's sarcoma-associated herpesvirus (human herpesvirus 8) infection of human fibroblast cells occurs through endocytosis. J. Virol. 77, 7978-7990.

Akula, S.M., Pramod, N.P., Wang, F.Z., and Chandran, B. (2001a). Human herpesvirus 8 envelope-associated glycoprotein B interacts with heparan sulfate-like moieties. Virology 284, 235-249. 10.1006/viro.2001.0921

Akula, S.M., Pramod, N.P., Wang, F.Z., and Chandran, B. (2002). Integrin alpha3beta1 (CD 49c/29) is a cellular receptor for Kaposi's sarcoma-associated herpesvirus (KSHV/HHV-8) entry into the target cells. Cell 108, 407-419.

Akula, S.M., Wang, F.Z., Vieira, J., and Chandran, B. (2001b). Human herpesvirus 8 interaction with target cells involves heparan sulfate. Virology 282, 245-255. 10.1006/viro.2000.0851

Alagiozoglou, L., Morris, L., Bredell, H., Martin, D.J., and Sitas, F. (2003). Human herpesvirus-8 antibodies and DNA in HIV-1 infected patients in South Africa. Epidemiology and infection 131, 1125-1129.

An, F.Q., Compitello, N., Horwitz, E., Sramkoski, M., Knudsen, E.S., and Renne, R. (2005). The latency-associated nuclear antigen of Kaposi's sarcoma-associated herpesvirus modulates cellular gene expression and protects lymphoid cells from p16 INK4A-induced cell cycle arrest. J. Biol. Chem. 280, 3862-3874. 10.1074/jbc.M407435200

An, F.Q., Folarin, H.M., Compitello, N., Roth, J., Gerson, S.L., McCrae, K.R., Fakhari, F.D., Dittmer, D.P., and Renne, R. (2006). Long-term-infected telomerase-immortalized endothelial cells: a model for Kaposi's sarcoma-associated herpesvirus latency *in vitro* and *in vivo*. J. Virol. 80, 4833-4846.

Ansari, M.Q., Dawson, D.B., Nador, R., Rutherford, C., Schneider, N.R., Latimer, M.J., Picker, L., Knowles, D.M., and McKenna, R.W. (1996). Primary body cavity-based AIDS-related lymphomas. American J. Clin. Pathol. 105, 221-229.

Antman, K., and Chang, Y. (2000). Kaposi's sarcoma. New Engl. J. Med. 342, 1027-1038. 10.1056/NEJM200004063421407

Ardavanis, A., Doufexis, D., Kountourakis, P., and Rigatos, G. (2008). A Kaposi's sarcoma complete clinical response after sorafenib administration. Ann Oncol 19, 1658-1659.

Arias, C., Weisburd, B., Stern-Ginossar, N., Mercier, A., Madrid, A.S., Bellare, P., Holdorf, M., Weissman, J.S., and Ganem, D. (2014). KSHV 2.0: a comprehensive annotation of the Kaposi's sarcoma-associated herpesvirus genome using next-generation sequencing reveals novel genomic and functional features. PLoS pathogens 10, e1003847. 10.1371/journal.ppat.1003847

Arvanitakis, L., Mesri, E.A., Nador, R.G., Said, J.W., Asch, A.S., Knowles, D.M., and Cesarman, E. (1996). Establishment and characterization of a primary effusion (body cavity-based) lymphoma cell line (BC-3) harboring kaposi's sarcoma-associated herpesvirus (KSHV/HHV-8) in the absence of Epstein-Barr virus. Blood 88, 2648-2654.

Bala, K., Bosco, R., Gramolelli, S., Haas, D.A., Kati, S., Pietrek, M., Havemeier, A., Yakushko, Y., Singh, V.V., Dittrich-Breiholz, O., *et al.* (2012). Kaposi's sarcoma herpesvirus K15 protein contributes to virus-induced angiogenesis by recruiting PLCgamma1 and activating NFAT1-dependent RCAN1 expression. PLoS pathogens 8, e1002927.

Balistreri, G., Viiliainen, J., Turunen, M., Diaz, R., Lyly, L., Pekkonen, P., Rantala, J., Ojala, K., Sarek, G., Teesalu, M., *et al.* (2016). Oncogenic Herpesvirus Utilizes Stress-Induced Cell Cycle Checkpoints for Efficient Lytic Replication. PLoS pathogens 12, e1005424. 10.1371/journal.ppat.1005424

Ballestas, M.E., Chatis, P.A., and Kaye, K.M. (1999). Efficient persistence of extrachromosomal KSHV DNA mediated by latency-associated nuclear antigen. Science 284, 641-644.

Ballestas, M.E., and Kaye, K.M. (2001). Kaposi's sarcoma-associated herpesvirus latency-associated nuclear antigen 1 mediates episome persistence through *cis*-acting terminal repeat (TR) sequence and specifically binds TR DNA. J. Virol. 75, 3250-3258.

Banerjee, S., Uppal, T., Strahan, R., Dabral, P., and Verma, S.C. (2016). The Modulation of Apoptotic Pathways by Gammaherpesviruses. Frontiers in microbiology 7, 585. 10.3389/fmicb.2016.00585

Barbera, A.J., Ballestas, M.E., and Kaye, K.M. (2004). The Kaposi's sarcoma-associated herpesvirus latency-associated nuclear antigen 1 N terminus is essential for chromosome association, DNA replication, and episome persistence. J. Virol. 78, 294-301.

Barbera, A.J., Chodaparambil, J.V., Kelley-Clarke, B., Luger, K., and Kaye, K.M. (2006). Kaposi's sarcoma-associated herpesvirus LANA hitches a ride on the chromosome. Cell cycle 5, 1048-1052. 10.4161/cc.5.10.2768

Baresova, P., Musilova, J., Pitha, P.M., and Lubyova, B. (2014). p53 tumour suppressor protein stability and transcriptional activity are targeted by Kaposi's sarcoma-associated herpesvirus-encoded viral interferon regulatory factor 3. Mol. Cell. Biol. 34, 386-399. 10.1128/MCB.01011-13

Beral, V., Peterman, T.A., Berkelman, R.L., and Jaffe, H.W. (1990). Kaposi's sarcoma among persons with AIDS: a sexually transmitted infection? Lancet 335, 123-128.

Bernaudin, J.F., Kambouchner, M., and Lacave, R. (2013). [Lymphatic vascular system, development and lymph formation. Review]. Revue de pneumologie clinique 69, 93-101. 10.1016/j.pneumo.2013.01.005

Birkmann, A., Mahr, K., Ensser, A., Yaguboglu, S., Titgemeyer, F., Fleckenstein, B., and Neipel, F. (2001). Cell surface heparan sulfate is a receptor for human herpesvirus 8 and interacts with envelope glycoprotein K8.1. J. Virol. 75, 11583-11593. 10.1128/JVI.75.23.11583-11593.2001

Bower, M., Collins, S., Cottrill, C., Cwynarski, K., Montoto, S., Nelson, M., Nwokolo, N., Powles, T., Stebbing, J., Wales, N., *et al.* (2008). British HIV Association guidelines for HIV-associated malignancies 2008. HIV medicine 9, 336-388. 10.1111/j.1468-1293.2008.00608.x

Branzei, D., and Foiani, M. (2008). Regulation of DNA repair throughout the cell cycle. Nat. Rev. Mol. Cell Biol. 9, 297-308. 10.1038/nrm2351

Brinkmann, M.M., Pietrek, M., Dittrich-Breiholz, O., Kracht, M., and Schulz, T.F. (2007). Modulation of host gene expression by the K15 protein of Kaposi's sarcoma-associated herpesvirus. J. Virol. 81, 42-58.

Cai, Q., Lan, K., Verma, S.C., Si, H., Lin, D., and Robertson, E.S. (2006). Kaposi's sarcoma-associated herpesvirus latent protein LANA interacts with HIF-1 alpha to upregulate RTA expression during hypoxia: Latency control under low oxygen conditions. J. Virol. 80, 7965-7975. 10.1128/JVI.00689-06

Cai, X., Lu, S., Zhang, Z., Gonzalez, C.M., Damania, B., and Cullen, B.R. (2005). Kaposi's sarcoma-associated herpesvirus expresses an array of viral microRNAs in latently infected cells. Proc Natl Acad Sci USA 102, 5570-5575. 10.1073/pnas.0408192102

Cancian, L., Hansen, A., and Boshoff, C. (2013). Cellular origin of Kaposi's sarcoma and Kaposi's sarcoma-associated herpesvirus-induced cell reprogramming. Trends in cell biology 23, 421-432. 10.1016/j.tcb.2013.04.001

Carbone, A., Cesarman, E., Gloghini, A., and Drexler, H.G. (2010). Understanding pathogenetic aspects and clinical presentation of primary effusion lymphoma through its derived cell lines. AIDS (London, England) 24, 479-490. 10.1097/QAD.0b013e3283365395

Carbone, A., Gloghini, A., Vaccher, E., Marchetti, G., Gaidano, G., and Tirelli, U. (2005). KSHV/HHV-8 associated lymph node based lymphomas in HIV seronegative subjects. Report of two cases with anaplastic large cell morphology and plasmablastic immunophenotype. J. Clin. Pathol. 58, 1039-1045. 10.1136/jcp.2005.026542

Carroll, P.A., Brazeau, E., and Lagunoff, M. (2004). Kaposi's sarcoma-associated herpesvirus infection of blood endothelial cells induces lymphatic differentiation. Virology 328, 7-18. 10.1016/j.virol.2004.07.008

Carroll, P.A., Kenerson, H.L., Yeung, R.S., and Lagunoff, M. (2006). Latent Kaposi's sarcoma-associated herpesvirus infection of endothelial cells activates

hypoxia-induced factors. J. Virol. 80, 10802-10812. 10.1128/JVI.00673-06
Casper, C., Nichols, W.G., Huang, M.L., Corey, L., and Wald, A. (2004). Remission of HHV-8 and HIV-associated multicentric Castleman disease with ganciclovir treatment. Blood 103, 1632-1634. 10.1182/blood-2003-05-1721
Cesarman, E., Chang, Y., Moore, P.S., Said, J.W., and Knowles, D.M. (1995). Kaposi's sarcoma-associated herpesvirus-like DNA sequences in AIDS-related body-cavity-based lymphomas. New Engl. J. Med. 332, 1186-1191. 10.1056/NEJM199505043321802
Cesarman, E., Nador, R.G., Bai, F., Bohenzky, R.A., Russo, J.J., Moore, P.S., Chang, Y., and Knowles, D.M. (1996). Kaposi's sarcoma-associated herpesvirus contains G protein-coupled receptor and cyclin D homologs which are expressed in Kaposi's sarcoma and malignant lymphoma. J. Virol. 70, 8218-8223.
Chadburn, A., Hyjek, E., Mathew, S., Cesarman, E., Said, J., and Knowles, D.M. (2004). KSHV-positive solid lymphomas represent an extra-cavitary variant of primary effusion lymphoma. The American journal of surgical pathology 28, 1401-1416.
Chakraborty, S., Veettil, M.V., and Chandran, B. (2012). Kaposi's Sarcoma Associated Herpesvirus Entry into Target Cells. Frontiers in microbiology 3, 6. 10.3389/fmicb.2012.00006
Chandriani, S., and Ganem, D. (2007). Host transcript accumulation during lytic KSHV infection reveals several classes of host responses. PloS one 2, e811.
Chandriani, S., and Ganem, D. (2010). Array-based transcript profiling and limiting-dilution reverse transcription-PCR analysis identify additional latent genes in Kaposi's sarcoma-associated herpesvirus. J. Virol. 84, 5565-5573. 10.1128/JVI.02723-09
Chang, H.H., and Ganem, D. (2013). A unique herpesviral transcriptional program in KSHV-infected lymphatic endothelial cells leads to mTORC1 activation and rapamycin sensitivity. Cell host and microbe 13, 429-440. 10.1016/j.chom.2013.03.009
Chang, Y., Cesarman, E., Pessin, M.S., Lee, F., Culpepper, J., Knowles, D.M., and Moore, P.S. (1994). Identification of herpesvirus-like DNA sequences in AIDS-associated Kaposi's sarcoma. Science 266, 1865-1869.
Chang, Y., Moore, P.S., Talbot, S.J., Boshoff, C.H., Zarkowska, T., Godden, K., Paterson, H., Weiss, R.A., and Mittnacht, S. (1996). Cyclin encoded by KS herpesvirus. Nature 382, 410. 10.1038/382410a0
Chaudhary, P.M., Jasmin, A., Eby, M.T., and Hood, L. (1999). Modulation of the NF-kappa B pathway by virally encoded death effector domains-containing proteins. Oncogene 18, 5738-5746. 10.1038/sj.onc.1202976
Chen, W., Hilton, I.B., Staudt, M.R., Burd, C.E., and Dittmer, D.P. (2010). Distinct p53, p53:LANA, and LANA complexes in Kaposi's Sarcoma--associated Herpesvirus Lymphomas. J. Virol. 84, 3898-3908. 10.1128/JVI.01321-09
Cheng, E.H., Nicholas, J., Bellows, D.S., Hayward, G.S., Guo, H.G., Reitz, M.S., and Hardwick, J.M. (1997). A Bcl-2 homolog encoded by Kaposi sarcoma-associated virus, human herpesvirus 8, inhibits apoptosis but does not heterodimerize with Bax or Bak. Proc Natl Acad Sci USA 94, 690-694.
Cheng, F., Pekkonen, P., Laurinavicius, S., Sugiyama, N., Henderson, S., Gunther, T., Rantanen, V., Kaivanto, E., Aavikko, M., Sarek, G., *et al.* (2011). KSHV-initiated notch activation leads to membrane-type-1 matrix metalloproteinase-dependent lymphatic endothelial-to-mesenchymal transition. Cell host and microbe 10, 577-590.
Cheng, F., Weidner-Glunde, M., Varjosalo, M., Rainio, E.M., Lehtonen, A., Schulz, T.F., Koskinen, P.J., Taipale, J., and Ojala, P.M. (2009). KSHV reactivation from latency requires Pim-1 and Pim-3 kinases to inactivate the latency-associated nuclear antigen LANA. PLoS pathogens 5, e1000324. 10.1371/journal.ppat.1000324
Child, E.S., and Mann, D.J. (2001). Novel properties of the cyclin encoded by Human Herpesvirus 8 that facilitate exit from quiescence. Oncogene 20, 3311-3322. 10.1038/sj.onc.1204447
Choi, J.K., Lee, B.S., Shim, S.N., Li, M., and Jung, J.U. (2000). Identification of the novel K15 gene at the rightmost end of the Kaposi's sarcoma-associated herpesvirus genome. J. Virol. 74, 436-446.
Choi, Y.B., and Nicholas, J. (2008). Autocrine and paracrine promotion of cell survival and virus replication by human herpesvirus 8 chemokines. J. Virol. 82, 6501-6513.
Connick, E., Kane, M.A., White, I.E., Ryder, J., and Campbell, T.B. (2004). Immune reconstitution inflammatory syndrome associated with Kaposi sarcoma during potent antiretroviral therapy. Clin. Infect Dis. 39, 1852-1855. 10.1086/426078
Coupland, S.E., Charlotte, F., Mansour, G., Maloum, K., Hummel, M., and Stein, H. (2005). HHV-8-associated T-cell lymphoma in a lymph node with concurrent peritoneal effusion in an HIV-positive man. The American journal of surgical pathology 29, 647-652.
Cronin, D.M., and Warnke, R.A. (2009). Castleman disease: an update on classification and the spectrum of associated lesions. Advances in anatomic pathology 16, 236-246. 10.1097/PAP.0b013e3181a9d4d3
da Silva, S.R., and de Oliveira, D.E. (2011). HIV, EBV and KSHV: viral cooperation in the pathogenesis of human malignancies. Cancer letters 305, 175-185. 10.1016/j.canlet.2011.02.007
Delgado, T., Carroll, P.A., Punjabi, A.S., Margineantu, D., Hockenbery, D.M., and Lagunoff, M. (2010). Induction of the Warburg effect by Kaposi's sarcoma herpesvirus is required for the maintenance of latently infected endothelial cells. Proc Natl Acad Sci USA 107, 10696-10701.
Delgado, T., Sanchez, E.L., Camarda, R., and Lagunoff, M. (2012). Global metabolic profiling of infection by an oncogenic virus: KSHV induces and requires lipogenesis for survival of latent infection. PLoS pathogens 8, e1002866.
Dillon, P.J., Gregory, S.M., Tamburro, K., Sanders, M.K., Johnson, G.L., Raab-Traub, N., Dittmer, D.P., and Damania, B. (2013). Tousled-like kinases modulate reactivation of gammaherpesviruses from latency. Cell

host and microbe 13, 204-214. 10.1016/j.chom.2012.12.005

Direkze, S., and Laman, H. (2004). Regulation of growth signalling and cell cycle by Kaposi's sarcoma-associated herpesvirus genes. Int. J. Exp. Pathol. 85, 305-319. 10.1111/j.0959-9673.2004.00407.x

Dittmer, D., Lagunoff, M., Renne, R., Staskus, K., Haase, A., and Ganem, D. (1998). A cluster of latently expressed genes in Kaposi's sarcoma-associated herpesvirus. J. Virol. 72, 8309-8315.

Dittmer, D.P. (2003). Transcription profile of Kaposi's sarcoma-associated herpesvirus in primary Kaposi's sarcoma lesions as determined by real-time PCR arrays. Cancer research 63, 2010-2015.

Domsic, J.F., Chen, H.S., Lu, F., Marmorstein, R., and Lieberman, P.M. (2013). Molecular basis for oligomeric-DNA binding and episome maintenance by KSHV LANA. PLoS pathogens 9, e1003672. 10.1371/journal.ppat.1003672

Du, M.Q., Liu, H., Diss, T.C., Ye, H., Hamoudi, R.A., Dupin, N., Meignin, V., Oksenhendler, E., Boshoff, C., and Isaacson, P.G. (2001). Kaposi sarcoma-associated herpesvirus infects monotypic (IgM lambda) but polyclonal naive B cells in Castleman disease and associated lymphoproliferative disorders. Blood 97, 2130-2136.

Dupin, N., Diss, T.L., Kellam, P., Tulliez, M., Du, M.Q., Sicard, D., Weiss, R.A., Isaacson, P.G., and Boshoff, C. (2000). HHV-8 is associated with a plasmablastic variant of Castleman disease that is linked to HHV-8-positive plasmablastic lymphoma. Blood 95, 1406-1412.

Ellis, M., Chew, Y.P., Fallis, L., Freddersdorf, S., Boshoff, C., Weiss, R.A., Lu, X., and Mittnacht, S. (1999). Degradation of p27(Kip) cdk inhibitor triggered by Kaposi's sarcoma virus cyclin-cdk6 complex. The EMBO journal 18, 644-653. 10.1093/emboj/18.3.644

Elsir, T., Smits, A., Lindstrom, M.S., and Nister, M. (2012). Transcription factor PROX1: its role in development and cancer. Cancer metastasis reviews 31, 793-805. 10.1007/s10555-012-9390-8

Ensoli, B., Sgadari, C., Barillari, G., Sirianni, M.C., Sturzl, M., and Monini, P. (2001). Biology of Kaposi's sarcoma. Eur J Cancer 37, 1251-1269.

Ensoli, B., and Sturzl, M. (1998). Kaposi's sarcoma: a result of the interplay among inflammatory cytokines, angiogenic factors and viral agents. Cytokine and growth factor reviews 9, 63-83.

Ensoli, B., and Sturzl, M. (1999). HHV-8 and multistep tumourigenesis. Trends Microbiol. 7, 310-312.

Field, N., Low, W., Daniels, M., Howell, S., Daviet, L., Boshoff, C., and Collins, M. (2003). KSHV vFLIP binds to IKK-gamma to activate IKK. Journal of cell science 116, 3721-3728. 10.1242/jcs.00691

Fielding, C.A., McLoughlin, R.M., Colmont, C.S., Kovaleva, M., Harris, D.A., Rose-John, S., Topley, N., and Jones, S.A. (2005). Viral IL-6 blocks neutrophil infiltration during acute inflammation. J Immunol 175, 4024-4029.

Flanagan, A.M., and Letai, A. (2008). BH3 domains define selective inhibitory interactions with BHRF-1 and KSHV BCL-2. Cell death and differentiation 15, 580-588. 10.1038/sj.cdd.4402292

Gaidano, G., and Carbone, A. (2001). Primary effusion lymphoma: a liquid phase lymphoma of fluid-filled body cavities. Advances in cancer research 80, 115-146.

Gallaher, A.M., Das, S., Xiao, Z., Andresson, T., Kieffer-Kwon, P., Happel, C., and Ziegelbauer, J. (2013). Proteomic screening of human targets of viral microRNAs reveals functions associated with immune evasion and angiogenesis. PLoS pathogens 9, e1003584.

Ganem, D. (2006). KSHV infection and the pathogenesis of Kaposi's sarcoma. Annual review of pathology 1, 273-296.

Garrigues, H.J., Rubinchikova, Y.E., Dipersio, C.M., and Rose, T.M. (2008). Integrin alphaVbeta3 Binds to the RGD motif of glycoprotein B of Kaposi's sarcoma-associated herpesvirus and functions as an RGD-dependent entry receptor. J. Virol. 82, 1570-1580. 10.1128/JVI.01673-07

Gastaldello, S., Hildebrand, S., Faridani, O., Callegari, S., Palmkvist, M., Di Guglielmo, C., and Masucci, M.G. (2010). A deneddylase encoded by Epstein-Barr virus promotes viral DNA replication by regulating the activity of cullin-RING ligases. Nature cell biology 12, 351-361. 10.1038/ncb2035

Gill, P.S., Tsai, Y.C., Rao, A.P., Spruck, C.H., 3rd, Zheng, T., Harrington, W.A., Jr., Cheung, T., Nathwani, B., and Jones, P.A. (1998). Evidence for multiclonality in multicentric Kaposi's sarcoma. Proc Natl Acad Sci USA 95, 8257-8261.

Glaunsinger, B., and Ganem, D. (2004). Highly selective escape from KSHV-mediated host mRNA shutoff and its implications for viral pathogenesis. J. Exp. Med. 200, 391-398. 10.1084/jem.20031881

Glenn, M., Rainbow, L., Aurade, F., Davison, A., and Schulz, T.F. (1999). Identification of a spliced gene from Kaposi's sarcoma-associated herpesvirus encoding a protein with similarities to latent membrane proteins 1 and 2A of Epstein-Barr virus. J. Virol. 73, 6953-6963.

Glesby, M.J., Hoover, D.R., Weng, S., Graham, N.M., Phair, J.P., Detels, R., Ho, M., and Saah, A.J. (1996). Use of antiherpes drugs and the risk of Kaposi's sarcoma: data from the Multicenter AIDS Cohort Study. J. Infect. Dis. 173, 1477-1480.

Godden-Kent, D., Talbot, S.J., Boshoff, C., Chang, Y., Moore, P., Weiss, R.A., and Mittnacht, S. (1997). The cyclin encoded by Kaposi's sarcoma-associated herpesvirus stimulates cdk6 to phosphorylate the retinoblastoma protein and histone H1. J. Virol. 71, 4193-4198.

Goto, H., Kojima, Y., Nagai, H., and Okada, S. (2013). Establishment of a CD4-positive cell line from an AIDS-related primary effusion lymphoma. International journal of hematology 97, 624-633. 10.1007/s12185-013-1339-3

Gradoville, L., Gerlach, J., Grogan, E., Shedd, D., Nikiforow, S., Metroka, C., and Miller, G. (2000). Kaposi's sarcoma-associated herpesvirus open reading frame 50/Rta protein activates the entire viral lytic cycle in the HH-B2 primary effusion lymphoma cell line. J. Virol. 74, 6207-6212.

Gramolelli, S., Weidner-Glunde, M., Abere, B., Viejo-Borbolla, A., Bala, K., Ruckert, J., Kremmer, E., and Schulz, T.F. (2015). Inhibiting the Recruitment of PLCgamma1 to Kaposi's Sarcoma Herpesvirus K15 Protein Reduces the Invasiveness and Angiogenesis of Infected Endothelial Cells. PLoS pathogens 11, e1005105. 10.1371/journal.ppat.1005105

Grundhoff, A., and Ganem, D. (2004). Inefficient establishment of KSHV latency suggests an additional role for continued lytic replication in Kaposi sarcoma pathogenesis. J. Clin. Invest. 113, 124-136. 10.1172/JCI17803

Guasparri, I., Keller, S.A., and Cesarman, E. (2004). KSHV vFLIP is essential for the survival of infected lymphoma cells. J. Exp. Med. 199, 993-1003. 10.1084/jem.20031467

Guerrero, A.D., Schmitz, I., Chen, M., and Wang, J. (2012). Promotion of Caspase Activation by Caspase-9-mediated Feedback Amplification of Mitochondrial Damage. Journal of clinical and cellular immunology 3. 10.4172/2155-9899.1000126

Gwack, Y., Byun, H., Hwang, S., Lim, C., and Choe, J. (2001a). CREB-binding protein and histone deacetylase regulate the transcriptional activity of Kaposi's sarcoma-associated herpesvirus open reading frame 50. J. Virol. 75, 1909-1917. 10.1128/JVI.75.4.1909-1917.2001

Gwack, Y., Hwang, S., Byun, H., Lim, C., Kim, J.W., Choi, E.J., and Choe, J. (2001b). Kaposi's sarcoma-associated herpesvirus open reading frame 50 represses p53-induced transcriptional activity and apoptosis. J. Virol. 75, 6245-6248. 10.1128/JVI.75.13.6245-6248.2001

Haas, D.A., Bala, K., Busche, G., Weidner-Glunde, M., Santag, S., Kati, S., Gramolelli, S., Damas, M., Dittrich-Breiholz, O., Kracht, M., *et al.* (2013). The inflammatory kinase MAP4K4 promotes reactivation of Kaposi's sarcoma herpesvirus and enhances the invasiveness of infected endothelial cells. PLoS pathogens 9, e1003737.

Hahn, A.S., Kaufmann, J.K., Wies, E., Naschberger, E., Panteleev-Ivlev, J., Schmidt, K., Holzer, A., Schmidt, M., Chen, J., Konig, S., *et al.* (2012). The ephrin receptor tyrosine kinase A2 is a cellular receptor for Kaposi's sarcoma-associated herpesvirus. Nature Med. 18, 961-966. 10.1038/nm.2805

Hansen, A., Henderson, S., Lagos, D., Nikitenko, L., Coulter, E., Roberts, S., Gratrix, F., Plaisance, K., Renne, R., Bower, M., *et al.* (2010). KSHV-encoded miRNAs target MAF to induce endothelial cell reprogramming. Genes and development 24, 195-205. 10.1101/gad.553410

Hellert, J., Weidner-Glunde, M., Krausze, J., Lunsdorf, H., Ritter, C., Schulz, T.F., and Luhrs, T. (2015). The 3D structure of Kaposi sarcoma herpesvirus LANA C-terminal domain bound to DNA. Proc Natl Acad Sci USA 112, 6694-6699. 10.1073/pnas.1421804112

Hellert, J., Weidner-Glunde, M., Krausze, J., Richter, U., Adler, H., Fedorov, R., Pietrek, M., Ruckert, J., Ritter, C., Schulz, T.F., *et al.* (2013). A structural basis for BRD2/4-mediated host chromatin interaction and oligomer assembly of Kaposi sarcoma-associated herpesvirus and murine gammaherpesvirus LANA proteins. PLoS pathogens 9, e1003640. 10.1371/journal.ppat.1003640

Hengge, U.R., Ruzicka, T., Tyring, S.K., Stuschke, M., Roggendorf, M., Schwartz, R.A., and Seeber, S. (2002). Update on Kaposi's sarcoma and other HHV8 associated diseases. Part 1: epidemiology, environmental predispositions, clinical manifestations, and therapy. The Lancet Infectious diseases 2, 281-292.

Hollingworth, R., Skalka, G.L., Stewart, G.S., Hislop, A.D., Blackbourn, D.J., and Grand, R.J. (2015). Activation of DNA Damage Response Pathways during Lytic Replication of KSHV. Viruses 7, 2908-2927. 10.3390/v7062752

Hong, Y.K., Foreman, K., Shin, J.W., Hirakawa, S., Curry, C.L., Sage, D.R., Libermann, T., Dezube, B.J., Fingeroth, J.D., and Detmar, M. (2004). Lymphatic reprogramming of blood vascular endothelium by Kaposi sarcoma-associated herpesvirus. Nature genetics 36, 683-685.

Hong, Y.K., Harvey, N., Noh, Y.H., Schacht, V., Hirakawa, S., Detmar, M., and Oliver, G. (2002). Prox1 is a master control gene in the program specifying lymphatic endothelial cell fate. Developmental dynamics: an official publication of the American Association of Anatomists 225, 351-357. 10.1002/dvdy.10163

Hosseinipour, M.C., Sweet, K.M., Xiong, J., Namarika, D., Mwafongo, A., Nyirenda, M., Chiwoko, L., Kamwendo, D., Hoffman, I., Lee, J., *et al.* (2014). Viral profiling identifies multiple subtypes of Kaposi's sarcoma. mBio 5, e01633-01614. 10.1128/mBio.01633-14

Hume, A.J., Finkel, J.S., Kamil, J.P., Coen, D.M., Culbertson, M.R., and Kalejta, R.F. (2008). Phosphorylation of retinoblastoma protein by viral protein with cyclin-dependent kinase function. Science 320, 797-799. 10.1126/science.1152095

Hwang, S., Gwack, Y., Byun, H., Lim, C., and Choe, J. (2001). The Kaposi's sarcoma-associated herpesvirus K8 protein interacts with CREB-binding protein (CBP) and represses CBP-mediated transcription. J. Virol. 75, 9509-9516. 10.1128/JVI.75.19.9509-9516.2001

Izumiya, Y., Lin, S.F., Ellison, T.J., Levy, A.M., Mayeur, G.L., Izumiya, C., and Kung, H.J. (2003). Cell cycle regulation by Kaposi's sarcoma-associated herpesvirus K-bZIP: direct interaction with cyclin-CDK2 and induction of G1 growth arrest. J. Virol. 77, 9652-9661.

Jackson, B.R., Noerenberg, M., and Whitehouse, A. (2014). A novel mechanism inducing genome instability in Kaposi's sarcoma-associated herpesvirus infected cells. PLoS pathogens 10, e1004098. 10.1371/journal.ppat.1004098

Janier, M., Morel, P., and Civatte, J. (1990). The Koebner phenomenon in AIDS-related Kaposi's sarcoma. Journal of the American Academy of Dermatology 22, 125-126.

Jarviluoma, A., Child, E.S., Sarek, G., Sirimongkolkasem, P., Peters, G., Ojala, P.M., and Mann, D.J. (2006). Phosphorylation of the cyclin-dependent kinase inhibitor p21Cip1 on serine 130 is essential for viral cyclin-mediated bypass of a p21Cip1-imposed G1 arrest. Mol. Cell. Biol. 26, 2430-2440.

Jarviluoma, A., Koopal, S., Rasanen, S., Makela, T.P., and Ojala, P.M. (2004). KSHV viral cyclin binds to p27KIP1 in primary effusion lymphomas. Blood 104, 3349-3354. 10.1182/blood-2004-05-1798

Jenner, R.G., Alba, M.M., Boshoff, C., and Kellam, P. (2001). Kaposi's sarcoma-associated herpesvirus latent and lytic gene expression as revealed by DNA arrays. J. Virol. 75, 891-902. 10.1128/JVI.75.2.891-902.2001

Johnson, N.C., Dillard, M.E., Baluk, P., McDonald, D.M., Harvey, N.L., Frase, S.L., and Oliver, G. (2008). Lymphatic endothelial cell identity is reversible and its maintenance requires Prox1 activity. Genes and development 22, 3282-3291. 10.1101/gad.1727208

Jones, T., Ramos da Silva, S., Bedolla, R., Ye, F., Zhou, F., and Gao, S.J. (2014). Viral cyclin promotes KSHV-induced cellular transformation and tumourigenesis by overriding contact inhibition. Cell cycle 13, 845-858. 10.4161/cc.27758

Jones, T., Ye, F., Bedolla, R., Huang, Y., Meng, J., Qian, L., Pan, H., Zhou, F., Moody, R., Wagner, B., *et al.* (2012). Direct and efficient cellular transformation of primary rat mesenchymal precursor cells by KSHV. J. Clin. Invest. 122, 1076-1081.

Judde, J.G., Lacoste, V., Briere, J., Kassa-Kelembho, E., Clyti, E., Couppie, P., Buchrieser, C., Tulliez, M., Morvan, J., and Gessain, A. (2000). Monoclonality or oligoclonality of human herpesvirus 8 terminal repeat sequences in Kaposi's sarcoma and other diseases. J. Natl. Cancer Inst. 92, 729-736.

Kaaya, E.E., Parravicini, C., Ordonez, C., Gendelman, R., Berti, E., Gallo, R.C., and Biberfeld, P. (1995). Heterogeneity of spindle cells in Kaposi's sarcoma: comparison of cells in lesions and in culture. J. Acquir. Immune Defic. Syndr. Hum. Retrovirol. 10, 295-305.

Kaldis, P., Ojala, P.M., Tong, L., Makela, T.P., and Solomon, M.J. (2001). CAK-independent activation of CDK6 by a viral cyclin. Molecular biology of the cell 12, 3987-3999.

Kaleeba, J.A., and Berger, E.A. (2006). Kaposi's sarcoma-associated herpesvirus fusion-entry receptor: cystine transporter xCT. Science 311, 1921-1924. 10.1126/science.1120878

Katano, H., Sato, Y., and Sata, T. (2001). Expression of p53 and human herpesvirus-8 (HHV-8)-encoded latency-associated nuclear antigen with inhibition of apoptosis in HHV-8-associated malignancies. Cancer 92, 3076-3084.

Kedes, D.H., and Ganem, D. (1997). Sensitivity of Kaposi's sarcoma-associated herpesvirus replication to antiviral drugs. Implications for potential therapy. J. Clin. Invest. 99, 2082-2086. 10.1172/JCI119380

Keller, S.A., Schattner, E.J., and Cesarman, E. (2000). Inhibition of NF-kappaB induces apoptosis of KSHV-infected primary effusion lymphoma cells. Blood 96, 2537-2542.

Kirsch, D.G., Doseff, A., Chau, B.N., Lim, D.S., de Souza-Pinto, N.C., Hansford, R., Kastan, M.B., Lazebnik, Y.A., and Hardwick, J.M. (1999). Caspase-3-dependent cleavage of Bcl-2 promotes release of cytochrome c. J. Biol. Chem. 274, 21155-21161.

Kledal, T.N., Rosenkilde, M.M., Coulin, F., Simmons, G., Johnsen, A.H., Alouani, S., Power, C.A., Luttichau, H.R., Gerstoft, J., Clapham, P.R., *et al.* (1997). A broad-spectrum chemokine antagonist encoded by Kaposi's sarcoma-associated herpesvirus. Science 277, 1656-1659.

Klotzbucher, A., Stewart, E., Harrison, D., and Hunt, T. (1996). The 'destruction box' of cyclin A allows B-type cyclins to be ubiquitinated, but not efficiently destroyed. The EMBO journal 15, 3053-3064.

Koon, H.B., Bubley, G.J., Pantanowitz, L., Masiello, D., Smith, B., Crosby, K., Proper, J., Weeden, W., Miller, T.E., Chatis, P., *et al.* (2005). Imatinib-induced regression of AIDS-related Kaposi's sarcoma. J Clin Oncol 23, 982-989.

Koon, H.B., Krown, S.E., Lee, J.Y., Honda, K., Rapisuwon, S., Wang, Z., Aboulafia, D., Reid, E.G., Rudek, M.A., Dezube, B.J., *et al.* (2014). Phase II trial of imatinib in AIDS-associated Kaposi's sarcoma: AIDS Malignancy Consortium Protocol 042. J Clin Oncol 32, 402-408.

Koopal, S., Furuhjelm, J.H., Jarviluoma, A., Jaamaa, S., Pyakurel, P., Pussinen, C., Wirzenius, M., Biberfeld, P., Alitalo, K., Laiho, M., *et al.* (2007). Viral oncogene-induced DNA damage response is activated in Kaposi sarcoma tumourigenesis. PLoS pathogens 3, 1348-1360.

Krishnan, H.H., Naranatt, P.P., Smith, M.S., Zeng, L., Bloomer, C., and Chandran, B. (2004). Concurrent expression of latent and a limited number of lytic genes with immune modulation and antiapoptotic function by Kaposi's sarcoma-associated herpesvirus early during infection of primary endothelial and fibroblast cells and subsequent decline of lytic gene expression. J. Virol. 78, 3601-3620.

Krown, S.E., Roy, D., Lee, J.Y., Dezube, B.J., Reid, E.G., Venkataramanan, R., Han, K., Cesarman, E., and Dittmer, D.P. (2012). Rapamycin with antiretroviral therapy in AIDS-associated Kaposi sarcoma: an AIDS Malignancy Consortium study. Journal of acquired immune deficiency syndromes 59, 447-454. 10.1097/QAI.0b013e31823e7884

Kumar, P., and Wood, C. (2013). Kaposi's sarcoma-associated herpesvirus transactivator Rta induces cell cycle arrest in G0/G1 phase by stabilizing and promoting nuclear localization of p27kip. J. Virol. 87, 13226-13238. 10.1128/JVI.02540-13

Laman, H., Coverley, D., Krude, T., Laskey, R., and Jones, N. (2001). Viral cyclin-cyclin-dependent kinase 6 complexes initiate nuclear DNA replication. Mol. Cell. Biol. 21, 624-635. 10.1128/MCB.21.2.624-635.2001

Laney, A.S., Cannon, M.J., Jaffe, H.W., Offermann, M.K., Ou, C.Y., Radford, K.W., Patel, M.M., Spira, T.J., Gunthel, C.J., Pellett, P.E., *et al.* (2007). Human herpesvirus 8 presence and viral load are associated with the progression of AIDS-associated Kaposi's sarcoma. AIDS (London, England) 21, 1541-1545. 10.1097/QAD.0b013e3282202b7d

Lee, J.S., Li, Q., Lee, J.Y., Lee, S.H., Jeong, J.H., Lee, H.R., Chang, H., Zhou, F.C., Gao, S.J., Liang, C., *et al.* (2009). FLIP-mediated autophagy regulation in cell death control. Nature cell biology 11, 1355-1362. 10.1038/ncb1980

Leidal, A.M., Cyr, D.P., Hill, R.J., Lee, P.W., and McCormick, C. (2012). Subversion of autophagy by Kaposi's sarcoma-associated herpesvirus impairs oncogene-induced senescence. Cell host and microbe 11, 167-180.

Li, M., Damania, B., Alvarez, X., Ogryzko, V., Ozato, K., and Jung, J.U. (2000). Inhibition of p300 histone acetyltransferase by viral interferon regulatory factor. Mol. Cell. Biol. 20, 8254-8263.

Li, M., Lee, H., Yoon, D.W., Albrecht, J.C., Fleckenstein, B., Neipel, F., and Jung, J.U. (1997). Kaposi's sarcoma-associated herpesvirus encodes a functional cyclin. J. Virol. 71, 1984-1991.

Li, S., Tan, M., Juillard, F., Ponnusamy, R., Correia, B., Simas, J.P., Carrondo, M.A., McVey, C.E., and Kaye, K.M. (2015). The Kaposi Sarcoma Herpesvirus Latency-associated Nuclear Antigen DNA Binding Domain Dorsal Positive Electrostatic Patch Facilitates DNA Replication and Episome Persistence. J. Biol. Chem. 290, 28084-28096. 10.1074/jbc.M115.674622

Li, X., Feng, J., and Sun, R. (2011). Oxidative stress induces reactivation of Kaposi's sarcoma-associated herpesvirus and death of primary effusion lymphoma cells. J. Virol. 85, 715-724. 10.1128/JVI.01742-10

Lin, R., Genin, P., Mamane, Y., Sgarbanti, M., Battistini, A., Harrington, W.J., Jr., Barber, G.N., and Hiscott, J. (2001). HHV-8 encoded vIRF-1 represses the interferon antiviral response by blocking IRF-3 recruitment of the CBP/p300 coactivators. Oncogene 20, 800-811. 10.1038/sj.onc.1204163

Liu, L., Eby, M.T., Rathore, N., Sinha, S.K., Kumar, A., and Chaudhary, P.M. (2002). The human herpes virus 8-encoded viral FLICE inhibitory protein physically associates with and persistently activates the Ikappa B kinase complex. J. Biol. Chem. 277, 13745-13751. 10.1074/jbc.M110480200

Long, E., Ilie, M., Hofman, V., Havet, K., Selva, E., Butori, C., Lacour, J.P., Nelson, A.M., Cathomas, G., and Hofman, P. (2009). LANA-1, Bcl-2, Mcl-1 and HIF-1alpha protein expression in HIV-associated Kaposi sarcoma. Virchows Archiv: an international journal of pathology 455, 159-170. 10.1007/s00428-009-0791-1

Lu, J., Jha, H.C., Verma, S.C., Sun, Z., Banerjee, S., Dzeng, R., and Robertson, E.S. (2014). Kaposi's sarcoma-associated herpesvirus-encoded LANA contributes to viral latent replication by activating phosphorylation of survivin. J. Virol. 88, 4204-4217. 10.1128/JVI.03855-13

Luftig, M.A. (2014). Viruses and the DNA Damage Response: Activation and Antagonism. Annual review of virology 1, 605-625. 10.1146/annurev-virology-031413-085548

Lukac, D.M., Kirshner, J.R., and Ganem, D. (1999). Transcriptional activation by the product of open reading frame 50 of Kaposi's sarcoma-associated herpesvirus is required for lytic viral reactivation in B cells. J. Virol. 73, 9348-9361.

Lukac, D.M., Renne, R., Kirshner, J.R., and Ganem, D. (1998). Reactivation of Kaposi's sarcoma-associated herpesvirus infection from latency by expression of the ORF 50 transactivator, a homolog of the EBV R protein. Virology 252, 304-312. 10.1006/viro.1998.9486

Ma, Q., Cavallin, L.E., Leung, H.J., Chiozzini, C., Goldschmidt-Clermont, P.J., and Mesri, E.A. (2013). A role for virally induced reactive oxygen species in Kaposi's sarcoma herpesvirus tumourigenesis. Antioxidants and redox signaling 18, 80-90. 10.1089/ars.2012.4584

Ma, T., Jham, B.C., Hu, J., Friedman, E.R., Basile, J.R., Molinolo, A., Sodhi, A., and Montaner, S. (2010). Viral G protein-coupled receptor up-regulates Angiopoietin-like 4 promoting angiogenesis and vascular permeability in Kaposi's sarcoma. Proc Natl Acad Sci USA 107, 14363-14368. 10.1073/pnas.1001065107

Ma, T., Patel, H., Babapoor-Farrokhran, S., Franklin, R., Semenza, G.L., Sodhi, A., and Montaner, S. (2015). KSHV induces aerobic glycolysis and angiogenesis through HIF-1-dependent upregulation of pyruvate kinase 2 in Kaposi's sarcoma. Angiogenesis 18, 477-488. 10.1007/s10456-015-9475-4

Mann, D.J., Child, E.S., Swanton, C., Laman, H., and Jones, N. (1999). Modulation of p27(Kip1) levels by the cyclin encoded by Kaposi's sarcoma-associated herpesvirus. The EMBO journal 18, 654-663. 10.1093/emboj/18.3.654

Maral, T. (2000). The Koebner phenomenon in immunosuppression-related Kaposi's sarcoma. Annals of plastic surgery 44, 646-648.

Martin, D., and Gutkind, J.S. (2008). Human tumour-associated viruses and new insights into the molecular mechanisms of cancer. Oncogene 27 Suppl 2, S31-42.

Masood, R., Cesarman, E., Smith, D.L., Gill, P.S., and Flore, O. (2002). Human herpesvirus-8-transformed endothelial cells have functionally activated vascular endothelial growth factor/vascular endothelial growth factor receptor. The American journal of pathology 160, 23-29. 10.1016/S0002-9440(10)64344-1

Miller, G., El-Guindy, A., Countryman, J., Ye, J., and Gradoville, L. (2007). Lytic cycle switches of oncogenic human gammaherpesviruses. Advances in cancer research 97, 81-109. 10.1016/S0065-230X(06)97004-3

Mishima, K., Watabe, T., Saito, A., Yoshimatsu, Y., Imaizumi, N., Masui, S., Hirashima, M., Morisada, T., Oike, Y., Araie, M., *et al.* (2007). Prox1 induces lymphatic endothelial differentiation via integrin alpha9 and other signaling cascades. Molecular biology of the cell 18, 1421-1429. 10.1091/mbc.E06-09-0780

Mocroft, A., Youle, M., Gazzard, B., Morcinek, J., Halai, R., and Phillips, A.N. (1996). Anti-herpesvirus treatment and risk of Kaposi's sarcoma in HIV infection. Royal Free/Chelsea and Westminster Hospitals Collaborative Group. AIDS (London, England) 10, 1101-1105.

Molden, J., Chang, Y., You, Y., Moore, P.S., and Goldsmith, M.A. (1997). A Kaposi's sarcoma-associated herpesvirus-encoded cytokine homolog (vIL-6) activates signaling through the shared gp130 receptor subunit. J. Biol. Chem. 272, 19625-19631.

Montaner, S., Sodhi, A., Molinolo, A., Bugge, T.H., Sawai, E.T., He, Y., Li, Y., Ray, P.E., and Gutkind, J.S. (2003). Endothelial infection with KSHV genes *in vivo* reveals that vGPCR initiates Kaposi's sarcomagenesis and can promote the tumourigenic potential of viral latent genes. Cancer cell 3, 23-36.

Moore, P.S. (2007). KSHV manipulation of the cell cycle and apoptosis. In Human Herpesviruses: Biology, Therapy, and Immunoprophylaxis, A. Arvin, G. Campadelli-Fiume, E. Mocarski, P.S. Moore, B. Roizman, R. Whitley, and K. Yamanishi, eds. (Cambridge).

Moore, P.S., Boshoff, C., Weiss, R.A., and Chang, Y. (1996). Molecular mimicry of human cytokine and cytokine response pathway genes by KSHV. Science 274, 1739-1744.

Mori, Y., Nishimoto, N., Ohno, M., Inagi, R., Dhepakson, P., Amou, K., Yoshizaki, K., and Yamanishi, K. (2000). Human herpesvirus 8-encoded interleukin-6 homologue (viral IL-6) induces endogenous human IL-6 secretion. Journal of medical virology 61, 332-335.

Mutlu, A.D., Cavallin, L.E., Vincent, L., Chiozzini, C., Eroles, P., Duran, E.M., Asgari, Z., Hooper, A.T., La Perle, K.M., Hilsher, C., *et al.* (2007). In vivo-restricted and reversible malignancy induced by human herpesvirus-8 KSHV: a cell and animal model of virally induced Kaposi's sarcoma. Cancer cell 11, 245-258.

Nador, R.G., Cesarman, E., Chadburn, A., Dawson, D.B., Ansari, M.Q., Sald, J., and Knowles, D.M. (1996). Primary effusion lymphoma: a distinct clinicopathologic entity associated with the Kaposi's sarcoma-associated herpes virus. Blood 88, 645-656.

Nakamura, H., Li, M., Zarycki, J., and Jung, J.U. (2001). Inhibition of p53 tumour suppressor by viral interferon regulatory factor. J. Virol. 75, 7572-7582. 10.1128/JVI.75.16.7572-7582.2001

Nakano, K., Isegawa, Y., Zou, P., Tadagaki, K., Inagi, R., and Yamanishi, K. (2003). Kaposi's sarcoma-associated herpesvirus (KSHV)-encoded vMIP-I and vMIP-II induce signal transduction and chemotaxis in monocytic cells. Archives of virology 148, 871-890. 10.1007/s00705-002-0971-7

Nealon, K., Newcomb, W.W., Pray, T.R., Craik, C.S., Brown, J.C., and Kedes, D.H. (2001). Lytic replication of Kaposi's sarcoma-associated herpesvirus results in the formation of multiple capsid species: isolation and molecular characterization of A, B, and C capsids from a gammaherpesvirus. J. Virol. 75, 2866-2878. 10.1128/JVI.75.6.2866-2878.2001

Nepka, C., Kanakis, D., Samara, M., Kapsoritakis, A., Potamianos, S., Karantana, M., and Koukoulis, G. (2012). An unusual case of Primary Effusion Lymphoma with aberrant T-cell phenotype in a HIV-negative, HBV-positive, cirrhotic patient, and review of the literature. CytoJournal 9, 16. 10.4103/1742-6413.97766

Ojala, P.M., Yamamoto, K., Castanos-Velez, E., Biberfeld, P., Korsmeyer, S.J., and Makela, T.P. (2000). The apoptotic v-Cyclin-CDK6 complex phosphorylates and inactivates Bcl-2. Nature cell biology 2, 819-825. 10.1038/35041064

Oksenhendler, E., Duarte, M., Soulier, J., Cacoub, P., Welker, Y., Cadranel, J., Cazals-Hatem, D., Autran, B., Clauvel, J.P., and Raphael, M. (1996). Multicentric Castleman's disease in HIV infection: a clinical and pathological study of 20 patients. AIDS (London, England) 10, 61-67.

Orenstein, J.M. (2008). Ultrastructure of Kaposi sarcoma. Ultrastructural pathology 32, 211-220. 10.1080/01913120802343871

Ottinger, M., Christalla, T., Nathan, K., Brinkmann, M.M., Viejo-Borbolla, A., and Schulz, T.F. (2006). Kaposi's sarcoma-associated herpesvirus LANA-1 interacts with the short variant of BRD4 and releases cells from a BRD4- and BRD2/RING3-induced G1 cell cycle arrest. J. Virol. 80, 10772-10786. 10.1128/JVI.00804-06

Park, J., Seo, T., Hwang, S., Lee, D., Gwack, Y., and Choe, J. (2000). The K-bZIP protein from Kaposi's sarcoma-associated herpesvirus interacts with p53 and represses its transcriptional activity. J. Virol. 74, 11977-11982.

Pattingre, S., Tassa, A., Qu, X., Garuti, R., Liang, X.H., Mizushima, N., Packer, M., Schneider, M.D., and Levine, B. (2005). Bcl-2 antiapoptotic proteins inhibit Beclin 1-dependent autophagy. Cell 122, 927-939. 10.1016/j.cell.2005.07.002

Paudel, N., Sadagopan, S., Chakraborty, S., Sarek, G., Ojala, P.M., and Chandran, B. (2012). Kaposi's sarcoma-associated herpesvirus latency-associated nuclear antigen interacts with multifunctional angiogenin to utilize its antiapoptotic functions. J. Virol. 86, 5974-5991. 10.1128/JVI.00070-12

Pekkonen, P., Jarviluoma, A., Zinovkina, N., Cvrljevic, A., Prakash, S., Westermarck, J., Evan, G.I., Cesarman, E., Verschuren, E.W., and Ojala, P.M. (2014). KSHV viral cyclin interferes with T-cell development and induces lymphoma through Cdk6 and Notch activation *in vivo*. Cell cycle 13, 3670-3684. 10.4161/15384101.2014.964118

Petre, C.E., Sin, S.H., and Dittmer, D.P. (2007). Functional p53 signaling in Kaposi's sarcoma-associated herpesvirus lymphomas: implications for therapy. J. Virol. 81, 1912-1922. 10.1128/JVI.01757-06

Petrova, T.V., Makinen, T., Makela, T.P., Saarela, J., Virtanen, I., Ferrell, R.E., Finegold, D.N., Kerjaschki, D., Yla-Herttuala, S., and Alitalo, K. (2002). Lymphatic endothelial reprogramming of vascular endothelial cells by the Prox-1 homeobox transcription factor. The EMBO journal 21, 4593-4599.

Pfeffer, S., Sewer, A., Lagos-Quintana, M., Sheridan, R., Sander, C., Grasser, F.A., van Dyk, L.F., Ho, C.K., Shuman, S., Chien, M., *et al.* (2005). Identification of microRNAs of the herpesvirus family. Nature methods 2, 269-276. 10.1038/nmeth746

Platt, G.M., Cannell, E., Cuomo, M.E., Singh, S., and Mittnacht, S. (2000). Detection of the human herpesvirus 8-encoded cyclin protein in primary effusion lymphoma-derived cell lines. Virology 272, 257-266. 10.1006/viro.2000.0343

Polizzotto, M.N., Uldrick, T.S., Hu, D., and Yarchoan, R. (2012). Clinical Manifestations of Kaposi Sarcoma Herpesvirus Lytic Activation: Multicentric Castleman Disease (KSHV-MCD) and the KSHV Inflammatory Cytokine Syndrome. Frontiers in microbiology 3, 73. 10.3389/fmicb.2012.00073

Polizzotto, M.N., Uldrick, T.S., Wang, V., Aleman, K., Wyvill, K.M., Marshall, V., Pittaluga, S., O'Mahony, D., Whitby, D., Tosato, G., *et al.* (2013). Human and viral interleukin-6 and other cytokines in Kaposi sarcoma herpesvirus-associated multicentric Castleman disease. Blood 122, 4189-4198. 10.1182/blood-2013-08-519959

Ponnusamy, R., Petoukhov, M.V., Correia, B., Custodio, T.F., Juillard, F., Tan, M., Pires de Miranda, M., Carrondo, M.A., Simas, J.P., Kaye, K.M., *et al.* (2015). KSHV but not MHV-68 LANA induces a strong bend upon binding to terminal repeat viral DNA. Nucleic acids research 43, 10039-10054. 10.1093/nar/gkv987

Poole, L.J., Zong, J.C., Ciufo, D.M., Alcendor, D.J., Cannon, J.S., Ambinder, R., Orenstein, J.M., Reitz, M.S., and Hayward, G.S. (1999). Comparison of genetic variability at multiple *loci* across the genomes of the major subtypes of Kaposi's sarcoma-associated herpesvirus reveals evidence for recombination and for two distinct types of open reading frame K15 alleles at the right-hand end. J. Virol. 73, 6646-6660.

Pyakurel, P., Montag, U., Castanos-Velez, E., Kaaya, E., Christensson, B., Tonnies, H., Biberfeld, P., and Heiden, T. (2006). CGH of microdissected Kaposi's sarcoma lesions reveals recurrent loss of chromosome Y in early and additional chromosomal changes in late tumour stages. AIDS (London, England) 20, 1805-1812. 10.1097/01.aids.0000244199.72887.3d

Rabkin, C.S., Janz, S., Lash, A., Coleman, A.E., Musaba, E., Liotta, L., Biggar, R.J., and Zhuang, Z. (1997). Monoclonal origin of multicentric Kaposi's sarcoma lesions. New Engl. J. Med. 336, 988-993. 10.1056/NEJM199704033361403

Radkov, S.A., Kellam, P., and Boshoff, C. (2000). The latent nuclear antigen of Kaposi sarcoma-associated herpesvirus targets the retinoblastoma-E2F pathway and with the oncogene Hras transforms primary rat cells. Nature Med. 6, 1121-1127. 10.1038/80459

Raghu, H., Sharma-Walia, N., Veettil, M.V., Sadagopan, S., and Chandran, B. (2009). Kaposi's sarcoma-associated herpesvirus utilizes an actin polymerization-dependent macropinocytic pathway to enter human dermal microvascular endothelial and human umbilical vein endothelial cells. J. Virol. 83, 4895-4911. 10.1128/JVI.02498-08

Ramalingam, D., Kieffer-Kwon, P., and Ziegelbauer, J.M. (2012). Emerging themes from EBV and KSHV microRNA targets. Viruses 4, 1687-1710. 10.3390/v4091687

Rappocciolo, G., Jenkins, F.J., Hensler, H.R., Piazza, P., Jais, M., Borowski, L., Watkins, S.C., and Rinaldo, C.R., Jr. (2006). DC-SIGN is a receptor for human herpesvirus 8 on dendritic cells and macrophages. J Immunol 176, 1741-1749.

Russo, J.J., Bohenzky, R.A., Chien, M.C., Chen, J., Yan, M., Maddalena, D., Parry, J.P., Peruzzi, D., Edelman, I.S., Chang, Y., *et al.* (1996). Nucleotide sequence of the Kaposi sarcoma-associated herpesvirus (HHV8). Proc Natl Acad Sci USA 93, 14862-14867.

Said, J.W., Shintaku, I.P., Asou, H., deVos, S., Baker, J., Hanson, G., Cesarman, E., Nador, R., and Koeffler, H.P. (1999). Herpesvirus 8 inclusions in primary effusion lymphoma: report of a unique case with T-cell phenotype. Archives of pathology and laboratory medicine 123, 257-260. 10.1043/0003-9985(1999)123<0257:HIIPEL>2.0.CO;2

Sakakibara, S., and Tosato, G. (2011). Viral interleukin-6: role in Kaposi's sarcoma-associated herpesvirus: associated malignancies. Journal of interferon and cytokine research: the official journal of the International Society for Interferon and Cytokine Research 31, 791-801. 10.1089/jir.2011.0043

Samaniego, F., Markham, P.D., Gendelman, R., Watanabe, Y., Kao, V., Kowalski, K., Sonnabend, J.A., Pintus, A., Gallo, R.C., and Ensoli, B. (1998). Vascular endothelial growth factor and basic fibroblast growth factor present in Kaposi's sarcoma (KS) are induced by inflammatory cytokines and synergize to promote vascular permeability and KS lesion development. The American journal of pathology 152, 1433-1443.

Samols, M.A., Hu, J., Skalsky, R.L., and Renne, R. (2005). Cloning and identification of a microRNA cluster within the latency-associated region of Kaposi's sarcoma-associated herpesvirus. J. Virol. 79, 9301-9305. 10.1128/JVI.79.14.9301-9305.2005

Sarek, G., Jarviluoma, A., and Ojala, P.M. (2006). KSHV viral cyclin inactivates $p27^{KIP1}$ through Ser10 and Thr187 phosphorylation in proliferating primary effusion lymphomas. Blood 107, 725-732. 10.1182/blood-2005-06-2534

Sarek, G., Kurki, S., Enback, J., Iotzova, G., Haas, J., Laakkonen, P., Laiho, M., and Ojala, P.M. (2007). Reactivation of the p53 pathway as a treatment modality for KSHV-induced lymphomas. J. Clin. Invest. 117, 1019-1028. 10.1172/JCI30945

Sarek, G., and Ojala, P.M. (2007). p53 reactivation kills KSHV lymphomas efficiently *in vitro* and *in vivo*: new hope for treating aggressive viral lymphomas. Cell cycle 6, 2205-2209. 10.4161/cc.6.18.4730

Sarid, R., Flore, O., Bohenzky, R.A., Chang, Y., and Moore, P.S. (1998). Transcription mapping of the Kaposi's sarcoma-associated herpesvirus (human herpesvirus 8) genome in a body cavity-based lymphoma cell line (BC-1). J. Virol. 72, 1005-1012.

Sarid, R., Sato, T., Bohenzky, R.A., Russo, J.J., and Chang, Y. (1997). Kaposi's sarcoma-associated herpesvirus encodes a functional bcl-2 homologue. Nature Med. 3, 293-298.

Schwartz, R.A. (2004). Kaposi's sarcoma: an update. J. Surg. Oncol. 87, 146-151. 10.1002/jso.20090

Shin, Y.C., Joo, C.H., Gack, M.U., Lee, H.R., and Jung, J.U. (2008). Kaposi's sarcoma-associated herpesvirus viral IFN regulatory factor 3 stabilizes hypoxia-inducible factor-1 alpha to induce vascular endothelial growth factor expression. Cancer research 68, 1751-1759. 10.1158/0008-5472.CAN-07-2766

Shin, Y.C., Nakamura, H., Liang, X., Feng, P., Chang, H., Kowalik, T.F., and Jung, J.U. (2006). Inhibition of the

ATM/p53 signal transduction pathway by Kaposi's sarcoma-associated herpesvirus interferon regulatory factor 1. J. Virol. 80, 2257-2266. 10.1128/JVI.80.5.2257-2266.2006

Simonelli, C., Spina, M., Cinelli, R., Talamini, R., Tedeschi, R., Gloghini, A., Vaccher, E., Carbone, A., and Tirelli, U. (2003a). Clinical features and outcome of primary effusion lymphoma in HIV-infected patients: a single-institution study. J Clin Oncol 21, 3948-3954. 10.1200/JCO.2003.06.013

Simonelli, C., Zanussi, S., Cinelli, R., Dal Maso, L., Di Gennaro, G., D'Andrea, M., Nasti, G., Spina, M., Vaccher, E., De Paoli, P., *et al.* (2003b). Impact of concomitant antiblastic chemotherapy and highly active antiretroviral therapy on human immunodeficiency virus (HIV) viremia and genotyping in HIV-infected patients with non-Hodgkin lymphoma. Clin. Infect Dis. 37, 820-827. 10.1086/377204

Sin, S.H., Kim, Y., Eason, A., and Dittmer, D.P. (2015). KSHV Latency Locus Cooperates with Myc to Drive Lymphoma in Mice. PLoS pathogens 11, e1005135. 10.1371/journal.ppat.1005135

Sivakumar, R., Sharma-Walia, N., Raghu, H., Veettil, M.V., Sadagopan, S., Bottero, V., Varga, L., Levine, R., and Chandran, B. (2008). Kaposi's sarcoma-associated herpesvirus induces sustained levels of vascular endothelial growth factors A and C early during *in vitro* infection of human microvascular dermal endothelial cells: biological implications. J. Virol. 82, 1759-1776. 10.1128/JVI.00873-07

Sodhi, A., Montaner, S., and Gutkind, J.S. (2004). Does dysregulated expression of a deregulated viral GPCR trigger Kaposi's sarcomagenesis? Faseb J 18, 422-427.

Sodhi, A., Montaner, S., Patel, V., Zohar, M., Bais, C., Mesri, E.A., and Gutkind, J.S. (2000). The Kaposi's sarcoma-associated herpes virus G protein-coupled receptor up-regulates vascular endothelial growth factor expression and secretion through mitogen-activated protein kinase and p38 pathways acting on hypoxia-inducible factor 1alpha. Cancer research 60, 4873-4880.

Song, J., Yoshida, A., Yamamoto, Y., Katano, H., Hagihara, K., Oka, S., Kimura, S., and Yoshizaki, K. (2004). Viral load of human herpesvirus 8 (HHV-8) in the circulatory blood cells correlates with clinical progression in a patient with HHV-8-associated solid lymphoma with aids-associated Kaposi's sarcoma. Leukemia and lymphoma 45, 2343-2347. 10.1080/10428190412331283242

Song, S.N., Tomosugi, N., Kawabata, H., Ishikawa, T., Nishikawa, T., and Yoshizaki, K. (2010). Down-regulation of hepcidin resulting from long-term treatment with an anti-IL-6 receptor antibody (tocilizumab) improves anemia of inflammation in multicentric Castleman disease. Blood 116, 3627-3634. 10.1182/blood-2010-03-271791

Soulier, J., Grollet, L., Oksenhendler, E., Cacoub, P., Cazals-Hatem, D., Babinet, P., d'Agay, M.F., Clauvel, J.P., Raphael, M., Degos, L., *et al.* (1995). Kaposi's sarcoma-associated herpesvirus-like DNA sequences in multicentric Castleman's disease. Blood 86, 1276-1280.

Staskus, K.A., Zhong, W., Gebhard, K., Herndier, B., Wang, H., Renne, R., Beneke, J., Pudney, J., Anderson, D.J., Ganem, D., *et al.* (1997). Kaposi's sarcoma-associated herpesvirus gene expression in endothelial (spindle) tumour cells. J. Virol. 71, 715-719.

Stine, J.T., Wood, C., Hill, M., Epp, A., Raport, C.J., Schweickart, V.L., Endo, Y., Sasaki, T., Simmons, G., Boshoff, C., *et al.* (2000). KSHV-encoded CC chemokine vMIP-III is a CCR4 agonist, stimulates angiogenesis, and selectively chemoattracts TH2 cells. Blood 95, 1151-1157.

Sturzl, M., Brandstetter, H., Zietz, C., Eisenburg, B., Raivich, G., Gearing, D.P., Brockmeyer, N.H., and Hofschneider, P.H. (1995). Identification of interleukin-1 and platelet-derived growth factor-B as major mitogens for the spindle cells of Kaposi's sarcoma: a combined *in vitro* and *in vivo* analysis. Oncogene 10, 2007-2016.

Sun, R., Lin, S.F., Staskus, K., Gradoville, L., Grogan, E., Haase, A., and Miller, G. (1999). Kinetics of Kaposi's sarcoma-associated herpesvirus gene expression. J. Virol. 73, 2232-2242.

Swanton, C., Mann, D.J., Fleckenstein, B., Neipel, F., Peters, G., and Jones, N. (1997). Herpes viral cyclin/Cdk6 complexes evade inhibition by CDK inhibitor proteins. Nature 390, 184-187. 10.1038/36606

Tolani, B., Gopalakrishnan, R., Punj, V., Matta, H., and Chaudhary, P.M. (2014). Targeting Myc in KSHV-associated primary effusion lymphoma with BET bromodomain inhibitors. Oncogene 33, 2928-2937. 10.1038/onc.2013.242

Tomlinson, C.C., and Damania, B. (2004). The K1 protein of Kaposi's sarcoma-associated herpesvirus activates the Akt signaling pathway. J. Virol. 78, 1918-1927.

Trus, B.L., Heymann, J.B., Nealon, K., Cheng, N., Newcomb, W.W., Brown, J.C., Kedes, D.H., and Steven, A.C. (2001). Capsid structure of Kaposi's sarcoma-associated herpesvirus, a gammaherpesvirus, compared to those of an alphaherpesvirus, herpes simplex virus type 1, and a betaherpesvirus, cytomegalovirus. J. Virol. 75, 2879-2890. 10.1128/JVI.75.6.2879-2890.2001

Uldrick, T.S., Wang, V., O'Mahony, D., Aleman, K., Wyvill, K.M., Marshall, V., Steinberg, S.M., Pittaluga, S., Maric, I., Whitby, D., *et al.* (2010). An interleukin-6-related systemic inflammatory syndrome in patients co-infected with Kaposi sarcoma-associated herpesvirus and HIV but without Multicentric Castleman disease. Clin. Infect Dis. 51, 350-358. 10.1086/654798

Van Dross, R., Yao, S., Asad, S., Westlake, G., Mays, D.J., Barquero, L., Duell, S., Pietenpol, J.A., and Browning, P.J. (2005). Constitutively active K-cyclin/cdk6 kinase in Kaposi sarcoma-associated herpesvirus-infected cells. J. Natl. Cancer Inst. 97, 656-666. 10.1093/jnci/dji113

Varjosalo, M., Bjorklund, M., Cheng, F., Syvanen, H., Kivioja, T., Kilpinen, S., Sun, Z., Kallioniemi, O., Stunnenberg, H.G., He, W.W., *et al.* (2008). Application of active and kinase-deficient kinome collection for identification of kinases regulating hedgehog signaling. Cell 133, 537-548. 10.1016/j.cell.2008.02.047

Veettil, M.V., Sadagopan, S., Sharma-Walia, N., Wang, F.Z., Raghu, H., Varga, L., and Chandran, B. (2008). Kaposi's sarcoma-associated herpesvirus forms a multimolecular complex of integrins (alphaVbeta5, alphaVbeta3, and alpha3beta1) and CD98-xCT during infection of human dermal microvascular endothelial cells, and CD98-xCT is essential for the postentry stage of infection. J. Virol. 82, 12126-12144. 10.1128/JVI.01146-08

Verschuren, E.W., Jones, N., and Evan, G.I. (2004). The cell cycle and how it is steered by Kaposi's sarcoma-associated herpesvirus cyclin. J. Gen. Virol. 85, 1347-1361. 10.1099/vir.0.79812-0

Viejo-Borbolla, A., Ottinger, M., Bruning, E., Burger, A., Konig, R., Kati, E., Sheldon, J.A., and Schulz, T.F. (2005). Brd2/RING3 interacts with a chromatin-binding domain in the Kaposi's Sarcoma-associated herpesvirus latency-associated nuclear antigen 1 (LANA-1) that is required for multiple functions of LANA-1. J. Virol. 79, 13618-13629. 10.1128/JVI.79.21.13618-13629.2005

Wang, H.W., Trotter, M.W., Lagos, D., Bourboulia, D., Henderson, S., Makinen, T., Elliman, S., Flanagan, A.M., Alitalo, K., and Boshoff, C. (2004a). Kaposi sarcoma herpesvirus-induced cellular reprogramming contributes to the lymphatic endothelial gene expression in Kaposi sarcoma. Nature genetics 36, 687-693. 10.1038/ng1384

Wang, L., and Damania, B. (2008). Kaposi's sarcoma-associated herpesvirus confers a survival advantage to endothelial cells. Cancer research 68, 4640-4648. 10.1158/0008-5472.CAN-07-5988

Wang, L., Dittmer, D.P., Tomlinson, C.C., Fakhari, F.D., and Damania, B. (2006). Immortalization of primary endothelial cells by the K1 protein of Kaposi's sarcoma-associated herpesvirus. Cancer research 66, 3658-3666. 10.1158/0008-5472.CAN-05-3680

Wang, L., Wakisaka, N., Tomlinson, C.C., DeWire, S.M., Krall, S., Pagano, J.S., and Damania, B. (2004b). The Kaposi's sarcoma-associated herpesvirus (KSHV/HHV-8) K1 protein induces expression of angiogenic and invasion factors. Cancer research 64, 2774-2781.

Wang, S.E., Wu, F.Y., Fujimuro, M., Zong, J., Hayward, S.D., and Hayward, G.S. (2003a). Role of CCAAT/enhancer-binding protein alpha (C/EBPalpha) in activation of the Kaposi's sarcoma-associated herpesvirus (KSHV) lytic-cycle replication-associated protein (RAP) promoter in cooperation with the KSHV replication and transcription activator (RTA) and RAP. J. Virol. 77, 600-623.

Wang, S.E., Wu, F.Y., Yu, Y., and Hayward, G.S. (2003b). CCAAT/enhancer-binding protein-alpha is induced during the early stages of Kaposi's sarcoma-associated herpesvirus (KSHV) lytic cycle reactivation and together with the KSHV replication and transcription activator (RTA) cooperatively stimulates the viral RTA, MTA, and PAN promoters. J. Virol. 77, 9590-9612.

Warburg, O. (1956a). On respiratory impairment in cancer cells. Science 124, 269-270.

Warburg, O. (1956b). On the origin of cancer cells. Science 123, 309-314.

Weber, K.S., Grone, H.J., Rocken, M., Klier, C., Gu, S., Wank, R., Proudfoot, A.E., Nelson, P.J., and Weber, C. (2001). Selective recruitment of Th2-type cells and evasion from a cytotoxic immune response mediated by viral macrophage inhibitory protein-II. Eur. J. Immunol. 31, 2458-2466.

Wei, F., Gan, J., Wang, C., Zhu, C., and Cai, Q. (2016). Cell Cycle Regulatory Functions of the KSHV Oncoprotein LANA. Frontiers in microbiology 7, 334. 10.3389/fmicb.2016.00334

Weich, H.A., Salahuddin, S.Z., Gill, P., Nakamura, S., Gallo, R.C., and Folkmann, J. (1991). AIDS-associated Kaposi's sarcoma-derived cells in long-term culture express and synthesize smooth muscle alpha-actin. The American journal of pathology 139, 1251-1258.

Wiener, Z., Hogstrom, J., Hyvonen, V., Band, A., Kallio, P., Holopainen, T., Dufva, O., Haglund, C., Kruuna, O., Oliver, G., *et al.* (2014). [Prox1 promotes expansion of the colorectal cancer stem cell population to fuel tumour growth and ischemia resistance]. Duodecim; laaketieteellinen aikakauskirja 130, 2208.

Wu, F.Y., Tang, Q.Q., Chen, H., ApRhys, C., Farrell, C., Chen, J., Fujimuro, M., Lane, M.D., and Hayward, G.S. (2002). Lytic replication-associated protein (RAP) encoded by Kaposi sarcoma-associated herpesvirus causes p21CIP-1-mediated G1 cell cycle arrest through CCAAT/enhancer-binding protein-alpha. Proc Natl Acad Sci USA 99, 10683-10688. 10.1073/pnas.162352299

Wu, F.Y., Wang, S.E., Tang, Q.Q., Fujimuro, M., Chiou, C.J., Zheng, Q., Chen, H., Hayward, S.D., Lane, M.D., and Hayward, G.S. (2003). Cell cycle arrest by Kaposi's sarcoma-associated herpesvirus replication-associated protein is mediated at both the transcriptional and post-translational levels by binding to CCAAT/enhancer-binding protein alpha and p21(CIP-1). J. Virol. 77, 8893-8914.

Wu, L., Lo, P., Yu, X., Stoops, J.K., Forghani, B., and Zhou, Z.H. (2000). Three-dimensional structure of the human herpesvirus 8 capsid. J. Virol. 74, 9646-9654.

Yamin, R., Kaynan, N.S., Glasner, A., Vitenshtein, A., Tsukerman, P., Bauman, Y., Ophir, Y., Elias, S., Bar-On, Y., Gur, C., *et al.* (2013). The viral KSHV chemokine vMIP-II inhibits the migration of Naive and activated human NK cells by antagonizing two distinct chemokine receptors. PLoS pathogens 9, e1003568. 10.1371/journal.ppat.1003568

Yang, T.Y., Chen, S.C., Leach, M.W., Manfra, D., Homey, B., Wiekowski, M., Sullivan, L., Jenh, C.H., Narula, S.K., Chensue, S.W., *et al.* (2000). Transgenic expression of the chemokine receptor encoded by human herpesvirus 8 induces an angioproliferative disease resembling Kaposi's sarcoma. J. Exp. Med. 191, 445-454.

Ye, F., Lattif, A.A., Xie, J., Weinberg, A., and Gao, S. (2012). Nutlin-3 induces apoptosis, disrupts viral latency and inhibits expression of angiopoietin-2 in Kaposi sarcoma tumour cells. Cell cycle 11, 1393-1399. 10.4161/cc.19756

Ye, F., Zhou, F., Bedolla, R.G., Jones, T., Lei, X., Kang, T., Guadalupe, M., and Gao, S.J. (2011). Reactive oxygen species hydrogen peroxide mediates Kaposi's sarcoma-associated herpesvirus reactivation from latency. PLoS pathogens 7, e1002054. 10.1371/journal.ppat.1002054

Yogev, O., Lagos, D., Enver, T., and Boshoff, C. (2014). Kaposi's sarcoma herpesvirus microRNAs induce metabolic transformation of infected cells. PLoS pathogens 10, e1004400. 10.1371/journal.ppat.1004400

Yoo, J., Kang, J., Lee, H.N., Aguilar, B., Kafka, D., Lee, S., Choi, I., Lee, J., Ramu, S., Haas, J., *et al.* (2010). Kaposin-B enhances the PROX1 mRNA stability during lymphatic reprogramming of vascular endothelial cells by Kaposi's sarcoma herpes virus. PLoS pathogens 6, e1001046. 10.1371/journal.ppat.1001046

Yoo, J., Lee, H.N., Choi, I., Choi, D., Chung, H.K., Kim, K.E., Lee, S., Aguilar, B., Kang, J., Park, E., *et al.* (2012). Opposing regulation of PROX1 by interleukin-3 receptor and NOTCH directs differential host cell fate reprogramming by Kaposi sarcoma herpes virus. PLoS pathogens 8, e1002770. 10.1371/journal.ppat.1002770

Yu, F., Harada, J.N., Brown, H.J., Deng, H., Song, M.J., Wu, T.T., Kato-Stankiewicz, J., Nelson, C.G., Vieira, J., Tamanoi, F., *et al.* (2007). Systematic identification of cellular signals reactivating Kaposi sarcoma-associated herpesvirus. PLoS pathogens 3, e44. 10.1371/journal.ppat.0030044

Zhu, Y., Haecker, I., Yang, Y., Gao, S.J., and Renne, R. (2013). gamma-Herpesvirus-encoded miRNAs and their roles in viral biology and pathogenesis. Curr. Opin. Virol. 3, 266-275. 10.1016/j.coviro.2013.05.013

10

Merkel Cell Polyomavirus: The New Kid on the Block

Jason Diaz[1], Margo MacDonald[2], and Jianxin You[2*]

[1]Department of Biology, School of Arts and Sciences, and [2]Department of Microbiology, Perelman School of Medicine, University of Pennsylvania, Philadelphia, PA 19104, USA

**Corresponding author: jianyou@pennmedicine.upenn.edu*

Abstract

Merkel cell polyomavirus (MCPyV or MCV) is the first polyomavirus clearly shown to be associated with human cancer. Nearly all healthy adults asymptomatically shed MCPyV from their skin. In elderly and immunosuppressed individuals, infection can lead to a highly lethal form of skin cancer, Merkel cell carcinoma (MCC). Excessive exposure to sunlight and ultraviolet (UV) radiation, immune suppression, and advanced age are the most important risk factors for MCPyV-associated MCC. The incidence of MCC has tripled over the past two decades, and the concern for MCC grows as the aging population with prolonged sun exposure increases. With the high prevalence of MCPyV infection and an increasing amount of MCC diagnoses, there is a need to better understand the biology of MCPyV and its oncogenic potential. This chapter summarizes recent discoveries in MCPyV molecular virology, host cellular tropism, cell culture infection and animal models, and the unique genetic features of MCPyV that make it the only polyomavirus discovered to date that is associated with a human cancer. MCPyV is more likely to induce MCC in immunocompromised patients, thus providing an excellent probe for examining the role of the host immune system in controlling virus-induced oncogenesis. The interplay between MCPyV infection and host immune response and how disruption of this virus-host interaction may contribute to MCC pathogenesis is also reviewed. Together, our current knowledge provides a platform for future mechanistic studies to fully elucidate the MCPyV infectious life cycle and the mechanisms by which MCPyV infection contributes to MCC pathogenesis.

Introduction

Merkel cell carcinoma (MCC) was first described by Cyril Toker in 1971 as a trabecular tumour of the dermis (Toker, 1972). MCC typically presents as a neuroendocrine carcinoma of the skin, and has historically been believed to arise from Merkel cells, a unique cell type of the skin bearing both epithelial and neurosecretory characteristics (Wang *et al.*, 2011). MCC incidence rates have tripled in the last decades, and although still a rare cancer, it is incredibly lethal, with a disease-associated mortality of 46% (Lemos *et al.*, 2010). Risk factors for MCC include advanced age, immunosuppression, and prolonged exposure to sunlight and UV radiation (Heath *et al.*, 2008; Chang and Moore, 2012).

Noting that immune suppression was a key risk factor for MCC, Patrick Moore and Yuan Chang at the University of Pittsburgh set out to determine whether an infectious agent might underlie MCC incidence (Feng *et al.*, 2008). After performing deep sequencing of the transcriptomes of MCC tumours, they then subtracted out known human transcript sequences to discover any foreign sequences highly expressed in these tumours. Using this Digital Transcriptome Subtraction method, they were able to isolate a highly

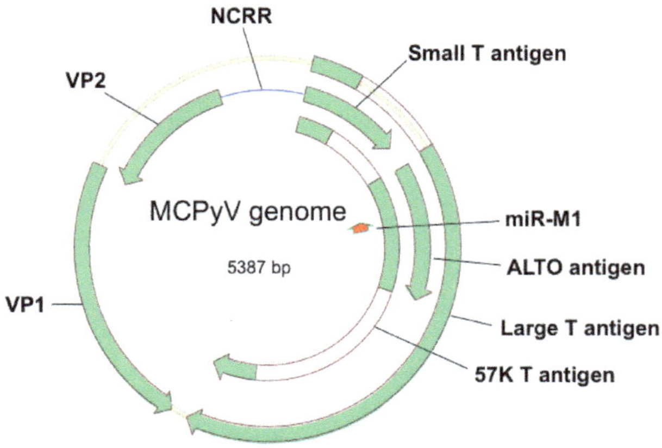

Figure 10.1. The structure of MCPyV genome. This schematic diagram shows the non-coding regulatory region (NCRR), early genes, late genes, and the virally encoded microRNA miR-M1.

expressed transcript with strong homology to the Large Tumour antigen (LT) of polyomaviruses (Feng *et al.*, 2007; Feng *et al.*, 2008). Using this transcript as a starting point they then identified a novel polyomavirus clonally integrated in these MCC tumours. They named this new polyomavirus Merkel cell polyomavirus (MCPyV or MCV) for its association with MCC.

Since that initial discovery in MCC tumours, it has been established that as many as 80% of all MCC tumours have clonally integrated MCPyV genomes (Feng *et al.*, 2008; Sastre-Garau *et al.*, 2009). Integration of the genome appears to occur before clonal expansion of the tumour, and persistent expression of the viral tumour antigens is required for MCC tumours to survive (Feng *et al.*, 2008; Houben *et al.*, 2010). These data strongly suggest that MCPyV plays an important oncogenic role in a large number of MCC tumours. Interestingly, it has also been shown that MCPyV infection of the skin is ubiquitous and largely asymptomatic in the general population (Kean *et al.*, 2009; Pastrana *et al.*, 2009; Tolstov *et al.*, 2009). Understanding the basic molecular virology of MCPyV and its role in oncogenesis has been a major interest in tumour virology since its discovery in 2008. More recent studies have called into question the cell of origin for MCPyV-associated MCC tumours, and there is a growing interest in understanding the mechanisms of persistent viral infection and the role of host immune responses in controlling MCPyV infection and associated tumourigenesis. In this chapter we review what has been established and what remains to be elucidated for the first polyomavirus linked to human cancer.

MCPyV molecular virology

MCPyV genome organization

MCPyV, like its family members SV40, JC virus and BK virus, is comprised of a small, circular double-stranded DNA genome surrounded by an icosahedral protein capsid (Gjoerup and Chang, 2010). The 5.3kb genome is organized into a multiply-spliced early "tumour antigen" region, a late region, and a non-coding regulatory region (NCRR) which separates these two coding regions (Figure 10.1). The NCRR contains the viral origin of replication (*Ori*) and the early and late promoters.

The early region encodes alternatively spliced tumour antigens, including Large Tumour antigen (LT), Small Tumour antigen (sT), 57-kilodalton tumour antigen (57kT), and the overprinting gene alternate LT ORF (ALTO) (Figure 10.1, right side). Both LT and sT share many characteristics of their SV40 counterparts, with some notable exceptions, whereas the function and physiological significance of both 57kT and ALTO remain to be elucidated (Gjoerup and Chang, 2010; Carter *et al.*, 2013; Spurgeon and Lambert, 2013).

LT antigen is a multifunctional protein with roles in manipulating the host cell cycle as well as driving replication of the viral genome. The N-terminal region of LT contains a conserved region 1 (CR1), a DnaJ domain that binds certain heat shock proteins, and an LxCxE motif that binds retinoblastoma protein (RB), all of which stimulate host cell proliferation through various mechanisms (Figure 10.2, top) (Shuda *et al.*, 2008). Flanking the RB binding site is the MCPyV Unique Region (MUR), a stretch of amino acids unique to MCPyV LT. The MUR has been reported to mediate LT binding to the endogenous vacuolar protein hVam6p, but the physiological significance of this interaction remains to be explored (Liu *et al.*, 2011). Additional functions encoded by the MUR remain to be discovered. Unlike other polyomavirus LT proteins

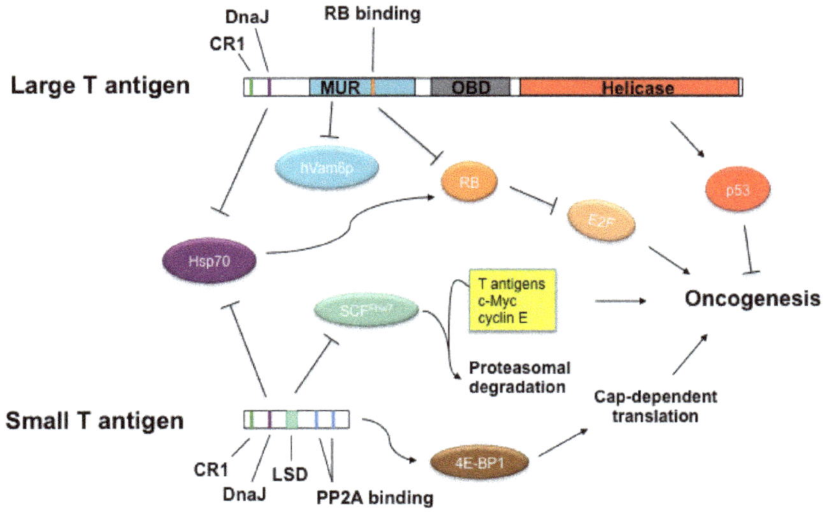

Figure 10.2. MCPyV oncoproteins interact with host cellular factors to drive tumour development. Shown are the domain structures of MCPyV LT and sT antigens. The two viral oncoproteins drive cellular transformation through interaction with various cellular targets. In the tumour-specific LTT mutants, the C-terminus of LT that activates p53 is omitted, thus releasing this anti-tumour break to allow oncogenic progression.

studied, MCPyV LT does not appear to directly bind the tumour suppressor p53, but nonetheless can inhibit p53-dependent reporter expression (Borchert *et al.*, 2014). The C-terminal region of LT contains an *Ori* Binding Domain (OBD), which is necessary for LT binding to the viral *Ori*, and a helicase domain, which acts as a hexameric complex to initiate unwinding of the viral DNA, stimulating replication of the viral genome (Kwun *et al.*, 2009; Diaz *et al.*, 2014) (Figure 10.2).

The sT protein shares portions of the LT N-terminal region, including the CR1 and DnaJ domains, but has a unique C-terminus (Figure 10.2). One of the key characteristics of this unique C-terminal domain is its ability to bind and inactivate PP2A phosphatases (Pallas *et al.*, 1990). Other polyomavirus sTs disrupt the Akt signalling pathway through PP2A inhibition, which leads to up-regulation of cellular proliferation (Sontag *et al.*, 1993; Rodriguez-Viciana *et al.*, 2006). MCPyV sT is uniquely able to more directly stimulate cellular proliferation by inducing hyper-phosphorylation of 4E-BP1 independent of PP2A binding (Shuda *et al.*, 2011). Indeed, unlike other polyomavirus sTs, MCPyV sT appears to play a central role in inducing oncogenesis, which we will discuss further below. sT also encodes a LT-stabilization domain (LSD), which binds the ubiquitin ligase SCFFbw7 and disrupts proteasomal degradation of both LT and certain cellular cell-cycle regulators (Kwun *et al.*, 2013) (Figure 10.2). Finally, MCPyV sT contains highly conserved iron-sulphur clusters in its unique C-terminal domain (Tsang *et al.*, 2016). These iron sulphur clusters are important for stimulating LT-mediated viral replication (Tsang *et al.*, 2016).

The late genes of MCPyV encode VP1 and VP2 (Figure 10.1, left side), which function as large and small subunits of the viral capsid, respectively. The major capsid protein VP1 is necessary and sufficient for the production of pseudovirions, while the minor protein VP2 seems to afford specificity in host cell targeting by unknown mechanisms (Tolstov *et al.*, 2009; Feng *et al.*, 2011; Neumann *et al.*, 2011; Schowalter *et al.*, 2011; Schowalter and Buck, 2013). These proteins recognize and bind sialic acid as well as heparin sulphate when attaching and entering target cells (Schowalter *et al.*, 2011). It is unclear how MCPyV targets specific cell types given that both sialic acid and heparin sulphate are ubiquitous, however other factors such as physical access may play a role. We will explore this further when we discuss MCPyV host cellular tropism.

MCPyV transcription and replication

The MCPyV NCRR contains the viral *Ori* flanked by the early and late promoter regions. Like other polyomaviruses, MCPyV early gene transcription begins soon after infection. The precise regulation of the MCPyV early promoter has not been studied. In related viruses like SV40, the presence of binding sites for ubiquitously expressed transcription factors like

Sp1, Sp2 and AP-1 allow the early genes to be expressed soon after the viral genome localizes to the nucleus; presumably the MCPyV early promoter functions in a similar manner (Dynan and Tjian, 1983a, b; White *et al.*, 2009). Many polyomaviruses, including SV40, encode a microRNA that autoregulates early gene transcription (Sullivan *et al.*, 2005). MCPyV encodes its own microRNA, termed miR-M1 (Figure 10.1), which was shown to downregulate expression of *LT*. This regulation of *LT* was critical for long-term MCPyV episome maintenance *in vitro* and could be important for establishing a persistent infection *in vivo* (Seo *et al.*, 2009; Theiss *et al.*, 2015).

After expression of the polyomavirus early genes *LT* and *sT*, viral genome amplification takes place. MCPyV LT binds to the viral *Ori* through its OBD, where it forms two hexameric complexes (Kwun *et al.*, 2009; Harrison *et al.*, 2011). LT must then be phosphorylated at a conserved threonine (T299) for activation of its helicase domain, which unwinds the viral DNA. Phosphorylation of the nearby T297 threonine disrupts LT binding to the viral *Ori*, and presumably acts as a break mechanism to limit both when and to what extent LT induces viral replication (Diaz *et al.*, 2014). MCPyV is unique in that sT can very robustly stimulate LT-mediated DNA replication. Both sT's LSD and the conserved iron-sulfur clusters are critical for this replication helper function (Kwun *et al.*, 2013; Tsang *et al.*, 2016).

When expressed at high levels in cells, *LT* mediates the formation of "replication factories" which appear as discrete foci within the host nucleus (Wang *et al.*, 2012; Li *et al.*, 2013; Tsang *et al.*, 2014;). This has been seen in both transient overexpression experiments as well as after infection of cells with MCPyV virions, but only when *LT* was expressed at a high level (Wang *et al.*, 2012; Li *et al.*, 2013; Tsang *et al.*, 2014; Liu *et al.*, 2016). These replication factories contain the viral genome and LT, as well as components of the host replication machinery, including the single-stranded DNA binding protein RPA70 and the DNA polymerase processivity factor PCNA (Wang *et al.*, 2012). Among polyomaviruses, MCPyV is unique in its recruitment of the host chromatin-binding factor BRD4 to these viral replication factories. BRD4 has been co-opted by a number of other DNA tumour viruses, including papillomaviruses and herpes viruses, for their replication and episome maintenance (You *et al.*, 2004;You *et al.*, 2006; Wang *et al.*, 2012; McBride and Jang, 2013; Wang *et al.*, 2013).

Interestingly, MCPyV replication also induces a robust host cellular DNA damage response (DDR). This has been evidenced by the phosphorylation of key mediators of both single- and double-stranded DNA damage response pathways, including H2AX, Ataxia telangiectasia mutated (ATM), Chk1 and Chk2 kinases (Li *et al.*, 2013). A number of these DDR proteins are recruited to viral replication factories. Inhibiting their function or reducing their expression through knockdown experiments attenuates viral replication, suggesting MCPyV has co-opted host DDR pathways for its replication (Tsang *et al.*, 2014). DDR activation appears to be most robustly stimulated by a region of the LT C-terminus; this activation stimulates p53 activity and can arrest the host cell cycle, providing an antagonistic activity against the proliferation induced by the N-terminal regions of LT (Li *et al.*, 2013) (Figure 10.2). Further highlighting the interplay of MCPyV infection and the host DDR, serine 816 of LT was shown to be phosphorylated by ATM kinase, although the functional significance of this phosphorylation remains unresolved (Li *et al.*, 2015). Although the host DDR has been shown to be manipulated by a number of DNA viruses, the specific molecular function of this DDR activation for MCPyV remain to be fully explored.

The MCPyV late genes are expressed only after viral replication (Feng *et al.*, 2011). As with the early promoter, the regulation of the MCPyV late promoter has not been studied; although once again the mechanism may be similar to that reported for other polyomaviruses, including a dependence on both viral DNA replication and robust *LT* expression (White *et al.*, 2009). The viral late genes *VP1* and *VP2* form capsids around the newly replicated viral genomes. How MCPyV virions exit cells, and whether this process is lytic or follows a less traumatic budding fashion, has not been resolved.

Viral genome integration and mechanisms of transformation

Similar to papillomavirus-associated cancers, MCPyV-induced MCC tumours typically have an integrated copy of the viral genome (Doorbar, 2006; Feng *et al.*, 2008; Doorbar *et al.*, 2012;). Although some viral proteins are expressed, these tumours do not support a productive viral life cycle. MCPyV-associated MCC tumours show a clonal integration pattern of the viral genome, suggesting that the integration event occurs early in oncogenesis, prior to the expansion of tumour cells. MCPyV-associated tumours typically express both of the major viral tumour antigens, LT and sT (Houben *et al.*, 2010; Shuda *et al.*, 2014). Each viral protein provides unique activities to drive tumourigenesis, and evidence supports both proteins playing a role in oncogenesis.

The MCPyV LT protein is typically expressed in a truncated form in MCC tumours (Shuda *et al.*, 2008). In addition to being integrated, a common feature of tumour-derived MCPyV genomes is the selection for mutations in the *LT* coding sequence that introduce premature stop codons which delete the C-terminal *Ori* binding and helicase domains. These tumour-specific mutations do not disrupt the expression of sT. The N-terminal portion of LT, which is still expressed in these tumours and referred to as LTT (tumour derived LT), retains the CR1, DnaJ and RB-binding motifs, allowing these tumour-specific LT mutants to robustly disrupt the host cell cycle. Indeed, an intact RB binding site is required to sustain growth of MCPyV-associated tumours, both *in vitro* and in xenograft models (Houben *et al.*, 2012). The selection for premature stop mutations is remarkably common in MCPyV-associated tumours, suggesting that retention of the replication domains in the C-terminus of LT is somehow antagonistic to tumourigenesis.

One selective pressure that may exist is the elimination of viral DNA replication after the genome has been integrated into the host DNA (Shuda *et al.*, 2008). Continuous LT-mediated replication from the integrated viral *Ori* could produce genotoxic levels of replication fork collisions and double-strand breaks in the host DNA; disrupting LT's OBD and helicase domains would alleviate this stress. Supporting this hypothesis, at least one tumour has been documented with an intact LT-coding sequence, but with a point mutation that disrupts a critical binding site recognized by the LT OBD; this point mutation was shown to make the viral *Ori* replication incompetent (Kwun *et al.*, 2009). In addition, other activities mediated by the C-terminal domain may be negatively selected for during tumourigenesis. For example, expression of just the OBD and helicase domains of MCPyV LT is sufficient to induce a DNA damage response, including activation of p53 and cell cycle arrest (Li *et al.*, 2013). This DDR induction could be due to non-specific helicase activity of MCPyV LT exerted on the host genome as even in the absence of viral *Ori* sequences MCPyV LT can form nuclear foci. This growth inhibitory property of the MCPyV LT C-terminal region may function as a barrier to malignant progression (Li *et al.*, 2013) (Figure 10.2). Another study further mapped LT's cell cycle arresting activity to the C-terminal 100 amino acids, indicating that at least some cell-cycle arresting activity exists outside of a replication context (Cheng *et al.*, 2013). None of these models are mutually exclusive: replicative stress, DNA damage responses and cell cycle arrest all pose challenges to oncogenesis, and together provide a reasonable rationale for the selection of truncated LT proteins which retain tumour-suppressor inhibiting activities while avoiding potentially antagonistic features residing in the C-terminal domain of LT.

A remarkable feature of MCPyV is the strong oncogenic potential of sT. In other polyomaviruses, sT generally plays a secondary role to LT in stimulating cellular proliferation and transformation. Contrary to other polyomaviruses, MCPyV LT alone is not sufficient to transform cells (Shuda *et al.*, 2011). MCPyV sT, however, has been suggested to transform immortalized rat fibroblasts in cell culture when expressed alone (Shuda *et al.*, 2011). sT's ability to induce hyper-phosphorylation of 4E-BP1 is a major driver of sT's oncogenic activity. Additionally, sT inhibits the E3 ubiquitin ligase SCFFbw7 to prevent proteasomal degradation of MCPyV LT and cell-cycle regulators like c-Myc and cyclin E, effectively stabilizing the expression of key viral and cellular proliferative proteins (Kwun *et al.*, 2013). Contrary to LT, no tumour mutations of sT have been reported, and sT is commonly expressed in MCPyV-associated tumours, again highlighting the important role this protein plays in MCPyV-derived cancers.

MCPyV host cellular tropism and the origin of MCC

While the oncogenic mechanisms of MCPyV have been actively studied since its discovery, much remains to be elucidated with respect to MCPyV natural infection. Studies of basic MCPyV virology have been hampered by the facts that MCPyV replicates poorly in the majority of cell lines tested thus far and its natural host cell up until very recently had not been described. Despite these challenges, many aspects of the viral life cycle have been successfully discovered in isolation. For example, expression of VP1 can produce pseudovirions capable of encapsidating small DNAs, including recombinant MCPyV genomes. The use of pseudovirions has allowed for the elucidation of key MCPyV binding factors, such as sialic acid and heparin sulfate, which mediate attachment and entry (Schowalter *et al.*, 2011; Neu *et al.*, 2012). Additionally, a number of cell lines, including the cervical cancer line C33A and the fibroblast line 293, have been shown to support early parts of the viral life cycle, including transcription and replication (Kwun *et al.*, 2009; Feng *et al.*, 2011; Wang *et al.*, 2012; Diaz *et al.*, 2014; Tsang *et al.*, 2014). Most of the studies cited in this chapter thus far have made use of such *in vitro* conditions to understand aspects of the viral life cycle in isolation. The lack of a robust cell culture system supporting the full MCPyV replication cycle has, however, limited our understanding of this important tumour virus.

Multiple lines of evidence point towards the skin being the major site of MCPyV replication in humans. Various deep sequencing experiments of adult skin have shown evidence of persistent and asymptomatic infection of MCPyV (Schowalter *et al.*, 2010; Foulongne *et al.*, 2012). Additionally, the cell lines which do support MCPyV replication have been either epithelial or fibroblast in origin (Kwun *et al.*, 2009; Feng *et al.*, 2011; Wang *et al.*, 2012). Finally, MCC is a tumour of the dermis and Merkel cells, the presumed cell of origin for MCC, are a resident of the epidermis. Following this line of reasoning, our lab examined the ability of various skin cell types to support MCPyV infection and recently reported that human dermal fibroblasts (HDFs) are a natural host cell of MCPyV (Liu *et al.*, 2016). In this study, we showed that both EGF and FGF were required for efficient infection of dermal fibroblasts; these factors may stimulate the expression of cellular factors necessary for infection, and the proliferative burst following growth factor treatment may produce an environment conducive to MCPyV infection and replication. These growth factors are also stimulated upon wounding (Werner and Grose, 2003), which may contribute to MCPyV persistent infection in the human skin. Another important factor supporting MCPyV infection is the induction of matrix metalloproteinases (MMPs) mediated by the WNT/β-catenin signalling pathway. These MMPs presumably provide MCPyV better access to its target cell-surface receptors. WNT signaling is critically important for the formation of hair follicles (Chen *et al.*, 2012), and we showed that MCPyV can efficiently infect the underlying dermal cells surrounding hair follicles (Liu *et al.*, 2016) (Figure 10.3). This finding matches well with the observation that MCPyV is frequently observed in eyebrow hair bulbs (Peretti *et al.*, 2014). Additionally, certain MCC risk factors, including UV exposure and aging, are known to up-regulate MMPs, once again linking risk of MCC incidence with MCPyV infection (Fisher *et al.*, 1996; Varani *et al.*, 2006; Gill and Parks, 2008; Cho *et al.*, 2009; Quan *et al.*, 2009; Quan and Fisher, 2015).

The identification of HDFs as a host cell for MCPyV may help resolve a long-standing debate in the MCC field regarding the cells of origin for MCC (Tilling and Moll, 2012). MCC has historically been thought to arise from Merkel cells due to its expression of cytokeratin 20, a unique marker of Merkel cells. A

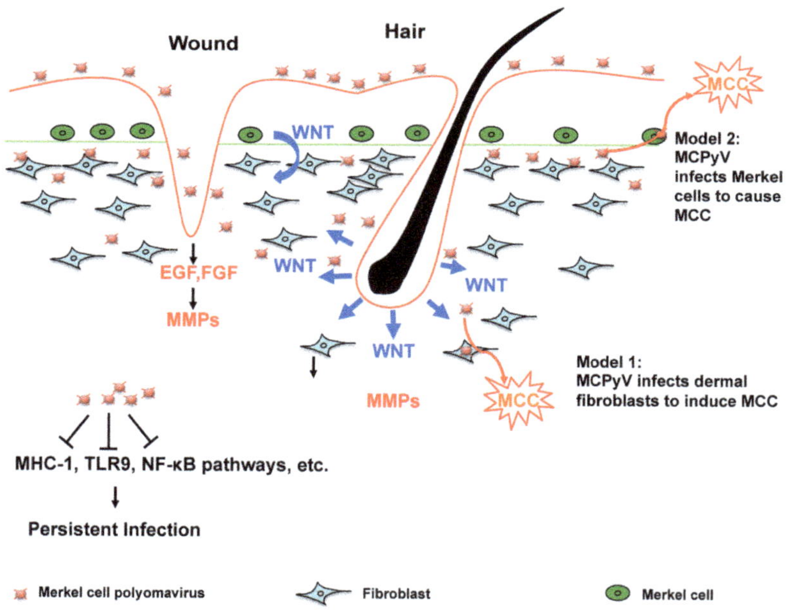

Figure 10.3. Molecular events contributing to MCPyV infection and MCC development. HDFs are a natural target cell of MCPyV infection in human skin. Induction of MMP genes by the WNT/β-catenin signaling pathway and growth factors, such as EGF and FGF, stimulates MCPyV infection. Skin damage caused by UV/ionizing radiation and wounding processes can induce EGF and FGF, and activate both WNT signaling and MMP expression, thus promoting viral infection. There are two working models for explaining how MCPyV-infected cells may give rise to MCC tumours. MCPyV infection of dermal fibroblasts could alter gene expression profile to induce MCC tumour development (Model 1). Alternatively, benign Merkel cells or progenitor cells located near the infected dermal fibroblasts could be infected as accidental bystanders, which represent a dead-end replication environment that promotes viral integration and transformation (Model 2). MCPyV may encode a number of immunomodulatory activities that play a role in establishing persistent infection.

number of recent findings, however, challenge this assumption. Merkel cells are post mitotic and do not have robust proliferative activity, making them less likely to support infection and tumourigenesis (Vaigot *et al.*, 1987; Moll *et al.*, 1996;). Additionally, Merkel cells are of epidermal origin while MCC tumours are thought to derive from the dermis (Morrison *et al.*, 2009; Van Keymeulen *et al.*, 2009; Calder and Smoller, 2010). MCC tumours have been reported to express markers common to pro-/pre-B cells, such as Paired box gene 5 (*PAX5*) and terminal deoxynucleotidyl transferase (TdT), suggesting that MCC tumours may derive from the B-cell lineage, although this has not been tested experimentally (Zur Hausen *et al.*, 2013). The finding that dermal fibroblasts support MCPyV infection provides exciting alternative hypotheses (Liu *et al.*, 2016) (Figure 10.3). The simplest model is that MCPyV infection of dermal fibroblasts can, over time, induce their transformation and up-regulation of genes commonly expressed in other cell types, including B-cells and Merkel cells (Figure 10.3, Model 1). This model agrees with the finding that MCC tumours are dermal in origin. An alternative hypothesis is that Merkel cells residing at the boundary of the epidermis and dermis may be infected as a non-productive bystander of dermal fibroblast infection (Figure 10.3, Model 2). In this model, the fact that Merkel cells do not support the full MCPyV life cycle may predispose this infection towards genome integration and transformation. These models – dermal fibroblast origin or bystander infection of Merkel cells – are not mutually exclusive, but both remain to be tested *in vivo* using animal and skin explant models (see below for a discussion of experimental models for studying MCPyV oncogenesis). The establishment of dermal fibroblasts as a natural host cell for MCPyV will open many new and exciting avenues of research and provides a tool for more thoroughly studying both MCPyV basic virology and its oncogenic potential.

Animal models for studying MCPyV oncogenic mechanism

An effective animal model is critical for studying the mechanisms underlying MCPyV-induced oncogenesis *in vivo*. Extensive effort has been devoted to developing animal models to examine the oncogenic activity of MCPyV oncogenes. In early studies, xenograft models utilizing mice injected with MCPyV-positive human MCC cells were used to examine the growth properties of these cells *in vivo* (Arora *et al.*, 2012; Dresang *et al.*, 2013). In a study led by Spurgeon *et al.*, a mouse model with conditional expression of both MCPyV sT and LTT antigens in the stratified epithelium of mice was developed (Spurgeon *et al.*, 2015). Mice expressing these antigens were observed to develop hyperplasia, hyperkeratosis, and a number of other skin abnormalities, but none of the mice developed MCC (Spurgeon *et al.*, 2015). A more recent transgenic mouse study specifically targeting expression of the oncogenic MCPyV sT antigen in the epidermis and epithelia of adult mice led to epithelial dysplasia, and a phenotype resembling squamous cell carcinoma *in situ* (Verhaegen *et al.*, 2015). Again, no MCC-like lesions were detected in this transgenic mouse model. In the latest report of a MCPyV sT animal model, ubiquitous high dose expression of MCPyV sT in multiple organs was found to be lethal in mice while low dose sT expression induced ear lobe dermal hyperkeratosis and hypergranulosis (Shuda *et al.*, 2015). While these attempts to establish transgenic mouse models for MCPyV oncogenes confirmed the tumourigenic capabilities of MCPyV sT and LTT *in vivo*, they did not generate lesions that fully recapitulate MCC pathogenesis (Shuda *et al.*, 2015; Spurgeon *et al.*, 2015; Verhaegen *et al.*, 2015). This could be partially attributed to the fact that most of these studies specifically targeted expression of the viral oncogenes in the epidermis while HDFs, rather than Merkel cells or other epithelial cells, are perhaps the primary target of MCPyV natural infection (Liu *et al.*, 2016). However, it is also possible that other unknown host somatic mutations are required to drive MCC oncogenesis. Therefore, better animal models are needed to investigate MCPyV-induced tumourigenesis *in vivo*. Our recent identification of HDFs as a target of MCPyV infection in human skin suggests that it is better to target expression of viral oncogenes to the dermis when generating animal models for studying how MCPyV oncogenes drive MCC development in live animals.

Furthermore, all of the reported MCPyV animal studies only expressed viral oncoproteins and did not try to recapitulate true MCPyV infection in the human skin. Since these animals have not actually been infected with MCPyV, the virus or its cancer-causing properties cannot be studied in the context of natural infection. As a result of the lack of an animal model exhibiting true MCPyV infection, it has not been possible to observe the viral life cycle or replication *in vivo*. An MCPyV infectable animal model is therefore needed to study the infectious life cycle of the virus. It is possible that only by monitoring MCPyV infection *in vivo* will we be able to elucidate the mechanisms of MCPyV-driven tumour formation and to determine how host and environmental risk factors contribute to this process. Because of the interactions of MCPyV with the immune system during MCC development described below, it is important to have an *in vivo* model in which the host immune response can be monitored in a more realistic manner than is allowed by *in vitro* studies. Building on the recent finding of HDFs as the target cells of MCPyV infection (Liu *et al.*, 2016), further research investigating the infectability of various animal fibroblasts will aid in identification of potential animal candidates for establishing a MCPyV infection animal model in the future.

Genetic basis for the oncogenic driver of MCPyV-associated MCC

Several recent studies have begun to define the genetic basis of MCC pathogenesis. These studies also provide insights for understanding the role of MCPyV in MCC pathogenesis. Delineating the differences in the causes of MCPyV-positive and MCPyV-negative MCCs will provide insights for determining the treatment options that work best for each type. It has been shown that UV radiation is the primary cause of the 20% of MCC cases that are MCPyV-negative (Harms *et al.*, 2015; Wong *et al.*, 2015; Goh *et al.*, 2016). Genetic studies have shown that, compared to MCPyV-positive MCCs, MCPyV-negative tumours display a much higher mutational burden. Specifically, these mutations were associated with a prominent UV-induced DNA damage signature exhibiting a high proportion of C>T and CC>TT mutations (Harms *et al.*, 2015; Wong *et al.*, 2015; Goh *et al.*, 2016). One study found that MCPyV-negative MCCs harboured a median of 1121 somatic single nucleotide variants (SSNVs) per tumour, while MCPyV-positive MCCs had a median of only 12.5 SSNVs per tumour (Goh *et al.*, 2016). Another study showed that MCPyV-negative MCCs exhibited an overall mutational burden of 10.09 ± 2.32 mutations/Mb while MCPyV-positive MCCs displayed a mutation burden of only 0.40 ± 0.09 mutations/Mb (Harms *et al.*, 2015). Both MCPyV-positive and MCPyV-negative MCC tumours are commonly found on sun-exposed regions of the body, such as the head, neck, and limbs (Wong *et al.*, 2015). However, the lower number of mutations in MCPyV-positive MCC, in combination with the lack of a UV mutation signature, indicate that UV plays a primary etiologic role in MCPyV-negative MCC tumourigenesis, but simply promotes tumour growth in MCPyV-positive MCC through local immunosuppressive effects on the tumour microenvironment, or possibly through promoting the mutations necessary for MCPyV integration and generation of the truncated viral LT antigen (Goh *et al.*, 2016). The lower number of genetic mutations found in the genome of MCPyV-positive MCCs compared to MCPyV-negative tumours and the lack of a definitive UV signature also support the idea that MCPyV oncogenes are the primary oncogenic drivers for these tumours.

It was also found that MCPyV-negative tumours contain a much higher number of cancer-promoting mutations compared to MCPyV-positive MCCs (Harms *et al.*, 2015; Wong *et al.*, 2015; Goh *et al.*, 2016). Some of the common mutations observed in MCPyV-negative MCCs include mutations in RB1, TP53 and PIK3CA, along with mutations in DNA damage repair and chromatin altering pathways (Harms *et al.*, 2015; Wong *et al.*, 2015; Goh *et al.*, 2016). Inactivating mutations in NOTCH indicate that it plays a tumour-suppressor role in MCC (Harms *et al.*, 2015). Activating mutations of HRAS were frequently observed in MCPyV-positive tumours, suggesting that these genetic mutations may cooperate with the viral oncogenes to drive the MCC tumourigenic progression (Harms *et al.*, 2015; Goh *et al.*, 2016;).

In multiple studies, ubiquitous inactivating mutations in RB1 were observed in MCPyV-negative tumours, but not in MCPyV-positive tumours (Harms *et al.*, 2015; Wong *et al.*, 2015; Goh *et al.*, 2016). This, in combination with the fact that the truncated form of MCPyV LT antigen interacts with and inactivates RB1, suggests that disruption of RB1 function is a critical event for all MCC tumourigenesis (Wong *et al.*, 2015). While there are many differences between MCPyV-positive and -negative MCCs, they both show high levels of T-cell infiltrating lymphocytes (TILs) (Wong *et al.*, 2015). The foreign, viral antigens expressed in MCPyV-positive MCCs are recognized by the host immune system and therefore make these tumours immunogenic. In addition, the high number of mutations observed in the virus-negative tumours may account for the higher tumour neoantigen burden observed in these tumours (Goh *et al.*, 2016). These features make both types of cancers excellent targets for potential therapeutic strategies encompassing immune checkpoint inhibitors (Wong *et al.*, 2015;Goh *et al.*, 2016).

MCPyV infection, host immune responses, and MCC

A complete understanding of how MCPyV drives MCC tumour development relies on elucidating the interplay between MCPyV infection and the host immune system. The role of MCPyV in MCC development, along with the fact that immune-suppressed individuals are at higher risk for developing this cancer, indicate that host immune function plays a critical role in controlling MCPyV-induced oncogenesis (Ma and Brewer, 2014). Additional evidence such as the occurrence of complete spontaneous tumour regression and presence of tumour reactive T cells also highlight the function of the immune system in preventing and eliminating MCC (Triozzi and Fernandez, 2013). Furthermore, better prognoses are generally observed in MCC patients with robust immune responses, and high intratumoural CD8+ and CD4+ lymphocyte infiltration predicts better survival (Paulson *et al.*, 2011; Sihto *et al.*, 2012; Sihto and Joensuu, 2012). However, immunosuppressed individuals only account for about 10% of MCC cases (Heath *et al.*, 2008); more than 90% of MCC patients with normal immune function fail to clear the MCC tumours that constantly express the highly antigenic, foreign MCPyV oncoproteins. MCC tumours continue to develop despite the presence of T cells recognizing MCPyV capsid proteins and oncoproteins that are constantly expressed in the tumours (Iyer *et al.*, 2011). These observations argue that immune evasion mechanisms exist for MCPyV-induced MCCs to escape immunological destruction. The facts that MCPyV infection is so prevalent in the general population (Foulongne *et al.*, 2012; Schowalter *et al.*, 2010; Tolstov *et al.*, 2009) and MCPyV-specific T-cell responses are clearly present in the blood of healthy individuals (Iyer *et al.*, 2011) also suggest that the virus has the ability to escape immune system-mediated clearance during latent infection (Figure 10.3, bottom).

Indeed, several recent studies have begun to reveal the potential roles of MCPyV proteins in modulation of the host immunological response. For example, in one study, 84% of MCC tumours exhibited downregulated expression of major histocompatibility complex class 1 (MHC-1) (Paulson *et al.*, 2014). MHC-1 is necessary for presenting peptides from intracellular proteins to CD8+ T cells (Hewitt, 2003), making its down-regulation an efficient mechanism for immune escape. MHC-1 expression was lower in MCPyV-positive tumours than in those that were MCPyV-negative (Paulson *et al.*, 2014), indicating that downregulation of MHC-1 may be a mechanism by which MCPyV viral proteins suppress recognition of MCPyV-derived antigens by CD8+ T cells. Another piece of evidence supporting the role of MCPyV in evading the immune response is its ability to downregulate Toll-like Receptor 9 (TLR9), a critical sensor of the host innate immune response that recognizes both viral and bacterial dsDNA (Shahzad *et al.*, 2013). This MCPyV function has been suggested to allow infected cells to escape host immune surveillance and to repress host innate immune responses during MCPyV-associated tumourigenesis (Shahzad *et al.*, 2013). MCPyV sT also down-regulates expression of cellular genes downstream of NF-κB pathways (Griffiths *et al.*, 2013). This could be a mechanism by which MCPyV subverts the host immune response to allow establishment of persistent infection in the host cells (Griffiths *et al.*, 2013) (Figure 10.3, bottom). In an analysis comparing gene expression profiles of MCPyV-negative and MCPyV-positive MCCs, Harms *et al.* showed that MCPyV-positive tumours had increased expression of immune response genes and were also enriched for peritumoural CD8+ T lymphocytes (Harms *et al.*, 2013), suggesting a potential role of viral proteins in modulating this cellular immune response.

MCPyV oncoproteins are not only specific and important for the growth of MCC cancer cells, but also foreign to the immune system. These key features of the viral oncogenes make them the idea targets of immunotherapy treatment for MCC. Despite the recent findings described above, many important questions remain in order to better understand the immune-evasion mechanisms employed by MCPyV to prevent host recognition. Once identified, the immuno-modulatory features of the viral proteins could also be subverted to induce antitumour immune response against MCPyV-positive tumour cells. Understanding how MCPyV modulates the host immune system could also suggest new targets for immune therapies.

Current and future therapeutic strategies

Although MCC is a highly aggressive carcinoma, there is currently no effective therapeutic regime available for curing this cancer. Early-stage, localized MCC is primarily treated with surgical excision (Brummer *et al.*, 2016). Adjuvant radiation therapy, or radiotherapy applied after the initial surgery, is also recommended for the primary tumour site because it has been linked to reduced local recurrence rates (Brummer *et al.*, 2016). MCC is generally radiosensitive, and adjuvant radiotherapy has been shown to improve local and regional recurrence rates, but this cancer frequently undergoes metastasis, increasing the probability that tumours may arise in areas that are harder to reach and to fully eradicate with radiotherapy (Rush *et al.*, 2011). Thus, chemotherapy has been applied to treat advanced stage MCC. Although MCC tumours are responsive to chemotherapy in the short term, the duration of the response is transient and many tumours often develop resistance to chemotherapy, which is associated with toxicity issues and lack of overall survival benefit in the studies completed so far (Saini and Miles, 2015; Brummer *et al.*, 2016; Cassler *et al.*, 2016). Additionally, chemotherapy has little benefit for MCC tumours because its immunosuppressive activity counteracts the cellular immune reaction to these tumours. Currently, there are very few viable options for patients with advanced or metastatic MCC (Cassler *et al.*, 2016).

Because of the inefficacy of these current treatments for advanced MCC, new therapies need to be explored. Considering the immunogenic properties of MCC, many studies are looking towards immunotherapy as a potential solution. For both primary and metastatic MCC tumours, patients with higher level of T cell infiltrates and increased expression of immune response markers showed higher rates of regression and better prognoses (Inoue *et al.*, 2000; Paulson *et al.*, 2011 Paulson *et al.*, 2013; Aldabagh *et al.*, 2014). This tight correlation between prognosis and immune function further supports the potential for immunotherapies in treating metastatic MCCs. As described above, studies have shown that MCCs, especially MCPyV-positive cases, evade immune response by down-regulating the expression of MHC-1, which is necessary for peptide recognition by T cells. Since this downregulation is reversible with interferon treatments, it has become a potential therapeutic target (Paulson *et al.*, 2014). Another promising immunotherapy approach for treating advanced MCCs targets the programmed cell death receptor 1/programmed cell death ligand 1 (PD-1/PD-L1) checkpoint. PD-L1 is often overexpressed in MCC tumours, especially in MCPyV-positive cases (Wong *et al.*, 2015). MCPyV-specific T-cells also express elevated levels of PD-1 (Afanasiev *et al.*, 2013). Ligation of PD-L1 with the PD-1 receptor on the surface of T cells activates an immune checkpoint, which inhibits the anti-tumour immune response (Keir *et al.*, 2008). Therefore, anti-PD-1 therapy is an attractive treatment option for MCC (Mantripragada and Birnbaum, 2015). In a recent study, patients treated with an anti-PD-1 antibody called pembrolizumab showed a response rate of 56% (Nghiem *et al.*, 2016). Although this immune checkpoint blockade strategy has demonstrated exciting potential for treating MCC tumours, these responses were not lasting, ranging in duration from 2.2 months to 9.7 months (Nghiem *et al.*, 2016).

While these recent studies have shown improvements in patients with various immunotherapies, the success has for the most part been minimal, further highlighting the need for new treatments for this cancer. With a disease-associated mortality of 46% and a rising incidence in the past two decades, it is important that new, more effective treatment options are explored (Saini and Miles, 2015). As described above, sequencing studies have revealed diverse mutational landscapes for MCPyV-positive and MCPyV-negative MCCs, suggesting that a precision medicine approach specifically targeting the viral or non-viral driver mutations in a particular MCC case may be an important option for treating MCC. In addition, since MCC is a virus-driven cancer, targeting MCPyV could be a viable strategy to treat the 80% of MCC cases that are MCPyV positive. Along this line, a DNA vaccine against the viral small T antigen is showing promise (Gomez *et al.*, 2013). Furthermore, MCPyV is more effective in driving MCC tumour development in immunocompromised populations, suggesting an important role of host immune repose in controlling the viral-induce carcinogenesis. It is therefore important to examine the immune escape mechanisms that allow the virus to maintain latent infection and to initiate oncogenic events (Figure 10.3, bottom). Mechanistic findings from these studies will offer important leads for developing viral-targeted therapeutic strategies to treat MCPyV-associated cancers.

Future Perspective

MCPyV is the only one of all currently known human polyomaviruses that has been associated with cancer. MCPyV infection occurs at a young age and is prevalent in the general population. During the course of its persistent infection, either due to prolonged sunlight exposure, viral genome amplification in the dermal cells, or infection of bystander Merkel cells, a dead-end host, MCPyV genomes become integrated into the host genome. Proliferation of cells carrying integrated viral genomes selects for LT truncation mutations that stimulate cellular transformation. Induction of MMP by aging and excessive UV exposure may favor tumour metastasis. Immune downregulation by viral proteins may allow MCPyV to establish a persistent infection; it may also play a key role in supporting the virally induced early stage MCC tumours to persist and expand.

Although MCC is a highly aggressive carcinoma, currently there is no effective therapeutic regime available for treating advanced stage cases. Recent discovery of HDFs as the major target of MCPyV infection in human skin provides an exciting opportunity for studying the infectious life cycle of this important oncogenic human polyomavirus. Identification of the target cells of MCPyV natural infection will also facilitate establishment of new animal models to investigate MCPyV biology and oncogenic mechanisms *in vivo*. MCPyV oncogenes are not only persistently expressed as foreign viral antigens in MCC tumours, but also required for the growth of the tumours cells. These key features make them the ideal target for developing novel immunotherapy to treat MCC tumours. Elucidation of the mechanism by which MCPyV escapes host immune surveillance to drive cellular transformation will facilitate the development of more effective virus-targeting immunotherapies for treating this highly aggressive skin cancer.

Acknowledgements

The authors would like to thank the members of our laboratory for helpful discussion and Dr. Wei Liu for figure preparation. This work was supported in part by the National Institutes of Health (NIH) Grants R01CA148768, R01CA142723, and R01CA187718.

References

Afanasiev, O.K., Yelistratova, L., Miller, N., Nagase, K., Paulson, K., Iyer, J.G., Ibrani, D., Koelle, D.M., and Nghiem, P. (2013). Merkel polyomavirus-specific T cells fluctuate with merkel cell carcinoma burden and express therapeutically targetable PD-1 and Tim-3 exhaustion markers. Clin Cancer Res 19, 5351-5360.

Aldabagh, B., Joo, J., and Yu, S.S. (2014). Merkel cell carcinoma: current status of targeted and future potential for immunotherapies. Semin Cutan Med Surg 33, 76-82.

Arora, R., Shuda, M., Guastafierro, A., Feng, H., Toptan, T., Tolstov, Y., Normolle, D., Vollmer, L.L., Vogt, A., Domling, A., *et al.* (2012). Survivin is a therapeutic target in Merkel cell carcinoma. Sci Transl Med 4, 133ra156.

Borchert, S., Czech-Sioli, M., Neumann, F., Schmidt, C., Wimmer, P., Dobner, T., Grundhoff, A., and Fischer, N. (2014). High-affinity Rb binding, p53 inhibition, subcellular localization, and transformation by wild-type or tumour-derived shortened Merkel cell polyomavirus large T antigens. J Virol 88, 3144-3160.

Brummer, G.C., Bowen, A.R., and Bowen, G.M. (2016). Merkel Cell Carcinoma: Current Issues Regarding Diagnosis, Management, and Emerging Treatment Strategies. Am J Clin Dermatol 17, 49-62.

Calder, K.B., and Smoller, B.R. (2010). New insights into merkel cell carcinoma. Adv Anat Pathol 17, 155-161.

Carter, J.J., Daugherty, M.D., Qi, X., Bheda-Malge, A., Wipf, G.C., Robinson, K., Roman, A., Malik, H.S., and Galloway, D.A. (2013). Identification of an overprinting gene in Merkel cell polyomavirus provides evolutionary insight into the birth of viral genes. Proc Natl Acad Sci USA 110, 12744-12749.

Cassler, N.M., Merrill, D., Bichakjian, C.K., and Brownell, I. (2016). Merkel Cell Carcinoma Therapeutic Update. Curr Treat Options Oncol 17, 36.

Chang, Y., and Moore, P.S. (2012). Merkel cell carcinoma: a virus-induced human cancer. Annu Rev Pathol 7, 123-144.

Chen, D., Jarrell, A., Guo, C., Lang, R., and Atit, R. (2012). Dermal beta-catenin activity in response to epidermal Wnt ligands is required for fibroblast proliferation and hair follicle initiation. Development 139, 1522-1533.

Cheng, J., Rozenblatt-Rosen, O., Paulson, K.G., Nghiem, P., and DeCaprio, J.A. (2013). Merkel cell polyomavirus large T antigen has growth-promoting and inhibitory activities. J Virol 87, 6118-6126.

Cho, S., Shin, M.H., Kim, Y.K., Seo, J.E., Lee, Y.M., Park, C.H., and Chung, J.H. (2009). Effects of infrared radiation and heat on human skin aging *in vivo*. J Investig Dermatol Symp Proc 14, 15-19.

Diaz, J., Wang, X., Tsang, S.H., Jiao, J., and You, J. (2014). Phosphorylation of large T antigen regulates merkel cell polyomavirus replication. Cancers (Basel) 6, 1464-1486.

Doorbar, J. (2006). Molecular biology of human papillomavirus infection and cervical cancer. Clin Sci (Lond) 110, 525-541.

Doorbar, J., Quint, W., Banks, L., Bravo, I.G., Stoler, M., Broker, T.R., and Stanley, M.A. (2012). The biology and life-cycle of human papillomaviruses. Vaccine 30 Suppl 5, F55-70.

Dresang, L.R., Guastafierro, A., Arora, R., Normolle, D., Chang, Y., and Moore, P.S. (2013). Response of Merkel cell polyomavirus-positive merkel cell carcinoma xenografts to a survivin inhibitor. PLoS One 8, e80543.

Dynan, W.S., and Tjian, R. (1983a). Isolation of transcription factors that discriminate between different promoters recognized by RNA polymerase II. Cell 32, 669-680.

Dynan, W.S., and Tjian, R. (1983b). The promoter-specific transcription factor Sp1 binds to upstream sequences in the SV40 early promoter. Cell 35, 79-87.

Feng, H., Kwun, H.J., Liu, X., Gjoerup, O., Stolz, D.B., Chang, Y., and Moore, P.S. (2011). Cellular and viral factors regulating Merkel cell polyomavirus replication. PLoS One 6, e22468.

Feng, H., Shuda, M., Chang, Y., and Moore, P.S. (2008). Clonal integration of a polyomavirus in human Merkel cell carcinoma. Science 319, 1096-1100.

Feng, H., Taylor, J.L., Benos, P.V., Newton, R., Waddell, K., Lucas, S.B., Chang, Y., and Moore, P.S. (2007). Human transcriptome subtraction by using short sequence tags to search for tumour viruses in conjunctival carcinoma. J Virol 81, 11332-11340.

Fisher, G.J., Datta, S.C., Talwar, H.S., Wang, Z.Q., Varani, J., Kang, S., and Voorhees, J.J. (1996). Molecular basis of sun-induced premature skin ageing and retinoid antagonism. Nature 379, 335-339.

Foulongne, V., Sauvage, V., Hebert, C., Dereure, O., Cheval, J., Gouilh, M.A., Pariente, K., Segondy, M., Burguière, A., Manuguerra, J.-C., *et al.* (2012). Human Skin Microbiota: High Diversity of DNA Viruses Identified on the Human Skin by High Throughput Sequencing. PLoS One 7, e38499.

Gill, S.E., and Parks, W.C. (2008). Metalloproteinases and their inhibitors: regulators of wound healing. Int J Biochem Cell Biol 40, 1334-1347.

Gjoerup, O., and Chang, Y. (2010). Chapter 1 - Update on Human Polyomaviruses and Cancer. In Adv Cancer Res, F.V.W. George, and K. George, eds. (Academic Press), pp. 1-51.

Goh, G., Walradt, T., Markarov, V., Blom, A., Riaz, N., Doumani, R., Stafstrom, K., Moshiri, A., Yelistratova, L., Levinsohn, J., *et al.* (2016). Mutational landscape of MCPyV-positive and MCPyV-negative Merkel cell carcinomas with implications for immunotherapy. Oncotarget 7, 3403-3415.

Gomez, B., He, L., Tsai, Y.C., Wu, T.C., Viscidi, R.P., and Hung, C.F. (2013). Creation of a Merkel cell polyomavirus small T antigen-expressing murine tumour model and a DNA vaccine targeting small T antigen. Cell Biosci 3, 29.

Griffiths, D.A., Abdul-Sada, H., Knight, L.M., Jackson, B.R., Richards, K., Prescott, E.L., Peach, A.H., Blair, G.E., Macdonald, A., and Whitehouse, A. (2013). Merkel cell polyomavirus small T antigen targets the NEMO adaptor protein to disrupt inflammatory signaling. J Virol 87, 13853-13867.

Harms, P.W., Patel, R.M., Verhaegen, M.E., Giordano, T.J., Nash, K.T., Johnson, C.N., Daignault, S., Thomas, D.G., Gudjonsson, J.E., Elder, J.T., *et al.* (2013). Distinct gene expression profiles of viral- and nonviral-associated merkel cell carcinoma revealed by transcriptome analysis. J Invest Dermatol 133, 936-945.

Harms, P.W., Vats, P., Verhaegen, M.E., Robinson, D.R., Wu, Y.M., Dhanasekaran, S.M., Palanisamy, N., Siddiqui, J., Cao, X., Su, F., *et al.* (2015). The Distinctive Mutational Spectra of Polyomavirus-Negative Merkel Cell Carcinoma. Cancer Res 75, 3720-3727.

Harrison, C.J., Meinke, G., Kwun, H.J., Rogalin, H., Phelan, P.J., Bullock, P.A., Chang, Y., Moore, P.S., and Bohm, A. (2011). Asymmetric assembly of Merkel cell polyomavirus large T-antigen origin binding domains at the viral origin. J Mol Biol 409, 529-542.

Heath, M., Jaimes, N., Lemos, B., Mostaghimi, A., Wang, L.C., Peñas, P.F., and Nghiem, P. (2008). Clinical characteristics of Merkel cell carcinoma at diagnosis in 195 patients: the AEIOU features. Journal of the American Academy of Dermatology 58, 375-381.

Hewitt, E.W. (2003). The MHC class I antigen presentation pathway: strategies for viral immune evasion. Immunology 110, 163-169.

Houben, R., Adam, C., Baeurle, A., Hesbacher, S., Grimm, J., Angermeyer, S., Henzel, K., Hauser, S., Elling, R., Brocker, E.B., *et al.* (2012). An intact retinoblastoma protein-binding site in Merkel cell polyomavirus large T antigen is required for promoting growth of Merkel cell carcinoma cells. Int J Cancer 130, 847-856.

Houben, R., Shuda, M., Weinkam, R., Schrama, D., Feng, H., Chang, Y., Moore, P.S., and Becker, J.C. (2010). Merkel cell polyomavirus-infected Merkel cell carcinoma cells require expression of viral T antigens. J Virol 84, 7064-7072.

Inoue, T., Yoneda, K., Manabe, M., and Demitsu, T. (2000). Spontaneous regression of merkel cell carcinoma: a comparative study of TUNEL index and tumour-infiltrating lymphocytes between spontaneous regression and non-regression group. J Dermatol Sci 24, 203-211.

Iyer, J.G., Afanasiev, O.K., McClurkan, C., Paulson, K., Nagase, K., Jing, L., Marshak, J.O., Dong, L., Carter, J., Lai, I., *et al.* (2011). Merkel cell polyomavirus-specific CD8(+) and CD4(+) T-cell responses identified in Merkel cell carcinomas and blood. Clin Cancer Res 17, 6671-6680.

Kean, J.M., Rao, S., Wang, M., and Garcea, R.L. (2009). Seroepidemiology of human polyomaviruses. PLoS Pathog 5, e1000363.

Keir, M.E., Butte, M.J., Freeman, G.J., and Sharpe, A.H. (2008). PD-1 and its ligands in tolerance and immunity. Annu Rev Immunol 26, 677-704.

Kwun, H.J., Guastafierro, A., Shuda, M., Meinke, G., Bohm, A., Moore, P.S., and Chang, Y. (2009). The Minimum Replication Origin of Merkel Cell Polyomavirus Has a Unique Large T-Antigen Loading Architecture and Requires Small T-Antigen Expression for Optimal Replication. J. Virol. 83, 12118-12128.

Kwun, H.J., Shuda, M., Feng, H., Camacho, C.J., Moore, P.S., and Chang, Y. (2013). Merkel cell polyomavirus small T antigen controls viral replication and oncoprotein expression by targeting the cellular ubiquitin ligase SCFFbw7. Cell Host Microbe 14, 125-135.

Lemos, B.D., Storer, B.E., Iyer, J.G., Phillips, J.L., Bichakjian, C.K., Fang, L.C., Johnson, T.M., Liegeois-Kwon, N.J., Otley, C.C., Paulson, K.G., *et al.* (2010). Pathologic nodal evaluation improves prognostic accuracy in Merkel cell carcinoma: analysis of 5823 cases as the basis of the first consensus staging system. J Am Acad Dermatol 63, 751-761.

Li, J., Diaz, J., Wang, X., Tsang, S.H., and You, J. (2015). Phosphorylation of Merkel Cell Polyomavirus Large Tumour Antigen at Serine 816 by ATM Kinase Induces Apoptosis in Host Cells. J Biol Chem 290, 1874-1884.

Li, J., Wang, X., Diaz, J., Tsang, S.H., Buck, C.B., and You, J. (2013). Merkel cell polyomavirus large T antigen disrupts host genomic integrity and inhibits cellular proliferation. J Virol 87, 9173-9188. PMCID: PMC3754048.

Liu, W., Yang, R., Payne, A.S., Schowalter, R.M., Spurgeon, M.E., Lambert, P.F., Xu, X., Buck, C.B., and You, J. (2016). Identifying the Target Cells and Mechanisms of Merkel Cell Polyomavirus Infection. Cell Host Microbe 19, 775-787.

Liu, X., Hein, J., Richardson, S.C., Basse, P.H., Toptan, T., Moore, P.S., Gjoerup, O.V., and Chang, Y. (2011). Merkel cell polyomavirus large T antigen disrupts lysosome clustering by translocating human Vam6p from the cytoplasm to the nucleus. J Biol Chem 286, 17079-17090.

Ma, J.E., and Brewer, J.D. (2014). Merkel cell carcinoma in immunosuppressed patients. Cancers (Basel) 6, 1328-1350.

Mantripragada, K., and Birnbaum, A. (2015). Response to Anti-PD-1 Therapy in Metastatic Merkel Cell Carcinoma Metastatic to the Heart and Pancreas. Cureus 7, e403.

McBride, A.A., and Jang, M.K. (2013). Current understanding of the role of the Brd4 protein in the papillomavirus lifecycle. Viruses 5, 1374-1394.

Moll, I., Zieger, W., and Schmelz, M. (1996). Proliferative Merkel cells were not detected in human skin. Arch Dermatol Res 288, 184-187.

Morrison, K.M., Miesegaes, G.R., Lumpkin, E.A., and Maricich, S.M. (2009). Mammalian Merkel cells are descended from the epidermal lineage. Dev Biol 336, 76-83.

Neu, U., Hengel, H., Blaum, B.S., Schowalter, R.M., Macejak, D., Gilbert, M., Wakarchuk, W.W., Imamura, A., Ando, H., Kiso, M., *et al.* (2012). Structures of Merkel cell polyomavirus VP1 complexes define a sialic acid binding site required for infection. PLoS Pathog 8, e1002738.

Neumann, F., Borchert, S., Schmidt, C., Reimer, R., Hohenberg, H., Fischer, N., and Grundhoff, A. (2011). Replication, Gene Expression and Particle Production by a Consensus Merkel Cell Polyomavirus (MCPyV) Genome. PLoS One 6, e29112.

Nghiem, P.T., Bhatia, S., Lipson, E.J., Kudchadkar, R.R., Miller, N.J., Annamalai, L., Berry, S., Chartash, E.K., Daud, A., Fling, S.P., *et al.* (2016). PD-1 Blockade with Pembrolizumab in Advanced Merkel-Cell Carcinoma. N Engl J Med.

Pallas, D.C., Shahrik, L.K., Martin, B.L., Jaspers, S., Miller, T.B., Brautigan, D.L., and Roberts, T.M. (1990). Polyoma small and middle T antigens and SV40 small t antigen form stable complexes with protein phosphatase 2A. Cell 60, 167-176.

Pastrana, D.V., Tolstov, Y.L., Becker, J.C., Moore, P.S., Chang, Y., and Buck, C.B. (2009). Quantitation of human seroresponsiveness to Merkel cell polyomavirus. PLoS Pathog 5, e1000578.

Paulson, K.G., Iyer, J.G., Blom, A., Warton, E.M., Sokil, M., Yelistratova, L., Schuman, L., Nagase, K., Bhatia, S., Asgari, M.M., *et al.* (2013). Systemic immune suppression predicts diminished Merkel cell carcinoma-specific survival independent of stage. J Invest Dermatol 133, 642-646.

Paulson, K.G., Iyer, J.G., Tegeder, A.R., Thibodeau, R., Schelter, J., Koba, S., Schrama, D., Simonson, W.T., Lemos, B.D., Byrd, D.R., *et al.* (2011). Transcriptome-wide studies of merkel cell carcinoma and validation of intratumoural CD8+ lymphocyte invasion as an independent predictor of survival. J Clin Oncol 29, 1539-1546.

Paulson, K.G., Tegeder, A., Willmes, C., Iyer, J.G., Afanasiev, O.K., Schrama, D., Koba, S., Thibodeau, R., Nagase, K., Simonson, W.T., *et al.* (2014). Downregulation of MHC-I expression is prevalent but reversible in Merkel cell carcinoma. Cancer Immunol Res 2, 1071-1079.

Peretti, A., Borgogna, C., Rossi, D., De Paoli, L., Bawadekar, M., Zavattaro, E., Boldorini, R., De Andrea, M., Gaidano, G., and Gariglio, M. (2014). Analysis of human beta-papillomavirus and Merkel cell polyomavirus infection in skin lesions and eyebrow hair bulbs from a cohort of patients with chronic lymphocytic leukaemia. Br J Dermatol 171, 1525-1528.

Quan, T., and Fisher, G.J. (2015). Role of Age-Associated Alterations of the Dermal Extracellular Matrix Microenvironment in Human Skin Aging: A Mini-Review. Gerontology 61, 427-434.

Quan, T., Qin, Z., Xia, W., Shao, Y., Voorhees, J.J., and Fisher, G.J. (2009). Matrix-degrading metalloproteinases in photoaging. J Investig Dermatol Symp Proc 14, 20-24.

Rodriguez-Viciana, P., Collins, C., and Fried, M. (2006). Polyoma and SV40 proteins differentially regulate PP2A to activate distinct cellular signaling pathways involved in growth control. Proc Natl Acad Sci USA 103, 19290-19295.

Rush, Z., Fields, R.C., Lee, N., and Brownell, I. (2011). Radiation therapy in the management of Merkel cell carcinoma: current perspectives. Expert Rev Dermatol 6, 395-404.

Saini, A.T., and Miles, B.A. (2015). Merkel cell carcinoma of the head and neck: pathogenesis, current and emerging treatment options. Onco Targets Ther 8, 2157-2167.

Sastre-Garau, X., Peter, M., Avril, M.F., Laude, H., Couturier, J., Rozenberg, F., Almeida, A., Boitier, F., Carlotti, A., Couturaud, B., *et al.* (2009). Merkel cell carcinoma of the skin: pathological and molecular evidence for a causative role of MCV in oncogenesis. J Pathol 218, 48-56.

Schowalter, R.M., and Buck, C.B. (2013). The Merkel cell polyomavirus minor capsid protein. PLoS Pathog 9, e1003558.

Schowalter, R.M., Pastrana, D.V., and Buck, C.B. (2011). Glycosaminoglycans and sialylated glycans sequentially facilitate Merkel cell polyomavirus infectious entry. PLoS Pathog 7, e1002161.

Schowalter, R.M., Pastrana, D.V., Pumphrey, K.A., Moyer, A.L., and Buck, C.B. (2010). Merkel cell polyomavirus and two previously unknown polyomaviruses are chronically shed from human skin. Cell Host Microbe 7, 509-515.

Seo, G.J., Chen, C.J., and Sullivan, C.S. (2009). Merkel cell polyomavirus encodes a microRNA with the ability to autoregulate viral gene expression. Virology 383, 183-187.

Shahzad, N., Shuda, M., Gheit, T., Kwun, H.J., Cornet, I., Saidj, D., Zannetti, C., Hasan, U., Chang, Y., Moore, P.S., *et al.* (2013). The T antigen locus of Merkel cell polyomavirus downregulates human Toll-like receptor 9 expression. J Virol 87, 13009-13019.

Shuda, M., Chang, Y., and Moore, P.S. (2014). Merkel cell polyomavirus-positive Merkel cell carcinoma requires viral small T-antigen for cell proliferation. J Invest Dermatol 134, 1479-1481.

Shuda, M., Feng, H., Kwun, H.J., Rosen, S.T., Gjoerup, O., Moore, P.S., and Chang, Y. (2008). T antigen mutations are a human tumour-specific signature for Merkel cell polyomavirus. Proc Natl Acad Sci USA 105, 16272-16277.

Shuda, M., Guastafierro, A., Geng, X., Shuda, Y., Ostrowski, S.M., Lukianov, S., Jenkins, F.J., Honda, K., Maricich, S.M., Moore, P.S., *et al.* (2015). Merkel Cell Polyomavirus Small T Antigen Induces Cancer and Embryonic Merkel Cell Proliferation in a Transgenic Mouse Model. PLoS One 10, e0142329.

Shuda, M., Kwun, H.J., Feng, H., Chang, Y., and Moore, P.S. (2011). Human Merkel cell polyomavirus small T antigen is an oncoprotein targeting the 4E-BP1 translation regulator. J Clin Invest 121, 3623-3634.

Sihto, H., Bohling, T., Kavola, H., Koljonen, V., Salmi, M., Jalkanen, S., and Joensuu, H. (2012). Tumour infiltrating immune cells and outcome of Merkel cell carcinoma: a population-based study. Clin Cancer Res 18, 2872-2881.

Sihto, H., and Joensuu, H. (2012). Tumour-infiltrating lymphocytes and outcome in Merkel cell carcinoma, a virus-associated cancer. Oncoimmunology 1, 1420-1421.

Sontag, E., Fedorov, S., Kamibayashi, C., Robbins, D., Cobb, M., and Mumby, M. (1993). The interaction of SV40 small tumour antigen with protein phosphatase 2A stimulates the map kinase pathway and induces cell proliferation. Cell 75, 887-897.

Spurgeon, M.E., Cheng, J., Bronson, R.T., Lambert, P.F., and DeCaprio, J.A. (2015). Tumourigenic activity of merkel cell polyomavirus T antigens expressed in the stratified epithelium of mice. Cancer Res 75, 1068-1079.

Spurgeon, M.E., and Lambert, P.F. (2013). Merkel cell polyomavirus: a newly discovered human virus with oncogenic potential. Virology 435, 118-130.

Sullivan, C.S., Grundhoff, A.T., Tevethia, S., Pipas, J.M., and Ganem, D. (2005). SV40-encoded microRNAs regulate viral gene expression and reduce susceptibility to cytotoxic T cells. Nature 435, 682-686.

Theiss, J.M., Gunther, T., Alawi, M., Neumann, F., Tessmer, U., Fischer, N., and Grundhoff, A. (2015). A Comprehensive Analysis of Replicating Merkel Cell Polyomavirus Genomes Delineates the Viral Transcription Program and Suggests a Role for mcv-miR-M1 in Episomal Persistence. PLoS Pathog 11, e1004974.

Tilling, T., and Moll, I. (2012). Which are the cells of origin in merkel cell carcinoma? J Skin Cancer 2012, 680410.

Toker, C. (1972). Trabecular carcinoma of the skin. Arch Dermatol 105, 107-110.

Tolstov, Y.L., Pastrana, D.V., Feng, H., Becker, J.C., Jenkins, F.J., Moschos, S., Chang, Y., Buck, C.B., and Moore, P.S. (2009). Human Merkel cell polyomavirus infection II. MCV is a common human infection that can be detected by conformational capsid epitope immunoassays. Int J Cancer 125, 1250-1256.

Triozzi, P.L., and Fernandez, A.P. (2013). The role of the immune response in merkel cell carcinoma. Cancers (Basel) 5, 234-254.

Tsang, S.H., Wang, R., Nakamaru-Ogiso, E., Knight, S.A., Buck, C.B., and You, J. (2016). The Oncogenic Small Tumour Antigen of Merkel Cell Polyomavirus Is an Iron-Sulfur Cluster Protein That Enhances Viral DNA Replication. J Virol 90, 1544-1556.

Tsang, S.H., Wang, X., Li, J., Buck, C.B., and You, J. (2014). Host DNA Damage Response Factors Localize to Merkel Cell Polyomavirus DNA Replication Sites to Support Efficient Viral DNA Replication. J Virol 88, 3285-3297. PMCID: PMC3957940.

Vaigot, P., Pisani, A., Darmon, Y.M., and Ortonne, J.P. (1987). The majority of epidermal Merkel cells are non-proliferative: a quantitative immunofluorescence analysis. Acta Derm Venereol 67, 517-520.

Van Keymeulen, A., Mascre, G., Youseff, K.K., Harel, I., Michaux, C., De Geest, N., Szpalski, C., Achouri, Y., Bloch, W., Hassan, B.A., *et al.* (2009). Epidermal progenitors give rise to Merkel cells during embryonic development and adult homeostasis. J Cell Biol 187, 91-100.

Varani, J., Dame, M.K., Rittie, L., Fligiel, S.E., Kang, S., Fisher, G.J., and Voorhees, J.J. (2006). Decreased collagen production in chronologically aged skin: roles of age-dependent alteration in fibroblast function and defective mechanical stimulation. Am J Pathol 168, 1861-1868.

Verhaegen, M.E., Mangelberger, D., Harms, P.W., Vozheiko, T.D., Weick, J.W., Wilbert, D.M., Saunders, T.L., Ermilov, A.N., Bichakjian, C.K., Johnson, T.M., *et al.* (2015). Merkel cell polyomavirus small T antigen is oncogenic in transgenic mice. J Invest Dermatol 135, 1415-1424.

Wang, T.S., Byrne, P.J., Jacobs, L.K., and Taube, J.M. (2011). Merkel cell carcinoma: update and review. Semin Cutan Med Surg 30, 48-56.

Wang, X., Helfer, C.M., Pancholi, N., Bradner, J.E., and You, J. (2013). Recruitment of Brd4 to the Human Papillomavirus Type 16 DNA Replication Complex Is Essential for Replication of Viral DNA. J. Virol. 87, 3871-3884. PMCID:PMC3624215.

Wang, X., Li, J., Schowalter, R.M., Jiao, J., Buck, C.B., and You, J. (2012). Bromodomain protein Brd4 plays a key role in Merkel cell polyomavirus DNA replication. PLoS Pathog 8, e1003021. PMCID:PMC3493480.

Werner, S., and Grose, R. (2003). Regulation of wound healing by growth factors and cytokines. Physiol Rev 83, 835-870.

White, M.K., Safak, M., and Khalili, K. (2009). Regulation of gene expression in primate polyomaviruses. J Virol 83, 10846-10856.

Wong, S.Q., Waldeck, K., Vergara, I.A., Schroder, J., Madore, J., Wilmott, J.S., Colebatch, A.J., De Paoli-Iseppi, R., Li, J., Lupat, R., *et al.* (2015). UV-Associated Mutations Underlie the Etiology of MCV-Negative Merkel Cell Carcinomas. Cancer Res 75, 5228-5234.

You, J., Croyle, J.L., Nishimura, A., Ozato, K., and Howley, P.M. (2004). Interaction of the bovine papillomavirus E2 protein with Brd4 tethers the viral DNA to host mitotic chromosomes. Cell 117, 349-360.

You, J., Srinivasan, V., Denis, G.V., Harrington, W.J., Jr., Ballestas, M.E., Kaye, K.M., and Howley, P.M. (2006). Kaposi's sarcoma-associated herpesvirus latency-associated nuclear antigen interacts with bromodomain protein Brd4 on host mitotic chromosomes. J Virol 80, 8909-8919. PMCID: PMC1563901.

Zur Hausen, A., Rennspiess, D., Winnepenninckx, V., Speel, E.J., and Kurz, A.K. (2013). Early B-cell differentiation in Merkel cell carcinomas: clues to cellular ancestry. Cancer Res.

11

DNA Tumour Viruses and the Host DNA Damage Response

Amy Y. Hafez and Micah A. Luftig*

Department of Molecular Genetics and Microbiology, Center for Virology, Duke University Medical Center, Durham, NC 27710 USA

**Corresponding author: micah.luftig@duke.edu*

Abstract

When viruses infect human cells, they must successfully resist host defences to faithfully replicate their genetic material. To do so, viruses interface with a number of cellular processes including the host DNA damage response. DNA viruses have a particularly precarious situation, in that they must infect cells without alerting the host DNA damage response pathway to recognize and launch a response against the viral DNA genome. This challenge has required DNA viruses to develop unique methods to allow the virus to interact with and evade the host DNA damage response. In this chapter, we focus on four DNA tumour viruses, Epstein-Barr virus (EBV), Kaposi sarcoma virus (KSHV), human papillomavirus (HPV), and adenovirus. These viruses modulate the DNA damage response by activating or deactivating sensors of DNA damage as well as key downstream effector molecules. Here we review the current literature discussing the integrated relationship between DNA tumour viruses and the host DNA damage response during lytic and latent viral infection.

Introduction

Our understanding of the intricacies of the DNA damage response (DDR) signalling pathway and how viruses interact with this host defence mechanism has greatly increased over the last few decades. Viruses have developed mechanisms to evade recognition by the host DDR, prevent downstream signalling consequences, and, in some cases, use the DDR to facilitate viral replication. In this chapter, we first discuss the sensors, kinases and effector kinases of the cellular DNA damage response and the major themes behind viral interaction with the DDR. Next, we examine the specific strategies DNA tumour viruses employ to interact with the DDR during lytic and latent life cycle stages. Lastly, we discuss emerging research on the distinction between viral and host DNA damage during infection.

Cellular DNA damage response

As cells grow and divide they experience constant threats to their genetic material. It has been estimated that each cell incurs 10,000 DNA lesions per day in the form of double-strand breaks (DSBs) and single-strand breaks (SSBs) (Lindahl, 1993). These lesions occur as a result of exposure to endogenous threats such as reactive oxygen species and exogenous environmental agents, for example ionizing radiation and radiomimetic chemicals that can damage cellular DNA (Figure 11.1). Furthermore, errors in normal cellular processes like DNA replication, V(D)J recombination and meiotic recombination may also result in DNA damage and lead to genetic mutations that can be passed down to subsequent generations (Hartlerode and Scully, 2009). If these mutations are not dealt with properly they can ultimately give rise to a number of diseases, including cancer. To maintain genetic integrity, cells have developed an integrative DNA damage response pathway that is tasked with sensing

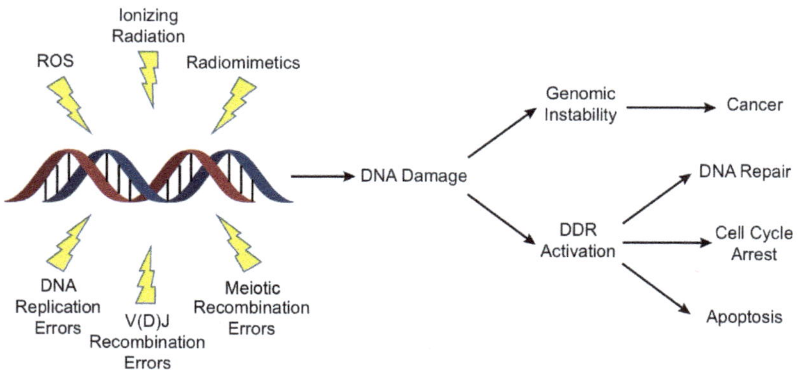

Figure 11.1. Sources of DNA damage. Schematic diagram of potential endogenous and exogenous sources of DNA damage that cause single-stranded DNA breaks and double-stranded DNA breaks. Abbreviation: ROS, reactive oxygen species.

damaged DNA, initiating a signalling cascade to alert necessary effector molecules, and controlling cell cycle checkpoints. Thus this damage response serves as an important innate tumour suppressor pathway. These DNA damage checkpoints have been found to function in activation of cell cycle arrest, DNA repair, activation of transcriptional programs, telomere length control, and induction of apoptosis (Figure 11.1) (Clarke *et al.*, 1993; Xu and Baltimore, 1996; Naito *et al.*, 1998; Cortez *et al.*, 1999; Martin *et al.*, 1999; Mills *et al.*, 1999; Gatei *et al.*, 2000; Lim *et al.*, 2000; Wu *et al.*, 2000; Zhao *et al.*, 2000).

In humans, initiation of the DDR is controlled by three major phosphoinositide 3-kinase-related protein kinases (PIKKs): ataxia telangiectasia mutated (ATM), ATM-Rad3-related protein (ATR), and DNA-dependent protein kinase (DNA-PK) (Sancar *et al.*, 2004; Ciccia and Elledge, 2010). ATM is known to primarily respond to DSBs, and has also been implicated in the general response to reactive oxygen species (ROS) and irradiation (Shiloh, 2003; Alexander and Walker, 2010; Guo *et al.*, 2010a; Guo *et al.*, 2010b; Cosentino *et al.*, 2011). Like ATM, DNA-PK is also a major kinase involved in DSB recognition primarily caused by genotoxic stresses or non-homologous end joining (NHEJ) during V(D)J recombination (Dobbs *et al.*, 2010; Nagasawa *et al.*, 2011). In contrast, ATR is predominantly activated by exposure of single-stranded DNA (ssDNA) ends and is crucial for controlling DNA replication (Zou, 2007; Shiotani and Zou, 2009a,b; Lopez-Contreras and Fernandez-Capetillo, 2010; Flynn and Zou, 2011). Activation of these DNA damage protein kinases leads to initiation of a signalling cascade resulting in phosphorylation of a number of downstream effector molecules, which leads to repair or in the case of irreparable damage, cellular arrest or apoptosis.

Following DSB formation, the MRE11-RAD50-NBS1 (MRN) complex is one of the first complexes to be recruited where it functions as a DNA damage sensor and assists in the recruitment of ATM to the damaged site (Figure 11.2a) (Falck *et al.*, 2005; Derheimer and Kastan, 2010). ATM is then activated by autophosphorylation at serine^{1981}, which induces its conformational change from a dimer to a monomer (Bakkenist and Kastan, 2003; So *et al.*, 2009). Upon initial activation, ATM then triggers a DDR signalling cascade that involves phosphorylation of histone variant H2AX at serine^{139} within minutes after DNA damage (Burma *et al.*, 2001). This signal for H2AX phosphorylation (γH2AX) is amplified and rapidly spreads over 500kb of the chromatin domains flanking the DNA damage (Meier *et al.*, 2007; Savic *et al.*, 2009). Mdc1 has been shown to facilitate this process by interacting with both ATM through its FHA domain, as well as directly binding to γH2AX, thus providing an mechanism for ATM to target nucleosomes containing γH2AX and recognize chromatin flanking the DNA damage site (Stewart *et al.*, 2003; Stucki *et al.*, 2005; Lou *et al.*, 2006). In addition to γH2AX phosphorylation, ATM signals to activate downstream effectors including Chk2, p53, BRCA1, 53BP1, and many more. Activation of H2AX triggers ubiquitinylation and SUMOylation cascades to promote recruitment of BRCA1 and 53BP1 and regulate DSB repair by homologous recombination (Huen *et al.*, 2007; Mailand *et al.*, 2007; Doil *et al.*, 2009; Stewart, 2009; Mattiroli *et al.*, 2012). Furthermore, ATM also phosphorylates Chk2 at threonine^{68} and p53 at serine^{15} to induce senescence or apoptosis (Figure 11.2a).

Like ATM, DNA-PK is also recruited to sites of DSBs. The KU70/KU80 heterodimer recruits and stabilizes DNA-PK at the DSB (Figure 11.2b) (Mahaney *et al.*, 2009). Artemis-mediated resectioning of the DSB

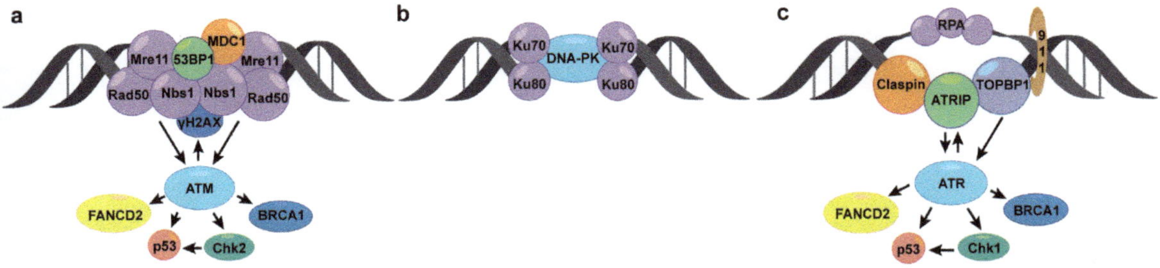

Figure 11.2. DNA damage response signalling pathways. **(A)** Schematic of the ATM DNA damage response pathway to double-stranded breaks. DNA damage is sensed by the MRN complex, which recruits ATM to the damage site. This allows activation of a downstream signalling cascade, which involves phosphorylation of γH2AX, activation of repair factors, including FANCD2, BRCA1 and 53BP1, and cell cycle arrest effectors, including Chk2 and p53. **(B)** Schematic representation of the DNA-PK DNA damage response pathway to double-stranded breaks. Ku70/80 recruits and stabilizes DNA-PK at the site of DNA damage to facilitate non-homologous end joining. **(C)** Schematic of the ATR DNA damage response pathway to single-stranded breaks. RPA coats single-stranded DNA and facilitates localization of ATR to DNA damage site. ATRIP forms a complex with ATR, binds RPA and interacts with the 911 complex to activate ATR. ATR goes on to activate downstream effectors, including p53, Chk1, and BRCA1.

occurs and the KU/DNA-PK complex recruits the XRCC4 adaptor and DNA ligase IV to allow for non-homologous end joining (NHEJ).

During each S phase, ATR is activated in response to ssDNA exposure, regulates firing of replication origins and initiates repair of stalled or collapsed replication forks (Nyberg *et al.*, 2002; Shechter *et al.*, 2004). Replication protein A (RPA) coats ssDNA and is important for the localization of ATR to the damage site (Figure 11.2c). ATR-interacting protein (ATRIP) forms a complex with ATR and binds directly to RPA-ssDNA. While RPA is sufficient for recruitment of ATR, the ATR-ATRIP complex must interact with the Rad9-Rad1-Hus1 (9-1-1) complex to activate ATR (Michael *et al.*, 2000; Stokes *et al.*, 2002; Byun *et al.*, 2005; MacDougall *et al.*, 2007). The 9-1-1 complex can then stimulate ATR signalling by bringing in topoisomerase-binding protein-1 (TOPBP1) to activate ATR (Furuya *et al.*, 2004; Delacroix *et al.*, 2007; Lee *et al.*, 2007). Once triggered, ATR goes on to activate downstream effectors, like p53, Chk1, and BRCA1 that are involved in induction of apoptosis, senescence, and DNA repair. Chk1 is activated by phosphorylation at serine^{317} and serine^{345} by means of Claspin, which is an adaptor molecule that brings ATR to Chk1 (Kumagai and Dunphy, 2000). Once Chk1 is phosphorylated it can target the cdc25 phosphatase cell cycle regulators that function by removing the inhibitory phosphorylation of cyclin-dependent kinases (Boutros *et al.*, 2006). ATR can also activate BRCA1, Werner syndrome ATP-dependent helicase and Bloom syndrome protein, which are all important proteins involved in regulating recombination and thus promote repair of DNA lesions (Figure 11.2c) (Tibbetts *et al.*, 2000; Pichierri *et al.*, 2003; Davies *et al.*, 2004; Li *et al.*, 2004).

While the ATM and ATR pathways are distinct, they do experience cross talk with each other. Most downstream effectors are phosphorylated by both ATM and ATR kinases, including BRCA1 and p53 (Siliciano *et al.*, 1997; Cortez *et al.*, 1999; Tibbetts *et al.*, 1999; Sarbassov *et al.*, 2004). Furthermore, ATR has been shown to phosphorylate H2AX in response to replication stress. This may recruit ATM to stressed replication forks (Ward and Chen, 2001). Although the ability for cross talk is available, evidence has been shown that ATM and ATR have unique functions and can act independent of each other (Brown and Baltimore, 2003; Chanoux *et al.*, 2009; Balmus *et al.*, 2012).

Viruses and the DNA damage response

The DDR can serve as an important cellular defence against viral pathogens. DNA viruses experience a significant challenge in preventing the DDR from recognizing replicating viral DNA genomes as damaged DNA. However, depending on the nature of viral interaction with the DDR, this can be beneficial or deleterious for viral replication. In the case of adenovirus, the DDR recognizes the DNA viral genome, preventing the virus from carrying out DNA cleavage and packaging. To overcome this barrier, adenovirus protein complex E4orf3/E1b55k degrades the hosts DDR sensing complex, MRN, to allow for DNA processing and prevent concatemerization (Figure 11.3a) (Stracker *et al.*, 2002). This is a classic example of the complex relationship between DNA viruses and the host DDR.

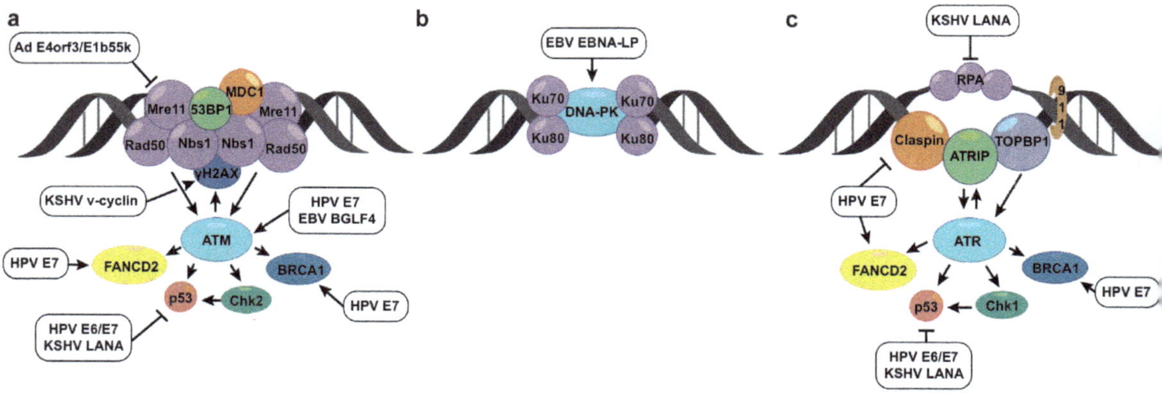

Figure 11.3. Viral interaction with the host DNA damage response pathway. Schematic representation indicating viruses and viral proteins that have been shown to interact with the (A) ATM, (B) DNA-PK and (C) ATR DNA damage response pathway. Abbreviations: Ad, adenovirus; EBV, Epstein-Barr virus; KSHV, Kaposi sarcoma-associated herpesvirus; and HPV, human papillomavirus.

Human DNA tumour viruses interface with the host DDR through a few distinct mechanisms. One mechanism is that viral oncoproteins can drive host cell proliferation to induce replicative stress thereby activating the DDR. As viruses are typically dependent on host cell machinery to replicate, triggering the host cell to proliferate boosts viral replication and diverts the DDR to the host DNA. During both latent and lytic phases of virus replication, a second mechanism of DDR engagement involves direct interaction of viral proteins with DDR machinery. This interaction can modulate the activity of the DDR such that it activates or inactivates effectors involved in the DDR to prevent senescence or apoptosis from occurring, which would ultimately suppress viral replication and promote DNA repair. Finally, human DNA tumour viruses can form viral DNA structures during replication that can be recognized by the DDR. During latency, DDR factors are crucial for maintaining the episomal DNA of large DNA tumour viruses. Furthermore, during lytic replication these viruses take advantage of DDR-mediated recombination and repair to process viral DNA and prepare it for packaging. Since the DDR can have positive and negative effects on viral DNA replication, the interaction between the virus and the DDR must be properly regulated. In this chapter, we will discuss recent advances in studying the DNA tumour viruses EBV, KSHV, HPV, and adenovirus and the strategies they employ to modulate the DDR pathway.

Viral replication and the DNA damage response

DNA tumour viruses, including herpesviruses, HPV, and adenovirus, are characterized by different life cycle phases; however, viruses have been shown to interact with the DDR pathway in each of these phases. All herpesviruses, including EBV and KSHV, exhibit a biphasic life cycle consisting of a lytic and latent stage. While these viruses typically establish latent infections, lytic reactivation can occur and has been shown to activate the DDR. The HPV life cycle is linked to the host keratinocyte and consists of three phases of replication; establishment, maintenance and productive, which are all affected by the DDR. The adenovirus life cycle is characterized by its DNA replication and can be divided into early and late phases. The early genes mostly function as regulatory proteins to modify host protein expression to facilitate viral DNA synthesis; activate viral genes, such as the virus-encoded DNA polymerase; and to evade the host immune defence and promote survival of the infected cell.

EBV and other herpesviruses encode an important protein kinase, called BGLF4, that functions as a regulator of lytic virus replication. This serine/threonine protein kinase is known for its cyclin-dependent kinase 1 (CDK1)-like activity and can phosphorylate both EBV latency and lytic proteins to regulate their transactivation and expression (Kato *et al.*, 2001; Kato *et al.*, 2003; Kawaguchi *et al.*, 2003; Yue *et al.*, 2005; Asai *et al.*, 2006; El-Guindy *et al.*, 2014). Interestingly, the Hayward group found that BGLF4 regulates the host DDR through the TIP60/ATM pathway. BGLF4 can phosphorylate TIP60 by acetylation activity. Activated TIP60 resulted in phosphorylation of ATM and downstream effectors (Li *et al.*, 2015). This TIP60/ATM activation by BGLF4 was found to be crucial for viral replication upon lytic induction as loss of TIP60 or ATM suppressed viral DNA replication (Li *et al.*, 2011). Recent advances made by utilizing an unbiased quantitative phosphoproteomic/mass spectrometry approach revealed that upon BGLF4 induction there is an increase in phosphorylation of cdc25c and PP1, which is known to regulate several aspects of the DDR (Kuntziger *et al.*, 2011; Li *et al.*, 2015). Taken together, BGLF4 phosphorylates TIP60 to activate the ATM pathway. Alternatively, BGLF4 can induce PP1 inhibition, which activates p38α that also triggers TIP60 phosphorylation to activate the DDR pathway (Li *et al.*, 2015). The Kenney group further identified activation of the ATM pathway as a key factor required for EBV lytic reactivation (Hagemeier *et al.*, 2012). ATM inhibition prevented efficient EBV lytic gene expression when in the presence of lytic reactivation-inducing stimuli including histone deacetylase inhibitors, transforming growth factor β, and B cell receptor engagement with anti-IgG. Specifically, expression of viral proteins BRLF1 and BRRF1 were observed to be ATM-dependent. Together these results suggest that ATM is important for a promoting lytic EBV reactivation.

Lytic reactivation in KSHV infected cells has been known to cause DNA double-stranded breaks, impair the non-homologous end joining repair pathway and activate the host DDR pathway (Xiao *et al.*, 2013; Jackson *et al.*, 2014). Recent work by the Grand laboratory has resulted in a detailed investigation of the activation of ATM, ATR, and DNA-PK DDR pathways during lytic replication. Interestingly, they find that KSHV lytic reactivation in B cells specifically activates the ATM and DNA-PK DDR kinases along with their downstream targets. To investigate the stages of the KSHV lytic lifecycle that are necessary for DDR activation, this group used ganciclovir, an antiviral drug that inhibits viral DNA synthesis without affecting expression of early viral lytic proteins (Matthews and Boehme, 1988; Hollingworth *et al.*, 2015). With treatment of ganciclovir immediate-early lytic proteins, RTA and ORF57 were expressed, while late envelope glycoprotein K8.1A was not expressed. Expression levels of DDR markers, phosphorylated H2AX and RPA32, were unchanged despite the lack of expression of late viral proteins. This suggests that DDR activation during lytic replication is mediated by expression of the early viral genes (Hollingworth *et al.*, 2015). Importantly, inhibition of the ATM pathway resulted in reduced production of infectious virus (Hollingworth *et al.*, 2015). The mechanistic link between KSHV lytic reactivation and the DDR pathway was further explored by the Ojala laboratory, who conducted a siRNA screen for novel regulators of KSHV reactivation. They found that depletion of MDM2, an E3 ubiquitin ligase, dramatically accelerated KSHV reactivation. MDM2 serves as a primary regulator of the p53 tumour suppressor pathway and functions to degrade p53 through its ubiquitin ligase activity (Haupt *et al.*, 1997). Further analysis investigating the mechanism behind MDM2 regulation of lytic reactivation lead to the discovery that during lytic reactivation cells activated p53 and p21 and accumulated DNA damage. Up-regulation of the p21 cell cycle regulator induced cell cycle arrest at G2, which is required to promote viral replication and transcription of viral lytic genes (Balistreri *et al.*, 2016). KSHV is not the only virus found to induce a G2/M cell cycle arrest during viral DNA amplification. Other DNA viruses found to induce arrest include HPV, adenoviruses, and parvoviruses (Davy and Doorbar, 2007). It is suggested that activation of G2/M cell cycle arrest facilitates viral replication to allow the virus to access cell cycle factors expressed at the G2 phase, including recombination and repair factors (Branzei and Foiani, 2008; Balistreri *et al.*, 2016). Furthermore, G2/M arrest can help the virus infected cell evade factors active in M phase that can impair virus propagation. Together, these findings suggest that KSHV-induced DNA damage during lytic reactivation ultimately requires DDR activation to allow the virus to propagate efficiently.

HPV infects squamous epithelial cells and causes cervical cancer (Walboomers *et al.*, 1999). The HPV life cycle is strongly associated with host cell differentiation (Longworth and Laimins, 2004). Upon infection of keratinocytes through a microwound in the stratified epithelium, HPV infects dividing basal cells. Infected daughter cells migrate into the suprabasal cell layer and undergo differentiation. E7 and E6 viral protein expression prevents exit from the cell cycle, thus promoting reentry of infected cells into S phase, and consequently promoting viral DNA synthesis. Cellular differentiation promotes the viral productive phase allowing for amplification of the viral genome (Moody and Laimins, 2010). The Moody group has previously established that HPV31-infected cells activate the ATM pathway to recruit repair factors, including Rad51, BRCA1, and P-RPA S33, to viral replication foci facilitating efficient replication (Gillespie *et al.*, 2012). Recently, the Moody group has reported that depletion of Rad51 and BRCA1, as well as inhibition of Rad51's recombinase activity, block productive replication but not viral genome maintenance (Chappell *et al.*, 2015). Overall, this indicates that HPV manipulates the host DDR to recruit DNA recombination repair factors to viral genomes and facilitate the productive phase.

While much progress has been made in the herpesvirus and HPV fields studying the effect of the DDR on lytic reactivation, recent foundational research in adenovirus has facilitated our understanding of this interplay. In a landmark paper from the Weitzman laboratory, they reveal that viruses can manipulate host chromatin to evade the host DDR. DNA virus genomes are compacted and packaged into virus particles along with small basic proteins. Adenovirus encodes a highly basic protein, called protein VII, which binds viral DNA and is incorporated in the virus particle (Vayda *et al.*, 1983; Chatterjee *et al.*, 1986). Protein VII localization was investigated and found to be present at both viral replication centres and in cellular chromatin stained for histone H1 and DAPI (Avgousti *et al.*, 2016). Protein VII uniquely altered cellular chromatin while other viral proteins had no effect on chromatin appearance. Importantly, mass spectrometry of nuclear fractions of infected cells expressing protein VII showed enrichment of HMGB1, 2 and 3. HMGB1 is a nuclear protein that when released serves as an extracellular danger signal that promotes detection by the immune system after infection (Lotze and Tracey, 2005; Lu *et al.*, 2012; Kang *et al.*, 2014). When protein VII is deleted, nuclear fractionation of infected cells resulted in lack of chromatin retention of HMGB1 and HMGB2. The Weitzman group conducted rigorous *in vivo* studies using a murine model of lipopolysaccharide-induced lung injury and found that expression of protein VII in transduced mouse lungs allowed for retention of mouse HMGB1. When mice were exposed to lipopolysaccharide to induce HMGB1 release, bronchoalveolar lavage fluid exhibited significantly less HMGB1 24 hours after lipopolysaccharide exposure and fewer neutrophils than mice expressing the empty vector. These findings suggest that protein VII plays an important role in allowing the virus to evade antiviral immune recognition by retaining HMGB1 in cellular chromatin (Avgousti *et al.*, 2016).

Latent viral infections and the DNA damage response

EBV-associated malignancies are characterized by latent infections in which a limited subset of viral proteins and small non-coding RNAs are expressed. The viral latency proteins include EBV nuclear antigens (EBNA 1, 2, 3A, 3B, 3C and LP), along with the latent membrane proteins (LMP1, 2A and 2B) (Yates *et al.*, 1985). EBNA-LP and EBNA2 are the first viral latent proteins to be expressed in infected B cells and are involved in transcriptional up-regulation of many viral and cellular genes. Viral genes, EBNA1, EBNA 3A, 3B, and 3C, along with cellular proto-oncogenes c-Myc, E2F1, and cyclin D2 are among the numerous genes up-regulated upon infection and their expression allows the infected B cells to enter the cell cycle, transitioning from the resting to proliferating state (Sinclair *et al.*, 1994; Kaiser *et al.*, 1999). The EBNA3 proteins are responsible for modulated EBNA2-dependent gene expression and can also regulate a number of other targets including repression of key tumour suppressors (Zhao *et al.*, 1996; Johannsen *et al.*, 1995; Skalska *et al.*, 2010; Maruo *et al.*, 2011; Skalska *et al.*, 2013). LMP1, a constitutively active mimic of the TNF receptor and B cell co-stimulatory molecule CD40, is then activated by EBNA2 and EBNA3C (Cahir McFarland *et al.*, 1999; Lin *et al.*, 2002). LMP1 activation can then trigger NFκB, p38, and JNK signalling pathways to promote proliferation and survival. LMP2 proteins act as a mimic of the B cell receptor and suppress endogenous B-cell receptor signalling and constitutively activates downstream tyrosine kinases for PI3K signalling and other pro-survival pathways (Miller, 1995). Expression of all viral latent proteins and non-coding RNAs is referred to as the latency III gene expression program. Latency III is established in EBV-associated AIDS lymphomas, post-transplant lymphoproliferative disorder, and *in vitro* production of LCLs (Rickinson, 2006). Alternatively, Hodgkin lymphoma and nasopharyngeal carcinoma express a restricted Latency II program, expressing EBNA1, LMP1, LMP2, and viral non-coding RNAs, and Burkitt lymphoma (BL) establish a latency I program of EBNA1 and non-coding RNAs (Babcock *et al.*, 2000).

KSHV-associated lymphomas establish two major reservoirs for infection, endothelial cells and B cells. Latency is established *in vitro* using human endothelial cell cultures and derived *in vivo* through primary effusion lymphomas (Lagunoff *et al.*, 2002). KSHV latency profile includes expression of v-Cyclin, viral FLICE inhibitory protein (vFLIP), latency-associated nuclear antigen (LANA), kaposins, and viral miRNAs (Staskus *et al.*, 1997; Chandriani and Ganem, 2010; Mesri *et al.*, 2010). LANA plays an important role in maintaining the viral episome (Ballestas *et al.*, 1999), inhibiting tumour suppressive host pathways (p53 and E2F-Rb) to allow the infected cell to evade innate host defences (Figure 11.3a) (Friborg *et al.*, 1999; Radkov *et al.*, 2000; Chipuk and Green, 2006; Chen *et al.*, 2010), and promoting telomerase expression (Verma *et al.*, 2004; Mesri *et al.*, 2010). v-Cyclin functions in the activation of cyclin-dependent kinase 6 and induces genomic instability (Chang *et al.*, 1996; Verschuren *et al.*, 2002) and vFLIP activates NFκB, promoting expression of anti-apoptotic molecules (Liu *et al.*, 2002; Guasparri *et al.*, 2004).

Like EBV and KSHV, the HPV genome is maintained as an episome. The viral proteins expressed are E1, E2, E6, and E7 (Moody and Laimins, 2010). While E1 and E2 are responsible for maintaining and replicating the viral episome, E1 has been found to be dispensable in some cases (Egawa *et al.*, 2012). E2 promotes stable episomal maintenance by tethering viral genomes to host mitotic chromosomes through chromatin adapter proteins (McBride, 2008). This allows for equal partitioning of HPV genomes to daughter cells after division (McBride, 2013). E6 and E7 function in promoting cell proliferation through antagonism of the host innate tumour suppressor pathway (p53 and Rb) (Figure 11.3a) (Dyson *et al.*, 1989; Munger *et al.*, 1989; Scheffner *et al.*, 1990; Cheng *et al.*, 1995; Dowhanick *et al.*, 1995). The Laimins group has extensively studied the effect of HPV on the DDR and found that the Fanconi anaemia (FA) pathway plays an important role in regulating the HPV life cycle and its interaction with the ATM and ATR DDR pathways. FA is a part of the DDR that mediates cross talk between the ATM and ATR pathways. The Laimins group reported that HPV activates the FA pathway, which results in an accumulation of FANCD2, a large nuclear protein responsible for recruiting downstream effector proteins involved in DNA repair (Figure 11.3a) (Kee and D'Andrea, 2010). Furthermore, studies show that FANCD2 colocalizes with other DNA repair factors and is preferentially recruited to viral genomes, as opposed to chromosomal DNA. Additionally, knockdown of FANCD2 was observed to reduce episomal maintenance in undifferentiated cells (Spriggs and Laimins, 2017). Other groups have also shown that loss of the FA pathway by knockdown of FANCA and FANCD2 stimulates E7 protein accumulation in infected human keratinocytes resulting in increased proliferation. These studies suggest that the FA pathway is an important regulator of HPV replication.

Viral gene expression induces a growth suppressive DDR

Many tumour viruses utilize the DDR to alter the host cellular environment to promote long-term proliferation, oncogenic transformation, and ultimately survival. B cell transformation efficiency when infected with EBV is low where only approximately 1% of infected B cells are immortalized (Henderson *et al.*, 1977; Sugden and Mark, 1977). This suggests that there must be host cell innate barriers at play to restrict virally induced proliferation. The DDR was identified by Nikitin *et al.* to induce a growth suppressive barrier responsible for ~10-fold restriction (Nikitin *et al.*, 2010). EBV infection of primary human B cells induced an ATM-mediated DDR, as indicated by phosphorylation of H2AX, ATM, Chk2, and formation of 53BP1 nuclear foci. These DDR marks did not co-localize with the viral genome, thus signifying that host DNA damage induced the DDR and not viral DNA damage. Furthermore, through fluorescent dye proliferation tracking studies it was reported that EBV induced a transient period of hyper-proliferation followed by a period of attenuated proliferation. Interestingly, this hyper-proliferative period was associated with activation of the DDR. Importantly, inhibition of ATM and Chk2 enhanced viral-mediated transformation, which suggests a mechanism where the DDR is an innate host tumour suppressor pathway responsible for restricting viral-mediated cell proliferation (Nikitin *et al.*, 2010).

The host gene expression of the hyper-proliferative B cells was studied in comparison to resting B cells and transformed LCLs to determine what distinct host factors may be responsible for inhibiting transformation. Microarray analysis revealed distinct transcriptomic changes among these cell populations, including enrichment for genes associated with cell cycle progression and the DDR in early proliferation (Nikitin *et al.*, 2010; Forte *et al.*, 2012; Price *et al.*, 2012). The rapidly proliferating EBV-infected B cells transitioned from expressing increased levels of EBNA-LP and EBNA2, to accumulating EBNA3 proteins and attenuated proliferation. As EBNA3 proteins begin accumulating they can act to attenuate EBNA2-regulated targets such as CD23 and c-Myc (Nikitin *et al.*, 2010). Overall, these studies support a model of heightened proliferative capacity due to c-Myc promoting replicative stress and DDR-mediated growth arrest. As cells progress beyond this phase, attenuated growth rates and c-Myc no longer activated the growth suppressive DDR.

Recently, McFadden *et al.* reported that growth arrest of hyper-proliferating EBV-infected B cells was associated with reduced level of mitochondrial respiration and decreased expression of genes involved in the TCA cycle and oxidative phosphorylation (McFadden *et al.*, 2016). Growth arrest was rescued by supplemented TCA cycle intermediates. Furthermore, arrested cells exhibited an increase in expression of p53 pathway gene targets, including sestrins leading to activation of AMPK, a reduction in mTOR signalling, and elevated autophagy, which was found to be important for maintaining hyper-proliferation, and ultimately survival during metabolic stress. Interestingly, glucose import and surface glucose transporter 1 (GLUT1) levels increased with long-term outgrowth, leading to increased metabolic processes, including glycolysis, oxidative phosphorylation, and suppression of basal autophagy (McFadden *et al.*, 2016). This study identified metabolic imbalance as a key instigator of DDR-mediated growth arrest during the hyper-proliferative period and subsequent barrier to B cell immortalization by EBV.

Suppression of KSHV-infected cells was first linked to the DDR by work conducted in the Ojala group investigating v-Cyclin in telomerase immortalized endothelial cells (Koopal *et al.*, 2007). v-Cyclin over-expression resulted in a decrease in cellular proliferation by triggering an S phase cell cycle arrest. These endothelial cells also experienced activation of phosphorylated H2AX, ATM, Chk2, and p53 DDR markers (Figure 11.3a). Inhibition of ATM in v-Cyclin over-expressing cells rescued the DDR activation and S phase growth arrest, however this resulted in abnormal mitosis (Koopal *et al.*, 2007). It is suggested that this abnormal mitosis is associated with v-Cyclin's ability to induce centrosome amplification in the absence of a functional DDR (Verschuren *et al.*, 2002). Work by the McCormick group found that the KSHV v-FLIP protein is capable of suppressing the v-Cyclin induced DDR and senescence (Leidal *et al.*, 2012).

DDR activation generates genomic instability

Expression of viral latent proteins from DNA tumour viruses can induce cell proliferation and promote growth transformation. However, this is often accompanied by persistent DDR due to irreparable DNA damage. Unrepaired DNA damage can lead to genomic instability and cancer development. Tumour viruses promote genomic instability and initiate tumourigenesis through diverse mechanisms including telomere dysfunction and induction of ROS.

Chromosomal aberrations are a hallmark of malignant transformation and while transformed LCLs are known to be karyotypically stable (Lacoste *et al.*, 2009), many EBV-associated malignancies have been observed to carry signs of chromosomal instability. The Masucci group has explored the link between EBV-associated malignancies and genomic instability in depth and has reported that EBV promotes genomic instability in Burkitt lymphoma (BL). This group conducted cytogenetic analyses in EBV negative and EBV positive BL lines and found that carrying EBV resulted in a significant increase in dicentric chromosomes, chromosome fragments and chromatid gaps. Furthermore, by examining telomeres using fluorescence *in situ* hybridization, EBV positive BL cells exhibited an overall increase in telomere size and telomeric fusion. Virally infected cells also revealed an increase in presence of phosphorylated H2AX (Kamranvar *et al.*, 2007). Expression of EBV nuclear antigen EBNA1, an EBV latency protein, was determined to be a mechanistic link between EBV viral proteins and chromosomal abnormalities. Stable or conditional expression of EBNA1 induced telomere fusion and heterogeneous length of telomeres (Kamranvar and Masucci, 2011). Interestingly, when telomere dysfunction was examined in primary human B cells infected with EBV, activation of the alternative lengthening of telomeres (ALT) pathway was found to be the preferred mechanism for lengthening telomeres (Kamranvar *et al.*, 2013). While most tumours use telomerase to extend telomeres, 10-15% of tumours employ ALT, a telomere recombination-based mechanism (Bryan *et al.*, 1995; Bryan *et al.*, 1997; Yeager *et al.*, 1999; Dunham *et al.*, 2000). Newly infected B cells exhibited hallmark signs of ALT activation, including telomere-associated promyelocytic leukemia nuclear bodies, telomeric-sister chromatid exchange, extra-chromosomal telomeres, telomere fusion, and telomere length heterogeneity. Importantly, early EBV infected B cells expressed low levels of TRF2, TRF1, POT1 and ATRX, shelterin complex proteins responsible for telomere protection (Kamranvar *et al.*, 2013).

Telomere dysfunction also occurs in KSHV-associated tumourigenesis. KSHV infection of immunocompromised patients is involved in the development of primary effusion lymphomas. The Hayward group found that LANA-expressing cell lines and primary effusion lymphoma lines both exhibited shortened telomeres, suggesting that LANA impacts telomere length. A proteomic screen for LANA-interacting proteins identified several telomere-associated proteins including TRF1, RPA1 and RPA2 (Figure 11.3c). Using chromatin immunoprecipitations, RPA1 and RPA2 were both found to bind telomeres in 293 and U2OS lines; however, when LANA was expressed in these cell lines there was a dramatic reduction in RPA1 and RPA2 binding at telomeres (Shamay *et al.*, 2012). This suggests a mechanism where LANA is responsible for inhibiting telomere binding by RPA to enhance telomere shortening (Figure 11.3c). This ultimately leads to LANA-mediated chromosome instability and may contribute to the development of KSHV-associated malignancies.

Additionally, HPV infection has been shown to increase genomic instability through expression of viral oncoprotein, E7. The Duensing group has shown that HPV-16 E7 expression triggered activation of the FA pathway, which resulted in recruitment of FANCD2 and FANCD1/BRCA1 to chromatin (Figure 11.3a). In FA-deficient cells, expression of HPV-16 E7 led to accelerated genome instability (Spardy *et al.*, 2007). Furthermore, HPV-16 E7 restores telomere length in a telomerase-independent method via ALT-associated promyelocytic leukemia bodies (APBs). APBs contained replicative stress proteins, including FANCD2, BRCA2, and MUS81, and knockdown of FANCD2 induced telomere dysfunction in ALT-dependent cells (Spardy *et al.*, 2008). Interestingly, the

Duensing group has also shown that cells expressing HPV-16 E7 can attenuate control of DNA damage checkpoints by accelerating proteolytic turnover of claspin, thus allowing cells to enter mitosis in the presence of DNA damage (Figure 11.3c). SCF$^{\beta TrCP}$-based claspin degradation machinery, including β-TrCP, Aurora A, and Polo-like kinase-1 were found to be deregulated and facilitated degradation of claspin to promote mitotic entry (Spardy *et al.*, 2009). These findings show HPV-16 E7 is a multifunctional protein that may induce genomic instability and contribute to extending the host cell life span to support viral genome propagation.

Reactive oxygen species facilitates DDR activation by DNA viruses

Oxidative stress caused by the up-regulation of ROS is among the many factors known to activate the DDR pathway. ROS activation upon oncogenic viral infection can cause oxidative DNA damage and results in increased mutagenesis and chromosomal alterations, ultimately promoting tumourigenesis. A number of human tumour viruses have been shown to activate ROS. Recent work from the Masucci laboratory has found that early after EBV infection, B cells exhibited an increased accumulation of ROS, which correlated with high levels of DNA damage (Chen *et al.*, 2016). Treatment with NACA, a ROS scavenger used to diminish the levels of ROS, resulted in a decrease in DNA damage along with down-regulation of the viral LMP1 latency protein, which is crucial for viral-mediated transformation. Furthermore, previous work has shown that the EBV protein EBNA1 can induce ROS levels leading to ATM-dependent DDR activation, thus suggesting a molecular source of damage for inducing telomere dysfunction (Gruhne *et al.*, 2009; Kamranvar and Masucci, 2011). It is suggested that increasing intracellular ROS levels may create a balance between activation of cellular antioxidant machinery and viral gene expression that inhibit the DDR to promote proliferation of infected cells. (Chen *et al.*, 2016).

Recently, ROS was also implicated in the DDR activation upon HPV infection. A few groups have shown that HPV, which initiates cellular transformation through expression of E6 and E7 proteins, can induce oxidative stress through E6/E7 activity. E6/E7-mediated ROS generation was observed to cause DNA damage by nicotinamide dinucleotide phosphate oxidases (NOXs), which produce superoxide, and promote tumourigenesis (Williams *et al.*, 2014; Marullo *et al.*, 2015). Together these studies have provided a new perspective on how oncogenic viral-induced oxidative stress may promote transformation.

Diversity of viral and host DNA damage response

DNA damage present in both viral and cellular genomes triggers DDR activation in infected cells. However, the nature of DDR signalling downstream of these two sources of damage may differ. Recently, a breakthrough study by the O'Shea laboratory provided evidence that viral and cellular genomes activate distinct DDR pathways in infected cells. They focused their investigation on the MRE11-RAD50-NBS1 (MRN) complex, which senses DNA damage and activates a global ATM-mediated DDR following adenovirus infection. In their recent report, they find that MRN binds to the adenoviral genome activating a localized ATM-mediated DDR. Typically, MRN recognition of cellular DNA damage initiates a global DDR where H2AX is phosphorylated over megabases of the genome flanking the damage site. In contrast to this signal amplification observed in cellular DDR pathways, damage to the viral genome did not induce H2AX phosphorylation across the cellular genome (Shah and O'Shea, 2015). Furthermore, they used wild type Adenovirus 5, which encodes early viral proteins, to infect human small airway epithelial cells and found that viral oncoproteins E1B-55K and E4-ORF3 inactivate MRN and allow viral replication, consistent with that found by Weitzman and colleagues (Figure 11.3a). The assembly of viral genomes within the nucleus triggered a unique global MRN-independent ATM response, but this global response was not found to impair viral replication. Lastly, they found that the localized MRN-ATM anti-viral DDR initiated at viral genomes affects viral replication but does not affect cellular DNA replication, thus providing evidence that the MRN-ATM anti-viral DDR selectively inhibits viral replication and maintains cellular proliferation (Shah and O'Shea, 2015). This discovery that viral and cellular genomes elicit distinct DDRs is particularly important as it begins to provide a mechanism for how the DDR could selectively target viral genomes while maintaining cellular viability.

Conclusion

DNA tumour viruses extensively engage the host DNA damage response. Many strategies are used by these viruses to activate, modulate, or suppress the DDR during latent and lytic stages, which allow the virus to establish a long-term infection and evade the host defence response. As discussed here, tumour viruses express oncoproteins to drive proliferation, which can lead to replicative stress and DNA damage, thus triggering the DDR. This activated DDR signalling cascade can result in one of three outcomes: DNA repair, apoptosis, or senescence. During latency, activation of the DDR mediated by viral oncoprotein expression typically results in strong cell cycle arrest. However, during lytic cycle, activation of the DDR often enhances viral replication. Therefore, viral interaction with the DDR can take on distinct roles depending on the stage of infection.

Lytic infection has been shown to promote activation of the DDR in response to viral replication, particularly through the ATM-mediated DDR signalling pathway. KSHV lytic reactivation specifically activates the ATM and DNA-PK pathways, which typically respond to double-stranded DNA breaks, while ATR is not activated by KSHV (Figure 11.3c). EBV also activates the ATM pathway by expression of BGLF4, which activates TIP60 allowing it to acetylate ATM and trigger downstream signal activation. Activation of the host DDR was required for efficient viral propagation and progeny virion formation. As viral replication requires host cell machinery, it is likely that efficient viral propagation requires activation of the DDR to allow the virus access to host cell cycle factors, including recombination and repair factors.

During latency, viral oncoprotein activation led to aberrant proliferation, which can induce replicative stress and result in activation of the host DDR and cell cycle arrest. This DDR activation upon latent EBV and KSHV infection was observed to cause a growth suppressive barrier to viral-mediated transformation. Viral oncoproteins have developed methods to regulate the DDR by triggering autophagy. Recent advances in this field have brought the importance of metabolism to the forefront of latent infection. Aberrant proliferation in EBV-infected B cells was found to induce metabolic stress characterized by an imbalance of glycolysis and oxidative phosphorylation and subsequently triggering cellular senescence.

Our understanding of the interplay between DNA tumour viruses and the host DDR has expanded rapidly. It is clear that viruses have developed a number of strategies to interact with the ATM pathway to promote viral replication, development of progeny virion and induce senescence. Furthermore, oncogenic stress has been shown to increase the incidence of downstream chromosomal abnormalities, including telomere dysfunction. However, to determine how and why the DDR is activated, we must define the upstream molecular sources of the damage and develop a better understanding of the genetic mechanism the DDR pathway employs to induce downstream telomere fusions, irreparable damage, and facilitate or inhibit virus replication. By developing a more specific understanding of the particular DDR factors involved and how they become activated, novel therapeutic strategies can be designed to target these factors and prevent tumourigenesis.

Acknowledgements

This work was supported by NIH R01-CA140337 to M.A.L. and R01-CA140337-S1 to A.Y.H. Additional funding came from the Duke CFAR, an NIH funded program, 5P30-AI064518, and NIH T32-CA009111 (A.Y.H.).

Conflicts of Interest

The authors declare no conflict of interest.

References

Alexander, A., and Walker, C.L. (2010). Differential localization of ATM is correlated with activation of distinct downstream signaling pathways. Cell Cycle 9, 3685-3686. 10.4161/cc.9.18.13253.

Asai, R., Kato, A., Kato, K., Kanamori-Koyama, M., Sugimoto, K., Sairenji, T., Nishiyama, Y., and Kawaguchi, Y. (2006). Epstein-Barr virus protein kinase BGLF4 is a virion tegument protein that dissociates from virions in a phosphorylation-dependent process and phosphorylates the viral immediate-early protein BZLF1. J Virol 80, 5125-5134. 10.1128/jvi.02674-05.

Avgousti, D.C., Herrmann, C., Kulej, K., Pancholi, N.J., Sekulic, N., Petrescu, J., Molden, R.C., Blumenthal, D., Paris, A.J., Reyes, E.D., *et al.* (2016). A core viral protein binds host nucleosomes to sequester immune danger signals. Nature 535, 173-177. 10.1038/nature18317.

Babcock, G.J., Hochberg, D., and Thorley-Lawson, A.D. (2000). The expression pattern of Epstein-Barr virus latent genes *in vivo* is dependent upon the differentiation stage of the infected B cell. Immunity 13, 497-506. S1074-7613(00)00049-2 [pii].

Bakkenist, C.J., and Kastan, M.B. (2003). DNA damage activates ATM through intermolecular autophosphorylation and dimer dissociation. Nature 421, 499-506.

Balistreri, G., Viiliainen, J., Turunen, M., Diaz, R., Lyly, L., Pekkonen, P., Rantala, J., Ojala, K., Sarek, G., Teesalu, M., *et al.* (2016). Oncogenic Herpesvirus Utilizes Stress-Induced Cell Cycle Checkpoints for Efficient Lytic Replication. PLoS Pathog 12, e1005424. 10.1371/journal.ppat.1005424.

Ballestas, M.E., Chatis, P.A., and Kaye, K.M. (1999). Efficient persistence of extrachromosomal KSHV DNA mediated by latency-associated nuclear antigen. Science 284, 641-644.

Balmus, G., Zhu, M., Mukherjee, S., Lyndaker, A.M., Hume, K.R., Lee, J., Riccio, M.L., Reeves, A.P., Sutter, N.B., Noden, D.M., *et al.* (2012). Disease severity in a mouse model of ataxia telangiectasia is modulated by the DNA damage checkpoint gene Hus1. Hum. Mol. Genet. 21, 3408-3420. 10.1093/hmg/dds173.

Boutros, R., Dozier, C., and Ducommun, B. (2006). The when and wheres of CDC25 phosphatases. Curr. Opin. Cell Biol.18, 185-191. 10.1016/j.ceb.2006.02.003.

Branzei, D., and Foiani, M. (2008). Regulation of DNA repair throughout the cell cycle. Nat. Rev. Mol. Cell Biol. 9, 297-308. 10.1038/nrm2351.

Brown, E.J., and Baltimore, D. (2003). Essential and dispensable roles of ATR in cell cycle arrest and genome maintenance. Genes Dev 17, 615-628. 10.1101/gad.1067403.

Bryan, T.M., Englezou, A., Dalla-Pozza, L., Dunham, M.A., and Reddel, R.R. (1997). Evidence for an alternative mechanism for maintaining telomere length in human tumors and tumor-derived cell lines. Nat Med 3, 1271-1274.

Bryan, T.M., Englezou, A., Gupta, J., Bacchetti, S., and Reddel, R.R. (1995). Telomere elongation in immortal human cells without detectable telomerase activity. The EMBO journal 14, 4240-4248.

Burma, S., Chen, B.P., Murphy, M., Kurimasa, A., and Chen, D.J. (2001). ATM phosphorylates histone H2AX in response to DNA double-strand breaks. J. Biol. Chem. 276, 42462-42467.

Byun, T.S., Pacek, M., Yee, M.C., Walter, J.C., and Cimprich, K.A. (2005). Functional uncoupling of MCM helicase and DNA polymerase activities activates the ATR-dependent checkpoint. Genes Dev 19, 1040-1052. 10.1074/jbc.C100466200

Cahir McFarland, E.D., Izumi, K.M., and Mosialos, G. (1999). Epstein-barr virus transformation: involvement of latent membrane protein 1-mediated activation of NF-kappaB. Oncogene 18, 6959-6964.

Chandriani, S., and Ganem, D. (2010). Array-based transcript profiling and limiting-dilution reverse transcription-PCR analysis identify additional latent genes in Kaposi's sarcoma-associated herpesvirus. J Virol 84, 5565-5573. 10.1128/jvi.02723-09.

Chang, Y., Moore, P.S., Talbot, S.J., Boshoff, C.H., Zarkowska, T., Godden, K., Paterson, H., Weiss, R.A., and Mittnacht, S. (1996). Cyclin encoded by KS herpesvirus. Nature 382, 410. 10.1038/382410a0.

Chanoux, R.A., Yin, B., Urtishak, K.A., Asare, A., Bassing, C.H., and Brown, E.J. (2009). ATR and H2AX cooperate in maintaining genome stability under replication stress. J Biol Chem 284, 5994-6003. 10.1074/jbc.M806739200.

Chappell, W.H., Gautam, D., Ok, S.T., Johnson, B.A., Anacker, D.C., and Moody, C.A. (2015). Homologous Recombination Repair Factors Rad51 and BRCA1 Are Necessary for Productive Replication of Human Papillomavirus 31. J Virol 90, 2639-2652. 10.1128/jvi. 02495-15.

Chatterjee, P.K., Vayda, M.E., and Flint, S.J. (1986). Adenoviral protein VII packages intracellular viral DNA throughout the early phase of infection. The EMBO journal 5, 1633-1644.

Chen, W., Hilton, I.B., Staudt, M.R., Burd, C.E., and Dittmer, D.P. (2010). Distinct p53, p53:LANA, and LANA complexes in Kaposi's Sarcoma--associated Herpesvirus Lymphomas. J Virol 84, 3898-3908. 10.1128/JVI.01321-09.

Chen, X., Kamranvar, S.A., and Masucci, M.G. (2016). Oxidative stress enables Epstein-Barr virus-induced B-cell transformation by posttranscriptional regulation of viral and cellular growth-promoting factors. Oncogene 35, 3807-3816. 10.1038/onc.2015.450.

Cheng, S., Schmidt-Grimminger, D.C., Murant, T., Broker, T.R., and Chow, L.T. (1995). Differentiation-dependent up-regulation of the human papillomavirus E7 gene reactivates cellular DNA replication in suprabasal differentiated keratinocytes. Genes Dev 9, 2335-2349.

Chipuk, J.E., and Green, D.R. (2006). Dissecting p53-dependent apoptosis. Cell Death Differ 13, 994-1002. 10.1038/sj.cdd.4401908.

Ciccia, A., and Elledge, S.J. (2010). The DNA damage response: making it safe to play with knives. Mol Cell 40, 179-204. 10.1016/j.molcel.2010.09.019.

Clarke, A.R., Purdie, C.A., Harrison, D.J., Morris, R.G., Bird, C.C., Hooper, M.L., and Wyllie, A.H. (1993). Thymocyte apoptosis induced by p53-dependent and independent pathways. Nature 362, 849-852. 10.1038/362849a0.

Cortez, D., Wang, Y., Qin, J., and Elledge, S.J. (1999). Requirement of ATM-dependent phosphorylation of brca1 in the DNA damage response to double-strand breaks. Science 286, 1162-1166.

Cosentino, C., Grieco, D., and Costanzo, V. (2011). ATM activates the pentose phosphate pathway promoting anti-oxidant defence and DNA repair. The EMBO journal 30, 546-555. 10.1038/emboj.2010.330.

Davies, S.L., North, P.S., Dart, A., Lakin, N.D., and Hickson, I.D. (2004). Phosphorylation of the Bloom's syndrome helicase and its role in recovery from S-phase arrest. Mol Cell Biol 24, 1279-1291.

Davy, C., and Doorbar, J. (2007). G2/M cell cycle arrest in the life cycle of viruses. Virology 368, 219-226. 10.1016/j.virol.2007.05.043.

Delacroix, S., Wagner, J.M., Kobayashi, M., Yamamoto, K., and Karnitz, L.M. (2007). The Rad9-Hus1-Rad1 (9-1-1) clamp activates checkpoint signaling via TopBP1. Genes Dev 21, 1472-1477. 10.1101/gad. 1547007.

Derheimer, F.A., and Kastan, M.B. (2010). Multiple roles of ATM in monitoring and maintaining DNA integrity. FEBS letters 584, 3675-3681. 10.1016/j.febslet. 2010.05.031.

Dobbs, T.A., Tainer, J.A., and Lees-Miller, S.P. (2010). A structural model for regulation of NHEJ by DNA-PKcs autophosphorylation. DNA Repair (Amst) 9, 1307-1314. 10.1016/j.dnarep.2010.09.019.

Doil, C., Mailand, N., Bekker-Jensen, S., Menard, P., Larsen, D.H., Pepperkok, R., Ellenberg, J., Panier, S., Durocher, D., Bartek, J., *et al.* (2009). RNF168 binds and amplifies ubiquitin conjugates on damaged chromosomes to allow accumulation of repair proteins. Cell 136, 435-446. 10.1016/j.cell.2008.12.041 S0092-8674(09)00004-X [pii].

Dowhanick, J.J., McBride, A.A., and Howley, P.M. (1995). Suppression of cellular proliferation by the papillomavirus E2 protein. J Virol 69, 7791-7799.

Dunham, M.A., Neumann, A.A., Fasching, C.L., and Reddel, R.R. (2000). Telomere maintenance by recombination in human cells. Nature genetics 26, 447-450. 10.1038/82586.

Dyson, N., Howley, P.M., Munger, K., and Harlow, E. (1989). The human papilloma virus-16 E7 oncoprotein is able to bind to the retinoblastoma gene product. Science 243, 934-937.

Egawa, N., Nakahara, T., Ohno, S., Narisawa-Saito, M., Yugawa, T., Fujita, M., Yamato, K., Natori, Y., and Kiyono, T. (2012). The E1 protein of human papillomavirus type 16 is dispensable for maintenance replication of the viral genome. J Virol 86, 3276-3283. 10.1128/jvi.06450-11.

El-Guindy, A., Lopez-Giraldez, F., Delecluse, H.J., McKenzie, J., and Miller, G. (2014). A locus encompassing the Epstein-Barr virus bglf4 kinase regulates expression of genes encoding viral structural proteins. PLoS Pathog 10, e1004307. 10.1371/journal.ppat.1004307.

Falck, J., Coates, J., and Jackson, S.P. (2005). Conserved modes of recruitment of ATM, ATR and DNA-PKcs to sites of DNA damage. Nature 434, 605-611. 10.1038/nature03442.

Flynn, R.L., and Zou, L. (2011). ATR: a master conductor of cellular responses to DNA replication stress. Trends in biochemical sciences 36, 133-140. 10.1016/j.tibs.2010.09.005.

Forte, E., Salinas, R.E., Chang, C., Zhou, T., Linnstaedt, S.D., Gottwein, E., Jacobs, C., Jima, D., Li, Q.J., Dave, S.S., *et al.* (2012). The Epstein-Barr virus (EBV)-induced tumor suppressor microRNA MiR-34a is growth promoting in EBV-infected B cells. J Virol 86, 6889-6898. 10.1128/JVI.07056-11 JVI.07056-11 [pii].

Friborg, J., Jr., Kong, W., Hottiger, M.O., and Nabel, G.J. (1999). p53 inhibition by the LANA protein of KSHV protects against cell death. Nature 402, 889-894..10.1038/47266.

Furuya, K., Poitelea, M., Guo, L., Caspari, T., and Carr, A.M. (2004). Chk1 activation requires Rad9 S/TQ-site phosphorylation to promote association with C-terminal BRCT domains of Rad4TOPBP1. Genes Dev 18, 1154-1164. 10.1101/gad.291104.

Gatei, M., Young, D., Cerosaletti, K.M., Desai-Mehta, A., Spring, K., Kozlov, S., Lavin, M.F., Gatti, R.A., Concannon, P., and Khanna, K. (2000). ATM-dependent phosphorylation of nibrin in response to radiation exposure. Nature genetics 25, 115-119. 10.1038/75508.

Gillespie, K.A., Mehta, K.P., Laimins, L.A., and Moody, C.A. (2012). Human papillomaviruses recruit cellular DNA repair and homologous recombination factors to viral replication centers. J Virol 86, 9520-9526. 10.1128/JVI.00247-12.

Gruhne, B., Sompallae, R., Marescotti, D., Kamranvar, S.A., Gastaldello, S., and Masucci, M.G. (2009). The Epstein-Barr virus nuclear antigen-1 promotes genomic instability via induction of reactive oxygen species. Proc Natl Acad Sci USA 106, 2313-2318. 0810619106 [pii] 10.1073/pnas.0810619106.

Guasparri, I., Keller, S.A., and Cesarman, E. (2004). KSHV vFLIP is essential for the survival of infected lymphoma cells. J Exp Med 199, 993-1003. 10.1084/jem.20031467.

Guo, Z., Deshpande, R., and Paull, T.T. (2010a). ATM activation in the presence of oxidative stress. Cell Cycle 9, 4805-4811. 10.4161/cc.9.24.14323.

Guo, Z., Kozlov, S., Lavin, M.F., Person, M.D., and Paull, T.T. (2010b). ATM activation by oxidative stress. Science 330, 517-521. 10.1126/science.1192912.

Hagemeier, S.R., Barlow, E.A., Meng, Q., and Kenney, S.C. (2012). The cellular ataxia telangiectasia-mutated kinase promotes epstein-barr virus lytic reactivation in response to multiple different types of lytic reactivation-inducing stimuli. J Virol 86, 13360-13370. 10.1128/JVI.01850-12.

Hartlerode, A.J., and Scully, R. (2009). Mechanisms of double-strand break repair in somatic mammalian cells. Biochem J 423, 157-168. 10.1042/bj20090942.

Haupt, Y., Maya, R., Kazaz, A., and Oren, M. (1997). Mdm2 promotes the rapid degradation of p53. Nature 387, 296-299. 10.1038/387296a0.

Henderson, E., Miller, G., Robinson, J., and Heston, L. (1977). Efficiency of transformation of lymphocytes by Epstein-Barr virus. Virology 76, 152-163.

Hollingworth, R., Skalka, G.L., Stewart, G.S., Hislop, A.D., Blackbourn, D.J., and Grand, R.J. (2015). Activation of DNA Damage Response Pathways during Lytic Replication of KSHV. Viruses 7, 2908-2927. 10.3390/v7062752.

Huen, M.S., Grant, R., Manke, I., Minn, K., Yu, X., Yaffe, M.B., and Chen, J. (2007). RNF8 transduces the DNA-damage signal via histone ubiquitylation and checkpoint protein assembly. Cell 131, 901-914. S0092-8674(07)01270-6 [pii] 10.1016/j.cell.2007.09.041.

Jackson, B.R., Noerenberg, M., and Whitehouse, A. (2014). A novel mechanism inducing genome instability in Kaposi's sarcoma-associated herpesvirus infected cells. PLoS Pathog 10, e1004098. 10.1371/journal.ppat.1004098.

Johannsen, E., Koh, E., Mosialos, G., Tong, X., Kieff, E., and Grossman, S.R. (1995). Epstein-Barr virus nuclear protein 2 transactivation of the latent membrane protein 1 promoter is mediated by J kappa and PU.1. J Virol 69, 253-262.

Kaiser, C., Laux, G., Eick, D., Jochner, N., Bornkamm, G.W., and Kempkes, B. (1999). The proto-oncogene c-myc is a direct target gene of Epstein-Barr virus nuclear antigen 2. J Virol 73, 4481-4484.

Kamranvar, S.A., Chen, X., and Masucci, M.G. (2013). Telomere dysfunction and activation of alternative lengthening of telomeres in B-lymphocytes infected by Epstein-Barr virus. Oncogene 32, 5522-5530. 10.1038/onc.2013.189.

Kamranvar, S.A., Gruhne, B., Szeles, A., and Masucci, M.G. (2007). Epstein-Barr virus promotes genomic instability in Burkitt's lymphoma. Oncogene 26, 5115-5123. 10.1038/sj.onc.1210324.

Kamranvar, S.A., and Masucci, M.G. (2011). The Epstein-Barr virus nuclear antigen-1 promotes telomere dysfunction via induction of oxidative stress. Leukemia 25, 1017-1025. 10.1038/leu.2011.35.

Kang, R., Chen, R., Zhang, Q., Hou, W., Wu, S., Cao, L., Huang, J., Yu, Y., Fan, X.G., Yan, Z., *et al.* (2014). HMGB1 in health and disease. Molecular aspects of medicine 40, 1-116. 10.1016/j.mam.2014.05.001.

Kato, K., Kawaguchi, Y., Tanaka, M., Igarashi, M., Yokoyama, A., Matsuda, G., Kanamori, M., Nakajima, K., Nishimura, Y., Shimojima, M., *et al.* (2001). Epstein-Barr virus-encoded protein kinase BGLF4 mediates hyperphosphorylation of cellular elongation factor 1delta (EF-1delta): EF-1delta is universally modified by conserved protein kinases of herpesviruses in mammalian cells. J Gen Virol 82, 1457-1463. 10.1099/0022-1317-82-6-1457.

Kato, K., Yokoyama, A., Tohya, Y., Akashi, H., Nishiyama, Y., and Kawaguchi, Y. (2003). Identification of protein kinases responsible for phosphorylation of Epstein-Barr virus nuclear antigen leader protein at serine-35, which regulates its coactivator function. J Gen Virol 84, 3381-3392. 10.1099/vir.0.19454-0.

Kawaguchi, Y., Kato, K., Tanaka, M., Kanamori, M., Nishiyama, Y., and Yamanashi, Y. (2003). Conserved protein kinases encoded by herpesviruses and cellular protein kinase cdc2 target the same phosphorylation site in eukaryotic elongation factor 1delta. J Virol 77, 2359-2368.

Kee, Y., and D'Andrea, A.D. (2010). Expanded roles of the Fanconi anemia pathway in preserving genomic stability. Genes Dev 24, 1680-1694. 10.1101/gad.1955310.

Koopal, S., Furuhjelm, J.H., Jarviluoma, A., Jaamaa, S., Pyakurel, P., Pussinen, C., Wirzenius, M., Biberfeld, P., Alitalo, K., Laiho, M., *et al.* (2007). Viral oncogene-induced DNA damage response is activated in Kaposi sarcoma tumorigenesis. PLoS Pathog 3, 1348-1360. 07-PLPA-RA-0224 [pii] 10.1371/journal.ppat.0030140.

Kumagai, A., and Dunphy, W.G. (2000). Claspin, a novel protein required for the activation of Chk1 during a DNA replication checkpoint response in Xenopus egg extracts. Mol Cell 6, 839-849.

Kuntziger, T., Landsverk, H.B., Collas, P., and Syljuasen, R.G. (2011). Protein phosphatase 1 regulators in DNA damage signaling. Cell Cycle 10, 1356-1362.

Lacoste, S., Wiechec, E., Dos Santos Silva, A.G., Guffei, A., Williams, G., Lowbeer, M., Benedek, K., Henriksson, M., Klein, G., and Mai, S. (2009). Chromosomal rearrangements after *ex vivo* Epstein-Barr virus (EBV) infection of human B cells. Oncogene. onc2009359 [pii]
10.1038/onc.2009.359.

Lagunoff, M., Bechtel, J., Venetsanakos, E., Roy, A.M., Abbey, N., Herndier, B., McMahon, M., and Ganem, D. (2002). De novo infection and serial transmission of Kaposi's sarcoma-associated herpesvirus in cultured endothelial cells. J Virol 76, 2440-2448.

Lee, J., Kumagai, A., and Dunphy, W.G. (2007). The Rad9-Hus1-Rad1 checkpoint clamp regulates interaction of TopBP1 with ATR. J Biol Chem 282, 28036-28044. 10.1074/jbc.M704635200.

Leidal, A.M., Cyr, D.P., Hill, R.J., Lee, P.W., and McCormick, C. (2012). Subversion of autophagy by Kaposi's sarcoma-associated herpesvirus impairs oncogene-induced senescence. Cell Host Microbe 11, 167-180. 10.1016/j.chom.2012.01.005.

Li, R., Liao, G., Nirujogi, R.S., Pinto, S.M., Shaw, P.G., Huang, T.C., Wan, J., Qian, J., Gowda, H., Wu, X., *et al.* (2015). Phosphoproteomic Profiling Reveals Epstein-Barr Virus Protein Kinase Integration of DNA Damage Response and Mitotic Signaling. PLoS Pathog 11, e1005346. 10.1371/journal.ppat.1005346.

Li, R., Zhu, J., Xie, Z., Liao, G., Liu, J., Chen, M.R., Hu, S., Woodard, C., Lin, J., Taverna, S.D., *et al.* (2011). Conserved herpesvirus kinases target the DNA damage response pathway and TIP60 histone acetyltransferase to promote virus replication. Cell Host Microbe 10, 390-400. 10.1016/j.chom.2011.08.013 S1931-3128(11)00284-8 [pii].

Li, W., Kim, S.M., Lee, J., and Dunphy, W.G. (2004). Absence of BLM leads to accumulation of chromosomal DNA breaks during both unperturbed and disrupted S phases. The Journal of cell biology 165, 801-812. 0.1083/jcb.200402095.

Lim, D.S., Kim, S.T., Xu, B., Maser, R.S., Lin, J., Petrini, J.H., and Kastan, M.B. (2000). ATM phosphorylates p95/nbs1 in an S-phase checkpoint pathway. Nature 404, 613-617. 10.1038/35007091.

Lin, J., Johannsen, E., Robertson, E., and Kieff, E. (2002). Epstein-Barr virus nuclear antigen 3C putative repression domain mediates coactivation of the LMP1 promoter with EBNA-2. J Virol 76, 232-242.

Lindahl, T. (1993). Instability and decay of the primary structure of DNA. Nature 362, 709-715. 10.1038/362709a0.

Liu, Y., Wang, X., Lo, A.K., Wong, Y.C., Cheung, A.L., and Tsao, S.W. (2002). Latent membrane protein-1 of Epstein-Barr virus inhibits cell growth and induces sensitivity to cisplatin in nasopharyngeal carcinoma cells. J Med Virol 66, 63-69.

Longworth, M.S., and Laimins, L.A. (2004). Pathogenesis of human papillomaviruses in differentiating epithelia. Microbiology and molecular biology reviews: MMBR 68, 362-372. 10.1128/mmbr.68.2.362-372.2004.

Lopez-Contreras, A.J., and Fernandez-Capetillo, O. (2010). The ATR barrier to replication-born DNA damage. DNA Repair (Amst) 9, 1249-1255. 10.1016/j.dnarep.2010.09.012.

Lotze, M.T., and Tracey, K.J. (2005). High-mobility group box 1 protein (HMGB1): nuclear weapon in the immune arsenal. Nature reviews Immunology 5, 331-342. 10.1038/nri1594.

Lou, Z., Minter-Dykhouse, K., Franco, S., Gostissa, M., Rivera, M.A., Celeste, A., Manis, J.P., van Deursen, J., Nussenzweig, A., Paull, T.T., *et al.* (2006). MDC1 maintains genomic stability by participating in the amplification of ATM-dependent DNA damage signals. Mol Cell 21, 187-200. 10.1016/j.molcel.2005.11.025.

Lu, B., Nakamura, T., Inouye, K., Li, J., Tang, Y., Lundback, P., Valdes-Ferrer, S.I., Olofsson, P.S., Kalb, T., Roth, J., *et al.* (2012). Novel role of PKR in inflammasome activation and HMGB1 release. Nature 488, 670-674. 10.1038/nature11290.

MacDougall, C.A., Byun, T.S., Van, C., Yee, M.C., and Cimprich, K.A. (2007). The structural determinants of checkpoint activation. Genes Dev 21, 898-903. 10.1101/gad.1522607.

Mahaney, B.L., Meek, K., and Lees-Miller, S.P. (2009). Repair of ionizing radiation-induced DNA double-strand breaks by non-homologous end-joining. Biochem J 417, 639-650. 10.1042/BJ20080413.

Mailand, N., Bekker-Jensen, S., Faustrup, H., Melander, F., Bartek, J., Lukas, C., and Lukas, J. (2007). RNF8 ubiquitylates histones at DNA double-strand breaks and promotes assembly of repair proteins. Cell 131, 887-900. S0092-8674(07)01271-8 [pii] 10.1016/j.cell.2007.09.040.

Martin, S.G., Laroche, T., Suka, N., Grunstein, M., and Gasser, S.M. (1999). Relocalization of telomeric Ku and SIR proteins in response to DNA strand breaks in yeast. Cell 97, 621-633.

Marullo, R., Werner, E., Zhang, H., Chen, G.Z., Shin, D.M., and Doetsch, P.W. (2015). HPV16 E6 and E7 proteins induce a chronic oxidative stress response via NOX2 that causes genomic instability and increased susceptibility to DNA damage in head and neck cancer cells. Carcinogenesis 36, 1397-1406. 10.1093/carcin/bgv126.

Maruo, S., Zhao, B., Johannsen, E., Kieff, E., Zou, J., and Takada, K. (2011). Epstein-Barr virus nuclear antigens 3C and 3A maintain lymphoblastoid cell growth by repressing $p16^{INK4A}$ and p14ARF expression. Proc Natl Acad Sci USA 108, 1919-1924. 1019599108 [pii] 10.1073/pnas.1019599108.

Matthews, T., and Boehme, R. (1988). Antiviral activity and mechanism of action of ganciclovir. Reviews of infectious diseases 10 Suppl 3, S490-494.

Mattiroli, F., Vissers, J.H., van Dijk, W.J., Ikpa, P., Citterio, E., Vermeulen, W., Marteijn, J.A., and Sixma, T.K. (2012). RNF168 ubiquitinates K13-15 on H2A/H2AX to drive DNA damage signaling. Cell 150, 1182-1195. 10.1016/j.cell.2012.08.005.

McBride, A.A. (2008). Replication and partitioning of papillomavirus genomes. Advances in virus research 72, 155-205. 10.1016/s0065-3527(08)00404-1.

McBride, A.A. (2013). The papillomavirus E2 proteins. Virology 445, 57-79. 10.1016/j.virol.2013.06.006.

McFadden, K., Hafez, A.Y., Kishton, R., Messinger, J.E., Nikitin, P.A., Rathmell, J.C., and Luftig, M.A. (2016). Metabolic stress is a barrier to Epstein-Barr virus-mediated B-cell immortalization. Proc Natl Acad Sci USA 113, E782-790. 10.1073/pnas.1517141113.

Meier, A., Fiegler, H., Munoz, P., Ellis, P., Rigler, D., Langford, C., Blasco, M.A., Carter, N., and Jackson, S.P. (2007). Spreading of mammalian DNA-damage response factors studied by ChIP-chip at damaged telomeres. The EMBO journal 26, 2707-2718. 10.1038/sj.emboj.7601719.

Mesri, E.A., Cesarman, E., and Boshoff, C. (2010). Kaposi's sarcoma and its associated herpesvirus. Nature reviews Cancer 10, 707-719. 10.1038/nrc2888.

Michael, W.M., Ott, R., Fanning, E., and Newport, J. (2000). Activation of the DNA replication checkpoint through RNA synthesis by primase. Science 289, 2133-2137.

Mills, K.D., Sinclair, D.A., and Guarente, L. (1999). MEC1-dependent redistribution of the Sir3 silencing protein from telomeres to DNA double-strand breaks. Cell 97, 609-620.

Moody, C.A., and Laimins, L.A. (2010). Human papillomavirus oncoproteins: pathways to transformation. Nature reviews Cancer 10, 550-560. 10.1038/nrc2886.

Munger, K., Werness, B.A., Dyson, N., Phelps, W.C., Harlow, E., and Howley, P.M. (1989). Complex formation of human papillomavirus E7 proteins with the retinoblastoma tumor suppressor gene product. Embo J 8, 4099-4105.

Nagasawa, H., Little, J.B., Lin, Y.F., So, S., Kurimasa, A., Peng, Y., Brogan, J.R., Chen, D.J., Bedford, J.S., and Chen, B.P. (2011). Differential role of DNA-PKcs phosphorylations and kinase activity in radiosensitivity and chromosomal instability. Radiation research 175, 83-89. 10.1667/rr2092.1.

Naito, T., Matsuura, A., and Ishikawa, F. (1998). Circular chromosome formation in a fission yeast mutant defective in two ATM homologues. Nature genetics 20, 203-206. 10.1038/2517.

Nikitin, P.A., Yan, C.M., Forte, E., Bocedi, A., Tourigny, J.P., White, R.E., Allday, M.J., Patel, A., Dave, S.S., Kim, W., *et al.* (2010). An ATM/Chk2-mediated DNA damage-responsive signaling pathway suppresses Epstein-Barr virus transformation of primary human B cells. Cell Host Microbe 8, 510-522. S1931-3128(10)00377-X [pii] 10.1016/j.chom.2010.11.004.

Nyberg, K.A., Michelson, R.J., Putnam, C.W., and Weinert, T.A. (2002). Toward maintaining the genome: DNA damage and replication checkpoints. Annual review of genetics 36, 617-656. 10.1146/annurev.genet.36.060402.113540.

Pichierri, P., Rosselli, F., and Franchitto, A. (2003). Werner's syndrome protein is phosphorylated in an ATR/ATM-dependent manner following replication arrest and DNA damage induced during the S phase of the cell cycle. Oncogene 22, 1491-1500. 10.1038/sj.onc.1206169.

Price, A.M., Tourigny, J.P., Forte, E., Salinas, R.E., Dave, S.S., and Luftig, M.A. (2012). Analysis of Epstein-Barr Virus-Regulated Host Gene Expression Changes through Primary B-Cell Outgrowth Reveals Delayed Kinetics of Latent Membrane Protein 1-Mediated NF-kappaB Activation. J Virol 86, 11096-11106. 10.1128/JVI.01069-12.

Radkov, S.A., Kellam, P., and Boshoff, C. (2000). The latent nuclear antigen of Kaposi sarcoma-associated herpesvirus targets the retinoblastoma-E2F pathway and with the oncogene Hras transforms primary rat cells. Nat Med 6, 1121-1127. 10.1038/80459.

Sancar, A., Lindsey-Boltz, L.A., Unsal-Kacmaz, K., and Linn, S. (2004). Molecular mechanisms of mammalian DNA repair and the DNA damage checkpoints. Annual review of biochemistry 73, 39-85. 10.1146/annurev.biochem.73.011303.073723.

Sarbassov, D.D., Ali, S.M., Kim, D.H., Guertin, D.A., Latek, R.R., Erdjument-Bromage, H., Tempst, P., and Sabatini, D.M. (2004). Rictor, a novel binding partner of mTOR, defines a rapamycin-insensitive and raptor-independent pathway that regulates the cytoskeleton.

Current biology: CB 14, 1296-1302. 10.1016/j.cub.2004.06.054.
Savic, V., Yin, B., Maas, N.L., Bredemeyer, A.L., Carpenter, A.C., Helmink, B.A., Yang-Iott, K.S., Sleckman, B.P., and Bassing, C.H. (2009). Formation of dynamic gamma-H2AX domains along broken DNA strands is distinctly regulated by ATM and MDC1 and dependent upon H2AX densities in chromatin. Mol Cell 34, 298-310. 10.1016/j.molcel.2009.04.012.
Scheffner, M., Werness, B.A., Huibregtse, J.M., Levine, A.J., and Howley, P.M. (1990). The E6 oncoprotein encoded by human papillomavirus types 16 and 18 promotes the degradation of p53. Cell 63, 1129-1136. 0092-8674(90)90409-8 [pii].
Shah, G.A., and O'Shea, C.C. (2015). Viral and Cellular Genomes Activate Distinct DNA Damage Responses. Cell 162, 987-1002. 10.1016/j.cell.2015.07.058.
Shamay, M., Liu, J., Li, R., Liao, G., Shen, L., Greenway, M., Hu, S., Zhu, J., Xie, Z., Ambinder, R.F., *et al.* (2012). A protein array screen for Kaposi's sarcoma-associated herpesvirus LANA interactors links LANA to TIP60, PP2A activity, and telomere shortening. J Virol 86, 5179-5191. 10.1128/jvi.00169-12.
Shechter, D., Costanzo, V., and Gautier, J. (2004). Regulation of DNA replication by ATR: signaling in response to DNA intermediates. DNA Repair (Amst) 3, 901-908. 10.1016/j.dnarep.2004.03.020.
Shiloh, Y. (2003). ATM and related protein kinases: safeguarding genome integrity. Nature reviews Cancer 3, 155-168. 10.1038/nrc1011.
Shiotani, B., and Zou, L. (2009a). ATR signaling at a glance. Journal of cell science 122, 301-304. 10.1016/j.molcel.2009.01.024.
Shiotani, B., and Zou, L. (2009b). Single-stranded DNA orchestrates an ATM-to-ATR switch at DNA breaks. Mol Cell 33, 547-558. 10.1016/j.molcel.2009.01.024.
Siliciano, J.D., Canman, C.E., Taya, Y., Sakaguchi, K., Appella, E., and Kastan, M.B. (1997). DNA damage induces phosphorylation of the amino terminus of p53. Genes Dev 11, 3471-3481.
Sinclair, A.J., Palmero, I., Peters, G., and Farrell, P.J. (1994). EBNA-2 and EBNA-LP cooperate to cause G0 to G1 transition during immortalization of resting human B lymphocytes by Epstein-Barr virus. The EMBO journal 13, 3321-3328.
Skalska, L., White, R.E., Franz, M., Ruhmann, M., and Allday, M.J. (2010). Epigenetic repression of p16(INK4A) by latent Epstein-Barr virus requires the interaction of EBNA3A and EBNA3C with CtBP. PLoS Pathog 6, e1000951. 10.1371/journal.ppat.1000951.
Skalska, L., White, R.E., Parker, G.A., Sinclair, A.J., Paschos, K., and Allday, M.J. (2013). Induction of p16(INK4a) is the major barrier to proliferation when Epstein-Barr virus (EBV) transforms primary B cells into lymphoblastoid cell lines. PLoS Pathog 9, e1003187. 10.1371/journal.ppat.1003187.
So, S., Davis, A.J., and Chen, D.J. (2009). Autophosphorylation at serine 1981 stabilizes ATM at DNA damage sites. The Journal of cell biology 187, 977-990. 10.1083/jcb.200906064.
Spardy, N., Covella, K., Cha, E., Hoskins, E.E., Wells, S.I., Duensing, A., and Duensing, S. (2009). Human papillomavirus 16 E7 oncoprotein attenuates DNA damage checkpoint control by increasing the proteolytic turnover of claspin. Cancer Res 69, 7022-7029. 10.1158/0008-5472.can-09-0925.
Spardy, N., Duensing, A., Charles, D., Haines, N., Nakahara, T., Lambert, P.F., and Duensing, S. (2007). The human papillomavirus type 16 E7 oncoprotein activates the Fanconi anemia (FA) pathway and causes accelerated chromosomal instability in FA cells. J Virol 81, 13265-13270. 10.1128/jvi.01121-07.
Spardy, N., Duensing, A., Hoskins, E.E., Wells, S.I., and Duensing, S. (2008). HPV-16 E7 reveals a link between DNA replication stress, fanconi anemia D2 protein, and alternative lengthening of telomere-associated promyelocytic leukemia bodies. Cancer Res 68, 9954-9963. ening of telomere-associated promyelocytic leukemia bodies." Cancer Res 68(23): 9954-9963.10.1158/0008-5472.CAN-08-0224.
Spriggs, C.C., and Laimins, L.A. (2017). FANCD2 Binds Human Papillomavirus Genomes and Associates with a Distinct Set of DNA Repair Proteins to Regulate Viral Replication. mBio 8. 10.1128/mBio.02340-16.
Staskus, K.A., Zhong, W., Gebhard, K., Herndier, B., Wang, H., Renne, R., Beneke, J., Pudney, J., Anderson, D.J., Ganem, D., *et al.* (1997). Kaposi's sarcoma-associated herpesvirus gene expression in endothelial (spindle) tumor cells. J Virol 71, 715-719.
Stewart, G.S. (2009). Solving the RIDDLE of 53BP1 recruitment to sites of damage. Cell Cycle 8, 1532-1538. 10.4161/cc.8.10.8351.
Stewart, G.S., Wang, B., Bignell, C.R., Taylor, A.M., and Elledge, S.J. (2003). MDC1 is a mediator of the mammalian DNA damage checkpoint. Nature 421, 961-966. 10.1038/nature01446
Stokes, M.P., Van Hatten, R., Lindsay, H.D., and Michael, W.M. (2002). DNA replication is required for the checkpoint response to damaged DNA in Xenopus egg extracts. The Journal of cell biology 158, 863-872. 10.1083/jcb.200204127.
Stracker, T.H., Carson, C.T., and Weitzman, M.D. (2002). Adenovirus oncoproteins inactivate the Mre11-Rad50-NBS1 DNA repair complex. Nature 418, 348-352. 10.1038/nature00863].
Stucki, M., Clapperton, J.A., Mohammad, D., Yaffe, M.B., Smerdon, S.J., and Jackson, S.P. (2005). MDC1 directly binds phosphorylated histone H2AX to regulate cellular responses to DNA double-strand breaks. Cell 123, 1213-1226. 10.1038/nature00863].
Sugden, B., and Mark, W. (1977). Clonal transformation of adult human leukocytes by Epstein-Barr virus. J Virol 23, 503-508.
Tibbetts, R.S., Brumbaugh, K.M., Williams, J.M., Sarkaria, J.N., Cliby, W.A., Shieh, S.Y., Taya, Y., Prives, C., and Abraham, R.T. (1999). A role for ATR in the DNA damage-induced phosphorylation of p53. Genes Dev 13, 152-157.
Tibbetts, R.S., Cortez, D., Brumbaugh, K.M., Scully, R., Livingston, D., Elledge, S.J., and Abraham, R.T. (2000). Functional interactions between BRCA1 and the checkpoint kinase ATR during genotoxic stress. Genes Dev 14, 2989-3002.

Vayda, M.E., Rogers, A.E., and Flint, S.J. (1983). The structure of nucleoprotein cores released from adenovirions. Nucleic Acids Res 11, 441-460.

Verma, S.C., Borah, S., and Robertson, E.S. (2004). Latency-associated nuclear antigen of Kaposi's sarcoma-associated herpesvirus up-regulates transcription of human telomerase reverse transcriptase promoter through interaction with transcription factor Sp1. J Virol 78, 10348-10359. 10.1128/jvi.78.19.10348-10359.2004.

Verschuren, E.W., Klefstrom, J., Evan, G.I., and Jones, N. (2002). The oncogenic potential of Kaposi's sarcoma-associated herpesvirus cyclin is exposed by p53 loss *in vitro* and *in vivo*. Cancer Cell 2, 229-241.

Walboomers, J.M., Jacobs, M.V., Manos, M.M., Bosch, F.X., Kummer, J.A., Shah, K.V., Snijders, P.J., Peto, J., Meijer, C.J., and Munoz, N. (1999). Human papillomavirus is a necessary cause of invasive cervical cancer worldwide. J. Pathol. 189, 12-19. 10.1002/(sici)1096-9896(199909)189:1<12::aid-path431>3.0.co;2-f.

Ward, I.M., and Chen, J. (2001). Histone H2AX is phosphorylated in an ATR-dependent manner in response to replicational stress. J Biol Chem 276, 47759-47762. 10.1074/jbc.C100569200.

Williams, V.M., Filippova, M., Filippov, V., Payne, K.J., and Duerksen-Hughes, P. (2014). Human papillomavirus type 16 E6* induces oxidative stress and DNA damage. J Virol 88, 6751-6761. 10.1128/jvi.03355-13.

Wu, X., Ranganathan, V., Weisman, D.S., Heine, W.F., Ciccone, D.N., O'Neill, T.B., Crick, K.E., Pierce, K.A., Lane, W.S., Rathbun, G., *et al.* (2000). ATM phosphorylation of Nijmegen breakage syndrome protein is required in a DNA damage response. Nature 405, 477-482. 10.1038/35013089.

Xiao, Y., Chen, J., Liao, Q., Wu, Y., Peng, C., and Chen, X. (2013). Lytic infection of Kaposi's sarcoma-associated herpesvirus induces DNA double-strand breaks and impairs non-homologous end joining. J Gen Virol 94, 1870-1875. 10.1099/vir.0.053033-0.

Xu, Y., and Baltimore, D. (1996). Dual roles of ATM in the cellular response to radiation and in cell growth control. Genes Dev 10, 2401-2410.

Yates, J.L., Warren, N., and Sugden, B. (1985). Stable replication of plasmids derived from Epstein-Barr virus in various mammalian cells. Nature 313, 812-815.

Yeager, T.R., Neumann, A.A., Englezou, A., Huschtscha, L.I., Noble, J.R., and Reddel, R.R. (1999). Telomerase-negative immortalized human cells contain a novel type of promyelocytic leukemia (PML) body. Cancer Res 59, 4175-4179.

Yue, W., Gershburg, E., and Pagano, J.S. (2005). Hyperphosphorylation of EBNA2 by Epstein-Barr virus protein kinase suppresses transactivation of the LMP1 promoter. J Virol 79, 5880-5885. 10.1128/jvi.79.9.5880-5885.2005.

Zhao, B., Marshall, D.R., and Sample, C.E. (1996). A conserved domain of the Epstein-Barr virus nuclear antigens 3A and 3C binds to a discrete domain of Jkappa. J Virol 70, 4228-4236.

Zhao, S., Weng, Y.C., Yuan, S.S., Lin, Y.T., Hsu, H.C., Lin, S.C., Gerbino, E., Song, M.H., Zdzienicka, M.Z., Gatti, R.A., *et al.* (2000). Functional link between ataxia-telangiectasia and Nijmegen breakage syndrome gene products. Nature 405, 473-477. 10.1038/35013083.

Zou, L. (2007). Single- and double-stranded DNA: building a trigger of ATR-mediated DNA damage response. Genes Dev 21, 879-885. 10.1101/gad.1550307.

12

Human Adenovirus: Tool to Reveal Mechanisms of Cellular Transformation and Vector to Treat Human Cancers

Ron T. Javier

Department of Molecular Virology and Microbiology, Baylor College of Medicine, Houston, Texas, USA

**Corresponding author: rjavier@bcm.edu*

Abstract

Human adenoviruses were discovered through their ability to cause acute respiratory infections in people. Shortly thereafter, this human pathogen was additionally shown to have the capacity to induce malignant tumours when inoculated into experimental animals. By representing the first known tumourigenic human virus, adenovirus attracted many talented scientists to study its oncogenic properties. The resulting work demonstrated that adenovirus is a powerful tool for exposing general molecular mechanisms that commonly trigger the development of cancers. Another advancement stemming from these investigations was the adenovirus vector, which arose from technologies that allowed manipulation of the adenovirus genome and production of cell lines that stably express adenovirus genes. Considering the tumourigenic properties of human adenovirus, it may be somewhat surprising that adenovirus vectors are now employed to combat human cancers. The goal of this chapter is to review the molecular mechanisms discovered for the adenovirus oncogenes and the promise of adenovirus vectors as effective cancer therapeutics.

Introduction

Human adenovirus was discovered by investigators seeking to identify new etiologic agents of acute respiratory infections. In 1953, Wallace Rowe and Robert Huebner attempted to identify the common cold virus, but instead isolated a new virus that caused degeneration of epithelial-like cells in explants of tonsil and adenoidal tissue surgically removed from children (Rowe *et al.*, 1953). Shortly thereafter, Maurice Hilleman isolated a new virus associated with an epidemic of acute febrile respiratory illnesses in Army recruits (Hilleman and Werner, 1954). Upon characterization, the two viruses proved to be related, and were named adenovirus after the adenoid tissue from which they were first isolated.

More than sixty human adenovirus serotypes have now been identified and classified into seven species (A to G) based on serology, haemagglutination, and genome sequence (Jones *et al.*, 2007; Robinson *et al.*, 2011). Human adenoviruses infect epithelial cells lining mucous membranes to cause primarily respiratory diseases, including pneumonia, croup, and bronchitis (Boyer *et al.*, 1959; Gray *et al.*, 2007; Metzgar *et al.*, 2007; Pavia, 2011), but are also associated with eye and bladder infections as well as infantile diarrhoea (Foy *et al.*, 1968; Numazaki *et al.*, 1973; Gary *et al.*, 1979; de Jong *et al.*, 1983; Anderson, 2010). Human adenoviruses are responsible for only a small proportion of acute respiratory infections in people, where they are linked to 5-10% of such infections in children, as well as to epidemic outbreaks in military recruits and patients living in nursing homes (Kunz and Ottolini, 2010). Certain adenovirus serotypes, such as species B adenovirus type 14, can produce life-threatening respiratory infections that require hospitalization in up to 40% of otherwise healthy people (CDC, 2007; Louie *et al.*, 2008 Kajon *et al.*, 2010; Carr *et al.*, 2011). Moreover, for neonates and immunosuppressed patients, including people with AIDS, adenovirus infections are responsible for severe and sometimes fatal complications, such as fulminant pneumonia, hepatitis, and encephalitis (Lion *et al.*, 2010; Tebruegge and Curtis, 2010). Current treatments for severe adenovirus-associated diseases are controversial and carry significant risks for adverse events (Tebruegge and Curtis, 2010).

The adenovirus virion is a 90-nm diameter, non-enveloped icosahedral particle (Liu *et al.*, 2010b). Distinctive fiber proteins project from each vertex of the icosahedral virion, and serve to bind attachment receptors on the surface of host cells (Zubieta *et al.*, 2005; Arnberg, 2009). An approximately 36-kb linear, double-stranded DNA (dsDNA) genome is encapsidated within the virion protein shell (Davison *et al.*, 2003). The viral genome codes for multiple transcription units, expression of which are classified into four kinetic classes: (1) immediate-early (*E1A*) expressed first after a viral infection, (2) early (E1B, *E2, E3,* and *E4*) expressed maximally before the onset of viral DNA replication, (3) intermediate (*IX, IVa2, L4* intermediate, and *E2*-late) expressed maximally at the onset of viral DNA replication, and (4) late (major late transcription unit; *L1-L5* messenger RNAs [mRNAs]) expressed maximally after the onset of viral DNA replication. These protein-encoding transcription units are transcribed by RNA polymerase II, and each typically codes for multiple polypeptides that mediate a common function. For example, in general, *E1A* activates transcription and induces quiescent host cells to progress into S-phase of the cell cycle (Berk, 2005); *E1B* blocks apoptosis (Chinnadurai, 2011); *E2* has a direct role in viral DNA replication (Stillman, 1983); *E3* modulates host immune responses to the viral infection (Lichtenstein *et al.*, 2004); and *major late transcription unit* encodes either structural proteins of the virion or proteins required for virion assembly and maturation (Akusjarvi, 2008). This organizational theme of adenovirus transcription units likely evolved to synchronize expression of different polypeptides required to carry out a single function during the viral life cycle. An exception is the *E4* transcription unit that encodes a more functionally diverse set of proteins that regulate transcription, RNA splicing, translation, mRNA nuclear export, DNA replication, and apoptosis (Weitzman, 2005).

The adenovirus genome also contains the non-coding viral associated I (*VAI)* RNA gene, which like cellular tRNA genes is transcribed by RNA polymerase III (Weinmann *et al.*, 1974). *VAI* expresses a highly structured 160-nt VAI RNA that functions to increase the translation of late viral mRNAs in adenovirus-infected cells (Ma and Mathews, 1996). During an adenovirus infection, a consequence of symmetrical transcription from complementary viral DNA strands is creation of dsRNA, which binds double-stranded RNA (dsRNA)-activated cellular protein kinase R (PKR) and stimulates its kinase activity. PKR is a key antiviral host factor and, when activated by dsRNA in virus-infected cells, it inhibits translation initiation by phosphorylating eukaryotic translation initiation factor eIF2α. The VAI RNA counteracts this host defense by competitively inhibiting dsRNA-induced activation of PKR by dsRNA. The approximately 200bp inverted terminal repeat (ITR) sequences located at both ends of the linear adenovirus genome function as the origins of viral DNA replication mediated by viral *E2* products (Davison *et al.*, 2003). In addition, a *cis*-acting DNA packaging sequence adjacent to the left ITR binds to the viral Iva2 protein to promote insertion of viral DNA into empty capsids and formation of infectious particles (Hearing *et al.*, 1987; Perez-Romero *et al.*, 2005; Christensen *et al.*, 2012).

A simplified overview of the adenovirus life cycle in its natural-host human cells centres around viral gene functions that modify infected cells to produce an optimal environment for viral replication. Most cells in a host are quiescent due to having exited the cell cycle. Therefore, early viral genes (*E1A, E1B, E4*) are required to stimulate quiescent host cells to reenter the cell cycle and progress into S phase. Doing so creates an environment that supplies large amounts of precursor nucleotides and host factors required for an upcoming massive burst in viral DNA and mRNA synthesis in infected cells (Berk, 2005; Chinnadurai, 2011). To support the high levels of late viral gene expression required to assemble as many as ten-thousand progeny virion particles per cell, two early viral genes (*E1B, E4*) additionally function to increase the cytoplasmic abundance of late viral mRNAs by blocking nuclear export of host mRNAs and selectively promoting nuclear export of late viral mRNAs (Flint and Gonzalez, 2003). In concert with the latter effect, an early and late viral gene (*E4, VAI*) function to increase translation of late viral mRNAs either by activating signalling cascades that stimulate protein synthesis or by specifically promoting translation of late viral mRNAs (Kitajewski *et al.*, 1986; O'Shea *et al.*, 2005a). During the entire adenovirus life cycle, multiple viral genes (*E1A, E1B,*

E3, E4, VAI) act in concert to counteract a variety of anti-viral host responses that would otherwise diminish or prevent viral gene expression, viral DNA replication, and production of progeny virions (Weitzman, 2005; Chahal *et al.*, 2012).

Adenovirus: The first known human virus with tumourigenic potential

From 1908 to 1910, Vilhelm Ellerman, Oluf Bang, and Peyton Rous identified the first tumour viruses, which were avian retroviruses that caused leukemia or solid sarcomas in chickens (Ellermann and Bang, 1908; Rous, 1911). Sixty-six years later, studies of avian Rous sarcoma virus by Harold Varmus and J. Michael Bishop would lead to the landmark discovery that cancer is a genetic disease caused in part by mutation and dysregulated expression of cellular genes (Stehelin *et al.*, 1976). By the early 1960s, researchers discovered that certain murine viruses, as well as simian virus SV40, also have the capacity to elicit tumours under experimental conditions (Rous and Beard, 1935; Bittner, 1936; Gross, 1951; Gross, 1953; Graffi, 1957; Friend, 1957; Moloney, 1960; Eddy *et al.*, 1962; Girardi *et al.*, 1962), leading to the strong suspicion that some human viruses may cause cancer in people. In 1962, John Trentin screened collections of human viruses for tumour induction in experimentally inoculated newborn hamsters. He found that subcutaneous inoculation of species A human adenovirus serotype 12 (Ad12) promoted the development of undifferentiated sarcomas at the site of infection (Trentin *et al.*, 1962). This was a seminal discovery to the fields of virology and cancer because it identified the first known oncogenic human virus.

Later studies showed that following subcutaneous inoculation of immune-competent rodents, all species A and B human adenoviruses can generate undifferentiated sarcomas, and two of seven tested species D human adenoviruses, serotypes 9 (Ad9) and 10 (Ad10), can elicit oestrogen-dependent mammary tumours, most of which histologically present as benign mammary fibroadenomas, though some malignant phyllodes tumours and solid sarcomas of the mammary glands are also produced (Javier *et al.*, 1991). Fibroadenomas and phyllodes tumours are mixed tumours composed of a cellular stroma embedded in a dense extracellular matrix and interspersed with benign glandular epithelium. *In situ* hybridization analyses identified Ad9 mRNA-expressing tumour cells as either mammary fibroblasts in the fibroadenomas or mammary myoepithelial cells in the phyllodes tumours and solid sarcomas (Javier *et al.*, 1991). In contrast, all species C human adenoviruses are nontumourigenic after inoculation into experimental animals (Berk, 2013).

It was also found that all species of human adenovirus, including non-tumourigenic species C like human adenovirus serotype 5 (Ad5), can induce oncogenic transformation in cultured primary rodent cells (Berk, 2013). This cell culture model for tumourigenesis studies was quickly adopted and continues to be widely used in research studies because animal experiments tend to be difficult and time consuming, as well as challenging when trying to decipher molecular mechanisms. The properties of normal cells and tumour cells differ in many important ways, and the alterations that manifest in tumour cells are referred to as transformed properties. The transformed phenotype includes changes in proliferation, growth, morphology, metabolism, and gene expression. These properties collectively define the cellular state of tumourigenic/oncogenic transformation. Abnormal cellular proliferation is a key transformed property that includes the capacity of cells to divide (i) in the absence of growth factors, (ii) after reaching confluency, (iii) in the absence of attachment to a solid substrate, and/or (iv) an unlimited number of times without undergoing permanent growth arrest (senescence). These divergences from normal cellular proliferation are termed growth factor-independent growth, loss of contact inhibition, anchorage-independent growth, and immortalization, respectively. Anchorage-independent growth most closely correlates with the capacity of transformed cells to form tumours in animals (Macpherson and Montagnier, 1964; Topp *et al.*, 1981).

While studies have failed to expose a reproducible link between human adenovirus infections and human cancers (Mackey *et al.*, 1976; Green *et al.*, 1979; Ibelgaufts *et al.*, 1982; Mende *et al.*, 2004; Gustafsson *et al.*, 2007; Kosulin *et al.*, 2007; Vasconcelos *et al.*, 2008; Miller *et al.*, 2009; Honkaniemi *et al.*, 2010; Wimmer *et al.*, 2010), the established tumourigenic or cellular transforming potential of adenovirus, together with the ease by which it is propagated in human cell lines and their genomes are mutated to study gene functions, attracted many researchers interested in exploiting this virus as a tool to discover mechanisms of carcinogenesis. Indeed, our fundamental understanding of molecular events that drive the development of human cancer is due in large part to seminal discoveries made from the study of viral oncoproteins (Javier and Butel, 2008), in particular the concept that cancer is a genetic disease caused by activating mutations in cellular proto-oncogenes, the discovery of the p53 tumour suppressor, and the discovery of functions for the retinoblastoma tumour suppressor.

Work on adenovirus also yielded important insights into our general understanding of gene expression, DNA replication, and regulation of cell-cycle progression and cellular proliferation. Adenovirus may be best known for its use in the discovery of mRNA splicing (Berget *et al.*, 1977; Chow *et al.*, 1977). More recently, adenoviruses have been productively exploited as vectors for vaccination, as well as for gene and cancer therapies (Wold and Toth, 2013; Capasso *et al.*, 2014; Crystal, 2014).

Identification of oncogenes encoded by human adenoviruses

Following the discovery of the tumourigenic potential of human adenovirus, efforts turned to identifying the viral genes responsible for provoking oncogenesis *in vivo* and cellular transformation *in vitro* (Berk, 2013; DiMaio and Fan, 2013; Wold and Ison, 2013). As a first step toward this goal, studies determined that adenovirus-induced rodent tumours and transformed rodent cells invariably fail to produce infectious viral particles, revealing distinct outcomes for human adenovirus infections of natural host human cells and experimental host rodent cells. Specifically, adenovirus-infected human cells are fully permissive for viral replication and support all steps of the viral life cycle -- virus entry; immediate early and early gene expression; DNA replication; intermediate and late gene expression; progeny virion production; and cell lysis to release progeny virions. By contrast, adenovirus-infected rodent cells are semi- or nonpermissive for human adenovirus replication and support only the early steps of the viral life cycle -- virus entry; immediate early and early gene expression; limited or no viral DNA replication; no intermediate or late gene expression; no progeny virion production; and no cell lysis and instead survival of infected cells. The latter abortive infection in rodent cells creates an opportunity for rare non-homologous integration of adenovirus DNA into the host cell genome. In rare cells where the integrated DNA includes one or more adenovirus oncogenes, stable oncogene expression drives uncontrolled proliferation and clonal expansion to generate tumours *in vivo* or transformed cell lines *in vitro*.

These observations set the stage to identify adenovirus oncogenes. This was first accomplished with species A and B adenoviruses that produce tumours in experimental animals and with species C adenoviruses that transform cultured cells. A key observation was that only a portion of the adenovirus genome was retained in some tumours and transformed cell lines. By examining a collection of such samples, investigators found that only the *E1A* and *E1B* transcription units, which are situated on the extreme left end of viral genome and together constitute the viral *E1* region, were invariably retained in tumours and transformed cell lines (Doerfler, 1968; Gallimore, 1974), suggesting that the viral *E1* region contains the oncogenic determinants of species A, B, and C adenoviruses. Moreover, examination of *E1A* and *E1B* mutant adenoviruses verified that both genes are indeed required for virus-induced transformation of rodents cells, as well as for productive viral replication in human cells (Jones and Shenk, 1979). Subsequent transfection experiments conducted with plasmids expressing individual adenovirus transcription units demonstrated that adenovirus *E1A* and *E1B* are both necessary and sufficient to promote full oncogenic transformation of primary rodent cells (Ruley, 1983; Jones, 1990). By contrast, *E1A* alone only induces partial cellular transformation (Houweling *et al.*, 1980), which lacks most features of fully transformed cells, including the capacity to form tumours in animals, whereas *E1B* alone fails to transform cells (Van den Elsen *et al.*, 1983). Therefore, *E1A* and *E1B* cooperate to induce a fully transformed state in rodent cells.

Interestingly, an acute effect of *E1A* expression alone in primary rodent cells is death by apoptosis for the majority of the cells (White, 1993; Bayley and Mymryk, 1994). Following this massive crisis period, rare E1A-expressing cells eventually emerge with a partially transformed phenotype that manifests as cell immortalization, which is the capacity to divide indefinitely (Houweling *et al.*, 1980; Moran *et al.*, 1986). Cell immortalization is a transformed phenotype because normal cells only divide a limited number of times before they stop dividing and undergo senescence. Furthermore, immortalization is among the transformed properties invariably acquired by cancer cells (Stewart and Weinberg, 2006).

Another important question that arose was why species A adenoviruses are highly tumourigenic whereas species C adenoviruses are nontumourigenic in experimental animals, despite the fact their *E1* regions display comparable transforming activity in cultured cells. An important insight was gleaned from the finding that cells transformed by species C viruses are tumourigenic when inoculated into immune-deficient animals but not when inoculated into immune-competent animals. By contrast, cells transformed by species A viruses are highly tumourigenic in both immune-deficient and immune-competent animals. These observations led to the hypothesis that species C virus-transformed cells fail to form tumours in immune-competent animals because these cells are more immunogenic than those transformed by a tumourigenic species A virus. In support of this idea, species A Ad12-transformed cells express very little cellular major histocompatibility complex (MHC) class I protein and are resistant to killing by cytotoxic T-cells, whereas species C Ad5-transformed cells express high levels of MHC class I and are sensitive to killing by cytotoxic T-cells (Bernards *et al.*, 1983). It was additionally found that the large E1A protein is responsible for these differences. This landmark study revealed that the E1A protein of tumourigenic Ad12 confers a tumourigenic phenotype because unlike Ad5 E1A, Ad12 E1A can downregulate MHC class 1 protein levels, thereby allowing Ad12-transformed cells to escape elimination by cytotoxic T-cells and to form tumours in immune-competent animals.

Based on studies of human adenovirus, mouse polyomavirus, and SV40, several important general concepts for small and medium-sized DNA tumour viruses emerged: (1) All or part of the viral genome must persist in transformed cells. (2) Viral DNA that persists must include a viral oncogene(s). (3) Viral DNA persists in cells by nonhomologous recombination with a host chromosome. (4) This type of integration event is by nature rare, which explains the inefficiency of transformation by these viruses. (5) More than one oncogene is typically encoded in the viral genome, and they cooperate to transform cells. (6) Viral oncogenes are also required for productive replication of the viruses in their natural host cells and are typically either immediate-early or early class viral genes. (7) The ability of transformed cells to avoid immune elimination is crucial for tumourigenesis *in vivo*. These general concepts do not exclude the possibility that virus-induced cancers may be triggered by a "hit and run" mechanism, where the virus initiates tumourigenesis but is not required to maintain the transformed state, resulting in eventual loss of viral sequences. In fact, evidence hints that certain adenovirus oncogenes may employ this strategy (Nevels *et al.*, 2001).

Unlike tumours elicited by species A and B human adenoviruses, mammary tumours induced by species D Ad9 invariably contain one or more integrated copies of the entire viral genome (Javier *et al.*, 1991). This observation failed to provide insights into the identity of the Ad9 oncogene(s). More informative was the fact that Ad9 induced mammary tumours express immediate early *E1A* and early *E4* mRNAs, but not early *E1B* or *E3* mRNAs, indicating that *E1A* and/or *E4* are candidates for encoding an Ad9 oncogene. Key to the identification of the primary oncogenic determinant of Ad9 was the observation that unlike Ad9, the closely related species D human adenovirus serotype 26 (Ad26) fails to elicit tumours of any kind in experimental animals (Javier *et al.*, 1992). Results with Ad9/Ad26 recombinant viruses identified a 5' segment of the *E4* transcription unit essential for Ad9-induced mammary tumourigenesis. The *E4* transcription unit contains six different open reading frames (ORFs; *E4orf1*, *E4orf2*, *E4orf3*, *E4orf4*, *E4orf6*, and *E4orf6/7*) (Javier and Shenk, 1996), and the crucial 5' *E4* segment encompassed *E4orf1*, *E4orf2*, and the 5' half of *E4orf3*. Subsequent studies with Ad9 viruses having deletion or point mutations within the 5' *E4* segment defined the minimal crucial Ad9 sequences to the *E4orf1* coding region and an adjacent small noncoding region situated between *E4orf1* and *E4orf2* (Javier, 1994; Thomas *et al.*, 2001). These mutant viruses lack detectable replication defects in permissive human cell lines. More importantly, E4orf1 protein is invariably expressed in Ad9-induced mammary tumours. The function of the small non-coding region has not been determined, though it was speculated to permit *E4orf1* mRNA expression in Ad9infected rodent mammary cells *in vivo*. Remarkably, substitution of the *E1* region in non-tumourigenic species C Ad5 with an Ad9 *E4orf1* expression cassette converts Ad5 into a tumourigenic replication-defective vector capable of inducing mammary tumours identical to those produced by Ad9. Moreover, an expression plasmid encoding Ad9 *E4orf1*, but not the six other Ad9 *E4* region-encoded ORFs, is sufficient to transform a rodent cell line (Javier, 1994). Interestingly, the same expression plasmid encoding *E4orf1* derived from species A-C adenoviruses fails to transform the rodent cell line, whereas expression plasmids encoding *E4orf1* derived from species A-D adenoviruses exhibit identical transforming activity in human cell lines (Weiss *et al.*, 1997b; Kumar *et al.*, 2014). The failure of *E4orf1* genes from species A-C adenoviruses to transform rodent cells stems from an expression defect specific to these host cells. In this regard, *E4orf1* protein is expressed in a rodent cell line stably containing the *E4orf1* plasmid of species D Ad9 but not species A-C adenoviruses, whereas E4orf1 protein is comparably expressed in human cell lines stably containing the *E4orf1* plasmid of species A-D adenoviruses. These observations suggest that species D Ad9 *E4orf1* uniquely functions as a tumourigenic determinant in rodents because unlike *E4orf1* genes from species A-C adenoviruses, it lacks an expression defect in rodent cells. Whether the expression defect of *E4orf1* genes from species A-C adenoviruses is at the mRNA or protein level has not been determined.

Given that the *E1* region is the major oncogenic determinant for species A-C adenoviruses, a possible role for *E1A* and/or *E1B* in species D Ad9-induced mammary tumourigenesis was explored (Thomas *et al.*, 1999). Transfection experiments were initially conducted to assess the capacity of an Ad9 *E1* region plasmid to transform low-passage rat embryo fibroblasts (REFs) (Jannun and Chinnadurai, 1987; Thomas *et al.*, 1999). Differing from the *E1* region of species A-C adenoviruses, species D Ad9 *E1* alone fails to induce transformed foci on REFs. Oncogenic mutant Ras alone also does not transform REFs, but does when combined with the Ad9 *E1* region or *E1A* from Ad9 or species C Ad5. In contrast, mutant Ras combined with *E1B* from Ad9 or species C Ad5 does not transform REFs. The findings suggest that the failure of the Ad9 E1 region alone to transform REFs is due to an Ad9 *E1B* expression defect in REFs, similar to the E1B expression defect observed in Ad9-induced rat mammary tumours (Javier *et al.*, 1991). Next, tumour induction by Ad9 mutant viruses having *E1A* or *E1B* deleted was compared to wild-type Ad9 in animals. Like wild-type Ad9, *E1A* and *E1B* mutant Ad9 viruses elicited mammary tumours at 100% frequency. However, mammary tumours elicited by the *E1A* mutant Ad9 were generally smaller than the comparably-sized mammary tumours produced by wild-type Ad9 and the *E1B* mutant Ad9, consistent with (i) expression of *E1A* but not *E1B* mRNA in wild-type Ad9-induced mammary tumours and (ii) cooperation of Ad9 *E1A* but not *E1B* with oncogenic Ras to transform REFs. More importantly, additionally considering the nontumourigenic phenotype of *E4orf1* mutant Ad9 viruses along with the similar tumourigenic phenotypes of wild-type Ad9, *E1B* mutant Ad9, and the *E1* region-deleted Ad5 vector that expresses Ad9 *E4orf1*, it can be concluded that Ad9 *E4orf1* is the major oncogenic determinant of Ad9. This finding is in striking contrast to species A-C adenoviruses where only *E1A* and *E1B* are the tumourigenic determinants. Furthermore, while Ad9 *E1A* increases the mass of wild-type Ad9-induced mammary tumours, Ad9 *E1A* becomes dispensable when the strong CMV promoter drives Ad9 *E4orf1* expression in an *E1* region-deleted Ad5 vector. The fact that subcutaneous inoculation of either species D Ad9 or species A-C adenoviruses produces oestrogen-dependent mammary tumours in only female animals or undifferentiated sarcomas in both female and male animals, respectively, further suggests that the *E4orf1* gene and the *E1* region also govern the tissue tropism of adenovirus-induced tumourigenesis. While the strict oestrogen dependence of Ad9 induced mammary tumourigenesis is not yet understood (Javier *et al.*, 1991), this dependence together with detection of oestrogen receptor mRNA expression in mammary tumour cells hints that Ad9 E4orf1 cooperates with ligand-activated oestrogen receptor, akin to the cooperation of two oncogenes, to induce neoplastic transformation of mammary gland cells *in vivo*.

With identification of the adenovirus oncogenes, attention next shifted to deciphering the molecular mechanisms whereby the products of these viral genes provoke tumourigenesis *in vivo* and cellular transformation *in vitro*.

Oncogenic mechanisms of adenovirus E1A

After infection of cells with human adenovirus, viral immediate-early *E1A* is the first transcription unit expressed due a strong enhancer element in the *E1A* promoter (Hearing and Shenk, 1983; Nevins *et al.*, 1979). Two major alternatively spliced E1A mRNAs, 12S and 13S, are produced by employing one of two 5′-splice sites separated by 138bp and the same 3′ splice site (Perricaudet *et al.*, 1979). The two polypeptides translated from the 12S and 13S E1A mRNAs are the 243-residue small E1A protein and the 289-residue large E1A protein, respectively. These proteins have identical amino acid sequences aside from an additional unique 46 residue peptide present in large E1A. Both proteins localize to the nucleus and function to regulate the transcription of both cellular and viral genes. Large and small E1A have a shared function, and large E1A additionally has a unique function mediated by the additional 46 residue peptide. The shared E1A function is to promote adenovirus-infected quiescent cells to reenter the cell cycle and progress into S phase, creating an optimal environment for viral replication by stimulating production of dNTPs for viral DNA replication, NTPs for viral transcription, and RNA processing proteins for alternative splicing and polyadenylation. The function unique to large E1A, on the other hand, is to activate transcription from its own promoter as well as the promoters of early viral genes *E1B*, *E2*, *E3*, and *E4* (Montell *et al.*, 1982; Montell *et al.*, 1984; Winberg and Shenk, 1984).

Conserved motifs in proteins often mediate important functions, so a search for such motifs in adenovirus E1A proteins was undertaken. Comparisons of the amino-acid sequences of different large E1A proteins revealed four conserved regions, termed CR1 (residues 40-80), CR2 (residues 121-139), CR3 (residues 140-186), and CR4 (residues 240-288) (Kimelman *et al.*, 1985; van Ormondt *et al.*, 1980; Pelka *et al.*, 2008). CR3 comprises the 46-residue peptide unique to large E1A and functions as a transcriptional activation domain (Shenk and Flint, 1991; Berk, 2005). CR3 mediates this function through an interaction with the

MED23 subunit of the human Mediator transcription complex (Boyer *et al.*, 1999; Stevens *et al.*, 2002). MED23 binds to activation domains of transcription factors, such as CR3 of E1A, to stimulate transcription of target genes by enhancing both pre-initiation complex assembly on promoters and transcriptional elongation (Cantin *et al.*, 2003; Wang *et al.*, 2005). CR3 additionally binds the cellular nuclear acetyltransferase p300, and also possibly acetyltransferase PCAF, to enhance large E1A-induced early viral gene activation (Pelka *et al.*, 2009a; Pelka *et al.*, 2009b). However, the observation that large and small E1A display equivalent oncogenic activity leads to the conclusion that CR3 is dispensable for E1A-induced cellular transformation (Braithwaite *et al.*, 1983; Spindler *et al.*, 1985; Zerler *et al.*, 1987).

Other studies revealed that CR4 at the E1A carboxyl-terminus surprisingly mediates anti-oncogenic functions as evidenced by E1A CR4 mutants exhibiting hyper-transformation and hyper-tumourigenic phenotypes in cooperation with oncogenic Ras (Subramanian *et al.*, 1989; Douglas *et al.*, 1991; Frisch, 1991; Chinnadurai, 1992; Quinlan and Douglas, 1992; Frisch, 1994). These phenotypes are due to the inability of CR4 E1A mutants to interact with and block the function of several different cellular factors, including the cellular transcriptional corepressor proteins CtBP1 and CtBP2, forkhead family transcription factors FOXK1 and FOXK2, and dual-specificity kinases DYRK1A and DYRK1B (Chinnadurai, 2002, Chinnadurai, 2007; Komorek *et al.*, 2010). In cells that express a CR4 E1A mutant, CtBP1/2 retains its normal function of blocking expression of genes that promote epithelial cell differentiation and apoptosis, thereby increasing cellular transformation, whereas in the presence of wild-type E1A, these CtBP functions are inactivated, thereby reducing cellular transformation. While less is known about the remaining CR4 targets, the FOXK1/2 transcription factors and the DYRK1A and DYRK1B kinases are postulated to activate genes that either promote cellular proliferation or prevent cellular differentiation, respectively, thereby conferring higher transforming activity to CR4 E1A mutants than wild-type E1A.

Mutational analyses of E1A identified two separate protein regions required for *E1A* to cooperate with *E1B* or activated Ras to induce full transformation of rodent cells. The first crucial region is defined by 80 amino-acid residues at the extreme aminoterminus of E1A encompassing unique sequences (residues 1-39) and adjacent conserved CR1 sequences (residues 40-80). The second crucial region is defined by conserved E1A CR2 sequences (residues 121-139). When E1A alone is transiently expressed in quiescent primary rodent cells, either one of these two crucial E1A regions is sufficient to induce G0 exit, passage through G1 phase, progression into S phase, and cellular DNA synthesis, whereas both E1A regions are required additionally to induce passage through the entire cell cycle and mitosis (Whyte *et al.*, 1988b; Howe *et al.*, 1990; Stein *et al.*, 1990; Wang *et al.*, 1993;). Based on an earlier landmark study showing that cellular transformation induced by mouse polyomavirus middle T depends on separate interactions with proto-oncogene products of cellular *Src* (gene captured and mutated in avian Rous sarcoma virus) and *PIK3CA* [encodes phosphatidylinositol-3-kinase (PI3K)], it was hypothesized that the crucial E1A protein regions likewise bind and activate cellular proto-oncogene products (Yee and Branton, 1985; Harlow *et al.*, 1986). The important cellular targets and functions discovered for the two crucial E1A regions are discussed below.

Oncogenic functions associated with E1A CR2.

The first identification of a cellular factor that mediates *E1A*-induced cellular transformation was reported by investigators in the laboratories of Ed Harlow and Robert Weinberg (Whyte *et al.*, 1988a). Following immunoprecipitation of E1A from extracts of adenovirus-infected cells, a 105kD cellular protein was recovered and identified as the retinoblastoma protein, which is the product of the *RB1* gene. Based on earlier studies, the RB1 protein represented a particularly exciting candidate to mediate E1A-induced cellular transformation because the results implicated a loss rather than gain of RB1 function in the development of the hereditary eye cancer retinoblastoma in children (Knudson, 1971, Vogel, 1979; Cavenee *et al.*, 1983; Abramson *et al.*, 1984; Knudson, 1984; Murphree and Benedict, 1984; Dryja *et al.*, 1986; Friend *et al.*, 1986; Fung *et al.*, 1987). The findings identified *RB1* as a cellular tumour suppressor gene as opposed to a proto-oncogene. Interestingly, an earlier study reported that inoculation of Ad12 into the eyes of newborn rats elicited retinoblastoma (Kobayashi and Mukai, 1973). Furthermore, mutations altering the conserved E1A amino-acid sequence leucine-x-cysteine-x-glutamic acid (LxCxE; x = any amino acid) within CR2 abolished not only E1A binding to RB1 but also E1A induced cellular transformation (Raychaudhuri *et al.*, 1991), providing direct evidence for a link between the E1A-RB1 interaction and oncogenesis. The interaction of E1A with the RB1 tumour suppressor represents a landmark discovery because it revealed for the first time that viral oncogenes can transform cells not only by activating cellular proto-oncogene products but also by inactivating cellular tumour suppressor proteins. The wide importance of this discovery is underscored by later studies showing that like adenovirus E1A, the otherwise unrelated SV40 Large T antigen and human papillomavirus (HPV) E7 oncoproteins have an E1A CR2-related LxCxE motif that likewise mediates binding to RB1 (DeCaprio, 2009). Moreover, introduction of the CR2 related element of SV40 Large T antigen into a transformation-defective E1A CR2 deletion mutant restored transforming activity (Moran, 1988).

Studies of DNA tumour virus oncogenes were instrumental in revealing how the RB1 tumour suppressor protein functions in normal cells. An important early step in allowing growth factor-stimulated quiescent cells to re-enter and progress through the cell cycle is the activation of cytosolic signalling pathways that stimulate transcription of the cyclin D gene (Matsushime *et al.*, 1991). Cyclin D is a regulatory subunit of cyclin dependent kinases (cdks), and as its levels rise in early G1 phase of the cell cycle, cyclin D associates with cdk4 and cdk6 to stimulate their kinase activity. In conjunction with subsequently activated cyclin E- and cyclin A-cdk2 complexes, cyclin D-*cdk4/6* hyper-phosphorylates and functionally inactivates the nuclear substrate RB1 (DeCaprio, 2009). Inactivation of RB1 allows cells to pass through an important mid-G1 phase cell-cycle checkpoint, the G1 restriction point, and to enter into S phase (Hatakeyama and Weinberg, 1995). This occurs because, in G0 and early-G1 phase prior to hyper-phosphorylation of RB1, it exists in an active, hypophosphorylated state that enforces the G1 restriction point to block S-phase progression. This inhibitory activity is due to RB1-mediated inhibition of E2F transcription factor family proteins E2F1, E2F2, and E2F3 (E2F-1/-2/-3). RB1 represses E2F-1/-2/-3 proteins by binding to their transcriptional activation domains. Notably, studies of adenovirus led to identification of E2F family proteins, which were named for their ability to mediate E1A-induced transcriptional activation of the adenovirus E2 promoter (Kovesdi *et al.*, 1986). In G0 and early-G1 phase, the E2F-1/-2/-3 proteins and histone-modifying enzymes that silence transcription, such as histone deacetylase (HDAC), exist in an HDAC:RB1:E2F complex that associates via the E2F DNA-binding domain with E2-Fbinding sites in the promoters of E2F-responsive cellular genes (Berk, 2005). The complex fails to activate transcription and instead inhibits transcription due to both RB1-mediated repression of E2F activity and HDAC-mediated transcriptional silencing. The transcriptional repression enforces the G1 restriction point that prevents S-phase progression because E2F-responsive cellular genes code for numerous critical S phase-specific factors such as DNA polymerases, topoisomerases, and enzymes of nucleotide metabolism required for cellular DNA replication (DeGregori *et al.*, 1995; Dick and Rubin, 2013). However, this repression is relieved by cyclin-cdk-mediated hyper-phosphorylation of RB1 that occurs at midG1 phase because this post-translational modification promotes RB1 to release both E2F and HDAC from the inhibitory HDAC:RB1:E2F complex, leaving free E2F proteins to associate with the promoter of E2F-responsive cellular genes and to activate their transcription, thereby driving passage of cells through the G1 restriction point and progression into S-phase (DeCaprio, 2009). Given that passage through the G1 restriction point commits cells to S-phase progression, completion of the cell cycle, and cell division, the RB1 tumour suppressor represents a critical negative regulator of inappropriate cellular proliferation (Hatakeyama and Weinberg, 1995).

As mentioned above, cellular transformation induced by E1A stems in part from sequestration of RB1 within inactive E1A:RB1 complexes and disruption of inhibitory HDAC:RB1:E2F complexes (Dick and Rubin, 2013). The E1A-RB1 interaction is primarily mediated by the CR2 LxCxE motif of E1A and a cleft region in the pocket domain of RB1 (Nevins, 1994; Liu and Marmorstein, 2007). The LxCxE motif is also commonly found in RB1-interacting cellular proteins, including histone-modifying enzymes that silence transcription, such as HDAC, where a LxCxE motif likewise mediates binding to the RB1 cleft region. As E2F proteins lack an LxCxE motif, the E2F transactivation domain mediates binding to RB1 via a region of the RB1 pocket domain formed by the A-B subdomain interface. This allows RB1 to interact simultaneously with HDAC and E2F. Thus, high-affinity LxCxE motif-mediated E1A binding to RB1 competitively displaces HDAC from HDAC:RB1:E2F complexes, and is also necessary but not sufficient to displace E2F. In addition E2F displacement requires specific E1A CR1 residues, which act by competitively inhibiting binding of the E2F transactivation domain to the RB1 A-B subdomain interface region (Lee *et al.*, 1998; Liu and Marmorstein, 2007). Note that the latter E1A CR1 residues are distinct from those that mediate CR1 functions discussed in the next section. E1A displacement of HDAC and E2F from HDAC:RB1:E2F complexes abrogates RB1-mediated transcriptional silencing of E2F-1/-2/-3-responsive cellular genes and additionally allows free E2F-1/-2/-3 to transcriptionally activate these genes. While the effects of E1A are similar to those produced by growth factor stimulation of cells, an important difference is that E1A permits cells to bypass the G1 restriction point in the absence of growth factor-induced RB1 hyper-phosphorylation. Consequently, E1A binding to RB1 triggers inappropriate, growth factor-independent transcriptional activation of E2F-responsive genes, cell-cycle progression, and cell proliferation, which together culminate in triggering oncogenic cellular transformation. The ground-breaking discoveries made from studies of adenovirus E1A laid the groundwork for our current detailed understanding of how the RB1 protein functions as a tumour suppressor and how its functional inactivation by mutation or interaction with viral oncoproteins provokes carcinogenesis.

E1A, SV40 Large T antigen, and HPV E7 additionally interact via their LxCxE motif with two additional retinoblastoma (RB) family proteins, p107 (RBL1) and p130 (RBL2) (DeCaprio, 2009). In contrast to RB1, p107 and p130 bind to E2F family proteins E2F4 and -5 (E2F-4/-5), which unlike E2F-1/-2/-3, lack the capacity to activate transcription. Rather, E2F-4/-5 repress transcription and do so when present in p107:E2F-4/-5 and p130:E2F-4/-5 complexes that function to maintain growth-arrested G0 cells from re-entering and progressing through the cell-cycle (Ghosh and Harter, 2003; Frolov and Dyson, 2004; Blais and Dynlacht, 2007; van den Heuvel and Dyson, 2008; Sha *et al.*, 2010). These inhibitory p107/p130 complexes bind to promoters of various genes, including some E2F-responsive genes, and repress transcription through recruitment of transcription-silencing chromatin remodelling proteins, including HDAC1, HDAC2, heterochromatin protein 1 (HP1), histone methyltransferase Suv39h1, and CtBP interacting protein. This recruitment is mediated by binding of an LxCxE motif in the chromatin remodeling proteins to an RB1-related cleft region in p107/p130. As with RB1, either growth factor-induced phosphorylation by cyclin-cdk or E1A binding in the absence of growth factors causes p107/p130 to release associated E2F-4/-5 and chromatin remodeling proteins (Cobrinik *et al.*, 1993; van den Heuvel and Dyson, 2008), thereby relieving the transcriptional repression by inhibitory p107/p130 complexes. This effect, together with RB1 inactivation, permits growth-arrested G0 cells to re-enter the cell cycle, pass through G1 phase, and progress into S phase (Ghosh and Harter, 2003). Unscheduled inactivation of p107 and p130 is also implicated in viral oncoprotein-induced cellular transformation, as evidenced by the fact that wild-type but not an LxCxE mutant of SV40 Large T antigen promotes transformation in RB1-null mouse embryo fibroblasts (Christensen and Imperiale, 1995; Zalvide *et al.*, 1998). Hence, the common targeting of RB family proteins by adenovirus E1A, SV40 Large T antigen, and HPV E7 provokes oncogenic transformation by causing cells arrested in either G0 or G1 to undergo unscheduled cell-cycle progression and proliferation. In addition to promoting oncogenesis, these activities are also beneficial to the productive replication of these viruses in their natural hosts by inducing infected quiescent G0 arrested or terminally differentiated epithelial cells to enter S phase, thereby creating an optimal cellular environment for viral DNA replication, viral late gene expression, and progeny virion production.

Oncogenic functions associated with the 80-residue amino-terminal E1A region spanning the unique amino-terminal sequences (residues 1-39) and CR1 (residues 40-80)

E1A immunoprecipitation analyses, like those that led to identification of cellular RB family proteins as E1A targets, revealed that the 80-residue amino-terminal E1A region that promotes S-phase progression and cellular transformation mediates binding to multiple cellular factors involved in the regulation of chromatin structure, implying that these oncogenic protein interactions likewise function to alter cellular gene expression. The first cellular targets identified for this crucial E1A region were the closely-related nuclear lysine acetyl-transferases p300 and CBP (Wang *et al.*, 1993; Eckner *et al.*, 1994; Arany *et al.*, 1995). Both p300 and CBP have histone acetyl-transferase (HAT) activity and function as global transcriptional co-activator proteins that increase gene expression by binding to transcription factors that recognize specific DNA sequences in the promoters of target genes (Goodman and Smolik, 2000). E1A binding to p300/CBP requires two separate E1A elements, one located within the unique amino-terminal region and the other located within CR1. Mutation of these elements abolishes the cellular transforming potential of E1A. Among the known cellular targets of the crucial 80 residue E1A region, p300/CBP seems most important for cellular transformation (Rasti *et al.*, 2005). Notably, SV40 Large T antigen and high risk HPV E7 likewise target p300/CBP (Bernat *et al.*, 2003; Ahuja *et al.*, 2005).

The mechanisms by which E1A binding to p300/CBP contributes to cellular transformation have recently emerged from studies investigating E1A-mediated global epigenetic reprogramming of cellular gene expression. Horwitz *et al.* showed that binding of small E1A to p300/CBP, which is a quantitative interaction in cells (Wang and Berk, 2002), drastically reduces total cellular levels of acetylation on histone H3 lysine 18 (H3K18) (Horwitz *et al.*, 2008). Dramatic H3K18 hypo-acetylation is also observed in cells expressing SV40 large T antigen. Given that p300/CBP mediates H3K18 acetylation and that H3K18 hypoacetylation and H3K18 hyper-acetylation

(H3K18ac) are markers for transcriptionally repressed or active DNA regions, respectively, the results suggest that small E1A blocks p300/CBP-mediated H3K18ac to cause genome-wide transcriptional silencing of cellular genes regulated by this modification.

Ferrari *et al.* extended these findings by showing that small E1A associates with the promoters of the majority of cellular genes in a specific temporal order either to activate or repress transcription (Ferrari *et al.*, 2008). Among genes whose transcription was altered by small E1A, 30% were activated, and this cluster was enriched for genes functioning in cell-cycle progression and cell growth. By contrast, the remaining 20% and 50% of the genes were repressed by small E1A, and these clusters were enriched for genes functioning either in inflammatory and pathogen responses or in cell differentiation, respectively. The same investigators subsequently reported that small E1A eliminates H3K18ac activity from genes that express cell type-specific functions and relocalizes this activity and p300/CBP to the 30% of E1A-activated genes that function to promote cell-cycle progression and cell growth (Ferrari *et al.*, 2012). These E1A-activated genes include the majority of genes repressed by RB1, but not p107 and p130, in the absence of either E1A expression or growth factors (Ferrari *et al.*, 2014), likely explaining why RB1 but not p107 and p130 is a tumour suppressor. Another interesting finding was that small E1A re-localized RB1 and p130 to E1A-repressed genes involved in inflammatory and pathogen responses or p107 to the E1A-repressed genes involved in differentiated cell functions. Results with small E1A mutants specifically unable to bind either p300/CBP or RB family proteins further demonstrated that E1A binding to p300/CBP is primarily responsible for both E1A-induced activation and repression of cellular genes, whereas E1A binding to RB family proteins instead increases the magnitude of each activity to biologically significant levels.

Related to the role of both p300/CBP and RB family proteins in E1A induced reprogramming of cellular gene expression, E1A is known to assemble a trimeric complex through simultaneous interactions with one RB family protein and either p300 or CBP, and this complex is required for transformation of primary rodent cells (Lee *et al.*, 1998; Liu and Marmorstein, 2007; Horwitz *et al.*, 2008; Ferreon *et al.*, 2009). Formation of the trimeric p300:E1A:RB1 complex also induces p300 to acetylate E1A on lysine residue K239 (Madison *et al.*, 2002), and also RB1 on lysine residues K873 and K874, with the latter modifications blocking cyclin-cdks from binding to and phosphorylating RB1 (Chan *et al.*, 2001). Prompted by these observations, Ferrari *et al.* recently provided an explanation for E1A-induced dual recruitment of p300/CBP and RB family proteins to certain repressed genes (Ferrari *et al.*, 2014). Specifically, they demonstrated that when E1A binds RB1 to displace E2Fs and activate E2F-responsive genes, it exploits the resulting E1A:RB1 complex in conjunction with p300 to repress transcription. E1A does so by recruiting p300 into the E1A:RB1 complex, thereby forming the trimeric p300:E1A:RB1 complex, where p300 acetylates RB1 on K873 and K874 to induce an inhibitory RB1 conformation that binds repressive chromatin-modifying HDAC enzymes that condense chromatin. Transcriptional repression and chromatin condensation mediated by the trimeric complex requires not only RB1 K873/K874 acetylation by p300 and RB1-associated HDAC activity but also E1A K239 acetylation by p300 and E1A binding to cellular p400 (discussed later), a SWI/SNF chromatin-remodelling ATPase. Furthermore, the trimeric complexes specifically target cellular genes bound to high levels of p300. Such genes are enriched for encoding components of the anti-viral TGFβ, TNF, and interleukin signalling pathways, which if expressed would be expected to inhibit adenovirus replication and oncogenic cellular transformation. Thus, the requirement for the trimeric p300:E1A:RB1 complex in cellular transformation is due to its ability to repress transcription of cellular genes that promote cell death or block cell proliferation. The findings support a model whereby E1A binding to p300/CBP and RB family proteins, separately or in combination, mediates defined epigenetic reprogramming that promotes cellular transformation by both activating and repressing cellular gene expression, and that also creates a favourable cellular environment for viral replication.

While p300/CBP generally functions as a transcriptional co-activator, it also represses transcription of some cellular genes. For example, Thimmapaya and colleagues reported that p300 in conjunction with the repressor Yin Yang 1 (YY1) and HDAC3 blocks transcription of the Myc oncogene and that E1A or SV40 Large T antigen via their interaction with p300 relieve this inhibition to stimulate cellular proliferation (Sankar *et al.*, 2008; Singhal *et al.*, 2008; Kadeppagari *et al.*, 2009). E1A binding to CBP also contributes to cellular transformation through a nontranscriptional mechanism (Turnell *et al.*, 2005). APC5 and APC7, which are components of the anaphase-promoting complex/cyclosome E3 (APC/C E3) ubiquitin ligase, bind to CBP through a motif related to the FxD/ExxxL sequence present in E1A CR1. This sequence allows E1A to displace CBP from the APC/C complex and thereby inactivate APC/C E3 ligase activity. The APC/C E3 ligase functions to ubiquitylate and promote degradation of the anaphase inhibitor Securin, which prevents sister chromatid separation. Given the known role of APC/C complex in controlling a mitotic checkpoint that maintains the fidelity of sister-chromatid separation, E1A-mediated inactivation activation of APC/C E3 by sequestration of CBP has been proposed to contribute to oncogenic transformation by promoting aberrant mitoses and genomic instability (Turnell and Mymryk, 2006).

An additional cellular target identified for the 80-residue amino-terminal E1A region is chromatin remodeler p400 (Deleu *et al.*, 2001; Fuchs *et al.*, 2001), which binds the adapter protein TRRAP. TRRAP associates with various chromatin-binding protein complexes with HAT activity, including the HAT protein TIP60 complex that contains p400 and the SAGA complex that contains the TRAPP-binding HAT protein GCN5 (Ikura *et al.*, 2000; Sterner and Berger, 2000; Nikiforov *et al.*, 2002). E1A residues 2635 within the crucial 80-residue amino terminal region bind not only p400 but also TRRAP and GCN5. These interactions are linked to E1A-induced transformation because the transformation-defective E1A deletion mutant missing residues 26-35 binds p300/CBP and RB1 but not p400, TRRAP, or GCN5 (Barbeau *et al.*, 1994; Lill *et al.*, 1997; McMahon *et al.*, 1998; Deleu *et al.*, 2001; Fuchs *et al.*, 2001; Lang and Hearing, 2003;). A key function of both TRRAP/TIP60 and TRRAP/GCN5 complexes is to activate Myc and its responsive genes by binding to and stabilizing the Myc protein (McMahon *et al.*, 1998; Frank *et al.*, 2001; Frank *et al.*, 2003; Patel *et al.*, 2004). Moreover, wild-type E1A but not the E1A mutant missing residues 26-35 enhances p400 binding to Myc, thereby reducing Myc ubiquitylation, stabilizing Myc, and stimulating the transcription of Myc-responsive genes (Patel *et al.*, 2004; Tworkowski *et al.*, 2008). E1A was also recently shown to associate with Myc (Vijayalingam *et al.*, 2016). These findings suggest that the crucial 80-residue amino-terminal E1A region contributes to E1A-induced cellular transformation through separate p300/CBP- and p400/TRRAP/GCN5-dependent mechanisms that share the common function of elevating Myc protein levels to activate Myc-responsive genes.

Adenovirus E1B

The primary transcript expressed from the adenovirus *E1B* gene undergoes alternative splicing to generate two major mRNAs, 22S and 13S, that code for the large and small E1B proteins (Bos *et al.*, 1981). Based on their molecular masses, the large and small E1B proteins are often referred to as E1B-55K or E1B1-9K, respectively. Either of these E1B proteins is sufficient independently to cooperate with E1A to promote oncogenic cellular transformation, though E1B-55K does so more potently and efficiently (Jochemsen *et al.*, 1982; Gallimore *et al.*, 1985). The first clue about how *E1B* gene products function in oncogenesis came from understanding the basis of inefficient partial transformation induced by *E1A* expression alone in rodent cells. As mentioned earlier, an acute effect of *E1A* expression in quiescent cells is CR1- and CR2-dependent unscheduled entry into and completion of the cell cycle, resulting in increased cellular proliferation. However, this unscheduled cellular proliferation is brief and soon followed by a dramatic crisis period characterized by massive cell death from which rare surviving, immortalized cells emerge. Notably, the characteristics of E1A induced cell death observed during the crisis period proved identical to apoptosis (Pilder *et al.*, 1984; White *et al.*, 1984; Babiss *et al.*, 1985; Rao *et al.*, 1992). Moreover, expression of either E1B1-9K or E1B-55K blocked the E1A induced apoptosis (Rowe *et al.*, 1953; Houweling *et al.*, 1980; White *et al.*, 1992; Teodoro and Branton, 1997), permitting efficient emergence of fully transformed, *E1A*- and *E1B*-expressing cells. During a productive adenovirus infection of human cells, the capacity of *E1B* to antagonize *E1A* induced apoptosis also prevents premature cell death that would otherwise diminish viral replication (White, 2001). In the context of either adenovirus-induced cellular transformation in rodent cells or productive adenovirus replication in human cells, unscheduled cell-cycle progression induced by *E1A* can be viewed as a double-edged sword, where one edge promotes cellular changes conducive to neoplastic transformation and optimal viral replication and the other edge triggers a detrimental host response that diminishes these activities. Thus, the central role of *E1B* in both tumourigenesis and viral replication is to counteract apoptosis triggered in part by a protective host response to abnormal *E1A*-induced cell-cycle progression.

Oncogenic mechanisms of the E1B-19K protein

The first insight into a mechanism by which E1B interferes with apoptosis was gleaned from the observation that E1B1-9K shares weak amino-acid sequence similarity with cellular Bcl-2 family proteins (White *et al.*, 1991), which are important regulators of apoptosis having either pro- or antiapoptotic activity. This finding led to the discovery that the antiapoptotic activity of E1B-19K stems from an ability to complex with and inactivate proapoptotic BCL2 family proteins Bak and Bax, which co-oligomerize and form pores in the outer mitochondrial membrane to release apoptogenic proteins, such as cytochrome c and Smac/Diablo, to stimulate caspase-family executioners of apoptosis (Cuconati *et al.*, 2002). While E1A expression is sufficient to promote apoptosis in rodent cells via CR1- and CR2-dependent unscheduled cell-cycle progression, apoptosis induced in human cells productively infected with an E1B1-9K mutant adenovirus additionally requires adenovirus DNA replication-induced activation of the cellular DNA damage checkpoint protein ATM (White *et al.*, 1991; Cuconati *et al.*, 2003), reminiscent of the capacity of E1A to sensitize cells to DNA damage-induced apoptosis (Attardi *et al.*, 2004). A consequence of combined E1A expression and ATM activation in adenovirus-infected human cells is the downregulation of cellular antiapoptotic BCL2 family protein MCL-1. Like apoptosis induced by UV- or γ-irradiation of cells (Nijhawan *et al.*, 2003), apoptosis triggered in E1B-19K mutant adenovirus-infected human cells also depends on downregulation of MCL-1 (Cuconati *et al.*, 2003). Moreover, identical to E1B-19K, antiapoptotic MCL-1 functions by binding to and inactivating proapoptotic BCL2 family proteins Bak and Bax. These findings reveal that E1B-19K functions as a viral mimic of antiapoptotic MCL-1, thereby permitting adenovirus to counteract an otherwise deleterious apoptotic response triggered by E1A and adenovirus infection-mediated MCL-1 down-regulation. The discovery that adenovirus E1B-19K functions as an antiapoptotic viral BCL-2 homolog created a new paradigm, which later revealed that viruses commonly encode such proteins that prevent detrimental premature death in infected cells.

Oncogenic mechanisms of the E1B-55K protein

A key insight into the mechanism by which E1B-55K blocks apoptosis in adenovirus E1 region-transformed cells came from the report that E1B-55K directly binds the important cellular protein p53 (Sarnow *et al.*, 1982), which had been discovered three years earlier through its interaction with SV40 large T-antigen (Lane and Crawford, 1979; Linzer and Levine, 1979). p53 was originally classified an oncoprotein because the tumour cell-derived *p53* cDNA used by many laboratories carried an unexpected cancer-associated point mutation that conferred weak oncogenic potential. Ten years after its initial discovery, however, *p53* was reclassified as a tumour suppressor gene when Bert Vogelstein and colleagues reported bi-allelic inactivation of *p53* in three human colon cancers (Baker *et al.*, 1989). It is now well established that *p53* is the most frequently mutated gene in human cancers and codes for a critical checkpoint protein that prevents oncogenic cellular transformation by permitting only normal, healthy cells to survive and proliferate (Levine, 2009; Carvajal and Manfredi, 2013; Surget *et al.*, 2013).

The p53 protein forms a tetrameric, sequence-specific transcription factor that activates or represses expression of target genes (Joerger and Fersht, 2008). It contains three protein domains involved in (i) DNA binding to response elements in target gene promoters, (ii) transactivation by recruitment of negative or positive transcriptional regulatory proteins, and (iii) oligomerization and post-translational modifications. In healthy cells, p53 is normally maintained at low levels in an inactive state (Levine, 2009). Downregulation and inhibition of p53 are largely mediated by the E3 ubiquitin ligase Mdm-2, which targets p53 for proteasomal degradation, blocks p53 transcriptional activity, and regulates p53 subcellular localization. A variety of stresses can induce p53 post-translational modifications, including phosphorylation, ubiquitination, acetylation, methylation, sumoylation, and neddylation, which relieve Mdm-2 inhibition and, in doing so, stabilize and activate p53.

A well-known stressor that induces p53 is DNA damage, which stimulates a response pathway that promotes p53 modification, stabilization, and activation (Levine, 2009). Activated p53 in turn can cause either cell-cycle arrest to allow cells time to repair minor DNA damage or apoptosis to kill cells having major DNA damage. When the function of the key checkpoint protein p53 is impaired, the consequence is unrepaired DNA damage and inappropriate cell-cycle progression, resulting in accumulation of mutations or genomic instability, either of which increases the risk of oncogenic cellular transformation. Abrogating this protective host response is also important during a productive viral infection because replicating viral DNA is recognized as damaged DNA and initiates a DNA damage response (Weitzman *et al.*, 2004). Another important stressor that can induce p53 is unscheduled RB1 inactivation, which allows E2F1 prematurely released from RB1 to stimulate transcription of the *p19ARF* gene (Zhang *et al.*, 1998). The *p19* gene product binds and inactivates Mdm-2 to stabilize and activate p53, leading to cell-cycle arrest or apoptosis (Stott *et al.*, 1998).

p53 induces apoptosis either by stimulating transcription of pro-apoptotic genes, like PUMA, Bid, Bax, Noxa, TRAILR2, CD95, Apaf1, and caspase-6 or by repressing transcription of anti-apoptotic genes, such as survivin (Fridman and Lowe, 2003). Though evidence suggests that p53 primarily induces apoptosis by transcriptional mechanisms, it can also promote the effect by transcription-independent mechanisms, such as binding and inhibition of antiapoptotic Bcl-XL and Bcl-2 or binding and activation of proapoptotic Bax and Bak (Speidel, 2010). In addition to inducing apoptosis, p53 also causes cell-cycle arrest by activating transcription of the *p21*$^{WAF1/CIP1}$ gene (el-Deiry *et al.*, 1993), which encodes the potent cyclin-dependent kinase inhibitor protein p21 that binds to cyclin-cdk2 and -cdk4/6 complexes and blocks their kinase activity. The consequence of p21-mediated cdk inhibition is to maintain RB1 in its active hypophosphorylated state that represses E2F and prevents cell-cycle progression.

Based on the regulation and function of p53 mentioned above, it is not surprising that the massive apoptotic cell death induced by E1A results from a dramatic increase in p53 protein levels caused by inappropriate E1A-mediated cell-cycle progression (Debbas and White, 1993; Lowe and Ruley, 1993). While E1B-19K primarily counteracts E1A-mediated p53-independent apoptotic pathways by inactivating pro-apoptotic Bak and Bax, E1B-55K blocks E1A-mediated p53-dependent apoptotic pathways by inactivating the p53 tumour suppressor. The redundant anti-apoptotic functions of E1B1-9K and E1B-55K underscore the critical importance of counteracting host antiviral apoptotic responses during an adenovirus infection.

Adenovirus E1B-55K directly binds to p53 and blocks its function through several distinct mechanisms. E1B-55K interacts with the p53 amino-terminal region, which consists of separate subdomains that bind to Mdm-2 or mediate transcriptional transactivation (Yew and Berk, 1992; Teodoro and Branton, 1997). The high affinity interaction of E1B-55K with p53 competitively inhibits Mdm2-mediated p53 degradation to further increase p53 levels induced by E1A (Levine, 1990), and additionally ablates the transcriptional transactivation function of p53 both by sterically hindering p53 binding to transcriptional coactivator proteins and by tethering p53 to the potent transcriptional repression domain present in E1B-55K (Yew and Berk, 1992; Yew *et al.*, 1994; Martin and Berk, 1998, 1999). The E1B-55K repression domain provides an absolutely essential function because two different E1B-55K mutants that lack a functional repression domain but retain the ability to bind p53 fail to inhibit p53 mediated transcriptional transactivation or to cooperate with E1A to promote cellular transformation (Teodoro and Branton, 1997). This finding indicates that adenovirus E1B-55K converts the p53 transcriptional transactivation function into a transcriptional repression function. When combined with the capacity of the p53-E1B-55K complex to stabilize p53 and to increase its binding more than 10-fold to cognate promoter elements of target genes, E1B-55K forms a highly abundant, repressive complex that potently abolishes expression of p53-responsive genes (Martin and Berk, 1998). In cells transformed by *E1A* and *E1B*, p53-E1B-55K complexes are also exported from the nucleus into the cytoplasm, where they accumulate to form large aggresomes (Zantema *et al.*, 1985; Liu *et al.*, 2005; Pennella *et al.*, 2010). While this cytoplasmic sequestration of p53 by E1B-55K would be anticipated to reduce p53 mediated transcriptional transactivation, it is insufficient to eliminate this activity or to promote cellular transformation in cooperation with E1A.

In human cells productively infected with adenovirus, E1B-55K also forms a complex with the E4orf6 protein expressed from the viral *E4* transcription unit (Sarnow *et al.*, 1984). The E1B-55K-E4orf6 complex binds p53 to promote its polyubiquitination and proteasomal degradation (Querido *et al.*, 2001). Interestingly, this viral complex similarly promotes degradation of important cellular dsDNA-break response proteins, including MRE11, RAD50, NBS1, DNA ligase IV, and Bloom helicase, which would otherwise interfere with adenovirus DNA replication by recognizing the termini of the linear viral genome and causing genome concatamerization (Weitzman *et al.*, 2004). The E1B-55K-E4orf6 complex targets these cellular factors for degradation through interaction with the cellular E3 ubiquitin ligase complex composed of Elongin B and C, cullin 5, and RBX1 (Querido *et al.*, 2001; Harada *et al.*, 2002; Stracker *et al.*, 2002; Schwartz *et al.*, 2008). Though its coding sequence generally is not retained in adenovirus-transformed rodent cells, the E4orf6 protein can bind and inhibit the function of p53 even in the absence of E1B-55K, and also cooperates with *E1A* or both *E1A* and *E1B* to induce or enhance cellular transformation, respectively (Moore *et al.*, 1996).

This cooperative interplay between adenovirus *E1A* and *E1B* represents a common theme among DNA tumour viruses as evidenced by the fact that the SV40 Large T-antigen and HPV E6 oncoproteins likewise bind and inactivate p53 in cells (Mietz *et al.*, 1992). The convergent functional evolution exposed by the common ability of four otherwise unrelated viral oncoproteins (adenovirus E1B-55K and E4orf6, SV40 Large T-antigen, and HPV E6) to abolish p53 function in cells underscores the central role of this activity in neoplastic cellular transformation induced by DNA tumour viruses.

Adenovirus E4orf1

The Ad9 *E4orf1* oncogenic determinant codes for a 125-residue polypeptide that is found at multiple cellular locations, including the plasma membrane (Frese *et al.*, 2003; Frese *et al.*, 2006), membrane vesicles (Chung *et al.*, 2007), cytoplasm (Weiss *et al.*, 1996), and nucleus (Chung *et al.*, 2007; Kumar *et al.*, 2014; Thai *et al.*, 2014). While the first known function of *E4orf1* was its role as the primary determinant of Ad9 induced tumourigenesis (Javier *et al.*, 1992; Javier, 1994; Weiss *et al.*, 1996; Thomas *et al.*, 1999; Thomas *et al.*, 2001), it was later also shown to be necessary for optimal replication of species C Ad5 in human epithelial cells (O'Shea *et al.*, 2005a; O'Shea *et al.*, 2005b; Thai *et al.*, 2014). Adenovirus *E4orf1* genes evolved from a cellular *dUTP pyrophosphatase* (*dUTPase*) gene (Weiss *et al.*, 1997b; Chung *et al.*, 2008), which codes for an essential homotrimeric enzyme of nucleotide metabolism that hydrolyzes dUTP to prevent its detrimental incorporation into replicating cellular DNA (Mol *et al.*, 1996). Nonetheless, E4orf1 lacks dUTPase catalytic activity, indicating functional divergence from dUTPase (Weiss *et al.*, 1997b), despite the fact that E4orf1 similarly forms a homotrimer and is predicted to share the conserved dUTPase protein fold (Chung *et al.*, 2008). Through divergent evolution, E4orf1 acquired several novel functions, including activation of cellular class IA PI3K, receptor tyrosine kinases (RTKs), and Myc (Frese *et al.*, 2003; O'Shea *et al.*, 2005a; Thai *et al.*, 2014), and also inactivation of certain cellular PDZ proteins to disrupt the tight junction (TJ) and cause a loss of apicobasal cell polarity (Latorre *et al.*, 2005). Among these E4orf1 functions, PI3K activation is currently the only known function required for E4orf1-induced cellular transformation and Ad9 induced mammary tumourigenesis (Javier *et al.*, 1991; Javier *et al.*, 1992; Javier, 1994; Thomas *et al.*, 2001; Frese *et al.*, 2003; Frese *et al.*, 2006; Kong *et al.*, 2014; Kumar *et al.*, 2014). In addition, both PI3K and Myc activation are known to enhance adenovirus virion production in permissive human cells (O'Shea *et al.*, 2005a; O'Shea *et al.*, 2005b; Thai *et al.*, 2014).

Studies of Ad9 E4orf1 mutant proteins expressed at wild-type E4orf1 levels but unable to promote Ad9-induced mammary tumours *in vivo* or to transform rodent cell lines *in vitro* proved instrumental in identifying functionally important E4orf1 protein elements (Lee *et al.*, 1997; Weiss *et al.*, 1997a; Chung *et al.*, 2007; Chung *et al.*, 2008). The work revealed two separate E4orf1 protein interaction elements, namely the PDZ domain-binding motif (PBM) and Domain 2 (D2), both of which are required for the oncogenic potential of E4orf1 (Lee *et al.*, 1997; Chung *et al.*, 2007). Notably, the E4orf1 PBM was the first known viral PBM (Javier and Rice, 2011). In the 125-residue polypeptide, E4orf1 PBM mutations cluster at the extreme carboxyl-terminus, whereas D2 mutations distribute over a wider central region of the protein. E4orf1 functions can depend on both the PBM and D2 (PI3K activation and transformation), only the PBM (TJ disruption and loss of apicobasal cell polarity), or only D2 (Myc activation and RTK activation). Both the PBM and D2 mediate E4orf1 binding to distinct subsets of cellular factors. D2 is required for E4orf1 binding to several unidentified cellular phosphoproteins (p70, p150, p190) (Chung *et al.*, 2007), though at least one D2 residue (D65) is required for binding to and activation of cellular Myc (Thai *et al.*, 2014). As its name implies, the PBM of E4orf1 instead mediates binding to a select group of cellular PDZ domain-containing proteins (Lee *et al.*, 1997; Glaunsinger *et al.*, 2000; Lee *et al.*, 2000; Glaunsinger *et al.*, 2001; Latorre *et al.*, 2005).

The PDZ domain is an evolutionarily conserved protein-protein interaction module approximately 90 amino-acid residues in length (Ye and Zhang, 2013). The name PDZ derives from the first three identified PDZ domain-containing proteins: **P**SD95, **D**lg, and **Z**O-1. PDZ domains usually bind target proteins via a PBM, which is short peptide sequence located at the extreme carboxyl terminus. Multiple types of PBMs having different consensus sequences are now recognized, and the first one recognized was the type I PBM with the consensus sequence -x-S/T-x-Φ-COOH, where x is any amino-acid residue and Φ is a hydrophobic residue. A functional type I PBM was discovered at the carboxyl-terminus of all adenovirus E4orf1 proteins, as well as the HTLV1 Tax oncoprotein and all high-risk HPV E6 oncoproteins (Lee *et al.*, 1997). The PDZ domain is frequently found in cytoplasmic and membrane adapter proteins that regulate a variety of cellular processes, such as cell signaling, cell-cell junction formation, and cellular polarity (Javier, 2008). In addition to often having multiple PDZ domains, PDZ proteins often contain a variety of other protein interaction domains, such as Src homology region-3 (SH3), Lin-2 and Lin-7 (L27), and leucine-rich repeat domains. The E4orf1 PBM interacts with five different cellular PDZ proteins, including Discs large 1 (Dlg1), Multi-PDZ domain Protein 1 (MUPP1), PALS1-Associated Tight Junction protein (PATJ), Membrane-Associated Guanylate kinase Inverted 1 (MAGI-1), and Zona Occludens protein 2 (ZO-2) (Glaunsinger *et al.*, 2000; Glaunsinger *et al.*, 2001; Latorre *et al.*, 2005; Lee *et al.*, 2000; Lee *et al.*, 1997). In polarized epithelial cells, Dlg1 localizes to the adherens junction (AJ), while MUPP1, PATJ, MAGI-1 and ZO-2 localize to the tight junction (TJ).

The fact that the dUTPase enzyme is a functional homo-trimer led to the discovery that E4orf1 likewise exists as a homo-trimer, and also differs from dUTPase by additionally existing as a monomer in cells (Chung *et al.*, 2008). Formation of the E4orf1 homo-trimer depends on a carboxyl-terminal trimerization (TRI) element consisting of residues V-K-I that overlap with the PBM. More importantly, the E4orf1 monomer and trimer interact with separate cellular proteins to differentially alter their subcellular localization and to exert distinct functions. For example, the E4orf1 monomer binds specifically to PDZ proteins MUPP1, PATJ, MAGI-1 and ZO-2 at the plasma membrane to mislocalize them in the cytoplasm (Glaunsinger *et al.*, 2000; Lee *et al.*, 2000; Glaunsinger *et al.*, 2001; Latorre *et al.*, 2005; Chung *et al.*, 2008), whereas the E4orf1 trimer binds specifically to the PDZ protein Dlg1 to promote its translocation from the cytoplasm to the plasma membrane (Frese *et al.*, 2006; Kong *et al.*, 2014; Kumar *et al.*, 2014). The distinct functions mediated by these differential changes in subcellular localization are E4orf1 monomer-specific disruption of the tight junction and loss of apicobasal cell polarity in epithelial cells and E4orf1 trimer-specific PI3K activation. The former function could be envisioned to facilitate the spread of progeny viruses (Latorre *et al.*, 2005). Findings further suggest that the D2 element only exists within the E4orf1 trimer. Supporting this idea, wild-type E4orf1, PBM mutant E4orf1, and D2 mutant E4orf1 proteins associated with membrane vesicles, whereas PBM/D2 or PBM/TRI double mutant E4orf1 proteins did not, suggesting that either the PBM or D2 is sufficient to target E4orf1 to membrane vesicles and that the ability of D2 to mediate this effect depends on E4orf1 trimerization (Chung *et al.*, 2007). Moreover, molecular modeling of E4orf1 to the crystal structure of human dUTPase homo-trimer predicts that E4orf1 homo-trimerization positions six of the seven D2 residues in close proximity at each of the three subunit interfaces, suggesting that the functional D2 element is formed when three E4orf1 monomers assemble into a homo-trimer.

PI3K is a lipid kinase discovered through its interaction with the middle T antigen oncoprotein of mouse polyomavirus (Whitman *et al.*, 1985). Functional PI3K is a heterodimer formed by binding of the p85 regulatory subunit to the p110 catalytic subunit (Cantley, 2002). In the cytoplasm, p85 in the PI3K heterodimer inhibits the lipid kinase activity of p110. Upon activation, RTKs and Ras bind to cytoplasmic PI3K and recruit it to the plasma membrane (Blume-Jensen and Hunter, 2001). These effects serve both to relieve the enzymatic inhibition of p110 in the PI3K heterodimer and to place PI3K into close contact with its lipid substrate phosphatidylinositol-4,5-bisphosphate (PIP2). By phosphorylating PIP2, PI3K creates the second messenger phosphatidylinositol 3,4,5-trisphosphate (PIP3), which in turn binds the PI3K effector proteins Akt and PDK1 to recruit them to the plasma membrane. At the membrane, PDK1 and the mTORC2 complex activate Akt by phosphorylating it on threonine 308 and serine 473 (Bhaskar and Hay, 2007), respectively. Activated Akt stimulates numerous downstream effectors involved in regulating important cellular processes. Consequently, this signalling cascade is commonly subverted by pathogenic human viruses to enhance virus-host interactions and viral disease (Cooray, 2004), and also represents one of the most frequently dysregulated pathways in human cancers (Yuan and Cantley, 2008).

The ability of E4orf1 to activate PI3K was discovered by screening E4orf1-expressing rodent cells for activated components of signalling pathways commonly involved in oncogenesis (Frese *et al.*, 2003). The results revealed that E4orf1 induces constitutive, growth factor-independent activation of class IA PI3K and the downstream Akt/mammalian target of rapamycin (mTOR) pathway, but not components of several other signaling pathways, such as βcatenin/TCF, JNK, NFκB, Notch, and Stat3. PI3K activation is also a conserved function for E4orf1 proteins encoded by species A-D adenoviruses (Frese *et al.*, 2003; Kumar *et al.*, 2014). The PI3K/Akt/mTOR pathway promotes cell survival, growth, and proliferation by inactivating pro-apoptotic Forkhead transcription factors, activating the protein synthesis and cell proliferation stimulator p70S6-kinase (S6K), and downregulating cyclin-dependent kinase inhibitor $p27^{Kip1}$. These activities were also detected in E4orf1-expressing cells and reversed by treatment with the PI3K inhibitor LY294002 (Frese *et al.*, 2003). Furthermore, the PI3K inhibitor LY294002 or the mTORC1 inhibitor rapamycin, each of which selectively blocks PI3K/Akt/mTOR signalling, abrogated both anchorage-independent growth and focus formation by E4orf1-expressing cells and eliminated the transformed growth properties of an Ad9 induced mammary tumour cell line. These findings directly implicated dysregulated PI3K/Akt/mTOR signalling in E4orf1-induced cellular transformation and Ad9-induced mammary tumourigenesis.

A subsequent study reported by O'Shea and colleagues showed that, in human cells productively infected with Ad5, stimulation of PI3K signalling by E4orf1 increases protein synthesis by activating mTOR

(O'Shea *et al.*, 2005a). mTOR is known to be activated by the Ras-family GTPase Rheb, which is negatively regulated by the GAP activity of the TSC1:TSC2 tumour suppressor complex. Findings showed that E4orf1 activates mTOR by inhibiting the TSC1:TSC2 complex. Activated mTOR in turn stimulates protein synthesis by phosphorylating and inhibiting 4EBP, which blocks the translation initiation by cap-binding translation initiation factor eIF4E, and also by phosphorylating and activating S6K to enhance ribosome biosynthesis and expression of translation factors (Zoncu *et al.*, 2011). Furthermore, E4orf1-induced PI3K activation also promotes quiescent Ad5-infected cells to progress into S phase independently of E1A-mediated RB1 inactivation.

The dual PBM and D2 dependence of E4orf1-induced PI3K activation suggested that one or more PBM- and D2-interacting proteins mediate this function. Frese *et al.* identified a crucial cellular target for the PBM (Frese *et al.*, 2006). This study showed that *Dlg1*$^{+/+}$ mouse embryo fibroblasts (MEF) but not *Dlg1*$^{-/-}$ mutant MEF are able to support E4orf1-induced PI3K activation and cellular transformation, indicating a requirement for PDZ protein Dlg1 for these E4orf1 functions. Moreover, the failure of *Dlg1*$^{-/-}$ mutant MEF to support E4orf1-induced PI3K activation is specific to a *Dlg1* deficiency because MUPP1$^{-/-}$ or MAGI-1$^{-/-}$ mutant MEF lack this defect.

Dlg1 is a suspected tumour suppressor and mammalian homolog of the *Drosophila* tumour suppressor discs large, and the *Dlg1* gene product is classified as membrane-associated guanylate kinase (MAGUK) protein. The fact that Ad9 E4orf1, high-risk HPV E6 and HTLV-1 Tax independently evolved to target Dlg1 provide compelling evidence implicating this PDZ protein in human cancers (Lee *et al.*, 1997; Roberts *et al.*, 2012; James and Roberts, 2016). An intriguing possibility is that the ability of high-risk HPV E6 and HTLV1 Tax, as well as influenza A virus NS1 that likewise targets Dlg1, to activate PI3K similarly depends on Dlg1 (Lee *et al.*, 1997; Liu *et al.*, 2001; Jeong *et al.*, 2005; Contreras-Paredes *et al.*, 2009; Bellon *et al.*, 2010; Liu *et al.*, 2010a; Saito *et al.*, 2010; Spangle and Munger, 2010, Golebiewski *et al.*, 2011; Oteiza and Mechti, 2011; Spangle *et al.*, 2012; Fan *et al.*, 2013; Spangle *et al.*, 2013). In addition to three PDZ domains, Dlg1 contains L27, SH3, and guanylate kinase-homology (GuK) domains, as well as alternative-splice insertion elements I1, I2, and I3 (Lue *et al.*, 1994; Javier, 2008; Roberts *et al.*, 2012). Dlg1 lacks catalytic activity and its domains/elements instead serve as protein-protein interaction modules, consistent with the established function of Dlg1 as an adaptor/scaffolding protein. Protein regions in Dlg1 required for the ability to support E4orf1-induced PI3K activation were identified by reconstituting Dlg1$^{-/-}$ mutant MEF with *Dlg1* deletion mutants missing one or more domain/element. Results demonstrated a requirement for several Dlg1 regions that included PDZ1-2, U3 (unique sequences between PDZ2 and PDZ3), SH3, and I3 (Frese *et al.*, 2006). The roles of these protein regions in E4orf1-induced PI3K activation were linked to an ability of Dlg1 either to bind E4orf1 (PDZ1-2) or to promote translocation of the Dlg1:E4orf1 complex to the plasma membrane (U3, SH3, and I3). These findings surprisingly revealed that the Dlg1 tumour suppressor also functions like the product of a proto-oncogene by translocating the Dlg1:E4orf1 complex to the plasma membrane to activate PI3K. An intriguing possibility is that E4orf1 both hijacks the Dlg1 oncogenic function to dysregulate PI3K signalling and inactivates the Dlg1 tumour suppressor function.

The mechanism by which the Dlg1:E4orf1 complex mediates PI3K activation was recently exposed through proteomic analyses that identified the PI3K regulatory subunits p85α and p85β as new cellular targets of E4orf1 (Kong *et al.*, 2014). This led to the discovery that E4orf1 in the Dlg1:E4orf1 complex separately also directly binds to PI3K, thereby forming a cytoplasmic Dlg1:E4orf1:PI3K ternary complex. Dlg1 in the ternary complex functions in an I3 element-dependent fashion to promote translocation of PI3K tethered to E4orf1 to the plasma membrane, while E4orf1 in the complex is presumed to stimulate PI3K p110 catalytic activity by relieving repression by the p85 regulatory subunit (Frese *et al.*, 2003). These effects place activated PI3K in contact with its membrane lipid substrates to allow production of PIP3, which in turn constitutively activates Akt and promotes oncogenic cellular transformation. This work was conducted in human mammary epithelial cells (MCF10A) stably expressing Ad9 E4orf1, but the ternary complex likewise mediates PI3K activation in MCF10A cells productively infected with Ad9. Moreover, the novel strategy of ternary complex-dependent PI3K dysregulation and oncogenic transformation demonstrated for species D Ad9 E4orf1 stably expressed in MCF10A cells is conserved for adenovirus species A-C E4orf1 proteins, as is ternary complex-dependent PI3K dysregulation in MCF10A cells productively infected with each of the respective adenovirus species (Kumar *et al.*, 2014).

Nonetheless, while PI3K activation is required for oncogenic cellular transformation induced by E4orf1, it may not be sufficient as evidenced by the fact that expression of a constitutively activated PI3K or PKB mutant fails to recapitulate E4orf1-induced cellular

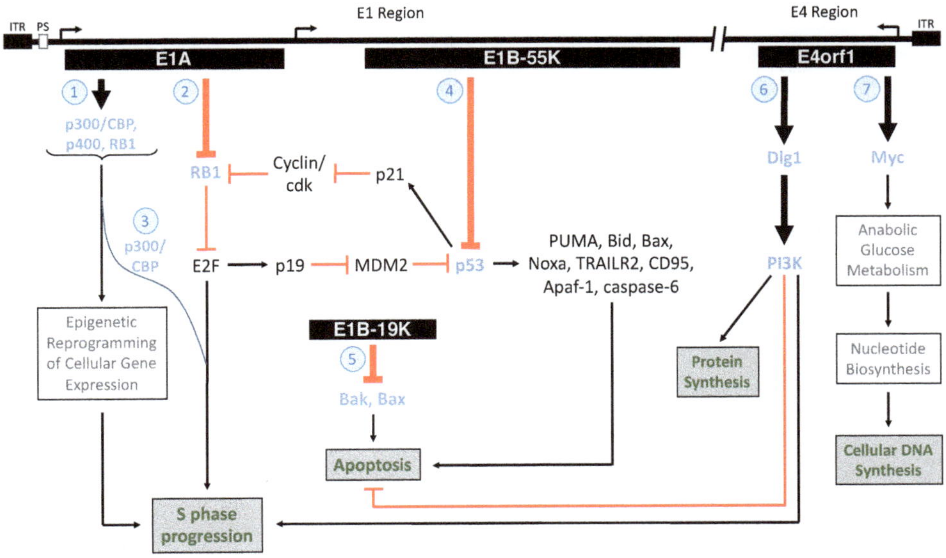

Figure 12.1. Key transforming activities for the primary oncogenic determinants (E1A, E1B, E4orf1) of tumourigenic adenoviruses. In step 1, the crucial 80-residue amino terminal region of E1A mediates binding to cellular p300/CBP and p400. These interactions cause global epigenetic reprogramming of cellular gene expression by activating genes that promote cell-cycle progression and cell growth and by repressing genes that promote inflammatory and pathogen responses or cell differentiation. In step 2, the crucial LxCxE motif in E1A CR2 mediates binding to the cellular RB1 tumour suppressor. The interaction releases E2F from repressive RB1 complexes, allowing free E2F to activate S phase-specific cellular genes. This interaction additionally contributes to global epigenetic reprogramming of cellular gene expression in step 1 through formation of trimeric p300:E1A:RB1 complexes that repress transcription of cellular genes that promote cell death or block cell proliferation. Conversely, in step 3, E1A-p300 complexes formed in step 1 additionally function in combination with free E2F in step 2 to activate transcription of E2F-responsive genes. These combined E1A activities cause unscheduled passage of cells through the restriction point, completion of the entire cell cycle, and increased cellular proliferation. However, the unscheduled E2F activation triggers a host response that activates p19, which in turn inactivates MDM2, thereby stabilizing and activating the cellular p53 tumour suppressor. In step 4, E1B-55K binds and inactivates p53 to repress transcription of the p21$^{WAF1/CIP1}$ gene and multiple pro-apoptotic cellular genes that would otherwise inhibit cell cycle progression or induce apoptosis, respectively. In step 5, E1B-19K binds and inactivates proapoptotic cellular proteins Bak and Bax, further ensuring the survival of E1A-expressing cells that express high levels of p53. Thus, *E1A and E1B* cooperate to induce cellular transformation and undifferentiated sarcomas in experimental animals. In step 6, the E4orf1 oncogenic determinant of Ad9 directly binds both Dlg1 and PI3K to form Dlg1:E4orf1:PI3K ternary complexes that constitutively activate PI3K, which promotes cellular transformation and mammary tumourigenesis in experimental animals by increasing protein synthesis, inducing S-phase progression, and blocking apoptosis. In step 7, E4orf1 also binds and activates cellular Myc in the nucleus to stimulate anabolic glucose metabolism and nucleic acid biosynthesis, thereby stimulating DNA synthesis, though it is not yet known whether this activity contributes to cellular transformation.

transformation of rodent cells (Frese *et al.*, 2003). Consequently, other E4orf1 functions in concert with PI3K activation may be required to transform cells. Other potentially oncogenic activities of E4orf1 are discussed below.

An exciting novel nuclear function for E4orf1 was recently reported by Thai *et al.* (2014). This group showed that Ad5 E4orf1 elevates nuclear Myc protein levels and also forms E4orf1:Myc complexes to activate Myc and stimulate transcription of specific Myc-responsive cellular glycolytic genes that promote anabolic glucose metabolism. In Ad5-infected cells, the resulting glucose metabolites shunted through the nonoxidative arm of the pentose phosphate pathway enhanced progeny virion production by increasing nucleotide biosynthesis required for optimal viral DNA replication. E4orf1-induced Myc upregulation and activation were shown to depend on one D2 residue (D65), which is additionally required for E4orf1-induced PI3K activation. Thus, it will be important to determine whether this important new E4orf1 function, in combination with PI3K activation, is required to promote oncogenic cellular transformation. Another interesting idea is that D2 generally functions to stabilize certain cellular targets given the reported E4orf1 D2-dependent Myc upregulation and E4orf1-mediated upregulation the PI3K subunits p110α, p85α, and p85β (Kong *et al.*, 2014; Thai *et al.*, 2014).

A more recent proteomics study by Kong *et al.* identified three RTKs -- epidermal growth factor receptor (EGFR), insulin receptor (InsR), and insulin-like growth factor 1 receptor (IGF1R) -- as new cellular targets of E4orf1 in MCF10A cells (Kong *et al.*, 2015). The results suggested that E4orf1 targets a Dlg1:EGFR complex that normally exists in MCF10A cells, and induces EGFR to associate with both InsR and IGF1R, resulting in formation of an E4orf1:Dlg1:EGFR:InsR:IGF1R quinary complex. Evidence further suggested that E4orf1 in the quinary complex promotes ligand-independent activation of the three RTKs, as well as EGFR-mediated stimulation of the downstream Ras/Mek/Erk pathway. Collectively, these quinary complex-mediated activities, in combination with ternary complex-mediated PI3K activation, function to promote constitutive Myc expression in MCF10A cells. This dysregulated Myc expression is postulated to enhance viral DNA replication and virion production in adenovirus-infected cells to ensure optimal Myc protein levels for formation of nuclear E4orf1:Myc complexes. Hence, it will be important to investigate a potential role for the quinary complex-dependent Myc upregulation in cellular transformation induced by E4orf1.

Model

The illustration presented in Figure 12.1 summarizes the known molecular mechanisms of the adenovirus E1A, E1B, and E4orf1 oncogenes described above.

Adenovirus vectors

Adenovirus vectors are used to correct genetic deficiencies (gene therapy), to induce protective immunity (vaccines), and to treat cancer, though clinical trials have employed these viral vectors primarily as cancer treatments (Wold and Toth, 2013). Adenovirus has many attractive features that prompted its use as a vector, including stable virions capable of infecting quiescent and dividing cells derived from many different tissues, ease of genome manipulation, virus propagation to high concentrations, well understood gene functions and life cycle, episomal maintenance, rare genome integration into the host chromosome, and safety proven in many clinical trials and by live virus vaccination of millions of soldiers.

There are two general classes of adenovirus vectors, replication competent and replication defective, which are employed either to exploit the lytic and immunogenic consequences of adenovirus replication (e.g., cancer therapy and vaccination) (Cerullo *et al.*, 2012) or to deliver transgenes (e.g., gene therapy) (Wold and Toth, 2013; Capasso *et al.*, 2014; Crystal, 2014), respectively. Replication-competent adenovirus vectors kill infected cells through lytic viral replication. This effect may be further amplified in tumours when progeny viruses released from lysed cells infect and kill neighbouring cells. In addition, the vast majority of replication-competent vectors used to combat cancer have been engineered with various genetic alterations to maintain robust lytic viral replication in cancer cells yet diminish infectivity and replication in normal cells.

Replication-defective vectors have some or all of the essential viral genes deleted. For example, first-generation vectors are missing essential viral *E1A* and *E1B*, and also non-essential *E3*. A transgene expression cassette not exceeding the total size of the deleted viral genes may be inserted into the vector genome. Due to deletion of essential *E1A* and *E1B* in first generation vectors, they must be propagated in complementing E1A- and E1B-expressing cell lines, such as the original HEK293 cells or the improved PER.C6 and N52.E6 cells (Graham *et al.*, 1977; Fallaux *et al.*, 1998; Schiedner *et al.*, 2000). However, viral gene expression by first-generation adenovirus vectors is leaky as high multiplicities of infection permit early and late viral gene expression that would otherwise depend on *E1A* (Yang *et al.*, 1994; Shimizu *et al.*, 2011). This viral gene expression, in combination with high vector doses generally required for effective gene therapy, stimulates detrimental host immune responses that can limit the usefulness of first-generation vectors. The host responses cause toxic side effects, shorten the time of transgene expression, and decrease effectiveness of readministered vectors (Nazir and Metcalf, 2005). Illustrating the potential toxicity, the death of one gene therapy patient was attributed to a cytokine storm triggered by administration of a high intravenous dose of a first generation vector (Raper *et al.*, 2002; Raper *et al.*, 2003). This tragic event prompted development of improved replication-defective vectors, such as the second-generation vectors (Koup *et al.*, 2009; Osada *et al.*, 2009), which are first-generation vectors additionally having viral *E4* or *E2*, or both deleted. Second generation vectors must be propagated on complementing cell lines that express *E1A*, *E1B*, and also *E2* and/or *E4* (Gorziglia *et al.*, 1999).

An advancement in replication-defective vectors was the development of helper-dependent or gutless adenovirus vectors, which have a large 37-kb cloning capacity because the vector genome contains only three small essential non-coding elements -- both terminal origins of DNA replication (ITRs) and the DNA packaging sequence (Brunetti-Pierri and Ng,

2011). Given the minimal DNA size requirement to package DNA into adenovirus particles, the genome of gutless vectors may require eukaryotic stuffer DNA in addition to the transgene for efficient assembly into particles. The propagation of gutless vectors depends on co-infection of cells with a helper adenovirus that supplies necessary viral early and late gene products. Using a helper adenovirus with the packaging sequence flanked by loxP sites and performing the co-infection in a cell line that conditionally expresses Cre recombinase allows for specific excision of the helper virus packaging sequence and selective assembly and enrichment of gutless vectors (Parks *et al.*, 1996; Hardy *et al.*, 1997). Further purification to remove helper virus is achieved by centrifugation on a caesium chloride density gradient or by anion exchange and size exclusion chromatography (Dormond and Kamen, 2011; Palmer and Ng, 2011). The absence of protein-encoding viral genes in gutless vectors reduces immunogenicity and toxicity and significantly lengthens transgene expression compared to both first- and second-generation vectors (Weaver *et al.*, 2009; Brunetti-Pierri and Ng, 2011; Palmer and Ng, 2011).

Adenovirus vectors to treat human cancers

Conditionally replication-competent vectors

Numerous replication-competent oncolytic adenovirus vectors are under evaluation as cancer treatments in experimental animal studies and clinical trials (Alemany, 2012; Cerullo *et al.*, 2012; Chen *et al.*, 2014; Cheng *et al.*, 2015; Larson *et al.*, 2015). Adenovirus vectors used to combat cancer are most often injected directly into tumours (Young *et al.*, 2006). By definition, oncolytic vectors replicate in and kill cancer cells, and the effect is enhanced by released progeny viruses infecting and destroying additional cancer cells in the tumour. Adenovirus replication and cell killing is also by nature more efficient in cancer cells than in normal cells. Cancer cells have mutations that dysregulate cellular gene expression and promote uncontrolled cell-cycle progression and proliferation. The consequence is that cancer cells perpetually exist in a state highly permissive for adenovirus replication, whereas normal human cells typically exist in a quiescent state and therefore are nonpermissive for adenovirus replication until viral gene expression induces a permissive state (Gonzalez *et al.*, 2006; Vaillancourt *et al.*, 2005). In addition, oncolytic adenovirus vectors are also commonly genetically modified to further increase selective killing of cancer cells. Direct killing of tumour cells by the viral vector unlikely represents the sole mechanism for eliminating cancer cells. For example, it has been proposed that anti-tumour immune responses elicited by virus-induced cell lysis play an even greater role in this process (Bell, 2014).

ONYX-015 is the Ad5 *E1B-55K* deletion mutant *dl1520* and the first oncolytic adenovirus vector investigated for the treatment of human cancer (Barker and Berk, 1987; Kirn, 2001). ONYX-015 replicates efficiently in most cancer cells and poorly in normal cells. Moreover, the efficient replication of ONYX-015 correlate with the presence of inactivating p53 mutations in cancer cells (Bischoff *et al.*, 1996). Given that adenovirus E1B-55K functions in part to inactivate p53, the findings were interpreted as indicating that cancer cells lacking functional p53 complement the E1B-55K defect of ONYX-015, thereby imparting a selective capacity to destroy p53-deficient cancer cells via lytic viral replication. However, this idea ultimately proved incorrect when the selective cancer cell-killing phenotype of ONYX015 was instead attributed to the capacity of cancer cells to complement the defect of ONYX015 for a separate E1B-55K function, namely increasing late viral gene expression by exporting nuclear, late viral mRNAs to the cytoplasm (O'Shea *et al.*, 2004). Thus, ONYX015 replicates efficiently in many types cancer cells and does so by a mechanism that is independent of a p53 functional deficiency. Nonetheless, multiple phase I and II clinical trials involving more than 300 patients have investigated ONYX015 as a treatment for several types of cancer (Wold and Toth, 2013; Larson *et al.*, 2015). In these studies, ONYX015 was injected repeatedly into tumours, either alone or in combination with chemotherapy. The vector replicated in tumour cells but not normal cells and, while toxicity was limited, reductions in tumour size were variable and generally modest. Nonetheless, stronger effects were generally obtained when treatment with ONYX-015 was combined with chemotherapy (Bressy and Benihoud, 2014). In a phase I/II clinical study, refractory colorectal metastases in the liver of 35 patients that had failed prior 5-fluorouracil/leucovorin chemotherapy were treated with ONYX-015 infused into the hepatic artery in conjunction with 5-fluorouracil/leucovorin chemotherapy (Reid *et al.*, 2002; Sze *et al.*, 2003; Reid *et al.*, 2005). While tumour masses acutely increased following vector infusion, they later regressed in 7 of 11 patients. It was speculated that the vector induced acute tumour enlargement by provoking an inflammatory response that subsequently contributed to tumour regression. Notably, the median survival time for patients with stable disease or partial responses was 19 months, and for other patients was 6 months. In a different phase I/II clinical study examining 21 patients with locally advanced adenocarcinoma of the pancreas or with metastatic disease, ONYX-015 treatment combined with gemcitabine chemotherapy resulted in partial regression of injected tumours in two patients, minor responses in two patients, and stable disease in six patients (Hecht *et al.*, 2003).

In phase I, II, and III clinical trials in China, the H101 (Oncorine) vector, having an E1B-55K defect similar to that of ONYX-015, produced significant anti-tumour responses, but no serious adverse events and little toxicity (Chen *et al.*, 2014). In 2005, H101 was approved in China for treatment of head and neck squamous cell carcinoma (HNSCC) (Jia and Kling, 2006). The strongest results were obtained when patients with advanced nasopharyngeal carcinoma were treated with H101 in combination with cisplatin/5-fluorouracil chemotherapy. This represents the first approval of any oncolytic vector for commercial use, and by 2009, more than 600 patients in clinical trials had been treated with H101 (Yu and Fang, 2007).

Another ONYX-015-related vector, Ad5-CD/TKrep, was evaluated in clinical trials for treatment of prostate cancer (Freytag *et al.*, 2007). This vector has a CD-TK fusion protein expression cassette inserted into an *E1B-55K* deletion site. CD-TK is a fusion of two prodrug-converting enzymes, CD *(Escherichia coli* cytosine deaminase) and TK (herpes simplex virus type 1 thymidine kinase) that convert non-toxic 5-fluorocytosine to toxic 5-fluorouracil or non-toxic ganciclovir/valganciclovir to toxic monophosphate forms, respectively. The products of the CD-TK enzymatic activities block DNA synthesis in infected tumour cells, as well as in uninfected tumour cells (bystander effect), and also sensitize the tumour cells to radiotherapy. An Ad5-CD/TKrep-related vector has an improved yeast CD-TK expression cassette inserted into an *E1B-55K* deletion site and an adenovirus death protein (ADP) expression cassette inserted into an E3 deletion site (Barton *et al.*, 2006). High ADP expression promotes efficient cell lysis at the late stage of an adenovirus infection to increase cell-cell spread of progeny vectors within tumours (Doronin *et al.*, 2000; Doronin *et al.*, 2003;). Together with radiation therapy and dual prodrug treatments, injection of these vectors into prostate tumours was safe and also increased patient survival compared to controls (Freytag *et al.*, 2002; Freytag *et al.*, 2003).

Fibre receptor proteins present in the adenovirus virion mediate binding to cellular attachment receptors on the surface of cells to initiate a viral infection. An infectivity-enhanced adenovirus vector is defined as having a genetically-modified fibre protein that interacts with novel cellular receptor to target the vector to new cell types (Yamamoto and Curiel, 2010; Pesonen *et al.*, 2011). The Ad5 fibre proteins normally mediate virion binding to cellular coxsackievirus and adenovirus receptor (CAR) to initiate a viral infection. However, many tumours types (e.g., ovary and bladder) express low CAR protein levels, which preclude the use of oncolytic Ad5 vectors for treating such cancers. In contrast, many cancer cells abundantly express αvβ integrins that interact with the peptide motif RGD, as well as desmoglein 2 that is the cellular receptor for the fibre protein of species B adenovirus type 3 (Ad3) (Krasnykh *et al.*, 2000; Tuve *et al.*, 2006; Wang *et al.*, 2011). These observations prompted construction of two different Ad5 vectors, one with a Ad5 fibre protein modified to contain the integrin-binding RGD motif and the other with a chimeric Ad5/Ad3 fibre protein able to bind desmoglein 2 (Krasnykh *et al.*, 1996). These infectivity-enhanced vectors, which are able to target cellular receptors highly expressed in many cancer cells, were additionally engineered with the CR2 *E1A* deletion in the oncolytic Ad5-D24 vector to create the Ad5-D24-RGD and Ad5/3Cox2L-D24 vectors. Considering that CR2 mutant E1A fails to inactivate RB1 and that RB1 is inactivated or the or RB1 pathway is defective in most cancer cells, Ad5 vectors carrying the CR2 *E1A* mutation are able to replicate lytically in the majority of cancer cells but not in normal cells. In clinical trials investigating these dual infectivity-enhanced/D24 Ad5 vectors, evidence for anti-cancer activity without significant adverse events has been reported (Kimball *et al.*, 2010; Pesonen *et al.*, 2010).

The secreted cytokine granulocyte-macrophage colony-stimulating factor (GMCSF) is an important growth factor and immune modulator of hematopoietic cells, including dendritic cells (DCs), and has emerged as the leading cytokine in the virotherapy field (Larson *et al.*, 2015). A *GM-CSF* expression cassette has been inserted into the *E3* deletion site of oncolytic adenovirus vectors, such as the Ad5 infectivity-enhanced and D24 vectors described above. Considering the immunosuppressive environment that typically exists within a tumour, GM-CSF expressed by an oncolytic vector replicating in cancer cells could be envisioned to overcome the immunosuppression by recruiting DC progenitors and inducing their differentiation into mature DCs, which after uptake and processing of tumour antigens, would migrate to local lymph nodes to stimulate antitumour responses. Phase I clinical trials with the GM-CSF-expressing Ad5/3D24-GMCSF vector (ONCOS-102) demonstrated that it was safe and stimulated an enhanced immune response with significant correlation between anti-viral and anti-tumour T-cells as well as progressive tumour infiltration of CD8 lymphocytes concomitant with systemic induction of tumour-specific CD8 T-cells (Kanerva *et al.*, 2013; Vassilev *et al.*, 2015).

Replication-defective vectors that express wild-type p53

The majority of human cancers carry inactivating *p53* mutations or defects in the *p53* pathway, and expression of wild-type p53 in *p53*-deficient cancer cells triggers apoptosis or blocks cell-cycle progression (Chen *et al.*, 2014). The restoration of functional *p53* in cancer cells additionally produces a bystander effect that inhibits tumour angiogenesis and stimulates anti-tumour immune responses (Nishizaki *et al.*, 1999; Waku *et al.*, 2000). Hence, it is not surprising that most clinical studies with adenovirus vectors have examined two closely related replication-defective, first-generation Ad5 vectors, Gendicine and Advexin, which have a wild-type *p53* expression cassette inserted into the *E1* region deletion (Wold and Toth, 2013). The main difference between Advexin and Gendicine is the use of a CMV or Rous sarcoma virus promoter, respectively, to drive *p53* expression.

Clinical trials have evaluated the Advexin vector for treating numerous types of cancers (Lane *et al.*, 2010). The findings revealed that Advexin is safe and had minor durable anti-tumour activity. For example, in a Phase I/II study, ten patients with chemoradiation-resistant advanced esophageal squamous cell carcinoma were treated every 28 days with intratumoural injections of Advexin on days 1 and 3 for up to five cycles (Shimada *et al.*, 2006). *p53* transgene expression was detected in tumour biopsies from all patients, and was accompanied by increased p21 and MDM2 mRNA levels in all but one case. Administration of multiple courses of Advexin was both feasible and well tolerated. Local tumour responses revealed stable disease in nine cases and progressive disease in one case, and three patients demonstrated absence of disease upon repeat biopsies. In a Phase II study, 13 patients with HNSCC of the oral cavity, oropharynx, larynx, or hypopharynx received surgery and perioperative Advexin injections in the primary tumour bed and the ipsilateral neck (Yoo *et al.*, 2009). Three patients received injections in the contralateral neck. The results suggest that the therapy was feasible and that the estimate for 1-year progression-free survival was 92%. Furthermore, two phase III trials examined Advexin treatment alone and Advexin treatment combined with methotrexate chemotherapy in advanced recurrent HNSCC (Nemunaitis *et al.*, 2009; Nemunaitis and Nemunaitis, 2011). Patients with tumours lacking p53 expression or expressing wild-type p53 showed an increase in survival time of 7.2 months, whereas patients with tumours expressing oncogenic mutant p53, presumably the dominant-negative inhibitor form of p53 that would inactivate wild-type p53 expressed by Advexin, showed a lower increase in survival time of 2.7 months. The findings indicate that expression of mutant p53 in recurrent HNSCC predicts reduced efficacy for Advexin treatment.

Notably, in 2003, Gendicine in combination with chemotherapy was approved by the State Food and Drug administration of China (SFDA) for the treatment of HNSCC, making it the first approved gene therapy product for commercial use anywhere (Pearson *et al.*, 2004). Reports of clinical trials in China demonstrate that Gendicine is safe and has significant anti-cancer activity, particularly in combination with radiotherapy and chemotherapy (Chen *et al.*, 2014). More than 7,000 patients have now been treated with this vector (Peng, 2005; Huang *et al.*, 2009; Ma *et al.*, 2009).

The replication-defective, first-generation Ad5 vector CTL-102 expresses the *E. coli* enzyme nitroreductase from the CMV promoter. Nitroreductase converts the weak alkylating agent CB1954 prodrug into a potent bifunctional alkylating agent that induces apoptosis in cells by crosslinking cellular DNA (Palmer *et al.*, 2003). The toxic product of this reaction is also cell permeable and provides a powerful bystander effect. In a phase I/II clinical trial evaluating CTL-102 inoculation into localized prostate cancers combined with prodrug treatment, minimal toxicity and preliminary evidence of antitumour activity, including an increase in time to 10% prostate-specific antigen (PSA) progression in 6 of 18 patients, were observed (Patel *et al.*, 2009).

The melanoma differentiation associated gene 7 (*mda-7*) expresses a secreted interleukin (IL)-10 family protein, also referred to as IL-24. MDA-7/IL-24 expression increases during terminal differentiation of normal cells, and is either low or absent in tumour cells. Furthermore, restoring expression of MDA-7/IL-24 in a wide variety of tumour cell types causes growth arrest and rapid cell death by apoptosis, and also sensitizes tumour cells to radiotherapy (Dent *et al.*, 2010). These findings prompted construction of a replication-defective, first-generation Ad5 vector expressing MDA7/IL24 (INGN-241) and its evaluation for antitumour activity. In a phase I clinical trial of patients with advanced carcinomas and melanomas, direct tumour inoculation of INGN-241 was found to be safe and had measurable tumouricidal activity as evidenced by the induction of apoptosis in a large percentage of the tumour volume with a measurable clinical response rate of approximately 44%, suggesting significant therapeutic value (Fisher *et al.*, 2003; Cunningham *et al.*, 2005; Tong *et al.*, 2005; Fisher *et al.*, 2007; Lebedeva *et al.*, 2007; Eager *et al.*, 2008).

Future Trends

The specific capacity of certain viruses to induce tumours in their natural hosts was not an evolutionarily selected trait. To survive, viruses must infect and replicate in their hosts, disseminate to a new host, and repeat the cycle. Given that in general tumour viruses do not productively replicate in their respective virus-induced cancers, tumourigenesis *per se* is an evolutionary dead-end that represents a rare unintended consequence of the replication strategy of a virus in its natural host cells. Like all viruses, human adenovirus is an obligate intracellular parasite. In this regard, adenovirus relies heavily on host factors present only during S phase of the cell cycle to create an optimal environment for viral DNA replication. However, in its natural host, adenovirus infects and replicates in respiratory and gastrointestinal tract epithelial cells that have exited the cell cycle and undergone terminal differentiation. As this cellular environment is not conducive to viral DNA replication, adenovirus evolved protein functions to drive quiescent cells from G0 to S phase by inactivating the cellular RB1 and p53 tumour suppressors and by hijacking the cellular p300/CBP transcriptional coactivator and p400 chromatin remodeller to reprogram global cellular gene expression through repression of genes that inhibit viral replication and through activation of genes that stimulate viral replication. To further increase viral DNA replication and also to prepare for the upcoming high-level late viral protein synthesis needed to assemble large numbers of progeny viruses, adenovirus evolved additional protein functions to increase nucleotide biosynthesis and protein translation by activating cellular Myc and PI3K, respectively. It should come as no surprise that a virus would evolve to target key host cell regulatory proteins in order to create an optimal environment for viral replication. Indeed, most of the cellular targets mentioned above are commonly dysregulated in nonviral-induced human cancers, providing an explanation for the tumourigenic potential of human adenovirus.

We still have much to learn from adenovirus, and future studies of its protein functions promise to reveal additional novel and widely important mechanisms of carcinogenesis. As examples, further investigations into the anti-oncogenic function of E1A CR4 has the potential to aid development of new therapeutic strategies to treat human cancers, while determining the crucial function of the D2 element in E4orf1-induced PI3K, MYC, and RTK activation may expose a new mechanism for oncogenic cellular transformation. Furthermore, considering the established oncogenic properties of human adenovirus under experimental conditions, including the ability to transform cultured human cells, and recent intriguing potential associations of adenovirus infections with cancer, it remains possible that future work will uncover an unexpected link of adenovirus to some types of cancer, perhaps induced by stable expression of adenovirus oncogenes or by a hit-and-run mechanism.

Considering that human adenovirus codes for several different oncogenes, the frequent use of this virus as a vector to treat human cancers may seem at first glance somewhat surprising. Nonetheless, results obtained with adenovirus vectors in clinical trials have been promising as evidenced by the approval of two different adenovirus vectors as drugs to treat human tumours in China. Continued progress in this field will clearly depend on increasing the efficacy of adenovirus vector treatments of cancers. Achieving this goal will require basic research to develop improved adenovirus vectors that target and kill tumour cells with higher efficiency and specificity, as well as new clinical trials to investigate the efficacy of combining adenovirus vectors with new types of chemotherapy, radiotherapy, and immunotherapy. An additional exciting avenue for future research with adenovirus vectors will be to determine whether direct killing of cancer cells or induction of an anti-tumour immune response plays a greater role in tumour cell elimination and patient survival.

References

Abramson, D.H., Ellsworth, R.M., Kitchin, F.D., and Tung, G. (1984). Second nonocular tumours in retinoblastoma survivors. Are they radiation-induced? Ophthalmology 91, 1351-1355.

Ahuja, D., Saenz-Robles, M.T., and Pipas, J.M. (2005). SV40 large T antigen targets multiple cellular pathways to elicit cellular transformation. Oncogene 24, 7729-7745.

Akusjarvi, G. (2008). Temporal regulation of adenovirus major late alternative RNA splicing. Front Biosci 13, 5006-5015.

Alemany, R. (2012). Chapter four--Design of improved oncolytic adenoviruses. Adv Cancer Res 115, 93-114.

Anderson, E.J. (2010). Prevention and treatment of viral diarrhea in pediatrics. Expert Rev Anti Infect Ther 8, 205-217.

Arany, Z., Newsome, D., Oldread, E., Livingston, D.M., and Eckner, R. (1995). A family of transcriptional adaptor proteins targeted by the E1A oncoprotein. Nature 374, 81-84.

Arnberg, N. (2009). Adenovirus receptors: implications for tropism, treatment and targeting. Rev Med Virol 19, 165-178.

Attardi, L.D., de Vries, A., and Jacks, T. (2004). Activation of the p53-dependent G1 checkpoint response in mouse embryo fibroblasts depends on the specific DNA damage inducer. Oncogene 23, 973-980.

Babiss, L.E., Ginsberg, H.S., and Darnell, J.E., Jr. (1985). Adenovirus E1B proteins are required for accumulation of late viral mRNA and for effects on cellular mRNA translation and transport. Mol Cell Biol 5, 2552-2558.

Baker, S.J., Fearon, E.R., Nigro, J.M., Hamilton, S.R., Preisinger, A.C., Jessup, J.M., vanTuinen, P., Ledbetter, D.H., Barker, D.F., Nakamura, Y., *et al.* (1989). Chromosome 17 deletions and p53 gene mutations in colorectal carcinomas. Science 244, 217-221.

Barbeau, D., Charbonneau, R., Whalen, S.G., Bayley, S.T., and Branton, P.E. (1994). Functional interactions within adenovirus E1A protein complexes. Oncogene 9, 359-373.

Barker, D.D., and Berk, A.J. (1987). Adenovirus proteins from both E1B reading frames are required for transformation of rodent cells by viral infection and DNA transfection. Virology 156, 107-121.

Barton, K.N., Paielli, D., Zhang, Y., Koul, S., Brown, S.L., Lu, M., Seely, J., Kim, J.H., and Freytag, S.O. (2006). Second-generation replication-competent oncolytic adenovirus armed with improved suicide genes and ADP gene demonstrates greater efficacy without increased toxicity. Mol Ther 13, 347-356.

Bayley, S.T., and Mymryk, J.S. (1994). Adenovirus e1a proteins and transformation (review). Int J Oncol 5, 425-444.

Bell, J. (2014). Oncolytic viruses: immune or cytolytic therapy? Mol Ther 22, 1231-1232.

Bellon, M., Baydoun, H.H., Yao, Y., and Nicot, C. (2010). HTLV-I Tax-dependent and -independent events associated with immortalization of human primary T lymphocytes. Blood 115, 2441-2448.

Berget, S.M., Moore, C., and Sharp, P.A. (1977). Spliced segments at the 5' terminus of adenovirus 2 late mRNA. Proc Natl Acad Sci USA 74, 3171-3175.

Berk, A.J. (2005). Recent lessons in gene expression, cell cycle control, and cell biology from adenovirus. Oncogene 24, 7673-7685.

Berk, A.J. (2013). Adenoviridae. In Fields Virology, D.M.H. Knipe, P. M., ed. (Philadelphia, PA: Wolters Kluwer Health/Lippincott Williams and Wilkins), pp. 1704-1731.

Bernards, R., Schrier, P.I., Houweling, A., Bos, J.L., van der Eb, A.J., Zijlstra, M., and Melief, C.J. (1983). Tumourigenicity of cells transformed by adenovirus type 12 by evasion of T-cell immunity. Nature 305, 776-779.

Bernat, A., Avvakumov, N., Mymryk, J.S., and Banks, L. (2003). Interaction between the HPV E7 oncoprotein and the transcriptional coactivator p300. Oncogene 22, 7871-7881.

Bhaskar, P.T., and Hay, N. (2007). The two TORCs and Akt. Dev Cell 12, 487-502.

Bischoff, J.R., Kirn, D.H., Williams, A., Heise, C., Horn, S., Muna, M., Ng, L., Nye, J.A., Sampson-Johannes, A., Fattaey, A., *et al.* (1996). An adenovirus mutant that replicates selectively in p53-deficient human tumour cells. Science 274, 373-376.

Bittner, J.J. (1936). Some Possible Effects of Nursing on the Mammary Gland Tumour Incidence in Mice. Science 84, 162.

Blais, A., and Dynlacht, B.D. (2007). E2F-associated chromatin modifiers and cell cycle control. Curr Opin Cell Biol 19, 658-662.

Blume-Jensen, P., and Hunter, T. (2001). Oncogenic kinase signalling. Nature 411, 355-365.

Bos, J.L., Polder, L.J., Bernards, R., Schrier, P.I., van den Elsen, P.J., van der Eb, A.J., and van Ormondt, H. (1981). The 2.2 kb E1b mRNA of human Ad12 and Ad5 codes for two tumour antigens starting at different AUG triplets. Cell 27, 121-131.

Boyer, G.S., Denny, F.W., Jr., and Ginsberg, H.S. (1959). Sequential cellular changes produced by types 5 and 7 adenoviruses in HeLa cells and in human amniotic cells; cytological studies aided by fluorescein-labelled antibody. J Exp Med 110, 827-844.

Boyer, T.G., Martin, M.E., Lees, E., Ricciardi, R.P., and Berk, A.J. (1999). Mammalian Srb/Mediator complex is targeted by adenovirus E1A protein. Nature 399, 276-279.

Braithwaite, A.W., Cheetham, B.F., Li, P., Parish, C.R., Waldron-Stevens, L.K., and Bellett, A.J. (1983). Adenovirus-induced alterations of the cell growth cycle: a requirement for expression of E1A but not of E1B. J Virol 45, 192-199.

Bressy, C., and Benihoud, K. (2014). Association of oncolytic adenoviruses with chemotherapies: an overview and future directions. Biochem Pharmacol 90, 97-106.

Brunetti-Pierri, N., and Ng, P. (2011). Helper-dependent adenoviral vectors for liver-directed gene therapy. Hum Mol Genet 20, R7-13.

Cantin, G.T., Stevens, J.L., and Berk, A.J. (2003). Activation domain-mediator interactions promote transcription preinitiation complex assembly on promoter DNA. Proc Natl Acad Sci USA 100, 12003-12008.

Cantley, L.C. (2002). The phosphoinositide 3-kinase pathway. Science 296, 1655-1657.

Capasso, C., Garofalo, M., Hirvinen, M., and Cerullo, V. (2014). The evolution of adenoviral vectors through genetic and chemical surface modifications. Viruses 6, 832-855.

Carr, M.J., Kajon, A.E., Lu, X., Dunford, L., O'Reilly, P., Holder, P., De Gascun, C.F., Coughlan, S., Connell, J., Erdman, D.D., *et al.* (2011). Deaths associated with human adenovirus-14p1 infections, Europe, 2009-2010. Emerging infectious diseases 17, 1402-1408.

Carvajal, L.A., and Manfredi, J.J. (2013). Another fork in the road--life or death decisions by the tumour suppressor p53. EMBO Rep 14, 414-421.

Cavenee, W.K., Dryja, T.P., Phillips, R.A., Benedict, W.F., Godbout, R., Gallie, B.L., Murphree, A.L., Strong, L.C., and White, R.L. (1983). Expression of recessive

alleles by chromosomal mechanisms in retinoblastoma. Nature 305, 779-784.
CDC (2007). Acute respiratory disease associated with adenovirus serotype 14--four states, 2006-2007. Morbidity Mortality Weekly Report 56, 1181-1184.
Cerullo, V., Koski, A., Vaha-Koskela, M., and Hemminki, A. (2012). Chapter eight--Oncolytic adenoviruses for cancer immunotherapy: data from mice, hamsters, and humans. Adv Cancer Res 115, 265-318.
Chahal, J.S., Qi, J., and Flint, S.J. (2012). The human adenovirus type 5 E1B 55 kDa protein obstructs inhibition of viral replication by type I interferon in normal human cells. PLoS pathogens 8, e1002853.
Chan, H.M., Krstic-Demonacos, M., Smith, L., Demonacos, C., and La Thangue, N.B. (2001). Acetylation control of the retinoblastoma tumour-suppressor protein. Nat Cell Biol 3, 667-674.
Chen, G.X., Zhang, S., He, X.H., Liu, S.Y., Ma, C., and Zou, X.P. (2014). Clinical utility of recombinant adenoviral human p53 gene therapy: current perspectives. Onco Targets Ther 7, 1901-1909.
Cheng, P.H., Wechman, S.L., McMasters, K.M., and Zhou, H.S. (2015). Oncolytic Replication of E1b-Deleted Adenoviruses. Viruses 7, 5767-5779.
Chinnadurai, G. (1992). Adenovirus E1a as a tumour-suppressor gene. Oncogene 7, 1255-1258.
Chinnadurai, G. (2002). CtBP, an unconventional transcriptional corepressor in development and oncogenesis. Mol Cell 9, 213-224.
Chinnadurai, G. (2007). Transcriptional regulation by C-terminal binding proteins. Int J Biochem Cell Biol 39, 1593-1607.
Chinnadurai, G. (2011). Opposing oncogenic activities of small DNA tumour virus transforming proteins. Trends Microbiol 19, 174-183.
Chow, L.T., Gelinas, R.E., Broker, T.R., and Roberts, R.J. (1977). An amazing sequence arrangement at the 5' ends of adenovirus 2 messenger RNA. Cell 12, 1-8.
Christensen, J.B., Ewing, S.G., and Imperiale, M.J. (2012). Identification and characterization of a DNA binding domain on the adenovirus IVa2 protein. Virology 433, 124-130.
Christensen, J.B., and Imperiale, M.J. (1995). Inactivation of the retinoblastoma susceptibility protein is not sufficient for the transforming function of the conserved region 2-like domain of simian virus 40 large T antigen. J Virol 69, 3945-3948.
Chung, S.H., Frese, K.K., Weiss, R.S., Prasad, B.V., and Javier, R.T. (2007). A New Crucial Protein-Interaction Element That Targets the Adenovirus E4-ORF1 Oncoprotein to Membrane Vesicles. J. Virol. 81, 4787-4797.
Chung, S.H., Weiss, R.S., Prasad, B.V.V., and Javier, R.T. (2008). Functionally distinct monomers and trimers produced by a viral oncoprotein. Oncogene 27, 1412–1420.
Cobrinik, D., Whyte, P., Peeper, D.S., Jacks, T., and Weinberg, R.A. (1993). Cell cycle-specific association of E2F with the p130 E1A-binding protein. Genes Dev 7, 2392-2404.
Contreras-Paredes, A., De la Cruz-Hernandez, E., Martinez-Ramirez, I., Duenas-Gonzalez, A., and Lizano, M. (2009). E6 variants of human papillomavirus 18 differentially modulate the protein kinase B/phosphatidylinositol 3-kinase (akt/PI3K) signaling pathway. Virology 383, 78-85.
Cooray, S. (2004). The pivotal role of phosphatidylinositol 3-kinase-Akt signal transduction in virus survival. J Gen Virol 85, 1065-1076.
Crystal, R.G. (2014). Adenovirus: the first effective *in vivo* gene delivery vector. Hum Gene Ther 25, 3-11.
Cuconati, A., Degenhardt, K., Sundararajan, R., Anschel, A., and White, E. (2002). Bak and Bax function to limit adenovirus replication through apoptosis induction. J Virol 76, 4547-4558.
Cuconati, A., Mukherjee, C., Perez, D., and White, E. (2003). DNA damage response and MCL-1 destruction initiate apoptosis in adenovirus-infected cells. Genes Dev 17, 2922-2932.
Cunningham, C.C., Chada, S., Merritt, J.A., Tong, A., Senzer, N., Zhang, Y., Mhashilkar, A., Parker, K., Vukelja, S., Richards, D., *et al.* (2005). Clinical and local biological effects of an intratumoural injection of mda-7 (IL24; INGN 241) in patients with advanced carcinoma: a phase I study. Mol Ther 11, 149-159.
Davison, A.J., Benko, M., and Harrach, B. (2003). Genetic content and evolution of adenoviruses. J Gen Virol 84, 2895-2908.
de Jong, J.C., Wigand, R., Kidd, A.H., Wadell, G., Kapsenberg, J.G., Muzerie, C.J., Wermenbol, A.G., and Firtzlaff, R.G. (1983). Candidate adenoviruses 40 and 41: fastidious adenoviruses from human infant stool. J Med Virol 11, 215-231.
Debbas, M., and White, E. (1993). Wild-type p53 mediates apoptosis by E1A, which is inhibited by E1B. Genes Dev 7, 546-554.
DeCaprio, J.A. (2009). How the Rb tumour suppressor structure and function was revealed by the study of Adenovirus and SV40. Virology 384, 274-284.
DeGregori, J., Kowalik, T., and Nevins, J.R. (1995). Cellular targets for activation by the E2F1 transcription factor include DNA synthesis- and G1/S-regulatory genes. Mol Cell Biol 15, 4215-4224.
Deleu, L., Shellard, S., Alevizopoulos, K., Amati, B., and Land, H. (2001). Recruitment of TRRAP required for oncogenic transformation by E1A. Oncogene 20, 8270-8275.
Dent, P., Yacoub, A., Hamed, H.A., Park, M.A., Dash, R., Bhutia, S.K., Sarkar, D., Wang, X.Y., Gupta, P., Emdad, L., *et al.* (2010). The development of MDA-7/IL-24 as a cancer therapeutic. Pharmacol Ther 128, 375-384.
Dick, F.A., and Rubin, S.M. (2013). Molecular mechanisms underlying RB protein function. Nat Rev Mol Cell Biol 14, 297-306.
DiMaio, D., and Fan, H. (2013). Viruses, Cell Transformation, and Cancer. In Fields Virology, D.M.H. Knipe, P. M., ed. (Philadelphia, PA: Wolters Kluwer Health/Lippincott Williams and Wilkins), pp. 154-189.
Doerfler, W. (1968). The fate of the DNA of adenovirus type 12 in baby hamster kidney cells. Proc Natl Acad Sci USA 60, 636-643.
Dormond, E., and Kamen, A.A. (2011). Manufacturing of adenovirus vectors: production and purification of

helper dependent adenovirus. Methods Mol Biol 737, 139-156.
Doronin, K., Toth, K., Kuppuswamy, M., Krajcsi, P., Tollefson, A.E., and Wold, W.S. (2003). Overexpression of the ADP (E3-11.6K) protein increases cell lysis and spread of adenovirus. Virology 305, 378-387.
Doronin, K., Toth, K., Kuppuswamy, M., Ward, P., Tollefson, A.E., and Wold, W.S. (2000). Tumour-specific, replication-competent adenovirus vectors overexpressing the adenovirus death protein. J Virol 74, 6147-6155.
Douglas, J.L., Gopalakrishnan, S., and Quinlan, M.P. (1991). Modulation of transformation of primary epithelial cells by the second exon of the Ad5 E1A12S gene. Oncogene 6, 2093-2103.
Dryja, T.P., Rapaport, J.M., Joyce, J.M., and Petersen, R.A. (1986). Molecular detection of deletions involving band q14 of chromosome 13 in retinoblastomas. Proc Natl Acad Sci USA 83, 7391-7394.
Eager, R., Harle, L., and Nemunaitis, J. (2008). Ad-MDA-7; INGN 241: a review of preclinical and clinical experience. Expert Opin Biol Ther 8, 1633-1643.
Eckner, R., Ewen, M.E., Newsome, D., Gerdes, M., DeCaprio, J.A., Lawrence, J.B., and Livingston, D.M. (1994). Molecular cloning and functional analysis of the adenovirus E1A-associated 300-kD protein (p300) reveals a protein with properties of a transcriptional adaptor. Genes Dev 8, 869-884.
Eddy, B.E., Borman, G.S., Grubbs, G.E., and Young, R.D. (1962). Identification of the oncogenic substance in rhesus monkey kidney cell culture as simian virus 40. Virology 17, 65-75.
el-Deiry, W.S., Tokino, T., Velculescu, V.E., Levy, D.B., Parsons, R., Trent, J.M., Lin, D., Mercer, W.E., Kinzler, K.W., and Vogelstein, B. (1993). WAF1, a potential mediator of p53 tumour suppression. Cell 75, 817-825.
Ellermann, V., and Bang, O. (1908). Experimentelle Leukamie bei Huhnern. Zentralbl Bakteriol Parasitenkd Infektionskr Hyg 46, 595-597.
Fallaux, F.J., Bout, A., van der Velde, I., van den Wollenberg, D.J., Hehir, K.M., Keegan, J., Auger, C., Cramer, S.J., van Ormondt, H., van der Eb, A.J., *et al.* (1998). New helper cells and matched early region 1-deleted adenovirus vectors prevent generation of replication-competent adenoviruses. Hum Gene Ther 9, 1909-1917.
Fan, S., Macken, C.A., Li, C., Ozawa, M., Goto, H., Iswahyudi, N.F., Nidom, C.A., Chen, H., Neumann, G., and Kawaoka, Y. (2013). Synergistic effect of the PDZ and p85beta-binding domains of the NS1 protein on virulence of an avian H5N1 influenza A virus. J. Virol. 87, 4861-4871.
Ferrari, R., Gou, D., Jawdekar, G., Johnson, S.A., Nava, M., Su, T., Yousef, A.F., Zemke, N.R., Pellegrini, M., Kurdistani, S.K., *et al.* (2014). Adenovirus small E1A employs the lysine acetylases p300/CBP and tumour suppressor Rb to repress select host genes and promote productive virus infection. Cell Host Microbe 16, 663-676.
Ferrari, R., Pellegrini, M., Horwitz, G.A., Xie, W., Berk, A.J., and Kurdistani, S.K. (2008). Epigenetic reprogramming by adenovirus e1a. Science 321, 1086-1088.
Ferrari, R., Su, T., Li, B., Bonora, G., Oberai, A., Chan, Y., Sasidharan, R., Berk, A.J., Pellegrini, M., and Kurdistani, S.K. (2012). Reorganization of the host epigenome by a viral oncogene. Genome Res 22, 1212-1221.
Ferreon, J.C., Martinez-Yamout, M.A., Dyson, H.J., and Wright, P.E. (2009). Structural basis for subversion of cellular control mechanisms by the adenoviral E1A oncoprotein. Proc Natl Acad Sci USA 106, 13260-13265.
Fisher, P.B., Gopalkrishnan, R.V., Chada, S., Ramesh, R., Grimm, E.A., Rosenfeld, M.R., Curiel, D.T., and Dent, P. (2003). mda-7/IL-24, a novel cancer selective apoptosis inducing cytokine gene: from the laboratory into the clinic. Cancer Biol Ther 2, S23-37.
Fisher, P.B., Sarkar, D., Lebedeva, I.V., Emdad, L., Gupta, P., Sauane, M., Su, Z.Z., Grant, S., Dent, P., Curiel, D.T., *et al.* (2007). Melanoma differentiation associated gene-7/interleukin-24 (mda-7/IL-24): novel gene therapeutic for metastatic melanoma. Toxicol Appl Pharmacol 224, 300-307.
Flint, S.J., and Gonzalez, R.A. (2003). Regulation of mRNA production by the adenoviral E1B 55-kDa and E4 Orf6 proteins. Curr Top Microbiol Immunol 272, 287-330.
Foy, H.M., Cooney, M.K., and Hatlen, J.B. (1968). Adenovirus type 3 epidemic associated with intermittent chlorination of a swimming pool. Arch Environ Health 17, 795-802.
Frank, S.R., Parisi, T., Taubert, S., Fernandez, P., Fuchs, M., Chan, H.M., Livingston, D.M., and Amati, B. (2003). MYC recruits the TIP60 histone acetyltransferase complex to chromatin. EMBO Rep 4, 575-580.
Frank, S.R., Schroeder, M., Fernandez, P., Taubert, S., and Amati, B. (2001). Binding of c-Myc to chromatin mediates mitogen-induced acetylation of histone H4 and gene activation. Genes Dev 15, 2069-2082.
Frese, K.K., Latorre, I.J., Chung, S.H., Caruana, G., Bernstein, A., Jones, S.N., Donehower, L.A., Justice, M.J., Garner, C.C., and Javier, R.T. (2006). Oncogenic function for the Dlg1 mammalian homolog of the Drosophila discs-large tumour suppressor. EMBO J 25, 1406-1417.
Frese, K.K., Lee, S.S., Thomas, D.L., Latorre, I.J., Weiss, R.S., Glaunsinger, B.A., and Javier, R.T. (2003). Selective PDZ protein-dependent stimulation of phosphatidylinositol 3-kinase by the adenovirus E4-ORF1 oncoprotein. Oncogene 22, 710-721.
Freytag, S.O., Khil, M., Stricker, H., Peabody, J., Menon, M., DePeralta-Venturina, M., Nafziger, D., Pegg, J., Paielli, D., Brown, S., *et al.* (2002). Phase I study of replication-competent adenovirus-mediated double suicide gene therapy for the treatment of locally recurrent prostate cancer. Cancer Res 62, 4968-4976.
Freytag, S.O., Stricker, H., Movsas, B., and Kim, J.H. (2007). Prostate cancer gene therapy clinical trials. Mol Ther 15, 1042-1052.
Freytag, S.O., Stricker, H., Pegg, J., Paielli, D., Pradhan, D.G., Peabody, J., DePeralta-Venturina, M., Xia, X.,

Brown, S., Lu, M., *et al.* (2003). Phase I study of replication-competent adenovirus-mediated double-suicide gene therapy in combination with conventional-dose three-dimensional conformal radiation therapy for the treatment of newly diagnosed, intermediate- to high-risk prostate cancer. Cancer Res 63, 7497-7506.

Fridman, J.S., and Lowe, S.W. (2003). Control of apoptosis by p53. Oncogene 22, 9030-9040.

Friend, C. (1957). Cell-free transmission in adult Swiss mice of a disease having the character of a leukemia. J Exp Med 105, 307-318.

Friend, S.H., Bernards, R., Rogelj, S., Weinberg, R.A., Rapaport, J.M., Albert, D.M., and Dryja, T.P. (1986). A human DNA segment with properties of the gene that predisposes to retinoblastoma and osteosarcoma. Nature 323, 643-646.

Frisch, S.M. (1991). Anti-oncogenic effect of adenovirus E1A in human tumour cells. Proc Natl Acad Sci USA 88, 9077-9081.

Frisch, S.M. (1994). E1a induces the expression of epithelial characteristics. J Cell Biol 127, 1085-1096.

Frolov, M.V., and Dyson, N.J. (2004). Molecular mechanisms of E2F-dependent activation and pRB-mediated repression. J Cell Sci 117, 2173-2181.

Fuchs, M., Gerber, J., Drapkin, R., Sif, S., Ikura, T., Ogryzko, V., Lane, W.S., Nakatani, Y., and Livingston, D.M. (2001). The p400 complex is an essential E1A transformation target. Cell 106, 297-307.

Fung, Y.K., Murphree, A.L., T'Ang, A., Qian, J., Hinrichs, S.H., and Benedict, W.F. (1987). Structural evidence for the authenticity of the human retinoblastoma gene. Science 236, 1657-1661.

Gallimore, P.H. (1974). Viral DNA in transformed cells. II. A study of the sequences of adenovirus 2 DNA in nine lines of transformed rat cells using specific fragments of the viral genome. J Mol Biol 89, 49-72.

Gallimore, P.H., Byrd, P.J., Whittaker, J.L., and Grand, R.J. (1985). Properties of rat cells transformed by DNA plasmids containing adenovirus type 12 E1 DNA or specific fragments of the E1 region: comparison of transforming frequencies. Cancer Res 45, 2670-2680.

Gary, G.W., Jr., Hierholzer, J.C., and Black, R.E. (1979). Characteristics of noncultivable adenoviruses associated with diarrhea in infants: a new subgroup of human adenoviruses. J Clin Microbiol 10, 96-103.

Ghosh, M.K., and Harter, M.L. (2003). A viral mechanism for remodeling chromatin structure in G0 cells. Mol Cell 12, 255-260.

Girardi, A.J., Sweet, B.H., Slotnick, V.B., and Hilleman, M.R. (1962). Development of tumours in hamsters inoculated in the neonatal period with vacuolating virus, SV-40. Proc Soc Exp Biol Med 109, 649-660.

Glaunsinger, B.A., Lee, S.S., Thomas, M., Banks, L., and Javier, R. (2000). Interactions of the PDZ-protein MAGI-1 with adenovirus E4-ORF1 and high-risk papillomavirus E6 oncoproteins. Oncogene 19, 5270-5280.

Glaunsinger, B.A., Weiss, R.S., Lee, S.S., and Javier, R. (2001). Link of the unique oncogenic properties of adenovirus type 9 E4-ORF1 to a select interaction with the candidate tumour suppressor protein ZO-2. EMBO J 20, 5578-5586.

Golebiewski, L., Liu, H., Javier, R.T., and Rice, A.P. (2011). The avian influenza virus NS1 ESEV PDZ binding motif associates with Dlg1 and Scribble to disrupt cellular tight junctions. J. Virol. 85, 10639-10648.

Gonzalez, R., Huang, W., Finnen, R., Bragg, C., and Flint, S.J. (2006). Adenovirus E1B 55-kilodalton protein is required for both regulation of mRNA export and efficient entry into the late phase of infection in normal human fibroblasts. J Virol 80, 964-974.

Goodman, R.H., and Smolik, S. (2000). CBP/p300 in cell growth, transformation, and development. Genes Dev 14, 1553-1577.

Gorziglia, M.I., Lapcevich, C., Roy, S., Kang, Q., Kadan, M., Wu, V., Pechan, P., and Kaleko, M. (1999). Generation of an adenovirus vector lacking E1, e2a, E3, and all of E4 except open reading frame 3. J Virol 73, 6048-6055.

Graffi, A. (1957). Chloroleukemia of mice. Ann N Y Acad Sci 68, 540-558.

Graham, F.L., Smiley, J., Russell, W.C., and Nairn, R. (1977). Characteristics of a human cell line transformed by DNA from human adenovirus type 5. J Gen Virol 36, 59-74.

Gray, G.C., McCarthy, T., Lebeck, M.G., Schnurr, D.P., Russell, K.L., Kajon, A.E., Landry, M.L., Leland, D.S., Storch, G.A., Ginocchio, C.C., *et al.* (2007). Genotype prevalence and risk factors for severe clinical adenovirus infection, United States 2004-2006. Clin Infect Dis 45, 1120-1131.

Green, M., Wold, W.S., Mackey, J.K., and Rigden, P. (1979). Analysis of human tonsil and cancer DNAs and RNAs for DNA sequences of group C (serotypes 1, 2, 5, and 6) human adenoviruses. Proc Natl Acad Sci USA 76, 6606-6610.

Gross, L. (1951). "Spontaneous" leukemia developing in C3H mice following inoculation in infancy, with AK-leukemic extracts, or AK-embrvos. Proc Soc Exp Biol Med 76, 27-32.

Gross, L. (1953). A filterable agent, recovered from Ak leukemic extracts, causing salivary gland carcinomas in C3H mice. Proc Soc Exp Biol Med 83, 414-421.

Gustafsson, B., Huang, W., Bogdanovic, G., Gauffin, F., Nordgren, A., Talekar, G., Ornelles, D.A., and Gooding, L.R. (2007). Adenovirus DNA is detected at increased frequency in Guthrie cards from children who develop acute lymphoblastic leukaemia. Br J Cancer 97, 992-994.

Harada, J.N., Shevchenko, A., Shevchenko, A., Pallas, D.C., and Berk, A.J. (2002). Analysis of the adenovirus E1B-55K-anchored proteome reveals its link to ubiquitination machinery. J Virol 76, 9194-9206.

Hardy, S., Kitamura, M., Harris-Stansil, T., Dai, Y., and Phipps, M.L. (1997). Construction of adenovirus vectors through Cre-lox recombination. J Virol 71, 1842-1849.

Harlow, E., Whyte, P., Franza, B.R., Jr., and Schley, C. (1986). Association of adenovirus early-region 1A proteins with cellular polypeptides. Mol Cell Biol 6, 1579-1589.

Hatakeyama, M., and Weinberg, R.A. (1995). The role of RB in cell cycle control. Prog Cell Cycle Res 1, 9-19.

Hearing, P., Samulski, R.J., Wishart, W.L., and Shenk, T. (1987). Identification of a repeated sequence element required for efficient encapsidation of the adenovirus type 5 chromosome. J Virol 61, 2555-2558.

Hearing, P., and Shenk, T. (1983). The adenovirus type 5 E1A transcriptional control region contains a duplicated enhancer element. Cell 33, 695-703.

Hecht, J.R., Bedford, R., Abbruzzese, J.L., Lahoti, S., Reid, T.R., Soetikno, R.M., Kirn, D.H., and Freeman, S.M. (2003). A phase I/II trial of intratumoural endoscopic ultrasound injection of ONYX-015 with intravenous gemcitabine in unresectable pancreatic carcinoma. Clin Cancer Res 9, 555-561.

Hilleman, M.R., and Werner, J.H. (1954). Recovery of new agent from patients with acute respiratory illness. Proc Soc Exp Biol Med 85, 183-188.

Honkaniemi, E., Talekar, G., Huang, W., Bogdanovic, G., Forestier, E., von Doblen, U., Engvall, M., Ornelles, D.A., Gooding, L.R., and Gustafsson, B. (2010). Adenovirus DNA in Guthrie cards from children who develop acute lymphoblastic leukaemia (ALL). Br J Cancer 102, 796-798.

Horwitz, G.A., Zhang, K., McBrian, M.A., Grunstein, M., Kurdistani, S.K., and Berk, A.J. (2008). Adenovirus small e1a alters global patterns of histone modification. Science 321, 1084-1085.

Houweling, A., van den Elsen, P.J., and van der Eb, A.J. (1980). Partial transformation of primary rat cells by the leftmost 4.5% fragment of adenovirus 5 DNA. Virology 105, 537-550.

Howe, J.A., Mymryk, J.S., Egan, C., Branton, P.E., and Bayley, S.T. (1990). Retinoblastoma growth suppressor and a 300-kDa protein appear to regulate cellular DNA synthesis. Proc Natl Acad Sci USA 87, 5883-5887.

Huang, P.I., Chang, J.F., Kirn, D.H., and Liu, T.C. (2009). Targeted genetic and viral therapy for advanced head and neck cancers. Drug Discov Today 14, 570-578.

Ibelgaufts, H., Jones, K.W., Maitland, N., and Shaw, J.F. (1982). Adenovirus-related RNA sequences in human neurogenic tumours. Acta Neuropathol 56, 113-117.

Ikura, T., Ogryzko, V.V., Grigoriev, M., Groisman, R., Wang, J., Horikoshi, M., Scully, R., Qin, J., and Nakatani, Y. (2000). Involvement of the TIP60 histone acetylase complex in DNA repair and apoptosis. Cell 102, 463-473.

James, C.D., and Roberts, S. (2016). Viral Interactions with PDZ Domain-Containing Proteins-An Oncogenic Trait? Pathogens 5.

Jannun, R., and Chinnadurai, G. (1987). Functional relatedness between the E1a and E1b regions of group C and group D human adenoviruses. Virus Res 7, 33-48.

Javier, R.T. (1994). Adenovirus type 9 E4 open reading frame 1 encodes a transforming protein required for the production of mammary tumours in rats. J. Virol. 68, 3917-3924.

Javier, R.T. (2008). Cell polarity proteins: common targets for tumourigenic human viruses. Oncogene 27, 7031-7046.

Javier, R.T., and Butel, J.S. (2008). The history of tumour virology. Cancer Res 68, 7693-7706.

Javier, R.T., Raska, K., Jr., Macdonald, G.J., and Shenk, T. (1991). Human adenovirus type 9-induced rat mammary tumours. J. Virol. 65, 3192-3202.

Javier, R.T., Raska, K., Jr., and Shenk, T. (1992). Requirement for the adenovirus type 9 E4 region in production of mammary tumours. Science 257, 1267-1271.

Javier, R.T., and Rice, A.P. (2011). Emerging theme: cellular PDZ proteins as common targets of pathogenic viruses. J Virol 85, 11544-11556.

Javier, R.T., and Shenk, T. (1996). Mammary tumours induced by human adenovirus type 9: a role for the viral early region 4 gene. Breast Cancer Res Treat 39, 57-67.

Jeong, S.J., Pise-Masison, C.A., Radonovich, M.F., Park, H.U., and Brady, J.N. (2005). Activated AKT regulates NF-kappaB activation, p53 inhibition and cell survival in HTLV-1-transformed cells. Oncogene 24, 6719-6728.

Jia, H., and Kling, J. (2006). China offers alternative gateway for experimental drugs. Nat Biotechnol 24, 117-118.

Jochemsen, H., Daniels, G.S., Hertoghs, J.J., Schrier, P.I., van den Elsen, P.J., and van der, E.A. (1982). Identification of adenovirus-type 12 gene products involved in transformation and oncogenesis. Virology 122, 15-28.

Joerger, A.C., and Fersht, A.R. (2008). Structural biology of the tumour suppressor p53. Annu Rev Biochem 77, 557-582.

Jones, M.S., 2nd, Harrach, B., Ganac, R.D., Gozum, M.M., Dela Cruz, W.P., Riedel, B., Pan, C., Delwart, E.L., and Schnurr, D.P. (2007). New adenovirus species found in a patient presenting with gastroenteritis. J Virol 81, 5978-5984.

Jones, N., and Shenk, T. (1979). Isolation of adenovirus type 5 host range deletion mutants defective for transformation of rat embryo cells. Cell 17, 683-689.

Jones, N.C. (1990). Transformation by the human adenoviruses. Semin Cancer Biol 1, 425-435.

Kadeppagari, R.K., Sankar, N., and Thimmapaya, B. (2009). Adenovirus transforming protein E1A induces c-Myc in quiescent cells by a novel mechanism. J Virol 83, 4810-4822.

Kajon, A.E., Lu, X., Erdman, D.D., Louie, J., Schnurr, D., George, K.S., Koopmans, M.P., Allibhai, T., and Metzgar, D. (2010). Molecular epidemiology and brief history of emerging adenovirus 14-associated respiratory disease in the United States. J Infect Dis 202, 93-103.

Kanerva, A., Nokisalmi, P., Diaconu, I., Koski, A., Cerullo, V., Liikanen, I., Tahtinen, S., Oksanen, M., Heiskanen, R., Pesonen, S., *et al.* (2013). Antiviral and antitumour T-cell immunity in patients treated with GM-CSF-coding oncolytic adenovirus. Clin Cancer Res 19, 2734-2744.

Kimball, K.J., Preuss, M.A., Barnes, M.N., Wang, M., Siegal, G.P., Wan, W., Kuo, H., Saddekni, S., Stockard, C.R., Grizzle, W.E., *et al.* (2010). A phase I study of a tropism-modified conditionally replicative adenovirus for recurrent malignant gynecologic diseases. Clin Cancer Res 16, 5277-5287.

Kimelman, D., Miller, J.S., Porter, D., and Roberts, B.E. (1985). E1a regions of the human adenoviruses and of the highly oncogenic simian adenovirus 7 are closely related. J Virol 53, 399-409.

Kirn, D. (2001). Clinical research results with dl1520 (Onyx-015), a replication-selective adenovirus for the treatment of cancer: what have we learned? Gene Ther 8, 89-98.

Kitajewski, J., Schneider, R.J., Safer, B., Munemitsu, S.M., Samuel, C.E., Thimmappaya, B., and Shenk, T. (1986). Adenovirus VAI RNA antagonizes the antiviral action of interferon by preventing activation of the interferon-induced eIF-2 alpha kinase. Cell 45, 195-200.

Knudson, A.G., Jr. (1971). Mutation and cancer: statistical study of retinoblastoma. Proc Natl Acad Sci USA 68, 820-823.

Knudson, A.G., Jr. (1984). Genetic predisposition to cancer. Cancer Detect Prev 7, 1-8.

Kobayashi, S., and Mukai, N. (1973). Retinoblastoma-like tumours induced in rats by human adenovirus. Invest Ophthalmol 12, 853-856.

Komorek, J., Kuppuswamy, M., Subramanian, T., Vijayalingam, S., Lomonosova, E., Zhao, L.J., Mymryk, J.S., Schmitt, K., and Chinnadurai, G. (2010). Adenovirus type 5 E1A and E6 proteins of low-risk cutaneous beta-human papillomaviruses suppress cell transformation through interaction with FOXK1/K2 transcription factors. J Virol 84, 2719-2731.

Kong, K., Kumar, M., Taruishi, M., and Javier, R.T. (2014). The human adenovirus E4-ORF1 protein subverts discs large 1 to mediate membrane recruitment and dysregulation of phosphatidylinositol 3-kinase. PLoS pathogens 10, e1004102.

Kong, K., Kumar, M., Taruishi, M., and Javier, R.T. (2015). Adenovirus E4-ORF1 Dysregulates Epidermal Growth Factor and Insulin/Insulin-Like Growth Factor Receptors To Mediate Constitutive Myc Expression. J. Virol. 89, 10774-10785.

Kosulin, K., Haberler, C., Hainfellner, J.A., Amann, G., Lang, S., and Lion, T. (2007). Investigation of adenovirus occurrence in pediatric tumour entities. J Virol 81, 7629-7635.

Koup, R.A., Lamoreaux, L., Zarkowsky, D., Bailer, R.T., King, C.R., Gall, J.G., Brough, D.E., Graham, B.S., and Roederer, M. (2009). Replication-defective adenovirus vectors with multiple deletions do not induce measurable vector-specific T cells in human trials. J Virol 83, 6318-6322.

Kovesdi, I., Reichel, R., and Nevins, J.R. (1986). Identification of a cellular transcription factor involved in E1A trans-activation. Cell 45, 219-228.

Krasnykh, V., Dmitriev, I., Navarro, J.G., Belousova, N., Kashentseva, E., Xiang, J., Douglas, J.T., and Curiel, D.T. (2000). Advanced generation adenoviral vectors possess augmented gene transfer efficiency based upon coxsackie adenovirus receptor-independent cellular entry capacity. Cancer Res 60, 6784-6787.

Krasnykh, V.N., Mikheeva, G.V., Douglas, J.T., and Curiel, D.T. (1996). Generation of recombinant adenovirus vectors with modified fibers for altering viral tropism. J Virol 70, 6839-6846.

Kumar, M., Kong, K., and Javier, R.T. (2014). Hijacking Dlg1 for oncogenic phosphatidylinositol 3-kinase activation in human epithelial cells is a conserved mechanism of human adenovirus E4-ORF1 proteins. J Virol 88, 14268-14277.

Kunz, A.N., and Ottolini, M. (2010). The role of adenovirus in respiratory tract infections. Curr Infect Dis Rep 12, 81-87.

Lane, D.P., Cheok, C.F., and Lain, S. (2010). p53-based cancer therapy. Cold Spring Harb Perspect Biol 2, a001222.

Lane, D.P., and Crawford, L.V. (1979). T antigen is bound to a host protein in SV40-transformed cells. Nature 278, 261-263.

Lang, S.E., and Hearing, P. (2003). The adenovirus E1A oncoprotein recruits the cellular TRRAP/GCN5 histone acetyltransferase complex. Oncogene 22, 2836-2841.

Larson, C., Oronsky, B., Scicinski, J., Fanger, G.R., Stirn, M., Oronsky, A., and Reid, T.R. (2015). Going viral: a review of replication-selective oncolytic adenoviruses. Oncotarget 6, 19976-19989.

Latorre, I.J., Roh, M.H., Frese, K.K., Weiss, R.S., Margolis, B., and Javier, R.T. (2005). Viral oncoprotein-induced mislocalization of select PDZ proteins disrupts tight junctions and causes polarity defects in epithelial cells. J Cell Sci 118, 4283-4293.

Lebedeva, I.V., Emdad, L., Su, Z.Z., Gupta, P., Sauane, M., Sarkar, D., Staudt, M.R., Liu, S.J., Taher, M.M., Xiao, R., *et al.* (2007). mda-7/IL-24, novel anticancer cytokine: focus on bystander antitumour, radiosensitization and antiangiogenic properties and overview of the phase I clinical experience (Review). Int J Oncol 31, 985-1007.

Lee, J.O., Russo, A.A., and Pavletich, N.P. (1998). Structure of the retinoblastoma tumour-suppressor pocket domain bound to a peptide from HPV E7. Nature 391, 859-865.

Lee, S.S., Glaunsinger, B., Mantovani, F., Banks, L., and Javier, R.T. (2000). Multi-PDZ domain protein MUPP1 is a cellular target for both adenovirus E4-ORF1 and high-risk papillomavirus type 18 E6 oncoproteins. J. Virol. 74, 9680-9693.

Lee, S.S., Weiss, R.S., and Javier, R.T. (1997). Binding of human virus oncoproteins to hDlg/SAP97, a mammalian homolog of the Drosophila discs large tumour suppressor protein. Proc Natl Acad Sci USA 94, 6670-6675.

Levine, A.J. (1990). The p53 protein and its interactions with the oncogene products of the small DNA tumour viruses. Virology 177, 419-426.

Levine, A.J. (2009). The common mechanisms of transformation by the small DNA tumour viruses: The inactivation of tumour suppressor gene products: p53. Virology 384, 285-293.

Lichtenstein, D.L., Toth, K., Doronin, K., Tollefson, A.E., and Wold, W.S. (2004). Functions and mechanisms of action of the adenovirus E3 proteins. Int Rev Immunol 23, 75-111.

Lill, N.L., Tevethia, M.J., Eckner, R., Livingston, D.M., and Modjtahedi, N. (1997). p300 family members associate with the carboxyl terminus of simian virus 40 large tumour antigen. J Virol 71, 129-137.

Linzer, D.I., and Levine, A.J. (1979). Characterization of a 54K dalton cellular SV40 tumour antigen present in SV40-transformed cells and uninfected embryonal carcinoma cells. Cell 17, 43-52.
Lion, T., Kosulin, K., Landlinger, C., Rauch, M., Preuner, S., Jugovic, D., Potschger, U., Lawitschka, A., Peters, C., Fritsch, G., *et al.* (2010). Monitoring of adenovirus load in stool by real-time PCR permits early detection of impending invasive infection in patients after allogeneic stem cell transplantation. Leukemia 24, 706-714.
Liu, H., Golebiewski, L., Dow, E.C., Krug, R.M., Javier, R.T., and Rice, A.P. (2010a). The ESEV PDZ-binding motif of the avian influenza A virus NS1 protein protects infected cells from apoptosis by directly targeting Scribble. J. Virol. 84, 11164-11174.
Liu, H., Jin, L., Koh, S.B., Atanasov, I., Schein, S., Wu, L., and Zhou, Z.H. (2010b). Atomic structure of human adenovirus by cryo-EM reveals interactions among protein networks. Science 329, 1038-1043.
Liu, X., and Marmorstein, R. (2007). Structure of the retinoblastoma protein bound to adenovirus E1A reveals the molecular basis for viral oncoprotein inactivation of a tumour suppressor. Genes Dev 21, 2711-2716.
Liu, Y., Shevchenko, A., Shevchenko, A., and Berk, A.J. (2005). Adenovirus exploits the cellular aggresome response to accelerate inactivation of the MRN complex. J Virol 79, 14004-14016.
Liu, Y., Wang, Y., Yamakuchi, M., Masuda, S., Tokioka, T., Yamaoka, S., Maruyama, I., and Kitajima, I. (2001). Phosphoinositide-3 kinase-PKB/Akt pathway activation is involved in fibroblast Rat-1 transformation by human T-cell leukemia virus type I tax. Oncogene 20, 2514-2526.
Louie, J.K., Kajon, A.E., Holodniy, M., Guardia-LaBar, L., Lee, B., Petru, A.M., Hacker, J.K., and Schnurr, D.P. (2008). Severe pneumonia due to adenovirus serotype 14: a new respiratory threat? Clin Infect Dis 46, 421-425.
Lowe, S.W., and Ruley, H.E. (1993). Stabilization of the p53 tumour suppressor is induced by adenovirus 5 E1A and accompanies apoptosis. Genes Dev 7, 535-545.
Lue, R.A., Marfatia, S.M., Branton, D., and Chishti, A.H. (1994). Cloning and characterization of hdlg: the human homologue of the Drosophila discs large tumour suppressor binds to protein 4.1. Proc Natl Acad Sci USA 91, 9818-9822.
Ma, G., Shimada, H., Hiroshima, K., Tada, Y., Suzuki, N., and Tagawa, M. (2009). Gene medicine for cancer treatment: commercially available medicine and accumulated clinical data in China. Drug Des Devel Ther 2, 115-122.
Ma, Y., and Mathews, M.B. (1996). Structure, function, and evolution of adenovirus-associated RNA: a phylogenetic approach. J Virol 70, 5083-5099.
Mackey, J.K., Rigden, P.M., and Green, M. (1976). Do highly oncogenic group A human adenoviruses cause human cancer? Analysis of human tumours for adenovirus 12 transforming DNA sequences. Proc Natl Acad Sci USA 73, 4657-4661.
Macpherson, I., and Montagnier, L. (1964). Agar Suspension Culture for the Selective Assay of Cells Transformed by Polyoma Virus. Virology 23, 291-294.
Madison, D.L., Yaciuk, P., Kwok, R.P., and Lundblad, J.R. (2002). Acetylation of the adenovirus-transforming protein E1A determines nuclear localization by disrupting association with importin-alpha. J Biol Chem 277, 38755-38763.
Martin, M.E., and Berk, A.J. (1998). Adenovirus E1B 55K represses p53 activation *in vitro*. J Virol 72, 3146-3154.
Martin, M.E., and Berk, A.J. (1999). Corepressor required for adenovirus E1B 55,000-molecular-weight protein repression of basal transcription. Mol Cell Biol 19, 3403-3414.
Matsushime, H., Roussel, M.F., Ashmun, R.A., and Sherr, C.J. (1991). Colony-stimulating factor 1 regulates novel cyclins during the G1 phase of the cell cycle. Cell 65, 701-713.
McMahon, S.B., Van Buskirk, H.A., Dugan, K.A., Copeland, T.D., and Cole, M.D. (1998). The novel ATM-related protein TRRAP is an essential cofactor for the c-Myc and E2F oncoproteins. Cell 94, 363-374.
Mende, Y., Schneider, P.M., Baldus, S.E., and Doerfler, W. (2004). PCR-screening of human esophageal and bronchial cancers reveals absence of adenoviral DNA sequences. Virus Res 104, 81-85.
Metzgar, D., Osuna, M., Kajon, A.E., Hawksworth, A.W., Irvine, M., and Russell, K.L. (2007). Abrupt emergence of diverse species B adenoviruses at US military recruit training centers. J Infect Dis 196, 1465-1473.
Mietz, J.A., Unger, T., Huibregtse, J.M., and Howley, P.M. (1992). The transcriptional transactivation function of wild-type p53 is inhibited by SV40 large T-antigen and by HPV-16 E6 oncoprotein. EMBO J 11, 5013-5020.
Miller, J.L., Harman, C., Weiner, C., and Baschat, A.A. (2009). Perinatal outcomes after second trimester detection of amniotic fluid viral genome in asymptomatic patients. J Perinat Med 37, 140-143.
Mol, C.D., Harris, J.M., McIntosh, E.M., and Tainer, J.A. (1996). Human dUTP pyrophosphatase: uracil recognition by a beta hairpin and active sites formed by three separate subunits. Structure 4, 1077-1092.
Moloney, J.B. (1960). Biological studies on a lymphoid-leukemia virus extracted from sarcoma 37. I. Origin and introductory investigations. J Natl Cancer Inst 24, 933-951.
Montell, C., Courtois, G., Eng, C., and Berk, A. (1984). Complete transformation by adenovirus 2 requires both E1A proteins. Cell 36, 951-961.
Montell, C., Fisher, E.F., Caruthers, M.H., and Berk, A.J. (1982). Resolving the functions of overlapping viral genes by site-specific mutagenesis at a mRNA splice site. Nature 295, 380-384.
Moore, M., Horikoshi, N., and Shenk, T. (1996). Oncogenic potential of the adenovirus E4orf6 protein. Proc Natl Acad Sci USA 93, 11295-11301.
Moran, E. (1988). A region of SV40 large T antigen can substitute for a transforming domain of the adenovirus E1A products. Nature 334, 168-170.
Moran, E., Zerler, B., Harrison, T.M., and Mathews, M.B. (1986). Identification of separate domains in the

adenovirus E1A gene for immortalization activity and the activation of virus early genes. Mol Cell Biol 6, 3470-3480.

Murphree, A.L., and Benedict, W.F. (1984). Retinoblastoma: clues to human oncogenesis. Science 223, 1028-1033.

Nazir, S.A., and Metcalf, J.P. (2005). Innate immune response to adenovirus. J Investig Med 53, 292-304.

Nemunaitis, J., Clayman, G., Agarwala, S.S., Hrushesky, W., Wells, J.R., Moore, C., Hamm, J., Yoo, G., Baselga, J., Murphy, B.A., *et al.* (2009). Biomarkers Predict p53 Gene Therapy Efficacy in Recurrent Squamous Cell Carcinoma of the Head and Neck. Clin Cancer Res 15, 7719-7725.

Nemunaitis, J., and Nemunaitis, J. (2011). Head and neck cancer: response to p53-based therapeutics. Head Neck 33, 131-134.

Nevels, M., Tauber, B., Spruss, T., Wolf, H., and Dobner, T. (2001). "Hit-and-run" transformation by adenovirus oncogenes. J Virol 75, 3089-3094.

Nevins, J.R. (1994). Cell cycle targets of the DNA tumour viruses. Curr Opin Genet Dev 4, 130-134.

Nevins, J.R., Ginsberg, H.S., Blanchard, J.M., Wilson, M.C., and Darnell, J.E., Jr. (1979). Regulation of the primary expression of the early adenovirus transcription units. J Virol 32, 727-733.

Nijhawan, D., Fang, M., Traer, E., Zhong, Q., Gao, W., Du, F., and Wang, X. (2003). Elimination of Mcl-1 is required for the initiation of apoptosis following ultraviolet irradiation. Genes Dev 17, 1475-1486.

Nikiforov, M.A., Chandriani, S., Park, J., Kotenko, I., Matheos, D., Johnsson, A., McMahon, S.B., and Cole, M.D. (2002). TRRAP-dependent and TRRAP-independent transcriptional activation by Myc family oncoproteins. Mol Cell Biol 22, 5054-5063.

Nishizaki, M., Fujiwara, T., Tanida, T., Hizuta, A., Nishimori, H., Tokino, T., Nakamura, Y., Bouvet, M., Roth, J.A., and Tanaka, N. (1999). Recombinant adenovirus expressing wild-type p53 is antiangiogenic: a proposed mechanism for bystander effect. Clin Cancer Res 5, 1015-1023.

Numazaki, Y., Kumasaka, T., Yano, N., Yamanaka, M., Miyazawa, T., Takai, S., and Ishida, N. (1973). Further study on acute hemorrhagic cystitis due to adenovirus type 11. N Engl J Med 289, 344-347.

O'Shea, C., Klupsch, K., Choi, S., Bagus, B., Soria, C., Shen, J., McCormick, F., and Stokoe, D. (2005a). Adenoviral proteins mimic nutrient/growth signals to activate the mTOR pathway for viral replication. EMBO J 24, 1211-1221.

O'Shea, C.C., Choi, S., McCormick, F., and Stokoe, D. (2005b). Adenovirus overrides cellular checkpoints for protein translation. Cell Cycle 4, 883-888.

O'Shea, C.C., Johnson, L., Bagus, B., Choi, S., Nicholas, C., Shen, A., Boyle, L., Pandey, K., Soria, C., Kunich, J., *et al.* (2004). Late viral RNA export, rather than p53 inactivation, determines ONYX-015 tumour selectivity. Cancer Cell 6, 611-623.

Osada, T., Yang, X.Y., Hartman, Z.C., Glass, O., Hodges, B.L., Niedzwiecki, D., Morse, M.A., Lyerly, H.K., Amalfitano, A., and Clay, T.M. (2009). Optimization of vaccine responses with an E1, E2b and E3-deleted Ad5 vector circumvents pre-existing anti-vector immunity. Cancer Gene Ther 16, 673-682.

Oteiza, A., and Mechti, N. (2011). The human T-cell leukemia virus type 1 oncoprotein tax controls forkhead box O4 activity through degradation by the proteasome. J. Virol. 85, 6480-6491.

Palmer, D.H., Milner, A.E., Kerr, D.J., and Young, L.S. (2003). Mechanism of cell death induced by the novel enzyme-prodrug combination, nitroreductase/CB1954, and identification of synergism with 5-fluorouracil. Br J Cancer 89, 944-950.

Palmer, D.J., and Ng, P. (2011). Rescue, amplification, and large-scale production of helper-dependent adenoviral vectors. Cold Spring Harb Protoc 2011, 857-866.

Parks, R.J., Chen, L., Anton, M., Sankar, U., Rudnicki, M.A., and Graham, F.L. (1996). A helper-dependent adenovirus vector system: removal of helper virus by Cre-mediated excision of the viral packaging signal. Proc Natl Acad Sci USA 93, 13565-13570.

Patel, J.H., Du, Y., Ard, P.G., Phillips, C., Carella, B., Chen, C.J., Rakowski, C., Chatterjee, C., Lieberman, P.M., Lane, W.S., *et al.* (2004). The c-MYC oncoprotein is a substrate of the acetyltransferases hGCN5/PCAF and TIP60. Mol Cell Biol 24, 10826-10834.

Patel, P., Young, J.G., Mautner, V., Ashdown, D., Bonney, S., Pineda, R.G., Collins, S.I., Searle, P.F., Hull, D., Peers, E., *et al.* (2009). A phase I/II clinical trial in localized prostate cancer of an adenovirus expressing nitroreductase with CB1954 [correction of CB1984]. Mol Ther 17, 1292-1299.

Pavia, A.T. (2011). Viral infections of the lower respiratory tract: old viruses, new viruses, and the role of diagnosis. Clin Infect Dis 52 Suppl 4, S284-289.

Pearson, S., Jia, H., and Kandachi, K. (2004). China approves first gene therapy. Nat Biotechnol 22, 3-4.

Pelka, P., Ablack, J.N., Fonseca, G.J., Yousef, A.F., and Mymryk, J.S. (2008). Intrinsic structural disorder in adenovirus E1A: a viral molecular hub linking multiple diverse processes. J Virol 82, 7252-7263.

Pelka, P., Ablack, J.N., Shuen, M., Yousef, A.F., Rasti, M., Grand, R.J., Turnell, A.S., and Mymryk, J.S. (2009a). Identification of a second independent binding site for the pCAF acetyltransferase in adenovirus E1A. Virology 391, 90-98.

Pelka, P., Ablack, J.N., Torchia, J., Turnell, A.S., Grand, R.J., and Mymryk, J.S. (2009b). Transcriptional control by adenovirus E1A conserved region 3 via p300/CBP. Nucleic Acids Res 37, 1095-1106.

Peng, Z. (2005). Current status of gendicine in China: recombinant human Ad-p53 agent for treatment of cancers. Hum Gene Ther 16, 1016-1027.

Pennella, M.A., Liu, Y., Woo, J.L., Kim, C.A., and Berk, A.J. (2010). Adenovirus E1B 55-kilodalton protein is a p53-SUMO1 E3 ligase that represses p53 and stimulates its nuclear export through interactions with promyelocytic leukemia nuclear bodies. J Virol 84, 12210-12225.

Perez-Romero, P., Tyler, R.E., Abend, J.R., Dus, M., and Imperiale, M.J. (2005). Analysis of the interaction of the adenovirus L1 52/55-kilodalton and IVa2 proteins with the packaging sequence *in vivo* and *in vitro*. J Virol 79, 2366-2374.

Perricaudet, M., Akusjarvi, G., Virtanen, A., and Pettersson, U. (1979). Structure of two spliced mRNAs from the transforming region of human subgroup C adenoviruses. Nature 281, 694-696.

Pesonen, S., Kangasniemi, L., and Hemminki, A. (2011). Oncolytic adenoviruses for the treatment of human cancer: focus on translational and clinical data. Mol Pharm 8, 12-28.

Pesonen, S., Nokisalmi, P., Escutenaire, S., Sarkioja, M., Raki, M., Cerullo, V., Kangasniemi, L., Laasonen, L., Ribacka, C., Guse, K., *et al.* (2010). Prolonged systemic circulation of chimeric oncolytic adenovirus Ad5/3-Cox2L-D24 in patients with metastatic and refractory solid tumours. Gene Ther 17, 892-904.

Pilder, S., Logan, J., and Shenk, T. (1984). Deletion of the gene encoding the adenovirus 5 early region 1b 21,000-molecular-weight polypeptide leads to degradation of viral and host cell DNA. J Virol 52, 664-671.

Querido, E., Blanchette, P., Yan, Q., Kamura, T., Morrison, M., Boivin, D., Kaelin, W.G., Conaway, R.C., Conaway, J.W., and Branton, P.E. (2001). Degradation of p53 by adenovirus E4orf6 and E1B-55K proteins occurs via a novel mechanism involving a Cullin-containing complex. Genes Dev 15, 3104-3117.

Quinlan, M.P., and Douglas, J.L. (1992). Immortalization of primary epithelial cells requires first- and second-exon functions of adenovirus type 5 12S. J Virol 66, 2020-2030.

Rao, L., Debbas, M., Sabbatini, P., Hockenbery, D., Korsmeyer, S., and White, E. (1992). The adenovirus E1A proteins induce apoptosis, which is inhibited by the E1B 19-kDa and Bcl-2 proteins. Proc Natl Acad Sci USA 89, 7742-7746.

Raper, S.E., Chirmule, N., Lee, F.S., Wivel, N.A., Bagg, A., Gao, G.P., Wilson, J.M., and Batshaw, M.L. (2003). Fatal systemic inflammatory response syndrome in a ornithine transcarbamylase deficient patient following adenoviral gene transfer. Mol Genet Metab 80, 148-158.

Raper, S.E., Yudkoff, M., Chirmule, N., Gao, G.P., Nunes, F., Haskal, Z.J., Furth, E.E., Propert, K.J., Robinson, M.B., Magosin, S., *et al.* (2002). A pilot study of *in vivo* liver-directed gene transfer with an adenoviral vector in partial ornithine transcarbamylase deficiency. Hum Gene Ther 13, 163-175.

Rasti, M., Grand, R.J., Mymryk, J.S., Gallimore, P.H., and Turnell, A.S. (2005). Recruitment of CBP/p300, TATA-binding protein, and S8 to distinct regions at the N terminus of adenovirus E1A. J Virol 79, 5594-5605.

Raychaudhuri, P., Bagchi, S., Devoto, S.H., Kraus, V.B., Moran, E., and Nevins, J.R. (1991). Domains of the adenovirus E1A protein required for oncogenic activity are also required for dissociation of E2F transcription factor complexes. Genes Dev 5, 1200-1211.

Reid, T., Galanis, E., Abbruzzese, J., Sze, D., Wein, L.M., Andrews, J., Randlev, B., Heise, C., Uprichard, M., Hatfield, M., *et al.* (2002). Hepatic arterial infusion of a replication-selective oncolytic adenovirus (dl1520): phase II viral, immunologic, and clinical endpoints. Cancer Res 62, 6070-6079.

Reid, T.R., Freeman, S., Post, L., McCormick, F., and Sze, D.Y. (2005). Effects of Onyx-015 among metastatic colorectal cancer patients that have failed prior treatment with 5-FU/leucovorin. Cancer Gene Ther 12, 673-681.

Roberts, S., Delury, C., and Marsh, E. (2012). The PDZ protein discs-large (DLG): the 'Jekyll and Hyde' of the epithelial polarity proteins. Febs J 279, 3549-3558.

Robinson, C.M., Singh, G., Henquell, C., Walsh, M.P., Peigue-Lafeuille, H., Seto, D., Jones, M.S., Dyer, D.W., and Chodosh, J. (2011). Computational analysis and identification of an emergent human adenovirus pathogen implicated in a respiratory fatality. Virology 409, 141-147.

Rous, P. (1911). A Sarcoma of the Fowl Transmissible by an Agent Separable from the Tumour Cells. J Exp Med 13, 397-411.

Rous, P., and Beard, J.W. (1935). The Progression to Carcinoma of Virus-Induced Rabbit Papillomas (Shope). J Exp Med 62, 523-548.

Rowe, W.P., Huebner, R.J., Gilmore, L.K., Parrott, R.H., and Ward, T.G. (1953). Isolation of a cytopathogenic agent from human adenoids undergoing spontaneous degeneration in tissue culture. Proc Soc Exp Biol Med 84, 570-573.

Ruley, H.E. (1983). Adenovirus early region 1A enables viral and cellular transforming genes to transform primary cells in culture. Nature 304, 602-606.

Saito, K., Saito, M., Taniura, N., Okuwa, T., and Ohara, Y. (2010). Activation of the PI3K-Akt pathway by human T cell leukemia virus type 1 (HTLV-1) oncoprotein Tax increases Bcl3 expression, which is associated with enhanced growth of HTLV-1-infected T cells. Virology 403, 173-180.

Sankar, N., Baluchamy, S., Kadeppagari, R.K., Singhal, G., Weitzman, S., and Thimmapaya, B. (2008). p300 provides a corepressor function by cooperating with YY1 and HDAC3 to repress c-Myc. Oncogene 27, 5717-5728.

Sarnow, P., Hearing, P., Anderson, C.W., Halbert, D.N., Shenk, T., and Levine, A.J. (1984). Adenovirus early region 1B 58,000-dalton tumour antigen is physically associated with an early region 4 25,000-dalton protein in productively infected cells. J Virol 49, 692-700.

Sarnow, P., Ho, Y.S., Williams, J., and Levine, A.J. (1982). Adenovirus E1b-58kd tumour antigen and SV40 large tumour antigen are physically associated with the same 54 kd cellular protein in transformed cells. Cell 28, 387-394.

Schiedner, G., Hertel, S., and Kochanek, S. (2000). Efficient transformation of primary human amniocytes by E1 functions of Ad5: generation of new cell lines for adenoviral vector production. Hum Gene Ther 11, 2105-2116.

Schwartz, R.A., Lakdawala, S.S., Eshleman, H.D., Russell, M.R., Carson, C.T., and Weitzman, M.D. (2008). Distinct requirements of adenovirus E1B-55K protein for degradation of cellular substrates. J Virol 82, 9043-9055.

Sha, J., Ghosh, M.K., Zhang, K., and Harter, M.L. (2010). E1A interacts with two opposing transcriptional pathways to induce quiescent cells into S phase. J Virol 84, 4050-4059.

Shenk, T., and Flint, J. (1991). Transcriptional and transforming activities of the adenovirus E1A proteins. Adv Cancer Res 57, 47-85.

Shimada, H., Matsubara, H., Shiratori, T., Shimizu, T., Miyazaki, S., Okazumi, S., Nabeya, Y., Shuto, K., Hayashi, H., Tanizawa, T., *et al.* (2006). Phase I/II adenoviral p53 gene therapy for chemoradiation resistant advanced esophageal squamous cell carcinoma. Cancer Sci 97, 554-561.

Shimizu, K., Sakurai, F., Machitani, M., Katayama, K., and Mizuguchi, H. (2011). Quantitative analysis of the leaky expression of adenovirus genes in cells transduced with a replication-incompetent adenovirus vector. Mol Pharm 8, 1430-1435.

Singhal, G., Kadeppagari, R.K., Sankar, N., and Thimmapaya, B. (2008). Simian virus 40 large T overcomes p300 repression of c-Myc. Virology 377, 227-232.

Spangle, J.M., Ghosh-Choudhury, N., and Munger, K. (2012). Activation of cap-dependent translation by mucosal human papillomavirus E6 proteins is dependent on the integrity of the LXXLL binding motif. J. Virol. 86, 7466-7472.

Spangle, J.M., and Munger, K. (2010). The human papillomavirus type 16 E6 oncoprotein activates mTORC1 signaling and increases protein synthesis. J. Virol. 84, 9398-9407.

Spangle, J.M., and Munger, K. (2013). The HPV16 E6 oncoprotein causes prolonged receptor protein tyrosine kinase signaling and enhances internalization of phosphorylated receptor species. PLoS Pathog 9, e1003237.

Speidel, D. (2010). Transcription-independent p53 apoptosis: an alternative route to death. Trends Cell Biol 20, 14-24.

Spindler, K.R., Eng, C.Y., and Berk, A.J. (1985). An adenovirus early region 1A protein is required for maximal viral DNA replication in growth-arrested human cells. J Virol 53, 742-750.

Stehelin, D., Varmus, H.E., Bishop, J.M., and Vogt, P.K. (1976). DNA related to the transforming gene(s) of avian sarcoma viruses is present in normal avian DNA. Nature 260, 170-173.

Stein, R.W., Corrigan, M., Yaciuk, P., Whelan, J., and Moran, E. (1990). Analysis of E1A-mediated growth regulation functions: binding of the 300-kilodalton cellular product correlates with E1A enhancer repression function and DNA synthesis-inducing activity. J Virol 64, 4421-4427.

Sterner, D.E., and Berger, S.L. (2000). Acetylation of histones and transcription-related factors. Microbiol Mol Biol Rev 64, 435-459.

Stevens, J.L., Cantin, G.T., Wang, G., Shevchenko, A., Shevchenko, A., and Berk, A.J. (2002). Transcription control by E1A and MAP kinase pathway via Sur2 mediator subunit. Science 296, 755-758.

Stewart, S.A., and Weinberg, R.A. (2006). Telomeres: cancer to human aging. Annu Rev Cell Dev Biol 22, 531-557.

Stillman, B.W. (1983). The replication of adenovirus DNA with purified proteins. Cell 35, 7-9.

Stott, F.J., Bates, S., James, M.C., McConnell, B.B., Starborg, M., Brookes, S., Palmero, I., Ryan, K., Hara, E., Vousden, K.H., *et al.* (1998). The alternative product from the human CDKN2A locus, p14(ARF), participates in a regulatory feedback loop with p53 and MDM2. EMBO J 17, 5001-5014.

Stracker, T.H., Carson, C.T., and Weitzman, M.D. (2002). Adenovirus oncoproteins inactivate the Mre11-Rad50-NBS1 DNA repair complex. Nature 418, 348-352.

Subramanian, T., La Regina, M., and Chinnadurai, G. (1989). Enhanced ras oncogene mediated cell transformation and tumourigenesis by adenovirus 2 mutants lacking the C-terminal region of E1a protein. Oncogene 4, 415-420.

Surget, S., Khoury, M.P., and Bourdon, J.C. (2013). Uncovering the role of p53 splice variants in human malignancy: a clinical perspective. Onco Targets Ther 7, 57-68.

Sze, D.Y., Freeman, S.M., Slonim, S.M., Samuels, S.L., Andrews, J.C., Hicks, M., Ahrar, K., Gupta, S., and Reid, T.R. (2003). Dr. Gary J. Becker Young Investigator Award: intraarterial adenovirus for metastatic gastrointestinal cancer: activity, radiographic response, and survival. J Vasc Interv Radiol 14, 279-290.

Tebruegge, M., and Curtis, N. (2010). Adenovirus infection in the immunocompromised host. Adv Exp Med Biol 659, 153-174.

Teodoro, J.G., and Branton, P.E. (1997). Regulation of p53-dependent apoptosis, transcriptional repression, and cell transformation by phosphorylation of the 55-kilodalton E1B protein of human adenovirus type 5. J Virol 71, 3620-3627.

Thai, M., Graham, N.A., Braas, D., Nehil, M., Komisopoulou, E., Kurdistani, S.K., McCormick, F., Graeber, T.G., and Christofk, H.R. (2014). Adenovirus E4ORF1-induced MYC activation promotes host cell anabolic glucose metabolism and virus replication. Cell metabolism 19, 694-701.

Thomas, D.L., Schaack, J., Vogel, H., and Javier, R. (2001). Several E4 region functions influence mammary tumourigenesis by human adenovirus type 9. J. Virol. 75, 557-568.

Thomas, D.L., Shin, S., Jiang, B.H., Vogel, H., Ross, M.A., Kaplitt, M., Shenk, T.E., and Javier, R.T. (1999). Early region 1 transforming functions are dispensable for mammary tumourigenesis by human adenovirus type 9. J. Virol. 73, 3071-3079.

Tong, A.W., Nemunaitis, J., Su, D., Zhang, Y., Cunningham, C., Senzer, N., Netto, G., Rich, D., Mhashilkar, A., Parker, K., *et al.* (2005). Intratumoural injection of INGN 241, a nonreplicating adenovector expressing the melanoma-differentiation associated gene-7 (mda-7/IL24): biologic outcome in advanced cancer patients. Mol Ther 11, 160-172.

Topp, W.C., Lane, D., and Pollack, R. (1981). Transformation by simian virus 40 and polyoma virus.

In DNA Tumour Viruses, J. Tooze, ed. (Cold Spring Harbor, NY: Cold Spring Harbor Laboratory), pp. 205-296.

Trentin, J.J., Yabe, Y., and Taylor, G. (1962). The quest for human cancer viruses. Science 137, 835-841.

Turnell, A.S., and Mymryk, J.S. (2006). Roles for the coactivators CBP and p300 and the APC/C E3 ubiquitin ligase in E1A-dependent cell transformation. Br J Cancer 95, 555-560.

Turnell, A.S., Stewart, G.S., Grand, R.J., Rookes, S.M., Martin, A., Yamano, H., Elledge, S.J., and Gallimore, P.H. (2005). The APC/C and CBP/p300 cooperate to regulate transcription and cell-cycle progression. Nature 438, 690-695.

Tuve, S., Wang, H., Ware, C., Liu, Y., Gaggar, A., Bernt, K., Shayakhmetov, D., Li, Z., Strauss, R., Stone, D., *et al.* (2006). A new group B adenovirus receptor is expressed at high levels on human stem and tumour cells. J Virol 80, 12109-12120.

Tworkowski, K.A., Chakraborty, A.A., Samuelson, A.V., Seger, Y.R., Narita, M., Hannon, G.J., Lowe, S.W., and Tansey, W.P. (2008). Adenovirus E1A targets p400 to induce the cellular oncoprotein Myc. Proc Natl Acad Sci USA 105, 6103-6108.

Vaillancourt, M.T., Atencio, I., Quijano, E., Howe, J.A., and Ramachandra, M. (2005). Inefficient killing of quiescent human epithelial cells by replicating adenoviruses: potential implications for their use as oncolytic agents. Cancer Gene Ther 12, 691-698.

Van den Elsen, P., Houweling, A., and Van der Eb, A. (1983). Expression of region E1b of human adenoviruses in the absence of region E1a is not sufficient for complete transformation. Virology 128, 377-390.

van den Heuvel, S., and Dyson, N.J. (2008). Conserved functions of the pRB and E2F families. Nat Rev Mol Cell Biol 9, 713-724.

van Ormondt, H., Maat, J., and Dijkema, R. (1980). Comparison of nucleotide sequences of the early E1a regions for subgroups A, B and C of human adenoviruses. Gene 12, 63-76.

Vasconcelos, G.M., Kang, M., Pombo-de-Oliveira, M.S., Schiffman, J.D., Lorey, F., Buffler, P., and Wiemels, J.L. (2008). Adenovirus detection in Guthrie cards from paediatric leukaemia cases and controls. Br J Cancer 99, 1668-1672.

Vassilev, L., Ranki, T., Joensuu, T., Jager, E., Karbach, J., Wahle, C., Partanen, K., Kairemo, K., Alanko, T., Turkki, R., *et al.* (2015). Repeated intratumoural administration of ONCOS-102 leads to systemic antitumour CD8+ T-cell response and robust cellular and transcriptional immune activation at tumour site in a patient with ovarian cancer. Oncoimmunology 4, e1017702.

Vijayalingam, S., Subramanian, T., Zhao, L.J., and Chinnadurai, G. (2016). The Cellular Protein Complex Associated with a Transforming Region of E1A Contains c-MYC. J Virol 90, 1070-1079.

Vogel, F. (1979). Genetics of retinoblastoma. Hum Genet 52, 1-54.

Waku, T., Fujiwara, T., Shao, J., Itoshima, T., Murakami, T., Kataoka, M., Gomi, S., Roth, J.A., and Tanaka, N. (2000). Contribution of CD95 ligand-induced neutrophil infiltration to the bystander effect in p53 gene therapy for human cancer. J Immunol 165, 5884-5890.

Wang, G., Balamotis, M.A., Stevens, J.L., Yamaguchi, Y., Handa, H., and Berk, A.J. (2005). Mediator requirement for both recruitment and postrecruitment steps in transcription initiation. Mol Cell 17, 683-694.

Wang, G., and Berk, A.J. (2002). In vivo association of adenovirus large E1A protein with the human mediator complex in adenovirus-infected and -transformed cells. J Virol 76, 9186-9193.

Wang, H., Li, Z.Y., Liu, Y., Persson, J., Beyer, I., Moller, T., Koyuncu, D., Drescher, M.R., Strauss, R., Zhang, X.B., *et al.* (2011). Desmoglein 2 is a receptor for adenovirus serotypes 3, 7, 11 and 14. Nat Med 17, 96-104.

Wang, H.G., Rikitake, Y., Carter, M.C., Yaciuk, P., Abraham, S.E., Zerler, B., and Moran, E. (1993). Identification of specific adenovirus E1A N-terminal residues critical to the binding of cellular proteins and to the control of cell growth. J Virol 67, 476-488.

Weaver, E.A., Nehete, P.N., Buchl, S.S., Senac, J.S., Palmer, D., Ng, P., Sastry, K.J., and Barry, M.A. (2009). Comparison of replication-competent, first generation, and helper-dependent adenoviral vaccines. PLoS One 4, e5059.

Weinmann, R., Raskas, H.J., and Roeder, R.G. (1974). Role of DNA-dependent RNA polymerases II and III in transcription of the adenovirus genome late in productive infection. Proc Natl Acad Sci USA 71, 3426-3439.

Weiss, R.S., Gold, M.O., Vogel, H., and Javier, R.T. (1997a). Mutant adenovirus type 9 E4 ORF1 genes define three protein regions required for transformation of CREF cells. J Virol 71, 4385-4394.

Weiss, R.S., Lee, S.S., Prasad, B.V., and Javier, R.T. (1997b). Human adenovirus early region 4 open reading frame 1 genes encode growth-transforming proteins that may be distantly related to dUTP pyrophosphatase enzymes. J. Virol. 71, 1857-1870.

Weiss, R.S., McArthur, M.J., and Javier, R.T. (1996). Human adenovirus type 9 E4 open reading frame 1 encodes a cytoplasmic transforming protein capable of increasing the oncogenicity of CREF cells. J. Virol. 70, 862-872.

Weitzman, M.D. (2005). Functions of the adenovirus E4 proteins and their impact on viral vectors. Front Biosci 10, 1106-1117.

Weitzman, M.D., Carson, C.T., Schwartz, R.A., and Lilley, C.E. (2004). Interactions of viruses with the cellular DNA repair machinery. DNA Repair (Amst) 3, 1165-1173.

White, E. (1993). Regulation of apoptosis by the transforming genes of the DNA tumour virus adenovirus. Proc Soc Exp Biol Med 204, 30-39.

White, E. (2001). Regulation of the cell cycle and apoptosis by the oncogenes of adenovirus. Oncogene 20, 7836-7846.

White, E., Cipriani, R., Sabbatini, P., and Denton, A. (1991). Adenovirus E1B 19-kilodalton protein overcomes the cytotoxicity of E1A proteins. J Virol 65, 2968-2978.

White, E., Grodzicker, T., and Stillman, B.W. (1984). Mutations in the gene encoding the adenovirus early region 1B 19,000-molecular-weight tumour antigen cause the degradation of chromosomal DNA. J Virol 52, 410-419.

White, E., Sabbatini, P., Debbas, M., Wold, W.S., Kusher, D.I., and Gooding, L.R. (1992). The 19-kilodalton adenovirus E1B transforming protein inhibits programmed cell death and prevents cytolysis by tumour necrosis factor alpha. Mol Cell Biol 12, 2570-2580.

Whitman, M., Kaplan, D.R., Schaffhausen, B., Cantley, L., and Roberts, T.M. (1985). Association of phosphatidylinositol kinase activity with polyoma middle-T competent for transformation. Nature 315, 239-242.

Whyte, P., Buchkovich, K.J., Horowitz, J.M., Friend, S.H., Raybuck, M., Weinberg, R.A., and Harlow, E. (1988a). Association between an oncogene and an anti-oncogene: the adenovirus E1A proteins bind to the retinoblastoma gene product. Nature 334, 124-129.

Whyte, P., Ruley, H.E., and Harlow, E. (1988b). Two regions of the adenovirus early region 1A proteins are required for transformation. J Virol 62, 257-265.

Wimmer, P., Tauber, B., Spruss, T., and Dobner, T. (2010). Adenovirus type 5 early encoded proteins of the E1 and E4 regions induce oncogenic transformation of primary rabbit cells. J Gen Virol 91, 1828-1833.

Winberg, G., and Shenk, T. (1984). Dissection of overlapping functions within the adenovirus type 5 E1A gene. EMBO J 3, 1907-1912.

Wold, W.S., and Ison, M.G. (2013). Adenoviruses. In Fields Virology, D.M. Knipe, and P.M. Howley, eds. (Philadelphia, PA: Lippincott Williams and Wilkins), pp. 1732–1767.

Wold, W.S., and Toth, K. (2013). Adenovirus vectors for gene therapy, vaccination and cancer gene therapy. Curr Gene Ther 13, 421-433.

Yamamoto, M., and Curiel, D.T. (2010). Current issues and future directions of oncolytic adenoviruses. Mol Ther 18, 243-250.

Yang, Y., Nunes, F.A., Berencsi, K., Furth, E.E., Gonczol, E., and Wilson, J.M. (1994). Cellular immunity to viral antigens limits E1-deleted adenoviruses for gene therapy. Proc Natl Acad Sci USA 91, 4407-4411.

Ye, F., and Zhang, M. (2013). Structures and target recognition modes of PDZ domains: recurring themes and emerging pictures. Biochem J 455, 1-14.

Yee, S.P., and Branton, P.E. (1985). Detection of cellular proteins associated with human adenovirus type 5 early region 1A polypeptides. Virology 147, 142-153.

Yew, P.R., and Berk, A.J. (1992). Inhibition of p53 transactivation required for transformation by adenovirus early 1B protein. Nature 357, 82-85.

Yew, P.R., Liu, X., and Berk, A.J. (1994). Adenovirus E1B oncoprotein tethers a transcriptional repression domain to p53. Genes Dev 8, 190-202.

Yoo, G.H., Moon, J., Leblanc, M., Lonardo, F., Urba, S., Kim, H., Hanna, E., Tsue, T., Valentino, J., Ensley, J., *et al*. (2009). A phase 2 trial of surgery with perioperative INGN 201 (Ad5CMV-p53) gene therapy followed by chemoradiotherapy for advanced, resectable squamous cell carcinoma of the oral cavity, oropharynx, hypopharynx, and larynx: report of the Southwest Oncology Group. Arch Otolaryngol Head Neck Surg 135, 869-874.

Young, L.S., Searle, P.F., Onion, D., and Mautner, V. (2006). Viral gene therapy strategies: from basic science to clinical application. J Pathol 208, 299-318.

Yu, W., and Fang, H. (2007). Clinical trials with oncolytic adenovirus in China. Curr Cancer Drug Targets 7, 141-148.

Yuan, T.L., and Cantley, L.C. (2008). PI3K pathway alterations in cancer: variations on a theme. Oncogene 27, 5497-5510.

Zalvide, J., Stubdal, H., and DeCaprio, J.A. (1998). The J domain of simian virus 40 large T antigen is required to functionally inactivate RB family proteins. Mol Cell Biol 18, 1408-1415.

Zantema, A., Fransen, J.A., Davis-Olivier, A., Ramaekers, F.C., Vooijs, G.P., DeLeys, B., and Van der Eb, A.J. (1985). Localization of the E1B proteins of adenovirus 5 in transformed cells, as revealed by interaction with monoclonal antibodies. Virology 142, 44-58.

Zerler, B., Roberts, R.J., Mathews, M.B., and Moran, E. (1987). Different functional domains of the adenovirus E1A gene are involved in regulation of host cell cycle products. Mol Cell Biol 7, 821-829.

Zhang, Y., Xiong, Y., and Yarbrough, W.G. (1998). ARF promotes MDM2 degradation and stabilizes p53: ARF-INK4a locus deletion impairs both the Rb and p53 tumour suppression pathways. Cell 92, 725-734.

Zoncu, R., Efeyan, A., and Sabatini, D.M. (2011). mTOR: from growth signal integration to cancer, diabetes and ageing. Nat Rev Mol Cell Biol 12, 21-35.

Zubieta, C., Schoehn, G., Chroboczek, J., and Cusack, S. (2005). The structure of the human adenovirus 2 penton. Mol Cell 17, 121-135.

www.ingramcontent.com/pod-product-compliance
Ingram Content Group UK Ltd.
Pitfield, Milton Keynes, MK11 3LW, UK
UKHW061959041026
15591UKWH00016B/1093

9 781910 190791